Categories and Representation Theory

With A Focus on 2-Categorical Covering Theory

Mathematical
Surveys
and
Monographs
Volume 271

Categories and Representation Theory

With A Focus on 2-Categorical Covering Theory

Hideto Asashiba

Translated by Hideto Asashiba

AMERICAN MATHEMATICAL SOCIETY
Providence, Rhode Island

2020 *Mathematics Subject Classification.* Primary 18-01, 18-02, 16-02, 16D90, 16G20, 16W22, 16W50.

For additional information and updates on this book, visit
www.ams.org/bookpages/surv-271

Library of Congress Cataloging-in-Publication Data

Names: Asashiba, Hideto, 1956- author, translator.
Title: Categories and representation theory : with a focus on 2-categorical covering theory / Hideto Asashiba ; translated by Hideto Asashiba.
Other titles: Ken to hyōgenron. English
Description: Providence, Rhode Island : American Mathematical Society, [2022] | Series: Mathematical surveys and monographs, 0076-5376 ; volume 271 | "Ken To Hyogenron by Hideto Asashiba ; Original Japanese language edition published by Saiensu-sha Co., Ltd." | Includes bibliographical references and index.
Identifiers: LCCN 2022021596 | ISBN 9781470464844 (paperback) | 9781470471507 (ebook)
Subjects: LCSH: Categories (Mathematics) | AMS: Category theory; homological algebra – Instructional exposition (textbooks, tutorial papers, etc.). | Category theory; homological algebra – Research exposition (monographs, survey articles). | Associative rings and algebras – Research exposition (monographs, survey articles). | Associative rings and algebras – Modules, bimodules and ideals – Module categories. | Associative rings and algebras – Representation theory of rings and algebras – Representations of quivers and partially ordered sets. | Associative rings and algebras – Rings and algebras with additional structure – Actions of groups and semigroups; invariant theory. | Associative rings and algebras – Rings and algebras with additional structure – Graded rings and modules.
Classification: LCC QA169 .A83 2022 | DDC 512/.62–dc23/eng20220823
LC record available at https://lccn.loc.gov/2022021596

∞ The paper used in this book is acid-free and falls within the guidelines
established to ensure permanence and durability.
Visit the AMS home page at `https://www.ams.org/`

10 9 8 7 6 5 4 3 2 1 27 26 25 24 23 22

Contents

Preface

This book is an English translation of the book **[12]** with the same title in Japanese published by Saiensu-sha on December 25, 2019. Most parts of chapters 4 and 5 were translated into English in **[13]** and appeared in the proceedings of ISCRA published by the Iranian Mathematical Society on August 27, 2020.

We remark that the 2-categories dealt with in this book are sometimes called *strict 2-categories*. There is a notion of a *bicategory* that is weaker than the notion of a 2-category. Roughly speaking, a bicategory is defined by replacing the equalities in the axioms of a 2-category by natural isomorphisms. Bicategories have more practical examples[1], but are more complicated to handle than 2-categories. It is known that every bicategory is equivalent to a 2-category as a bicategory (see Leinstar **[32]** for details).

The following are added to this translation. In Appendix A, we added Proposition A.4.4 to give an example of how to treat 2-categories consisting of linear categories. In the Japanese version, due to lack of space, some propositions in Sect. 2.3 were turned into exercises or simply cited without proofs. To make up for this, the translation includes answers to those questions and omitted proofs in Appendix B. Namely, in Sect. B.1, we show that the category of *framed* algebras (see Remark B.1.6) is equivalent to a subcategory of the category of finite linear categories, which gives an answer to Exercise 2.3.9. In Sect. B.2, we give answers to Exercises 2.3.15, 2.3.21, 2.3.23, and 2.3.24 with some explanations in order to turn exercises to propositions. In Sect. B.3, the Krull-Schmidt Theorem (Theorem 2.3.16) is proved using the notion of radical maps. In Sect. B.4, we give two proofs of Theorem 2.3.19, which is a special form of the Morita Theorem. The first proof is elementary; it only requires knowledge in Chapter 2 and a fundamental fact on tensor products over an algebra. The second proof explains a relationship between the first one and the Hom-tensor adjoint. After that we extend this theorem to (one half of) the full version of the Morita theorem by generalizing the second proof. The material in Sect. B.1–B.4 above covers the omitted parts in Sect. 2.3. Sect. B.5 is devoted to the definition of left Kan extensions and its relationships with left adjoint functors, and provides background for the beginning sentences in Sect. 7.1.

[1] Such as a bicategory of all small categories whose 1-morphisms are bimodules over small categories, whose compositions are given by tensor products, and whose 2-morphisms are bimodule morphisms between bimodules.

Finally, I would like to thank Professor Goro C. Kato for recommending and encouraging me to publish the translation. I would also like to thank the AMS for giving me permission to publish this book with the additional part, Professor Eriko Hironaka for her kind consultation, and my son Rihito Asashiba for some advice on translation.

November, 2021 Hideto Asashiba

Introduction

Category theory is used in representation theory of algebras in a variety of ways, just as it is in other areas of mathematics. The connection is twofold. On the one hand, many kinds of categories, functors, and natural transformations appear naturally in the study of representation theory. On the other, the general theorey of categories can also be applied to give new insights into representation theory. For instance, an algebra A itself can be regarded as a linear category with a single object and from A we can define the category $\operatorname{Mod} A$ of (right) A-modules, the stable module category $\underline{\operatorname{Mod}}A$, and the derived category $\mathscr{D}(\operatorname{Mod} A)$. We can then apply the general theories of abelian categories. Other categories such as the comodule category over a coalgebra, differential graded categories, A_∞-categories, 2-categories, and $(\infty, 1)$-categories also appear. In this book, we explain how categories are used in the representation theory of algebras by focusing on 2-categorical covering theory. Throughout this book, we set G and $\Bbbk$ to be a group and a field (or just a commutative ring in many cases), respectively.

Covering theory. Gabriel [**23**], Riedtmann [**41**], and Bongartz–Gabriel [**16**] introduced the use of covering theory into representation theory of algebras. This gave a way to reduce representations of algebras containing oriented cycles in their quivers[1] to those of algebras or linear categories containing fewer (or no) oriented cycles in their quivers and, in particular, the representation theory of selfinjective algebras (typical algebras containing oriented cycles in their quivers) was developed (see [**42**], [**43**] for example). Paper [**23**] introduced the notion of general covering functors, the most important ones in applications being Galois coverings. These are essentially the canonical covering functor from a linear category $\mathscr{C}$ with an action of a group G to its orbit category $\mathscr{C}/G$.

For a given algebra A, a linear category $\mathscr{C}$ with a G-action is called a *covering* of A if $\mathscr{C}/G \cong A$. There are several ways to compute coverings $\mathscr{C}$ of A. For selfinjective algebras, coverings $\mathscr{C}$ are given in many cases by so-called repetitive categories (Hughes–Waschbüsch [**30**]) of other algebras that are much easier to handle. For general algebras, we can use a way to construct a "universal covering"[2] $\tilde{A}$ by using a presentation of A by a quiver with (minimal) relations (Waschbüsch [**46**], Martinez-Villa–de la Peña [**36**]). Other coverings $\mathscr{C}$ are given as orbit categories $\tilde{A}/H$ of $\tilde{A}$ by a group H. The most general way that includes the above cases is given by forming the smash product $\mathscr{C} := A\#G$ by giving a G-grading to A (see e.g., E. Green [**26**], Cibils–Marcos [**19**]), which is explained in this book.

[1] Representations of algebras without oriented cycles in their quivers are much simpler than others.

[2] This may vary by presentations of A by quivers and relations.

In a later section of Ch. 2, we explain a way to construct algebras by quivers that can be seen as geometrical objects. Seen this way, coverings of algebras are viewed as algebraic versions of the usual coverings in geometry. In many cases, a quiver defining a covering has infinitely many vertices, in which case the defined algebra does not have an identity element. We usually deal with such an algebra as a linear category. Therefore we extend our range of consideration to the linear categories. Ch. 3 gives an overview of the classical covering theory by Gabriel, and in Chs. 5 and 7, we introduce the improved covering theory ([**5**], [**7**]), thus removing the assumptions that become obstructions for its applications. This requires families of natural isomorphisms and 2-categorical considerations. In particular in Ch. 5, we show that the orbit category construction and the smash product construction extends to 2-equivalences that are 2-quasi-inverses to each other.

Pseudo-actions of a group. In this book, especially in the main chapter (Ch. 5), we deal with actions of a group as pseudo-actions in one of their most generalized forms. The reasons for this are listed as follows: (1) In many actual settings when we consider an "action" of a group G on a category $\mathscr{C}$, elements a of G act on it as autoequivalences F rather than automorphisms, i.e., F does not have an inverse, but only a quasi-inverse F^-: $\mathbb{1}_{\mathscr{C}} \cong F^- \circ F \neq \mathbb{1}_{\mathscr{C}}$. In this case, there exist no actions X of G on $\mathscr{C}$ satisfying the condition $(*)$ $X(a) = F, X(a^{-1}) = F^-$. Indeed, if it exists, then we have to have $X(a^{-1}a) = X(a^{-1}) \circ X(a)$, but the left-hand side is $X(1) = \mathbb{1}_{\mathscr{C}}$, and the right-hand side is $F^- \circ F \neq \mathbb{1}_{\mathscr{C}}$, which gives a contradiction. However, when G is a cyclic group generated by an element a of G, there exists a pseudo-action X satisfying the condition $(*)$ (see Proposition 5.4.15). Note that for a pseudo-action X, the endo-functor $X(a)$ of $\mathscr{C}$ necessarily turns out to be an equivalence, but does not need to be an automorphism for each $a \in G$. (2) Pseudo-action has been defined by Deligne and others. However, there seems to be no evidence that this is a special case of a pseudofunctor. We discuss this relationship in detail in this book (Sect. 5.1). (3) Because in this way, the pseudo-action of the group can be regarded as a simple example of a pseudofunctor (lax functor, or colax functor), it will make it easier for the readers to understand pseudofunctors and lax functors later on if they become familiar with pseudo-actions first. (4) If we generalize the Cohen-Montgomery duality to our 2-categorical setting that uses pseudo-actions, we can give a strictification of transforming the pseudo-action of the group into an action. (5) Since group actions are simpler in theory and pseudo-actions can eventually be reduced to group actions by strictification, many papers have a tendency to keep the discussion within group actions. Therefore, it seems that there are few opportunities for exposure to pseudo-actions. For these reasons, we adopted the pseudo-action formulation, although it has not yet been published as a paper.

Contents of the chapters. We now describe the contents of each chapter. In Ch. 1, we define the basic terms of category theory, such as categories, functors, and natural transformations, products, coproducts, kernels, and cokernels. Due to lack of space, reviews of limits and colimits are kept to a minimum, and more detailed elementary topics are left to other books. Next, we define algebras and linear categories, and explain how the algebra itself can be regarded as a linear category.

In Ch. 2, we first show that considering a representation of an algebra is equivalent to considering a module over it. Based on this result, we define modules over a linear category as its representations from the beginning. Given an algebra A viewed as a linear category $\mathscr{C}$, we show that the category of modules over A and the category of modules over $\mathscr{C}$ are equivalent. Finally, we explain how to construct algebras and linear categories using quivers, and show that the category of representations of a bound quiver and the category of modules over the corresponding linear category are isomorphic.

In Ch. 3, we introduce coverings (and covering functors) as an example of a situation in which algebras ought to be considered as linear categories, and review classical covering theory in the representation theory of algebras. In studying the category $\operatorname{Mod} A$ of modules over A, we can replace A by the basic algebra B of A (see Definition 2.3.18) because $\operatorname{Mod} A$ and $\operatorname{Mod} B$ are equivalent as categories. Here, B is *skeletal* (also called *basic*) as a category, namely, different objects are non-isomorphic to each other. Thanks to this property, the covering theory of algebras can proceed in terms of equalities between functors. However, when constructing a covering theory for general categories (especially module categories and derived categories) that are not necessarily skeletal, the equalities between functors must be replaced by natural isomorphisms (or natural transformations), thus 2-categorical considerations become inevitable.

In Ch. 4, we summarize the basics of 2-categories necessary for the later use. We introduce string diagrams to precisely distinguish vertical and horizontal compositions of 2-morphisms, and apply the interchange law visually and effortlessly. Thereafter, we use string diagrams for proofs involving 2-morphisms. We also introduce a way to handle objects and 1-morphisms using string diagrams, a method that is used in the exposition of adjoint systems. There does not seem to be many books that explain in detail string diagrams and their actual use. Therefore, this is one of the features of this book. In preparation for Ch. 5, the definition of 2-equivalences using modifications is explained.

In Ch. 5, we generalize classical covering theory using 2-categories, and develop a 2-categorical general theory of the Cohen-Montgomery duality. In the original paper [**5**], we gave a proof of the existence of a 2-equivalence between a 2-category of linear categories with action of G and the 2-category of G-graded linear categories. In this book, we extend the statement to prove the existence of a 2-equivalence between a 2-category of linear categories with a *pseudo-action* of G and the same 2-category of G-graded linear categories as above. The most important point here is that it is necessary to weaken the definition of degree-preserving functors from the usual one in order to prove this 2-equivalence, as is done in [**5**], which deals with group actions. Using this result, it is possible to transform a category with pseudo-action of G into a category with (free) action of G that is G-equivariantly equivalent to the original one. It is also possible to transform a degree-preserving functor between G-graded linear categories into a strictly degree-preserving functor that is G-homogeneously equivalent to the original one.

In Ch. 6, we give ways to compute orbit categories and smash products in practice that were defined theoretically in the previous chapter. Here we restrict ourselves to just the case of actions of G, not of general pseudo-actions of G.

In Ch. 7, we investigate relationships between the module categories $\operatorname{Mod}\mathscr{C}$ and $\operatorname{Mod}\mathscr{C}/G$ of a linear categories $\mathscr{C}$ and $\mathscr{C}/G$, respectively, where $\mathscr{C}$ is a covering of

$\mathscr{C}/G$. The main tool for this investigation is the *pushdown* functor $P_{\bullet}\colon \operatorname{Mod}\mathscr{C} \to \operatorname{Mod}\mathscr{C}/G$ that is defined from a covering functor $P\colon \mathscr{C} \to \mathscr{C}/G$. This is defined as a left adjoint to the *pullup* functor $P^{\bullet}\colon \operatorname{Mod}\mathscr{C}/G \to \operatorname{Mod}\mathscr{C}, M \mapsto M \circ P^{\mathrm{op}}$ ($M \in (\operatorname{Mod}\mathscr{C}/G)_0$). Since this left adjoint can be constructed as a tensor product over a small category, we explain the unique existence of tensor products over a small category in general. This fact is extended to tensor products over a light category in the appendix. Moreover, we prove the fact that the pushdown functor induces a precovering functor between categories of finitely generated modules, which is an important point in applying covering theory to representation theory.

In the final chapter, we outline a way to extend a covering theory that only deals with actions (or pseudo-actions) of a group so far, to one that deals with actions (or lax actions) of a small category ([**8**], [**9**]).

Set theory for the foundation of category theory. Finally, we explain how we treat set theory for the foundation of category theory. Various axiomatic systems of set theory have been proposed. Here we adopt ZFC, and denote the class of all sets by SET. The main theme of the book is to give a 2-functor between 2-categories $\mathbf{C}$ and $\mathbf{D}$, where the objects of $\mathbf{C}$ (resp. $\mathbf{D}$) form a collection $\mathbf{C}_0$ (resp. $\mathbf{D}_0$) of linear categories with (pseudo-)action of the group G (resp. G-graded linear categories). Both of the collections $\mathbf{C}_0$ and $\mathbf{D}_0$ should be taken as widely as possible. However, until this book was written, we did not know how wide. In that case, it is safer to restrict to the class of small categories to avoid the paradox of set theory, and therefore we constructed the theory in that setting. However, even if we start from a small category $\mathscr{C}$, its module category $\operatorname{Mod}\mathscr{C}$ turns out to be not small anymore, and we cannot treat $\operatorname{Mod}\mathscr{C}$ in this setting.

Universes. The term small set can be interpreted in two ways. First, in the case where an element of the class SET is called a small set, and a second in the case where after fixing a (Grothendieck) universe $\mathfrak{U}$ containing the set $\mathbb{N}$ of all natural numbers, and call the elements of $\mathfrak{U}$ a small set (or, more precisely a $\mathfrak{U}$-small set). In the latter case, a subset of $\mathfrak{U}$ is called a $\mathfrak{U}$-class. Note that even a $\mathfrak{U}$-class is an element of SET. We started writing this book under the former understanding of small sets, but, we have decided to take the latter one for this book. In this case, a category $\mathscr{C}$ is called a $\mathfrak{U}$-*small category* if its object set $\mathscr{C}_0$ and the local morphism set $\mathscr{C}(x,y)$ for each $x, y \in \mathscr{C}_0$ are $\mathfrak{U}$-small sets; and is called a $\mathfrak{U}$-*light category* if $\mathscr{C}_0$ is a $\mathfrak{U}$-class and $\mathscr{C}(x,y)$ is a $\mathfrak{U}$-small set for each $x, y \in \mathscr{C}_0$. The latter is the usual one appearing in the literature. This makes it impossible to handle classes not in SET, but we can safely build theories within the range of SET, and it is possible to handle a wide variety of phenomena. Here we additionally assume the *axiom of universe* stating that for any set there exists a universe containing it as an element. This axiom is known to be independent of the ZFC. From this standpoint, in order to treat the module category[3] $\operatorname{Mod}\mathscr{C}$ stated above, we could extend the universe $\mathfrak{U}$ to a universe $\mathfrak{U}'$ that contains the set consisting of all objects and morphisms of $\operatorname{Mod}\mathscr{C}$ as an element because then $\operatorname{Mod}\mathscr{C}$ turns out to be a $\mathfrak{U}'$-small category. For the time being, we can handle it in this way. But if we would allow the universe to be extended accordingly, then for example when considering $\operatorname{Mod}(\operatorname{Mod}\mathscr{C})$, we also need a further extension, and it would be necessary for us to extend universes indefinitely. As in this case, we would have to extend a universe even when the set

[3]when $\mathscr{C}$ is a small $\mathfrak{U}$-category, $\operatorname{Mod}\mathscr{C}$ turns out to be a $\mathfrak{U}$-class.

in question is only slightly larger, and when we consider a set, we always would have to be aware of for which universe it is small. Thus, it would be convenient if we could fix one universe $\mathfrak{U}$ and, on that basis, if we could guarantee that what we want to construct fits within the range of SET. In what follows, we omit the prefix "$\mathfrak{U}$-" because we fix one universe $\mathfrak{U}$.

Levy's hierarchy. When looking for a better way to deal with the set theoretic paradox, a method was found to use a hierarchy of SET by the k-classes ($0 \le k \in \mathbb{Z}$) defined by Levy **[33]**. Since this was just a preprint and the proofs were not written at all, we verified all the statements. Most of them were easy to check, but it remained to verify whether the existence of ΨA, which played an essential role in his preprint, can be proved or not within the range of SET. The definition of ΨA itself is implicitly written to use the Knaster-Tarski theorem. However, since the setting of the statement does not satisfy the assumption of this theorem, it cannot be applied simply. Therefore, we were very careful to prove the existence of this ΨA. In proving this, we made full use of the axiom of universe, although at first we intended to assume only the existence of a universe containing the set $\mathbb{N}$. Once that the verification was complete, we decided to use it as our foundation of category theory, which is another feature of this book. This allows us to compute the level (called the *level of moderation*) of a category in question, and it is now possible to construct a theory always within the range of SET. This is very useful. For example, as noted above, we can prove the unique existence of tensor products even over a light category (see Appendix A.6). By the same argument, it should be possible to prove this statement over a k-moderate category ($k \ge 1$). In this book, these contents are summarized in the appendix, which can be referred to at any time. Since the category whose objects are the light categories and whose morphisms are the functors between them turns out to be a 2-moderate category, and not a light category anymore, we know that it is defined without any problems and is safe to use. By further adding the natural transformations between the functors in it as the 2-morphisms, we can define the 2-category $\mathbf{CAT}$. Similarly it is also possible to define the 2-category $\Bbbk$-$\mathbf{CAT}$ of light $\Bbbk$-categories and $\Bbbk$-linear functors between them. As we started writing this book, we could only handle the 2-category $\Bbbk$-$\mathbf{Cat}$ of small linear categories, but now that we have much more freedom, it has become possible to include even the light categories in both $\mathbf{C}_0$ and $\mathbf{D}_0$ (see Remark 5.8.15)[4]. This has also simplified the definition of categories with pseudo-action of G.

[4]Of course, we can also include the k-moderate linear categories for a fixed $k \ge 1$.

Acknowledgments

I would like to express my appreciation to Professor Hiroyuki Nakaoka for his helpful comments on the draft and for pointing out many typographical errors, and to Professor Teruyuki Yorioka for his comments on the draft of the appendix on set theory. I would like to thank my family for their support, which made it easier for me to write this book; and I dedicate this book to my father-in-law, Shohjiroh Tozaki, who passed away while I was working on the manuscript. Finally, I would like to thank Mr. Ryohei Ohmizo, editor of Saiensu-sha, for encouraging me to write this book and patiently waiting for me to complete it, and Mr. Kohsuke Hirase for his careful proofreading of the book, despite a major correction in the proofreading process.

October, 2019 Hideto Asashiba

Preliminaries and Conventions

As axioms of sets we take ZFC (the Zermelo–Fraenkel axioms with the axiom of choice) and add one more axiom (U), the *axiom of universe*, which states that any set is an element of some (Grothendieck) universe. We set SET to be the class of all sets. Throughout this book, we fix an infinite universe $\mathfrak{U}$ (i.e. a Grothendieck universe $\mathfrak{U} \ni \mathbb{N}$). Elements (resp. subsets) of $\mathfrak{U}$ are called *$\mathfrak{U}$-small sets* (resp. *$\mathfrak{U}$-classes*), which we simply call *small sets* (resp. *classes*) if there seems to be no confusion. A small set is equal to 0-class, a class is equal to 1-class, and a k-class ($\in \mathsf{SET}$) is defined in Levy's paper **[33]** for all $k \geq 2$.

A category $\mathscr{C}$ is called *small* (resp. *light*, *k-moderate*) if its object set $\mathscr{C}_0$ is a small set (resp. a class, k-class) and $\mathscr{C}(x, y)$ are small (resp. small, k-class) for all $x, y \in \mathscr{C}_0$. (A 1-moderate categories are simply called *moderate categories.*) We usually deal with light categories.

In the sequel, we fix a commutative ring $\Bbbk$ whose base set is small, and all linear spaces and linear categories are considered over $\Bbbk$. Unless otherwise stated, the base set of M is assumed to be small for all $\Bbbk$-modules M.

Conventions. Throughout this book, we use the following conventions.

(1) "a is A, b is B, $\cdots$, and c is C" or "$a, b, \cdots c$ are A, B, C, respectively" will be written as "a (resp. b, $\ldots$, c) is A (resp. B, $\ldots$, C)."

(2) The set of integers is denoted by $\mathbb{Z}$, and non-negative integers are called *natural numbers*, the set of which is denoted by $\mathbb{N}$. Thus $0 \in \mathbb{N}$ in this book.

(3) For a set S, we write $(x \in S)$ to mean "for each $x \in S$". We write $(\exists x \in S)$ to mean that there exists some $x \in S$.

(4) For $\Bbbk$-modules V, W, we write $V \leq W$ to mean that V is a submodule of W.

(5) For a family $(X_i)_{i \in I}$ of sets indexed by a set I, we denote by

$$\bigsqcup_{i \in I} X_i := \bigcup_{i \in I} \{(i, x) \mid x \in X_i\}$$

their *disjoint union*. If $(X_i)_{i \in I}$ is disjoint from the beginning, namely if $X_i \cap X_j = \emptyset$ $(i \neq j, i, j \in I)$, then we identify as $\bigsqcup_{i \in I} X_i = \bigcup_{i \in I} X_i$.

(6) If we consider a finite family $(X, Y, \ldots, Z)$ (n sets) of sets without using an indexed set, then we set

$$X \sqcup Y \sqcup \cdots \sqcup Z := \bigsqcup_{i=0}^{n-1} S_i$$

by setting $S_0 := X, S_1 := Y, \ldots, S_{n-1} := Z$. For instance, $X \sqcup Y := \{(0, x) \mid x \in X\} \cup \{(1, y) \mid y \in Y\}$.

(7) For elements i, j of a set I, we set $\delta_{i,j}$ to be the Kronecker's delta:

$$\delta_{i,j} = \begin{cases} 1 & (i = j), \\ 0 & (i \neq j). \end{cases}$$

(8) Let I be a set, $(A_i)_{i \in I}$ a family of abelian groups. For each $i \in I$, zero of A_i is different from each other and we have to write it 0_i to distinguish them. But we identify those zeros and denote them by the symbol 0, by which the symbol $\delta_{i,j}$ becomes useful. For instance, choose one $j \in I$ and take $a \in A_j$. Then we write the notation $(\delta_{i,j} a)_{i \in I} \in \prod_{i \in I} A_i$ in the following sense: For each $i \in I$, we have

$$\delta_{i,j} a = \begin{cases} a & (i = j), \\ 0_i & (i \neq j). \end{cases}$$

If $i \neq j$, then since originally $a \in A_j$, we have to have $0a = 0_j$, but we read it $0_j = 0 = 0_i$ because $0a = \delta_{i,j} a \in A_i$.

(9) The mapping $x \mapsto f(x)$ is denoted by $f(\text{-})$ or by $f(?)$. In particular, for the mapping of two variables $(x, y) \mapsto f(x, y)$ we write $f(\text{-}, ?)$ or $f(?, \text{-})$ to emphasize that we have two kinds of variables. For instance, $x \mapsto 2x$ is denoted by $2 \cdot (\text{-})$, by $2(\text{-})$, or by $2?$, and $(M, N) \mapsto (\operatorname{Mod} \mathscr{C})(M, N)$ is denoted by $(\operatorname{Mod} \mathscr{C})(\text{-}, ?)$[5].

(10) We denote by $\mathbf{1}$ a discrete category with a unique object $*$.

[5]Note that this is not garbled.

CHAPTER 1

Categories

1.1. Monoids and categories

DEFINITION 1.1.1. A *monoid* is a sequence of the following data that satisfies the axioms below:

Data:

- A non-empty set G,
- A map $\mu\colon G \times G \to G$, $\mu(x, y) =: xy$ $(x, y \in G)$,
- An elefment 1 of G.

Axioms:

- $(xy)z = x(yz)$ $(x, y, z \in G)$,
- $x1 = x = 1x$ $(x \in G)$.

The set G (resp. μ, 1) above is called the *base set* (resp. the *operation*, the *unit*) of this monoid. A monoid whose base set is small is called a *small monoid*. All monoids are assumed to be small monoids unless otherwise stated. A monoid with $G = \{1\}$ is called a *trivial monoid* (see Remark 1.1.15).

EXERCISE 1.1.2. Let $(G, \mu, 1)$ be a monoid. Define a map $\mu^{\mathrm{op}}\colon G \times G \to G$ by $\mu^{\mathrm{op}}(x, y) := \mu(y, x)$ $(x, y \in G)$. Then show that the triple $(G, \mu^{\mathrm{op}}, 1)$ turns out to be a monoid that is called the *opposite monoid* of $(G, \mu, 1)$.

DEFINITION 1.1.3. A *group* is a sequence of the following data that satisfies the axioms below:

Data:

- a monoid $(G, \mu, 1)$,
- a map $\iota\colon G \to G, \iota(x) =: x^{-1}$ $(x \in G)$.

Axioms:

- $x^{-1}x = 1 = xx^{-1}$ $(x \in G)$.

We call x^{-1} above the *inverse* of x. A group is called a *small group* if its base set is small. In what follows, all groups are assumed to be small groups unless otherwise stated.

EXAMPLE 1.1.4. Let $+$ be the addition of $\mathbb{Z}$, and $-\colon \mathbb{Z} \to \mathbb{Z}$ the function $x \mapsto -x$ $(x \in \mathbb{Z})$. Then $\mathbb{Z} := (\mathbb{Z}, +, 0, -)$ is a group.

EXERCISE 1.1.5. Let $(G, \mu, 1, \iota)$ be a group. Then show that $(G, \mu^{\mathrm{op}}, 1, \iota)$ is also a group, which is called the *opposite group* of $(G, \mu, 1, \iota)$.

First, we give a general definition of quivers and categories.

DEFINITION 1.1.6. A *quiver* is a sequence $Q = (Q_0, Q_1, s, t)$ of the following data:

- sets Q_0, Q_1 (not necessarily small sets or classes, i.e., $Q_0, Q_1 \in \mathsf{SET}$)
- maps $s, t \colon Q_1 \to Q_0$.

For each $x, y \in Q_0$ we set

$$Q(x,*) := \{\alpha \in Q_1 \mid s(\alpha) = x\}, Q(*,y) := \{\alpha \in Q_1 \mid t(\alpha) = y\}, \text{ and}$$
$$Q(x,y) := Q(x,*) \cap Q(*,y).$$

Q is called a *finite quiver* if both Q_0 and Q_1 are finite sets, and is said to be *locally finite* if $Q(x,*) \cup Q(*,x)$ is finite for all $x \in Q_0$. The quiver Q is illustrated as an oriented graph by drawing each element $\alpha \in Q_1$ as an arrow $s(\alpha) \xrightarrow{\alpha} t(\alpha)$ (if Q_0, Q_1 are actually small enough to draw). For this reason, we call elements of Q_0 (resp. Q_1) *vertices* (resp. *arrows*) of Q; and for each $\alpha \in Q_1$, $s(\alpha)$ and $t(\alpha)$ are called the *source* and the *target* of α, respectively. Note that if $Q = (Q_0, Q_1, s, t)$ is a quiver, then so is (Q_0, Q_1, t, s), which is called the *opposite quiver* of Q, and is denoted by Q^{op}.

REMARK 1.1.7 (A remark on giving quivers). When a quiver Q is given as $Q = (Q_0, Q_1, s, t)$ as in the definition, we obtain the data $(Q_0, (Q(x,y))_{x,y \in Q_0})$ that satisfies the following condition:

$$(x,y) \neq (x',y') \Rightarrow Q(x,y) \cap Q(x',y') = \emptyset, \quad (x,y,x',y' \in Q_0). \tag{1.1}$$

Indeed, if $Q(x,y) \cap Q(x',y') \neq \emptyset$, then there exists some $a \in Q(x,y) \cap Q(x',y')$, which causes $(x,y) = (x',y')$ because $x = s(a) = x'$, and $y = t(a) = y'$. It is also possible to reconstruct the original data of the quiver from this. Namely, $Q_1 = \bigsqcup_{x,y \in Q_0} Q(x,y)$ and s, t are determined as follows: By the condition (1.1), for each $a \in Q_1$ there exists a unique $(x,y) \in Q_0 \times Q_0$ such that $a \in Q(x,y)$, and then $s(a) = x, t(a) = y$. The two operations above are inverses of each other. Thus, a quiver may be defined as the following data satisfying the axiom below.

Data:

- a set Q_0,
- a family of sets $(Q(x,y))_{x,y \in Q_0}$.

Axiom: (1.1).
We often use this form when we define categories later.

DEFINITION 1.1.8. A *category* is a sequence of the following data that satisfies the axioms below.

Data:

- A quiver $\mathscr{C} := (\mathscr{C}_0, \mathscr{C}_1, \mathrm{dom}, \mathrm{cod})$,
- A family of maps $\circ := (\circ_{x,y,z} \colon \mathscr{C}(y,z) \times \mathscr{C}(x,y) \to \mathscr{C}(x,z))_{(x,y,z) \in \mathscr{C}_0 \times \mathscr{C}_0 \times \mathscr{C}_0}$,
- A family $\mathbb{1} := (\mathbb{1}_x)_{x \in \mathscr{C}_0}$, $\mathbb{1}_x \in \mathscr{C}(x,x)$ of elements of $\mathscr{C}_1$.

Axioms:

(associativity) $(h \circ g) \circ f = h \circ (g \circ f)$
$(x,y,z,w \in \mathscr{C}_0,\ (h,g,f) \in \mathscr{C}(z,w) \times \mathscr{C}(y,z) \times \mathscr{C}(x,y))$,

(unitality) $f \circ \mathbb{1}_x = f = \mathbb{1}_y \circ f\ (x,y \in \mathscr{C}_0,\ f \in \mathscr{C}(x,y))$.

In a category $\mathscr{C}$, we call an element of $\mathscr{C}_0$ (resp. $\mathscr{C}_1$) an *object* of (resp. a *morphism* in) $\mathscr{C}$, and $\mathscr{C}_0$ (resp. $\mathscr{C}_1$, $\mathscr{C}(x,y)$ $(x,y \in \mathscr{C}_0)$) the *object set* (resp. the *morphism set*, the *local morphism set* (from x to y)) of $\mathscr{C}$. The morphism $\mathbb{1}_x$ $(x \in \mathscr{C})$ is called the *identity* of x.

A category $\mathscr{C}$ is said to be *locally small* if $\mathscr{C}(x,y)$ is small for all $x, y \in \mathscr{C}_0$. A locally small category $\mathscr{C}$ is called a *light category* (resp. a *small category*) if $\mathscr{C}_0$ is a class (resp. a small set) (cf. Definition A.3.1).

REMARK 1.1.9. As defined above, this book deals only with those categories $\mathscr{C}$ for which $\mathscr{C}_0 \in \mathsf{SET}$ and $\mathscr{C}(x,y) \in \mathsf{SET}$ for all $x, y \in \mathscr{C}_0$. Those categories are usually called *small categories.* Although the exact name of a small category above is a $\mathfrak{U}$-small category, the meaning of "small" is always assumed to be defined by the fixed universe $\mathfrak{U}$, so that "$\mathfrak{U}$-" is omitted.

EXAMPLE 1.1.10. If $\mathscr{C} = (\mathscr{C}_0, \mathscr{C}_1, \mathrm{dom}, \mathrm{cod}, \circ, \mathbb{1})$ is a category, then so is

$$\mathscr{C}^{\mathrm{op}} := (\mathscr{C}_0, \mathscr{C}_1, \mathrm{cod}, \mathrm{dom}, \circ^{\mathrm{op}}, \mathbb{1}),$$

which is called the *opposite category* of $\mathscr{C}$, where $\circ^{\mathrm{op}} \colon \mathscr{C}^{\mathrm{op}}(y,x) \times \mathscr{C}^{\mathrm{op}}(z,y) \to \mathscr{C}^{\mathrm{op}}(z,x)$, $(x,y,z \in \mathscr{C}_0)$ is defined by $f \circ^{\mathrm{op}} g := g \circ f$, $(f \in \mathscr{C}^{\mathrm{op}}(y,x) = \mathscr{C}(x,y), g \in \mathscr{C}^{\mathrm{op}}(z,y) = \mathscr{C}(y,z))$

EXERCISE 1.1.11. Show that $\mathscr{C}^{\mathrm{op}}$ is a category in the above.

DEFINITION 1.1.12. Let $\mathscr{C}$ be a category, and let $x, y \in \mathscr{C}_0$, $f \in \mathscr{C}(x,y)$.

(1) f is called an *isomorphism* if there exists some $g \in \mathscr{C}(y,x)$ such that $g \circ f = \mathbb{1}_x$ and $f \circ g = \mathbb{1}_y$.
(2) When an isomorphism $f \in \mathscr{C}(x,y)$ exists, we say that x and y are *isomorphic*, and denote it by $x \cong y$.

REMARK 1.1.13. In Definition 1.1.12, g is uniquely determined for each isomorphism f. Indeed, for any $h \in \mathscr{C}(y,x)$ that has the same property as g, the equality $(g \circ f) \circ h = g \circ (f \circ h)$ shows that $\mathbb{1}_z \circ h = g \circ \mathbb{1}_y$, and hence $h = g$. This g is called the *inverse* of f, and is denoted by f^{-1}.

DEFINITION 1.1.14. A category is called a *sigleton* if its object set consists of exactly one element. Sometimes the unique element is denoted by $*$.

REMARK 1.1.15. A sigleton can be identified with a monoid. In this sense, a category can be regarded as a monoid *with multiple objects* (a "multiobjectificaiton" of a monoid). We denote by $\mathbf{1}$ the trivial monoid viewed as a category: $\mathbf{1} = (\{*\}, \{\mathbb{1}_*\}, \mathrm{dom}, \mathrm{cod}, \circ, \mathbb{1})$.

Indeed,

- If $\mathscr{C} = (\mathscr{C}_0, \mathscr{C}_1, \mathrm{dom}, \mathrm{cod}, \circ, \mathbb{1})$ is a singleton, and $\mathscr{C}_0 = \{*\}$, then we have $\mathscr{C}_1 = \mathscr{C}(*,*)$, and
$$G_{\mathscr{C}} := (\mathscr{C}_1, \circ_{*,*,*}, \mathbb{1}_*)$$
become a monoid.
- Conversely, if $(G, \mu, 1)$ is a monoid, then the maps $\mathrm{dom}, \mathrm{cod} \colon G \to \{*\}$ are uniquely defined by $\mathrm{dom}(\alpha) := * =: \mathrm{cod}(\alpha)$ $(\alpha \in G)$, and
$$\mathscr{C}_G := (\{*\}, G, \mathrm{dom}, \mathrm{cod}, (\mu)_{(x,y,z) \in \{*\}\times\{*\}\times\{*\}}, (1)_{x \in \{*\}})$$
becomes a singleton.
- The two constructions above are inverses of each other.

EXAMPLE 1.1.16. An arbitrary set $S \in \mathsf{SET}$ can be regarded as a category. Namely, we can identify it with the category $\mathscr{C}$ defined as follows, which is called the *discrete category* defined by S.

- $\mathscr{C}_0 := S$.
- For each $x, y \in S$, $\mathscr{C}(x,y) := \begin{cases} \{\mathbb{1}_x\} & (x = y); \\ \emptyset & (x \neq y). \end{cases}$
- Let $x, y, z \in \mathscr{C}_0$. We define a map
$$\circ_{x,y,z} \colon \mathscr{C}(y,z) \times \mathscr{C}(x,y) \to \mathscr{C}(x,z)$$
as follows: When $x = y = z$, $\circ_{x,y,z}$ is uniquely determined as $\mathbb{1}_x \circ \mathbb{1}_x := \mathbb{1}_x$. Otherwise, we may set $\circ_{x,y,z}$ to be the trivial map because $\mathscr{C}(y,z) \times \mathscr{C}(x,y) = \emptyset$.

In particular, the discrete category defined by the set $\{*\}$ with only one element turns out to be $\mathbf{1}$, the trivial monoid viewed as a category.

EXAMPLE 1.1.17. Any preordered set $(S, \leq)$ can be regarded as a category. Namely, it can be identified with the category $\mathscr{C}$ defined as follows:

- $\mathscr{C}_0 := S$.
- For each $x, y \in S$ $\mathscr{C}(x,y) := \begin{cases} \{(y,x)\} & (y \geq x); \\ \emptyset & (\text{otherwise}). \end{cases}$
- Let $x, y, z \in \mathscr{C}_0$. Then we define a map
$$\circ_{x,y,z} \colon \mathscr{C}(y,z) \times \mathscr{C}(x,y) \to \mathscr{C}(x,z)$$
as follows: If $z \geq y \geq x$, then both $\mathscr{C}(y,z) \times \mathscr{C}(x,y)$ and $\mathscr{C}(x,z)$ are singletons, the map is uniquely determined $\circ_{x,y,z}$. Namely, we set $(z,y) \circ (y,x) := (z,x)$. Otherwise, we have $\mathscr{C}(y,z) \times \mathscr{C}(x,y) = \emptyset$, and hence $\circ_{x,y,z}$ should be the trivial map.

For instance, let S be a set consisting of integers, and define a preorder $x \leq y$ by the condition that x divides y. We denote by $\mathscr{C}(S, |)$ the category defined as above from these data. For example, the category $\mathscr{C}(\{-1,1\}, |)$ is presented by the following quiver and the composition defined above:

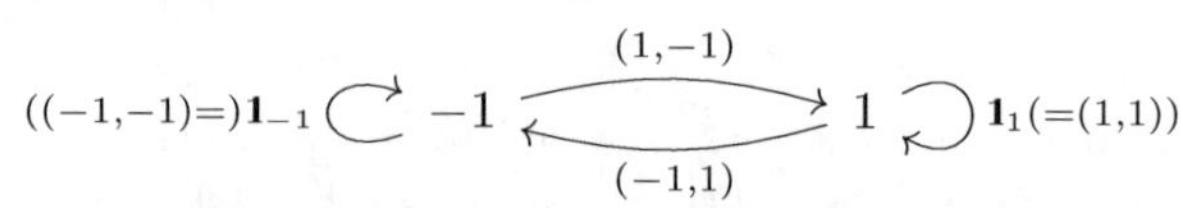

EXAMPLE 1.1.18. The category $\mathbf{Set} = (\mathbf{Set}_0, \mathbf{Set}_1, \mathrm{dom}, \mathrm{cod}, \circ, \mathbb{1})$ defined as follows is called the *category of small sets*.

- $\mathbf{Set}_0$ is the set of all small sets (i.e., $\mathbf{Set}_0 := \mathfrak{U}$).
- $\mathbf{Set}_1$ is the set of all maps between small sets.
- For each $f \colon X \to Y$ in $\mathbf{Set}_1$, $\mathrm{dom}(f) := X, \mathrm{cod}(f) := Y$.
- For each $A, B, C \in \mathbf{Set}_0$, we define the map
$$\circ_{A,B,C} \colon \mathbf{Set}(B,C) \times \mathbf{Set}(A,B) \to \mathbf{Set}(A,C)$$
as the usual composition of maps between sets.
- $\mathbb{1} := (\mathbb{1}_A)_{A \in \mathbf{Set}_0}$, where for each $A \in \mathbf{Set}_0$, $\mathbb{1}_A \colon A \to A$ is the identity map of A.

As is seen from the definition, $\mathbf{Set}$ is a light category.

1.1.a. Monomorphisms and epimorphisms. By characterizing the properties of monomorphisms and epimorphisms in the category **Set** of small sets in terms of categories, these concepts are defined in general categories as follows:

DEFINITION 1.1.19. Let $\mathscr{C}$ be a category, and $f\colon x \to y$ a morphism in $\mathscr{C}$.

(1) The morphism f is called a *monomorphism* if $f \circ a = f \circ b$ implies $a = b$ for all $a, b\colon z \to x$ and all $z \in \mathscr{C}_0$.
(2) The morphism f is called an *epimorphism* if $a \circ f = b \circ f$ implies $a = b$ for all $a, b\colon y \to z$ and all $z \in \mathscr{C}_0$.

If we reverse all the directions of arrows (i.e., considering in the opposite category) (1) turns out to be (2) and (2) turns out to be (1). In such a case, we say that (1) and (2) are *dual* to each other.

REMARK 1.1.20. Let $f\colon x \to y$ be a morphism in a category $\mathscr{C}$. Then the conditions required to be a monomorphism (resp. an epimorphism) are restated in terms of maps as follows.

(1) The morphism f is a monomorphism if and only if the map
$$\mathscr{C}(z, f)\colon \mathscr{C}(z, x) \to \mathscr{C}(z, y),\ a \mapsto f \circ a$$
is a monomorphism (between sets) for all $z \in \mathscr{C}_0$.
(2) The morphism f is an epimorphism if and only if the map
$$\mathscr{C}(f, z)\colon \mathscr{C}(y, z) \to \mathscr{C}(x, z),\ a \mapsto a \circ f$$
is a monomorphism (between sets) for all $z \in \mathscr{C}_0$.
(3) Note in the above that both a monomorphism and an epimorphism in a category become a monomorphism (an injection) in terms of maps.

Then, what does it mean to become an epimorphism (a surjection) in terms of maps? To answer this question, we prepare the following terminologies.

DEFINITION 1.1.21. Let $f\colon x \to y$ be a morphism in a category $\mathscr{C}$.

(1) The morphism f is called a *section* if there exists a morphism $g\colon y \to x$ such that $g \circ f = \mathbb{1}_x$. This is also called a *split monomorphism*.
(2) Dually, The morphism f is called a *retraction* if there exists a morphism $g\colon y \to x$ such that $f \circ g = \mathbb{1}_y$. This is also called a *split epimorphism*.

The following proposition gives an answer to the question above.

PROPOSITION 1.1.22. *Let $f\colon x \to y$ be a morphism in a category $\mathscr{C}$.*

(1) *The morphism f is a section if and only if the map*
$$\mathscr{C}(f, z)\colon \mathscr{C}(y, z) \to \mathscr{C}(x, z),\ a \mapsto a \circ f$$
is an epimorphism (between sets) for all $z \in \mathscr{C}_0$.
(2) *The morphism f is a retraction if and only if the map*
$$\mathscr{C}(z, f)\colon \mathscr{C}(z, x) \to \mathscr{C}(z, y),\ a \mapsto f \circ a$$
is an epimorphism (between sets) for all $z \in \mathscr{C}_0$.

PROOF. Since both can be proved similarly, we will show only (1).

($\Rightarrow$). Assume that f is a section. Then we have $g \circ f = \mathbb{1}_x$ for some $g\colon y \to x$. Then for each $z \in \mathscr{C}_0$, the map
$$\mathscr{C}(g, z)\colon \mathscr{C}(x, z) \to \mathscr{C}(y, z),\ a \mapsto a \circ g$$

satisfies $\mathscr{C}(f,z)\circ\mathscr{C}(g,z) = \mathbb{1}_{\mathscr{C}(x,z)}$ because $(\mathscr{C}(f,z)\circ\mathscr{C}(g,z))(a) = a\circ g\circ f = a$ for all $a \in \mathscr{C}(x,z)$. Thus $\mathscr{C}(g,z)$ is an epimorphism (between sets).

($\Leftarrow$). Assume that $\mathscr{C}(f,z)$ is an epimorphism (between sets) for all $z \in \mathscr{C}_0$. Apply this fact to $z = x$ to have that $\mathscr{C}(f,x)\colon \mathscr{C}(y,x) \to \mathscr{C}(x,x)$ is an epimorphism (between sets). Then there exists a morphism $g \in \mathscr{C}(y,x)$ such that $\mathscr{C}(f,x)(g) = \mathbb{1}_x$ because $\mathbb{1}_x \in \mathscr{C}(x,x)$. Here since $\mathscr{C}(f,x)(g) = g\circ f$, we have $g \circ f = \mathbb{1}_x$, and f is a section. □

1.1.b. Products and coproducts. Next, we take up direct products and disjoint unions in the category **Set** of small sets. By characterizing their properties in terms of categories, these concepts are defined in general categories as follows:

DEFINITION 1.1.23. Let $\mathscr{C}$ be a category, I a set, and $(x_i)_{i\in I}$ a family of objects in $\mathscr{C}$.

(1) A *product* of the family $(x_i)_{i\in I}$ is a sequence of the following data that has the property below.

Data:
- an object x in $\mathscr{C}$ (called a *product object*),
- a family $\pi := (\pi_i\colon x \to x_i)_{i\in I}$ of morphisms in $\mathscr{C}$ (called a *projection family*).

Universality of products: For each $y \in \mathscr{C}_0$, each family $\rho := (\rho_i\colon y \to x_i)_{i\in I}$ of morphisms in $\mathscr{C}$ uniquely factors through π. Namely, there exists a unique morphism $f\colon y \to x$ in $\mathscr{C}$ such that the diagram

(1.2)
$$\begin{array}{ccc} y & \xrightarrow{\quad f \quad} & x \\ & {\scriptstyle \rho_i}\searrow \quad \swarrow{\scriptstyle \pi_i} & \\ & x_i & \end{array}$$

commutes for all $i \in I$.

We also call this property the *universality* of the projection family π, and this f the *canonical* morphism determined by this universality.

(2) A *coproduct* of the family $(x_i)_{i\in I}$ is a sequence of the following data that has the property below.

Data:
- an object x in $\mathscr{C}$ (called a *coproduct object*),
- a family $\sigma := (\sigma_i\colon x_i \to x)_{i\in I}$ of morphisms in $\mathscr{C}$ (called an *injection family*).

Universality of coproducts: For each $y \in \mathscr{C}_0$, each family $\tau := (\tau_i\colon x_i \to y)_{i\in I}$ of morphisms in $\mathscr{C}$ uniquely factors through σ. Namely, there exists a unique morphism $f\colon x \to y$ in $\mathscr{C}$ such that the diagram

(1.3)
$$\begin{array}{ccc} & x_i & \\ & {\scriptstyle \sigma_i}\swarrow \quad \searrow{\scriptstyle \tau_i} & \\ x & \xrightarrow{\quad f \quad} & y \end{array}$$

commutes for all $i \in I$.

We also call this property the *universality* of the injection family σ, and this f the *canonical* morphism determined by this universality.

EXERCISE 1.1.24. In the category **Set** of small sets, show that direct products of sets give products, and disjoint unions of sets give coproducts. Namely, when I is a small set and $(X_i)_{i\in I}$ is a family of small sets, show the following:

(1) For each $i \in I$ let π_i be the i-th standard projections

$$\pi_i \colon \prod_{j\in I} X_j \to X_i, \ (x_j)_{j\in I} \mapsto x_i.$$

Then the pair $(\prod_{i\in I} X_i, (\pi_i)_{i\in I})$ turns out to be a product of $(X_i)_{i\in I}$.

(2) For each $i \in I$ let σ_i be the i-th standard injections

$$\sigma_i \colon X_i \to \bigsqcup_{i\in I} X_i, \ x \mapsto (i, x).$$

Then the pair $(\bigsqcup_{i\in I} X_i, (\sigma_i)_{i\in I})$ turns out to be a coproduct of $(X_i)_{i\in I}$.

LEMMA 1.1.25. *Let $\mathscr{C}$ be a category, I a set, and $(x_i)_{i\in I}$ a family of objects in $\mathscr{C}$.*

(1) *If both $(x, (\pi_i)_{i\in I})$ and $(y, (\rho_i)_{i\in I})$ are products of $(x_i)_{i\in I}$, then the canonical morphism f (the uniquely existing morphism f) in the diagram* (1.2) *is an isomorphism. In this sense, the two projection families π and ρ are (canonically) isomorphic. In particular, the product objects x and y are (canonically) isomorphic. By this reason, we denote the product object of $(x_i)_{i\in I}$ by $\prod_{i\in I} x_i$.*

(2) *Dually, if both $(x, (\sigma_i)_{i\in I})$ and $(y, (\tau_i)_{i\in I})$ are coproducts of $(x_i)_{i\in I}$, then the canonical morphism f (the uniquely existing morphism f) in the diagram* (1.3) *is an isomorphism. In this sense, the two injection families σ and τ are (canonically) isomorphic. In particular, the coproduct objects x and y are (canonically) isomorphic. By this reason, we denote the coproduct object of $(x_i)_{i\in I}$ by $\coprod_{i\in I} x_i$.*

PROOF.

(1) Let $(x, (\pi_i)_{i\in I})$ and $(y, (\rho_i)_{i\in I})$ products of $(x_i)_{i\in I}$. Since $(x, (\pi_i)_{i\in I})$ is a product of $(x_i)_{i\in I}$, there exists a unique morphism $f \colon y \to x$ such that $\rho_i = \pi_i \circ f$ for all $i \in I$. Since also $(y, (\rho_i)_{i\in I})$ is a product of $(x_i)_{i\in I}$, there exists a unique morphism $g \colon x \to y$ such that $\pi_i = \rho_i \circ g$ for all $i \in I$. These form the following commutative diagram:

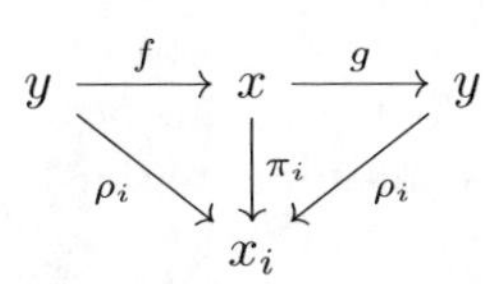

$(i \in I)$.

Hence we have $\rho_i = \rho_i \circ (g \circ f)$ for all $i \in I$. Further, we also have $\rho_i = \rho_i \circ \mathbb{1}_y$ for all $i \in I$. Thus by the universality of $(\rho_i)_{i\in I}$ we have $g \circ f = \mathbb{1}_y$. By the symmetry of the argument we have$f \circ g = \mathbb{1}_x$. Therefore f is an isomorphism.

(2) This is proved by making the argument above in $\mathscr{C}^{\mathrm{op}}$.

□

REMARK 1.1.26. The notions of (1) and (2) in Definition 1.1.23 are also dual to each other. The universalities of this definition can be rephrased in terms of maps as follows.

(1) The universality of a product is equivalent to saying that for each $y \in \mathscr{C}_0$ the map

$$(1.4)\qquad (\mathscr{C}(y,\pi_i))_{i\in I}\colon \mathscr{C}(y, \prod_{i\in I} x_i) \to \prod_{i\in I}\mathscr{C}(y,x_i),\ f \mapsto (\pi_i \circ f)_{i\in I}$$

to the direct product of sets is a bijection. We denote the inverse of this map by

$$(\mathscr{C}(y,\pi_i))_{i\in I}{}^{-1}\colon \prod_{i\in I}\mathscr{C}(y,x_i) \to \mathscr{C}(y, \prod_{i\in I} x_i),\ (f_i)_{i\in I} \mapsto [f_i]_{i\in I}.$$

Therefore if we set $f_i := \pi_i \circ f$ $(i \in I)$, then we have $f = [f_i]_{i\in I}$, which is called the *column vector presentation* of f with respect to the projection family $(\pi_i)_{i\in I}$. The canonical morphism $[f_i]_{i\in I}$ is characterized by the formula $f_i = \pi_i \circ [f_i]_{i\in I}$. In particular, when I is a finite set $I = \{1, 2, \dots, n\}$ for some integer $n \geq 1$, we denote $[f_i]_{i\in I}$ by $[f_i]_{i=1}^n$ or $\begin{bmatrix} f_1 \\ f_2 \\ \vdots \\ f_n \end{bmatrix}$.

(2) The universality of a coproduct is equivalent to saying that for each $y \in \mathscr{C}_0$, the map

$$(1.5)\qquad (\mathscr{C}(\sigma_i,y))_{i\in I}\colon \mathscr{C}(\coprod_{i\in I} x_i, y) \to \prod_{i\in I}\mathscr{C}(x_i,y),\ f \mapsto (f\circ\sigma_i)_{i\in I}$$

to the direct product of sets is a bijection. We denote the inverse of this map by

$$(\mathscr{C}(\sigma_i,y))_{i\in I}{}^{-1}\colon \prod_{i\in I}\mathscr{C}(x_i,y) \to \mathscr{C}(\coprod_{i\text{ß}I} x_i, y),\ (f_i)_{i\in I} \mapsto {}^t[f_i]_{i\in I},$$

where ^{t}[-] denotes the transpose. Therefore if we set $f_i := f \circ \sigma_i$ $(i \in I)$, then we have $f = {}^t[f_i]_{i\in I}$, which is called the *row vector presentation* of f with respect to the injection family $(\sigma_i)_{i\in I}$. The canonical morphism ${}^t[f_i]_{i\in I}$ is characterized by the formula $f_i = {}^t[f_i]_{i\in I} \circ \sigma_i$. In particular, when I is a finite set $I = \{1, 2, \dots, n\}$ for some integer $n \geq 1$, we denote ${}^t[f_i]_{i\in I}$ by ${}^t[f_i]_{i=1}^n$ or $[f_1, f_2 \dots, f_n]$.

Note that in the formulas (1.4) and (1.5), when we look at them from left to right, the product notation $\prod$ comes out of the second variable position (the covariant position) as $\prod$ (unchanged) in (1.4), but the coproduct notation $\coprod$ comes out of the first variable position (the contravariant position) as $\prod$ (reversed up side down) in (1.5). In both cases what comes out is the $\prod$, which denotes the direct product of sets.

By combining bijections in formulas (1.4) and (1.5), we immediately obtain the following.

PROPOSITION 1.1.27. *Let $\mathscr{C}$ be a category, I, J sets, and $(x_i)_{i\in I}, (y_j)_{j\in J}$ families of objects in $\mathscr{C}$. Furthermore, let $(\sigma_i\colon x_i \to \coprod_{i\in I} x_i)_{i\in I}, (\pi_j\colon \prod_{j\in J} y_j \to y_j)_{j\in J}$ be an injection family of the coproduct of $(x_i)_{i\in I}$ and a projection family of the product of $(y_j)_{j\in J}$, respectively. Then the map*

$$(\mathscr{C}(\sigma_i,\pi_j))_{j,i}\colon \mathscr{C}(\coprod_{i\in I} x_i, \prod_{j\in J} y_j) \to \prod_{j\in J}\prod_{i\in I}\mathscr{C}(x_i,y_j), f \mapsto (\pi_j\circ f\circ\sigma_i)_{j\in J, i\in I}$$

is bijective[1]. *We denote the inverse* $(\mathscr{C}(\sigma_i, \pi_j))_{j,i}{}^{-1}$ *by*

$$\prod_{j\in J}\prod_{i\in I}\mathscr{C}(x_i, y_j) \to \mathscr{C}(\coprod_{i\in I} x_i, \prod_{j\in J} y_j), (f_{j,i})_{j\in J, i\in I} \mapsto [f_{j,i}]_{j\in J, i\in I}.$$

Therefore if we set $f_{j,i} := \pi_j \circ f \circ \sigma_i$ $(i \in I, j \in J)$, *then it holds that* $f = [f_{j,i}]_{j\in J, i\in I}$. *We call this the* matrix presentation *of* f *with respect to* $(\sigma_i)_{i\in I}, (\pi_j)_{j\in J}$. *The canonical morphism* $[f_{j,i}]_{j\in J, i\in I}$ *is characterized by the formula*

$$f_{j,i} = \pi_j \circ [f_{j,i}]_{j\in J, i\in I} \circ \sigma_i \qquad (i \in I, j \in J).$$

1.2. Monoid homomorphisms and functors

DEFINITION 1.2.1. Let $G = (G, \mu, 1)$ and $G' = (G', \mu, 1)$ be monoids. Then a map $f\colon G \to G'$ satisfying the axioms below is called a *monoid homomorphism* from G to G'.

Axioms:

- $f(1) = 1$,
- $f(ab) = f(a)f(b)$ $(a, b \in G)$.

DEFINITION 1.2.2. Let $Q := (Q_0, Q_1, s, t)$ and $Q' := (Q'_0, Q'_1, s', t')$ be quivers. Then a *quiver morphism* (resp. a *contravariant quiver morphism*) is a pair of the following data that satisfies the axiom below, and we denote it by $(f_0, f_1)\colon Q \to Q'$:

Data:

- a map $f_0\colon Q_0 \to Q'_0$,
- a map $f_1\colon Q_1 \to Q'_1$.

Axiom:

- If $\alpha\colon x \to y$ in Q, then $f_1(\alpha)\colon f_0(x) \to f_0(y)$ (resp. $f_1(\alpha)\colon f_0(y) \to f_0(x)$) in Q'.

If there seems to be no confusion, then we denote f_0, f_1 and (f_0, f_1) simply by f. The axiom of quiver morphisms can be rephrased that both of the following diagrams commute:

$$\begin{CD} Q_1 @>s>> Q_0 \\ @V{f_1}VV @VV{f_0}V \\ Q'_1 @>>{s'}> Q'_0 \end{CD} \qquad \begin{CD} Q_1 @>t>> Q_0 \\ @V{f_1}VV @VV{f_0}V \\ Q'_1 @>>{t'}> Q'_0 \end{CD}$$

---

[1] Using the matrix presentation for a morphism from the coproduct to the product given above, we will later give the matrix presentation for a morphism from the coproduct to the *coproduct* in the module category. When objects x, y, z are presented as coproducts, in order to make the composite of matrix presentations of two morphisms $x \xrightarrow{f} y$, $y \xrightarrow{g} z$ the same as the usual product of matrices, it is enough to set $\pi_j \circ f \circ \sigma_i$ above to be the (j, i)-entry of the matrix (see Lemma 2.4.9). (This is caused by writing the composite of f and g "from right to left" as $g \circ f$.) By this reason, we wrote J on the left and I on the right on the right-hand side. To keep the compatibility with this matrix presentation we used the column vector notation in (1.4) and the row vector notation in (1.5).

DEFINITION 1.2.3. Let $\mathscr{C} := (\mathscr{C}_0, \mathscr{C}_1, \mathrm{dom}, \mathrm{cod}, \circ, \mathbb{1})$ and $\mathscr{C}' := (\mathscr{C}'_0, \mathscr{C}'_1, \mathrm{dom}', \mathrm{cod}', \circ, \mathbb{1})$ be categories. A *functor* (resp. a *contravariant functor*) from $\mathscr{C}$ to $\mathscr{C}'$ is a quiver morphism (resp. a contravariant quiver morphism) $F\colon \mathscr{C} \to \mathscr{C}'$ that satisfies the axioms below.

Axioms:

- $F(\mathbb{1}_x) = \mathbb{1}_{F(x)}$ $(x \in \mathscr{C}_0)$,
- $F(g \circ f) = F(g) \circ F(f)$ (resp. $F(g \circ f) = F(f) \circ F(g)$)
  $((g, f) \in \mathscr{C}(y, z) \times \mathscr{C}(x, y), x, y \in \mathscr{C}_0)$.

By abuse of notation we often denote F_0, F_1, and (F_0, F_1) simply by F. When contrasted with the contravariant functors, functors are called *covariant functors*. We can regard a contravariant functor $\mathscr{C} \to \mathscr{C}'$ as a covariant functor $\mathscr{C}^{\mathrm{op}} \to \mathscr{C}'$.

REMARK 1.2.4. A functor between singleton categories is nothing but a homomorphism between monoids. It is also clear from the definition that a functor preserves isomorphisms, i.e., if $F\colon \mathscr{C} \to \mathscr{C}'$ is a functor and $f\colon x \to y$ is an isomorphism in $\mathscr{C}$, then so is $F(f)\colon F(x) \to F(y)$ in $\mathscr{C}'$.

EXERCISE 1.2.5. Show that the following statements hold for a locally small category $\mathscr{C}$ and $x \in \mathscr{C}_0$.

(1) A functor $\mathscr{C}(x, \text{-})\colon \mathscr{C} \to \mathbf{Set}$ is defined as follows:

For each $y \in \mathscr{C}_0$, $y \mapsto \mathscr{C}(x, y)$, and for each morphism $y \xrightarrow{f} z$ in $\mathscr{C}$, $f \mapsto \mathscr{C}(x, f)\colon \mathscr{C}(x, y) \to \mathscr{C}(x, z)$, where $\mathscr{C}(x, f)(g) := f \circ g$ $(g \in \mathscr{C}(x, y))$. This functor is called the *representable* (covariant) functor represented by x.

(2) A functor $\mathscr{C}(\text{-}, x)\colon \mathscr{C}^{\mathrm{op}} \to \mathbf{Set}$ is defined as follows:

For each $y \in \mathscr{C}_0$, $y \mapsto \mathscr{C}(y, x)$, and for each morphism $z \xrightarrow{f} y$ in $\mathscr{C}$, $f \mapsto \mathscr{C}(f, x)\colon \mathscr{C}(y, x) \to \mathscr{C}(z, x)$, where $\mathscr{C}(f, x)(g) := g \circ f$ $(g \in \mathscr{C}(y, x))$. This functor is called the *representable* contravariant functor represented by x.

(3) If $F\colon \mathscr{C} \to \mathscr{D}$ is a functor, then a functor $F^{\mathrm{op}}\colon \mathscr{C}^{\mathrm{op}} \to \mathscr{D}^{\mathrm{op}}$ is defined by $F^{\mathrm{op}}(x) := F(x), F^{\mathrm{op}}(f) := F(f)$, $(x \in \mathscr{C}_0, f \in \mathscr{C}_1)$.

Now we define the direct product of categories, and introduce functors of two variables.

DEFINITION 1.2.6.

(1) For categories $\mathscr{C}$ and $\mathscr{D}$ we define a category $\mathscr{C} \times \mathscr{D}$ as follows, which is called the *direct product* of $\mathscr{C}$ and $\mathscr{D}$.

- The object set is given by

$$(\mathscr{C} \times \mathscr{D})_0 := \mathscr{C}_0 \times \mathscr{D}_0.$$

- For any $(x, y), (x', y') \in (\mathscr{C} \times \mathscr{D})_0$, we set

$$(\mathscr{C} \times \mathscr{D})((x, y), (x', y')) := \mathscr{C}(x, x') \times \mathscr{D}(y, y').$$

- For each pair of composable morphisms $(x, y) \xrightarrow{(f,g)} (x', y') \xrightarrow{(f',g')} (x'', y'')$, we define

$$(f', g') \circ (f, g) := (f' \circ f, g' \circ g)\colon (x, y) \to (x'', y'').$$

- For each $(x,y) \in (\mathscr{C}\times\mathscr{D})_0$, the identity morphism is given by $\mathbb{1}_{(x,y)} = (\mathbb{1}_x, \mathbb{1}_y)$.

(2) A functor from the direct product of categories to a category is called a *functor of two variables*.

REMARK 1.2.7. In the setting of Definition 1.2.6, since each morphism $(f,g)\colon (x,y) \to (x',y')$ in $\mathscr{C}\times\mathscr{D}$ can be written as $(\mathbb{1}_x, g)\circ(f,\mathbb{1}_y) = (f,g) = (f,\mathbb{1}_y)\circ(\mathbb{1}_x,g)$, a functor of two variables $F\colon \mathscr{C}\times\mathscr{D}\to\mathscr{E}$ satisfies the equality $F(\mathbb{1}_x,g)\circ F(f,\mathbb{1}_y) = F(f,\mathbb{1}_y)\circ F(\mathbb{1}_x,g)$, i.e., the following diagram commutes:

$$\begin{array}{ccc} F(x,y) & \xrightarrow{F(x,g)} & F(x,y') \\ {\scriptstyle F(f,y)}\downarrow & & \downarrow{\scriptstyle F(f,y')} \\ F(x',y) & \xrightarrow[F(x',g)]{} & F(x',y'). \end{array}$$

We also define the direct product of functors.

DEFINITION 1.2.8. Let $F\colon \mathscr{C}\to\mathscr{D}$ and $F'\colon \mathscr{C}'\to\mathscr{D}'$ be functors. Then we define a functor $F\times F'\colon \mathscr{C}\times\mathscr{C}'\to\mathscr{D}\times\mathscr{D}'$ by $(F\times F')(x,x') := (F(x),F'(x'))$ and $(F\times F')(f,f') := (F(f),F'(f'))$ for all $(x,x'),(y,y')\in(\mathscr{C}\times\mathscr{C}')_0$ and for all $(f,f')\in(\mathscr{C}\times\mathscr{C}')((x,x'),(y,y')))$. It is easy to verify that this is a functor.

DEFINITION 1.2.9. Let $F\colon \mathscr{C}\to\mathscr{D}$ be a functor (resp. a quiver morphism) between categories (resp. quivers). Then for each $x,y\in\mathscr{C}_0$, a map $F_{y,x}\colon \mathscr{C}(x,y)\to\mathscr{D}(F(x),F(y))$ is defined by $f\mapsto F(f)$ for all $f\in\mathscr{C}(x,y)$.

(1) F is said to be *full* if $F_{y,x}$ is surjective for all $x,y\in\mathscr{C}_0$.
(2) F is said to be *faithful* if $F_{y,x}$ is injective for all $x,y\in\mathscr{C}_0$.
(3) F is said to be *dense* if for each $y\in\mathscr{D}_0$ there exists an $x\in\mathscr{C}_0$ such that $F(x)\cong y$ in $\mathscr{D}$. (This notion is only defined if $\mathscr{D}$ is a category.)

Usually F is said to be *fully faithful* if it is full and faithful.

In this connection, we give the definitions of subcategories and full subcategories.

DEFINITION 1.2.10. Let $\mathscr{C},\mathscr{D}$ be categories (resp. quivers). Then $\mathscr{D}$ is called a *subcategory* (resp. subquiver) of $\mathscr{C}$ if for each $i=0,1$, we have $\mathscr{D}_i\subseteq\mathscr{C}_i$, and the pair $F=(F_0,F_1)$ of the inclusions $F_i\colon\mathscr{D}_i\to\mathscr{C}_i$ is a functor (resp. a quiver morphism) $\mathscr{D}\to\mathscr{C}$. This functor is called an *inclusion functor*. When it is the case, $\mathscr{D}$ is called a *full subcategory* (resp. *full subquiver*) of $\mathscr{C}$ if F is full. Obviously, this is equivalent to saying that $\mathscr{D}(x,y)=\mathscr{C}(x,y)$ for all $x,y\in\mathscr{D}_0$.

1.3. Natural transformations

In order to consider relationships between functors, we use the concept of natural transformations.

DEFINITION 1.3.1. Let $\mathscr{C}$ and $\mathscr{D}$ be categories, and $F,F'\colon\mathscr{C}\to\mathscr{D}$ functors. A *natural transformation* from F to F' is a family $\alpha := (\alpha_x)_{x\in\mathscr{C}_0}\in\prod_{x\in\mathscr{C}_0}\mathscr{D}(F(x), F'(x))$ (we sometimes write αx for α_x) of morphisms that satisfies the axiom below and is denoted by $\alpha\colon F\Rightarrow F'$:

Axiom: The following diagram commutes for all morphisms $f\colon x \to y$ in $\mathscr{C}$:

$$\begin{array}{ccc} F(x) & \xrightarrow{\alpha_x} & F'(x) \\ {\scriptstyle F(f)}\downarrow & & \downarrow{\scriptstyle F'(f)} \\ F(y) & \xrightarrow[\alpha_y]{} & F'(y) \end{array} .$$

Here, the α above is called a *natural isomorphism* if α_x is an isomorphism in $\mathscr{D}$ for all $x \in \mathscr{C}_0$. Functors F and F' are said to be *isomorphic* if there exists a natural isomorphism $\alpha\colon F \Rightarrow F'$, which is denoted by $F \cong F'$ (see Remark 1.3.9 (1)).

EXAMPLE 1.3.2. Let $F\colon \mathscr{C} \to \mathscr{D}$ be a functor between categories. Then $\mathbb{1}_F := (\mathbb{1}_{F(x)})_{x\in\mathscr{C}_0}$ is a natural transformation $F \Rightarrow F$. This is called the *identity natural transformation* of F

EXERCISE 1.3.3. Let $\mathscr{C}$ be a locally small category, and $f\colon x \to y$ a morphism in $\mathscr{C}$. Prove the following.

(1) Define $\mathscr{C}(f,\text{-})\colon \mathscr{C}(y,\text{-}) \Rightarrow \mathscr{C}(x,\text{-})$ by

$$\mathscr{C}(f,\text{-}) := (\mathscr{C}(f,z)\colon \mathscr{C}(y,z) \to \mathscr{C}(x,z))_{z\in\mathscr{C}_0}.$$

Then this is a natural transformation.

(2) Define $\mathscr{C}(\text{-},f)\colon \mathscr{C}(\text{-},x) \Rightarrow \mathscr{C}(\text{-},y)$ by

$$\mathscr{C}(\text{-},f) := (\mathscr{C}(z,f)\colon \mathscr{C}(z,x) \to \mathscr{C}(z,y))_{z\in\mathscr{C}_0}.$$

Then this is a natural transformation.

EXERCISE 1.3.4. Let $\mathscr{C}$ be a category, I a set, and $(x_i)_{i\in I}$ a family of objects of $\mathscr{C}$. Show that we obtain the natural isomorphisms

$$\begin{cases} (\mathscr{C}(\text{-},\pi_i))_{i\in I} := ((\mathscr{C}(z,\pi_i))_{i\in I})_{z\in\mathscr{C}_0}\colon \mathscr{C}(\text{-},\prod_{i\in I} x_i) \xrightarrow{\sim} \prod_{i\in I}\mathscr{C}(\text{-},x_i), \\ (\mathscr{C}(\sigma_i,\text{-}))_{i\in I} := ((\mathscr{C}(\sigma_i,z))_{i\in I})_{z\in\mathscr{C}_0}\colon \mathscr{C}(\coprod_{i\in I} x_i,\text{-}) \xrightarrow{\sim} \prod_{i\in I}\mathscr{C}(x_i,\text{-}), \end{cases} \tag{1.6}$$

by the formulas (1.4), (1.5) in Remark 1.1.26.

DEFINITION 1.3.5. For a diagram

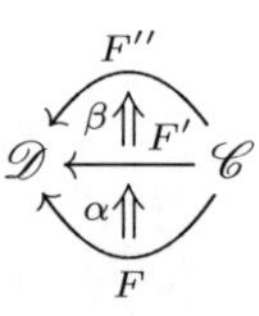

of categories, functors and natural transformations, $\beta \bullet \alpha := (\beta_x \circ \alpha_x\colon F(x) \to F''(x))_{x\in\mathscr{C}_0}$ turns out to be a natural transformation $F \Rightarrow F''$, which is called the *vertical composite* of α and β.

Next, we define composites of a functor and a natural transformation.

DEFINITION 1.3.6. For a diagram

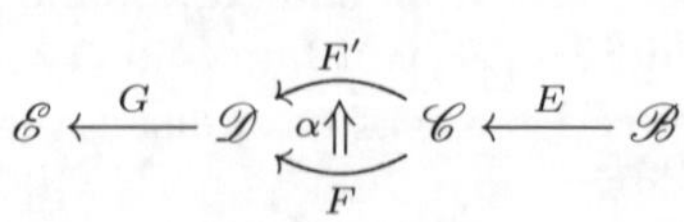

of categories, functors and natural transformations, the composites $\alpha \circ E$ and $G \circ \alpha$ of a functor and a natural transformation are defined as follows:

$$\alpha \circ E := (\alpha_{E(x)} : F(E(x)) \to F'(E(x)))_{x \in \mathscr{B}_0} : F \circ E \Rightarrow F' \circ E,$$
$$G \circ \alpha := (G(\alpha_x) : G(F(x)) \to G(F'(x)))_{x \in \mathscr{C}_0} : G \circ F \Rightarrow G \circ F'.$$

As is easily seen, these turn out to be natural transformations.

DEFINITION 1.3.7. For a diagram

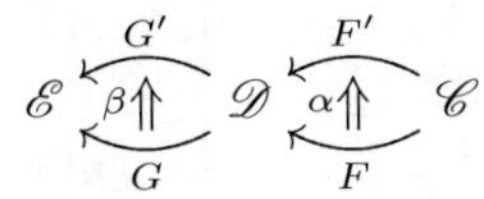

of categories, functors and natural transformations, we see that the following holds by Definitions 1.3.5 and 1.3.6:

$$(\beta \circ F') \bullet (G \circ \alpha) = (G' \circ \alpha) \bullet (\beta \circ F)$$

We set this common natural transformation to be $\beta \circ \alpha$, and call it the *horizontal composite* of α and β. From the definition, in particular we have $\mathbb{1}_G \circ \alpha = G \circ \alpha$ and $\beta \circ \mathbb{1}_F = \beta \circ F$.

DEFINITION 1.3.8. Let $\mathscr{C}$ and $\mathscr{D}$ be categories. We define the *functor category* $\mathscr{F} := \mathrm{Fun}(\mathscr{C}, \mathscr{D})$ from $\mathscr{C}$ to $\mathscr{D}$ as a category as follows. This $\mathscr{F}$ is often denoted by $\mathscr{D}^{\mathscr{C}}$.

- $\mathscr{F}_0 := \{F \mid F : \mathscr{C} \to \mathscr{D}$ is a functor$\}$.
- For each $F, F' \in \mathscr{F}_0$, $\mathscr{F}(F, F') := \{\alpha : F \Rightarrow F'$ is a natural transformation$\}$.
- The composition of $\mathscr{F}$ is given by the vertical composition of natural transformations.
- For each $F \in \mathscr{F}_0$, we set $\mathbb{1}_F$ to be the identity natural transformation of F defined in Example 1.3.2.

REMARK 1.3.9. In the definition above,

(1) A natural transformation α is a natural isomorphism if and only if α is an isomorphism in the category $\mathscr{F}$.
(2) We remark that if $\mathscr{C}$ is a small category, then $\mathscr{F}$ is a usual category (a light category), but if $\mathscr{C}$ is a light category and not a small category, then $\mathscr{F}$ turns out to be a 2-*moderate* category (see Proposition A.3.4).

1.4. Isomorphisms and equivalences of categories

DEFINITION 1.4.1. Let $\mathscr{C}, \mathscr{D}$ be categories and $F : \mathscr{C} \to \mathscr{D}$ a functor.

(1) F is called an *isomorphism* if there exists a functor $G : \mathscr{D} \to \mathscr{C}$ such that $G \circ F = \mathbb{1}_{\mathscr{C}}, F \circ G = \mathbb{1}_{\mathscr{D}}$. The functor G is called the *inverse* of F.
(2) F is called an *equivalence* if there exists a functor $G : \mathscr{D} \to \mathscr{C}$ such that $G \circ F \cong \mathbb{1}_{\mathscr{C}}, F \circ G \cong \mathbb{1}_{\mathscr{D}}$. The functor G is called a *quasi-inverse* of F.

REMARK 1.4.2.

(1) In statement (1) above, G is uniquely determined by F (if it exists). Indeed, let G' be another inverse of F. Then $(G' \circ F) \circ G = G' \circ (F \circ G)$ implies $G = G'$ by the property of inverses. By this reason, we denote the inverse of F by F^{-1}.

(2) However, in (2) above, G is not necessarily uniquely determined by F, as we see in the following example.

EXAMPLE 1.4.3. Let $\mathscr{C} := \mathscr{C}(\{-1,1\}, |)$ be the category defined in Example 1.1.17, and $\mathscr{D} := \mathbf{1}$ the singleton discrete category. Define a functor $F\colon \mathscr{C} \to \mathscr{D}$ by $F(x) := 1, F(f) := \mathbb{1}_1$ $(x \in \mathscr{C}_0, f \in \mathscr{C}_1)$. Define also a functor $G_i\colon \mathscr{D} \to \mathscr{C}$ by $G(1) := i, G(\mathbb{1}_1) := \mathbb{1}_i$ $(i = \pm 1)$. Then both G_{-1} and G_1 are quasi-inverses of F but $G_{-1} \neq G_1$.

The following theorem characterizes a functor F as an equivalence by the properties of F itself. See also Remark 4.4.6 for the proof.

THEOREM 1.4.4. *The following are equivalent for a functor $F\colon \mathscr{C} \to \mathscr{D}$ between categories:*

(1) *F is an equivalence.*
(2) *F is fully faithful and dense.*

PROOF. (1) $\Rightarrow$ (2). Let $G\colon \mathscr{D} \to \mathscr{C}$ be a quasi-inverse of F. Then there exist natural isomorphisms $\eta\colon \mathbb{1}_{\mathscr{C}} \Rightarrow G \circ F$ and $\varepsilon\colon F \circ G \Rightarrow \mathbb{1}_{\mathscr{D}}$. First of all, F is dense because there exist $G(u) \in \mathscr{C}_0$ and an isomorphism $\varepsilon_u\colon F(G(u)) \to u$ for all $u \in \mathscr{D}_0$.

In order to show that F is fully faithful, let $x, y \in \mathscr{C}_0$ and consider the map $F_{y,x}\colon \mathscr{C}(x,y) \to \mathscr{D}(F(x), F(y))$ induced from F. It is enough to show that this is a bijection. To do so we show that $F_{y,x}^{-1}$ is given by the map F' defined by

$$F'\colon \mathscr{D}(F(x), F(y)) \to \mathscr{C}(x,y), F'(g) := \eta_y^{-1} \circ G(g) \circ \eta_x \ (g \in \mathscr{D}(F(x), F(y)).$$

Since for each $f \in \mathscr{C}(x,y)$, the naturality of η yields a commutative diagram

$$\begin{array}{ccc} x & \xrightarrow{\quad f \quad} & y \\ {\scriptstyle \eta_x}\downarrow & & \downarrow{\scriptstyle \eta_y} \\ G(F(x)) & \xrightarrow[G(F(f))]{} & G(F(y)), \end{array}$$

we have $f = \eta_y^{-1} \circ G(F(f)) \circ \eta_x = (F' \circ F_{y,x})(f)$. Thus $F' \circ F_{y,x} = \mathbb{1}_{\mathscr{C}(x,y)}$. In particular, since $F_{y,x}$ is injective, F is faithful. Note here that the similar argument shows that G is also faithful. Next, in order to show $F_{y,x} \circ F' = \mathbb{1}_{\mathscr{D}(F(x),F(y))}$, take an arbitrary $g \in \mathscr{D}(F(x), F(y))$. Apply the commutativity of the diagram above to $f := F'(g) = \eta_y^{-1} \circ G(g) \circ \eta_x$ to obtain $G(g) = \eta_y \circ f \circ \eta_x^{-1} = G(F(f))$. Here since G is faithful, we have $g = F(f) = (F_{y,x} \circ F')(g)$, and hence $F_{y,x} \circ F' = \mathbb{1}_{\mathscr{D}(F(x),F(y))}$. Accordingly, $F_{y,x}^{-1} = F'$.

(2) $\Rightarrow$ (1). It suffices to construct a functor $G\colon \mathscr{D} \to \mathscr{C}$ and natural isomorphisms $\varepsilon\colon F \circ G \Rightarrow \mathbb{1}_{\mathscr{D}}$ and $\varepsilon'\colon G \circ F \Rightarrow \mathbb{1}_{\mathscr{C}}$. For each $u \in \mathscr{D}_0$, since F is dense, there exists an $x \in \mathscr{C}_0$ and an isomorphism $\varepsilon_u\colon F(x) \to u$ in $\mathscr{D}$. We then set $G(u) := x$, and we obtain an isomorphism $\varepsilon_u\colon F(G(u)) \to u$. Next for each $u, v \in \mathscr{D}_0$ take any $g \in \mathscr{D}(u,v)$. Then since $\varepsilon_v^{-1} \circ g \circ \varepsilon_u \in \mathscr{D}(F(G(u)), F(G(v)))$ and F is fully faithful, there exists a unique $f \in \mathscr{D}(G(u), G(v))$ such that $F(f) = \varepsilon_v^{-1} \circ g \circ \varepsilon_u$. We then set $G(g) := f$. The above defines a quiver morphism $G\colon \mathscr{D} \to \mathscr{C}$. It is easy to see that this becomes a functor.

By the construction above, we have $F(G(g)) = \varepsilon_v^{-1} \circ g \circ \varepsilon_u$ for all $u, v \in \mathscr{D}_0$, $g \in \mathscr{D}(u,v)$. Thus $\varepsilon := (\varepsilon_u)_{u \in \mathscr{D}_0}\colon F \circ G \Rightarrow \mathbb{1}_{\mathscr{D}}$ is a natural isomorphism.

Finally, we construct a natural isomorphism $\varepsilon'\colon G \circ F \Rightarrow \mathbb{1}_{\mathscr{C}}$. For each $x \in \mathscr{C}_0$ apply the construction above to $u := F(x)$ to have that $G(F(x)) \in \mathscr{C}_0$ and that

$\varepsilon_{F(x)} \in \mathscr{D}(F(G(F(x))), F(x))$ is an isomorphism. Again by the fact that F is fully faithful, there exists a unique $\varepsilon'_x \in \mathscr{C}(G(F(x)), x)$ such that $F(\varepsilon'_x) = \varepsilon_{F(x)}$. We see that this ε'_x is also an isomorphism from the facts that F is fully faithful and that $\varepsilon_{F(x)}$ is an isomorphism. Here, we set $\varepsilon' := (\varepsilon'_x)_{x\in\mathscr{C}_0} \colon G \circ F \Rightarrow \mathbb{1}_{\mathscr{C}}$. Then the naturality of ε yields the equalities

$$\begin{aligned} F(f \circ \varepsilon'_x) &= F(f) \circ F(\varepsilon'_x) \\ &= F(f) \circ \varepsilon_{F(x)} \\ &= \varepsilon_{F(y)} \circ F(G(F(f))) \\ &= F(\varepsilon'_y) \circ F(G(F(f))) \\ &= F(\varepsilon'_y \circ G(F(f))) \end{aligned}$$

for all $x, y \in \mathscr{C}_0$ and $f \in \mathscr{C}(x, y)$, which shows the naturality of ε' because F is faithful. □

It is clear from the definition that an isomorphism $F \colon \mathscr{C} \to \mathscr{D}$ between categories is an equivalence. In the following, we will clarify the difference between isomorphisms and equivalences by listing the conditions for equivalences to become isomorphisms. First, we mention the following fact. As the proof is easy, we let this be an exercise (it is proved in the same way as the proof of the fact that a homomorphism between groups has an inverse homomorphism if and only if it is bijective).

EXERCISE 1.4.5. The following are equivalent for a functor $F = (F_0, F_1) \colon \mathscr{C} \to \mathscr{D}$ between categories:

(1) F is an isomorphsim.
(2) Both F_0 and F_1 are bijective.

The following proposition describes the difference between isomorphisms and equivalences:

PROPOSITION 1.4.6. *The following are equivalent for a functor* $F = (F_0, F_1)\colon \mathscr{C} \to \mathscr{D}$ *between categories:*

(1) *F is an isomorphism.*
(2) *F is an equivalence, and F_0 is bijective.*

PROOF. (1) $\Rightarrow$ (2). This is clear from Exercise 1.4.5 above.

(2) $\Rightarrow$ (1). Assume statement (2). Then it is enough to show that F_1 is bijective by Exercise 1.4.5. By Theorem 1.4.4, F is fully faithful. Thus, for each $x, y \in \mathscr{C}_0$, $F_{y,x} \colon \mathscr{C}(x, y) \to \mathscr{D}(F(x), F(y))$ is bijective. Therefore so is

$$F_1 = \bigsqcup_{x,y\in\mathscr{C}_0} F_{y,x} \colon \mathscr{C}_1 = \bigsqcup_{x,y\in\mathscr{C}_0} \mathscr{C}(x, y) \to \bigsqcup_{x,y\in\mathscr{C}_0} \mathscr{D}(F(x), F(y)) = \mathscr{D}_1.$$

(The last equality follows from the fact that F_0 is bijective.) □

DEFINITION 1.4.7. Let $\mathscr{C}$ be a category. The relation $x \cong y$ between elements $x, y \in \mathscr{C}_0$ is easily seen to be an equivalence relation on the set $\mathscr{C}_0$. The equivalence class of an x with respect to this relation is called the *isomorphism class*, or shortly the *isoclass* of x. A full subcategory $\mathscr{C}'$ of $\mathscr{C}$ is called a *skeleton* if $\mathscr{C}'_0$ forms a complete set of representatives of the isoclasses of objects in $\mathscr{C}$. $\mathscr{C}$ is called *basic* or *skeletal* if distinct objects in $\mathscr{C}$ are non-isomorphic. Obviously, a skeleton of $\mathscr{C}$ is basic.

LEMMA 1.4.8. *Let $F: \mathscr{C} \to \mathscr{D}$ be a functor between categories, and assume that both $\mathscr{C}$ and $\mathscr{D}$ are basic. Then the following are equivalent:*

(1) *F is an isomorphism.*
(2) *F is an equivalence.*

PROOF. Since the implication (1) $\Rightarrow$ (2) is obvious, we only show the implication (2) $\Rightarrow$ (1). Assume that F is an equivalence. Then it is enough to show that F_0 is bijective by Proposition 1.4.6. Let F' be a quasi-inverse of F. For each $y \in \mathscr{D}_0$, since $F(F'(y)) \cong y$ in the basic category $\mathscr{D}$, we have $F(F'(y)) = y$. Thus F_0 is surjective. Further for each $x, x' \in \mathscr{C}_0$, if $F(x) = F(x')$, then $x \cong F'(F(x)) = F'(F(x')) \cong x'$ in the basic category $\mathscr{C}$, and hece $x = x'$. Therefore F_0 is bijective. □

PROPOSITION 1.4.9. *Let $\mathscr{C}$ be a category.*

(1) *If $\mathscr{C}'$ is a skeleton of $\mathscr{C}$, then $\mathscr{C}'$ and $\mathscr{C}$ are equivalent.*
(2) *If both $\mathscr{C}'$ and $\mathscr{C}''$ are skeletons of $\mathscr{C}$, then $\mathscr{C}'$ and $\mathscr{C}''$ are isomorphic.*

PROOF.

(1) The inclusion functor $\mathscr{C}' \to \mathscr{C}$ is fully faithful and dense. Therefore the assertion follows by Theorem 1.4.4.

(2) Let $F: \mathscr{C}' \to \mathscr{C}, G: \mathscr{C}'' \to \mathscr{C}$ be the inclusion functors, which are equivalences by assumption. By Theorem 1.4.4, G has a quasi-inverse $G': \mathscr{C} \to \mathscr{C}''$. Then it is enough to show that $G' \circ F: \mathscr{C}' \to \mathscr{C}''$ is an equivalence by Lemma 1.4.8. Now since both F and G' are fully faithful, so is $G' \circ F$. Further, for each $x \in \mathscr{C}_0''$, since $\mathscr{C}'$ is a skeleton of $\mathscr{C}$, there exists $y \in \mathscr{C}_0'$ such that $F(y) \cong G(x)$. Then $(G' \circ F)(y) \cong (G' \circ G)(x) \cong x$. Therefore $G' \circ F$ is also dense. Hence the assertion follows by Theorem 1.4.4.

□

From the above, we immediately obtain the following.

PROPOSITION 1.4.10. *Let $\mathscr{C}$ and $\mathscr{D}$ be categories with skeletons $\mathscr{C}'$ and $\mathscr{D}'$, respectively. Then $\mathscr{C}$ and $\mathscr{D}$ are equivalent if and only if $\mathscr{C}'$ and $\mathscr{D}'$ are isomorphic.*

1.5. Algebras and linear categories

DEFINITION 1.5.1. A monoid $(A, \mu, 1)$ satisfying the axiom below is called an *algebra*.

Axioms:

- A is a vector space.
- $\mu: A \times A \to A$ is bilinear (This μ is called the *multiplication* of A).

Let A be an algebra. A is called a *small algebra* if A is small as a monoid. In the sequel we assume that all algebras are small algebras unless otherwise stated. The dimension of A as a vector space is called the *dimension* of A. A is said to be *finite-dimensional* if its dimension is finite. Finally, A is called a *zero algebra* if $A = \{0\}$[2].

[2]As is easily seen, this condition is equivalent to the fact that $1 = 0$, and hence is also equivalent to saying that A is a trivial monoid.

REMARK 1.5.2. For each $c \in \Bbbk$ we have $(c1)a = ca = a(c1)$. Indeed, since μ is bilinear, we have

$$(c1)a = \mu(c1, a) = c\mu(1, a) = ca.$$
$$a(c1) = \mu(a, c1) = c\mu(a, 1) = ca.$$

REMARK 1.5.3.

(1) The definition above is also applied to the case that $\Bbbk$ is a commutative ring. In particular, when $\Bbbk = \mathbb{Z}$, the notion of algebras coincide with that of rings.

(2) When $\Bbbk$ is a field, the structure of an algebra can be examined by using linear algebra thanks to its vector space structure.

EXAMPLE 1.5.4. The following triples become algebras.

(1)
- $\Bbbk$,
- The multiplication of $\Bbbk$,
- $1 \in \Bbbk$.

(2)
- $\Bbbk[x] :=$ (the vector space consisting of the polynomials in one variable x with coefficients in $\Bbbk$).
- The usual multiplication of polynomials,
- $1 \in \Bbbk$.

(3) Let V be a vector space.
- $\operatorname{End}_{\Bbbk}(V)$:=(the vector space consisting of the linear transformations of V),
- The composition of linear transformations,
- $\mathbb{1}_V$ (the identity map).

(4) Let n be a positive integer.
- $M_n(\Bbbk)$:=(the vector space consisting of the square matrices of size n over $\Bbbk$).
- The usual multiplication of matrices,
- E_n (the identity matrix).

(4′) Let n be a positive integer, and A an algebra.
- $M_n(A)$:=(the vector space consisting of square matrices of size n over A).
- The usual multiplication of matrices,
- E_n (the identity matrix).

(5) Let n be a positive integer, and A an algebra.
- $T_n(A) :=$ (the vector space consisting of the lower triangular matrices in $M_n(A)$),
- The usual multiplication of matrices,
- E_n.

(6) Let $A = (A, \mu, 1)$ be an algebra.
- A,
- $\mu^{\rm op}\colon A \times A \to A, \mu^{\rm op}(a, b) := \mu(b, a)$ $((a, b) \in A \times A)$,
- 1.

This algebra is denoted by $A^{\rm op}$, and is called the *opposite* algebra of A.

DEFINITION 1.5.5. A $\Bbbk$-*linear category* (or $\Bbbk$-category for short) is a category $\mathscr{C} = (\mathscr{C}_0, \mathscr{C}_1, \mathrm{dom}, \mathrm{cod}, \circ, \mathbb{1})$ that satisfies the axioms below.

Axioms:

- $\mathscr{C}(x,y)$ is a vector space for all $x,y \in \mathscr{C}_0$,
- $\circ_{x,y,z} \colon \mathscr{C}(y,z) \times \mathscr{C}(x,y) \to \mathscr{C}(x,z)$ is bilinear for all $x,y,z \in \mathscr{C}_0$.

EXAMPLE 1.5.6.

(0) The category

$$\mathrm{MOD}\,\Bbbk = ((\mathrm{MOD}\,\Bbbk)_0, (\mathrm{MOD}\,\Bbbk)_1, \mathrm{dom}, \mathrm{cod}, \circ, \mathbb{1})$$

consisting of the whole vector spaces (not necessarily small vector spaces) and linear maps between them is a linear category, where

$(\mathrm{MOD}\,\Bbbk)_0 :=$ (the set of all [not necessarily small] vector spaces),
$(\mathrm{MOD}\,\Bbbk)_1 :=$ (the set of all linear maps between vector spaces),
for an element $f \colon X \to Y$ in $(\mathrm{MOD}\,\Bbbk)_1$, we set

$$\mathrm{dom}(f) := X,\ \mathrm{cod}(f) := Y,$$

for elements $X \xrightarrow{f} Y \xrightarrow{g} Z$ in $(\mathrm{MOD}\,\Bbbk)_1$, we set

$$(g \circ f)(x) := g(f(x))\ (x \in X),$$

and

for an element X in $(\mathrm{MOD}\,\Bbbk)_0$, we set $\mathbb{1}_X :=$ the identity map $X \to X$.

Indeed, each $(\mathrm{MOD}\,\Bbbk)(X,Y) = \mathrm{Hom}_\Bbbk(X,Y)$ is a vector space with respect to the following addition and the scalar multiplication, and the composition is bilinear: for each $f,g \in \mathrm{Hom}_A(X,Y), c \in \Bbbk$,

$$\begin{cases} (f+g)(x) := f(x) + g(x), \\ \quad (cf)(x) := c(f(x)) \qquad (x \in X). \end{cases}$$

(1) The full subcategory (see Definition 1.2.10) $\mathrm{Mod}^{\mathfrak{U}}\,\Bbbk$ of $\mathrm{MOD}\,\Bbbk$ consisting of small vector spaces is a linear category. We usually omit $\mathfrak{U}$ and denote it by $\mathrm{Mod}\,\Bbbk$.

(2) The full subcategory $\mathrm{mod}\,\Bbbk$ of $\mathrm{Mod}\,\Bbbk$ consisting of finite-dimensional vector spaces is a linear category.

(3) When $\mathscr{C} = (\mathscr{C}_0, \mathscr{C}_1, \mathrm{dom}, \mathrm{cod}, \circ, \mathbb{1})$ is a linear category, the opposite category $\mathscr{C}^{\mathrm{op}}$ of $\mathscr{C}$ is a linear category,

REMARK 1.5.7.

(1) Note that we have $(\mathrm{Mod}\,\Bbbk)_0 \subsetneq (\mathrm{MOD}\,\Bbbk)_0$. In this book, we work almost always in the range of the class SET of all sets, and usually deal with $\mathrm{Mod}\,\Bbbk$ instead of $\mathrm{MOD}\,\Bbbk$ that is too big; and we fix a universe $\mathfrak{U}$ and will not change it. Therefore the notation $\mathrm{Mod}\,\Bbbk$ is not confusing. Note, however, that if we replace the universe $\mathfrak{U}$ by another universe ($\mathfrak{U} \subsetneq$) $\mathfrak{U}'$, then we have $(\mathrm{Mod}^{\mathfrak{U}}\,\Bbbk)_0 \subsetneq (\mathrm{Mod}^{\mathfrak{U}'}\,\Bbbk)_0$.

(2) When we discuss topics using hierarchy of sets (e.g., the Yoneda Lemma 2.5.8), we use the following notation. For each integer $k \geq 0$, the full subcategory of $\mathrm{MOD}\,\Bbbk$ consisting of vector spaces V such that the base set of V is a k-class (see Definition A.2.5) is denoted by $\mathrm{Mod}^k\,\Bbbk$. Therefore, in particular, we have $\mathrm{Mod}^0\,\Bbbk = \mathrm{Mod}\,\Bbbk$ and $(\mathrm{Mod}^k\,\Bbbk)_0 \subsetneq (\mathrm{Mod}^{k+1}\,\Bbbk)_0$ for all $k \geq 0$.

(3) Each linear category that is a singleton can be identified with an algebra. In this sense, a linear category can be regarded as an algebra with multiple objects.

EXERCISE 1.5.8. Let I be a small set and $M_i \in (\mathrm{Mod}\,\Bbbk)_0$ $(i \in I)$. We denote the direct product and the direct sum of $(M_i)_{i\in I}$ by $\prod_{i\in I} M_i$ and by $\coprod_{i\in I} M_i$, respectively[3]. For each $j \in I$ we call the linear map $\pi_j \colon \prod_{i\in I} M_i \to M_j$, $(m_i)_{i\in I} \mapsto m_j$ the *j-th projection*, and the linear map $\sigma_j \colon M_j \to \coprod_{i\in I} M_i$, $m \mapsto (\delta_{ij}\, m)_{i\in I}$ the *j-th injection*. Then in the category $\mathrm{Mod}\,\Bbbk$ show that $(\prod_{i\in I} M_i, (\pi_i)_{i\in I})$ and $(\coprod_{i\in I} M_i, (\sigma_i)_{i\in I})$ turn out to be the product and the coproduct of $(M_i)_{i\in I}$, respectively. We call them the *standard direct product* and the *standard direct sum* of $(M_i)_{i\in I}$, respectively.

1.5.a. Kernels and cokernels of morphisms in a linear category. By characterizing the properties of the kernel and the cokernel of a morphism (linear map) in the category $\mathrm{Mod}\,\Bbbk$ of small vector spaces in terms of categories, these concepts are defined for linear categories in general as follows:

DEFINITION 1.5.9. Let $\mathscr{C}$ be a linear category and $f \colon x \to y$ a morphism in $\mathscr{C}$.

(1) A *kernel* of f is a pair of the following data that satisfies the axioms below.

Data:

- an object u in $\mathscr{C}$ (this is called a *kernel object*),
- a morphism $a \colon u \to x$ (this is called a *kernel morphism*).

Axioms:

(a) $fa = 0$;

(b) a is universal with respect to the property above, i.e., for each $b \colon v \to x$ with $fb = 0$, there exists a unique $c \colon v \to u$ such that $b = ac$.

(2) A *cokernel* of f is a pair of the following data that satisfies the axioms below.

Data:

- an object u in $\mathscr{C}$ (this is called a *cokernel object*),
- a morphism $a \colon y \to u$ of $\mathscr{C}$ (this is called a *cokernel morphism*).

Axioms:

(a) $af = 0$;

(b) a is universal with respect to the property above, i.e., for each $b \colon y \to v$ with $bf = 0$, there exists a unique $c \colon u \to v$ such that $b = ca$.

REMARK 1.5.10. Let $\mathscr{C}$ be a linear category, $f \colon x \to y$ a morphism in $\mathscr{C}$. The property for a morphism to be a kernel (resp. a cokernel) of f is expressed in terms of linear maps as follows:

(1) A morphism $a \colon u \to x$ in $\mathscr{C}$ is a kernel morphism of f if and only if for each $z \in \mathscr{C}_0$ the sequence

$$0 \to \mathscr{C}(z,u) \xrightarrow{\mathscr{C}(z,a)} \mathscr{C}(z,x) \xrightarrow{\mathscr{C}(z,f)} \mathscr{C}(z,y)$$

is exact in $\mathrm{Mod}\,\Bbbk$, i.e., $\mathscr{C}(z,a)$ is a kernel morphism of $\mathscr{C}(z,f)$.

---

[3]See Example 2.1.7

(2) A morphism $a\colon y \to u$ in $\mathscr{C}$ is a cokernel of f if and only if for each $z \in \mathscr{C}_0$ the sequence

$$0 \to \mathscr{C}(u,z) \xrightarrow{\mathscr{C}(a,z)} \mathscr{C}(y,z) \xrightarrow{\mathscr{C}(f,z)} \mathscr{C}(x,z)$$

is exact in $\operatorname{Mod}\Bbbk$, i.e., $\mathscr{C}(a,z)$ is a cokernel morphism of $\mathscr{C}(f,z)$.

The kernel and the cokernel of a morphism are uniquely determined up to isomorphisms in the following sense. Since the proof is easy, we leave it for an exercise.

PROPOSITION 1.5.11. *Let $\mathscr{C}$ be a linear category, $f\colon x \to y$ a morphism in $\mathscr{C}$.*

(1) *If both (u,a) and (v,b) are kernels of f, then there exists an isomorphism $c\colon v \to u$ such that $b = ac$. Therefore we denote "the" kernel object and "the" kernel morphism of f by $\operatorname{Ker} f$ and by $\ker f\colon \operatorname{Ker} f \to x$, respectively. Furthermore the kernel morphism $\ker f$ is a monomorphism* (see Definition 1.1.19).

(2) *If both (u,a) and (v,b) are cokernels of f, then there exists an isomorphism $c\colon u \to v$ such that $b = ca$. Therefore we denote "the" cokernel object and "the" cokernel morphism of f by $\operatorname{Coker} f$ and by $\operatorname{coker} f\colon y \to \operatorname{Coker} f$, respectively. Furthermore the cokernel morphism $\operatorname{coker} f$ is an epimorphism.*

EXERCISE 1.5.12. Prove Proposition 1.5.11 (cf. the proof of 1.1.25).

1.5.b. Finite products and finite coproducts in a linear category. In this subsection, we examine products and coproducts of a family $(x_i)_{i\in I}$ of objects of a linear category $\mathscr{C}$ in the case that I is finite. Therefore, we may assume that $I = \{1, 2, \dots, n\}$ for some positive integer n. We fix these notations throughout this section, and let $x \in \mathscr{C}_0$. In this case, we denote $\coprod_{i\in I} x_i$ (resp. $\prod_{i\in I} x_i$) by $\coprod_{i=1}^n x_i$ or $x_1 \amalg x_2 \amalg \cdots \amalg x_n$ (resp. by $\prod_{i=1}^n x_i$ or $x_1 \times x_2 \times \cdots \times x_n$).

DEFINITION 1.5.13. A (finite) *direct sum system* is a family $(x_i \xrightarrow{\sigma_i} x \xrightarrow{\pi_i} x_i)_{i=1}^n$ of morphisms in $\mathscr{C}$ that satisfies the following:

(1) $\pi_i \circ \sigma_j = \delta_{i,j} \mathbb{1}_{x_i}$ $(1 \le i, j \le n)$ (cf. Convensions (8)); and
(2) $\sum_{i=1}^n \sigma_i \circ \pi_i = \mathbb{1}_x$.

PROPOSITION 1.5.14. *The following are equivalent:*

(1) *There exists a sequence $(\pi_i\colon x \to x_i)_{i=1}^n$ of morphisms in $\mathscr{C}$ such that $(x, (\pi_i)_{i=1}^n)$ is a product of $(x_i)_{i=1}^n$.*
(2) *There exists a sequence $(\sigma_i\colon x_i \to x)_{i=1}^n$ of morphisms in $\mathscr{C}$ such that $(x, (\sigma_i)_{i=1}^n)$ is a coproduct of $(x_i)_{i=1}^n$.*
(3) *There exists a direct sum system $(x_i \xrightarrow{\sigma_i} x \xrightarrow{\pi_i} x_i)_{i=1}^n$.*

By this fact, the existence of $\prod_{i=1}^n x_i$ and that of $\coprod_{i=1}^n x_i$ are equivalent, and the product and the coproduct of $(x_i)_{i=1}^n$ coincide: $\prod_{i=1}^n x_i \cong \coprod_{i=1}^n x_i$, which is called the direct sum *of $(x_i)_{i=1}^n$.*

PROOF. (1) $\Rightarrow$ (3). Assume that there exists a sequence $(\pi_i\colon x \to x_i)_{i=1}^n$ of morphisms in $\mathscr{C}$ such that $(x, (\pi_i)_{i=1}^n)$ is a product of $(x_i)_{i=1}^n$. For each $1 \le j \le n$ take

$$\sigma_j := [\delta_{i,j} \mathbb{1}_{x_i}]_{i=1}^n \in \mathscr{C}(x_j, x)$$

as the canonical morphism corresponding to the element $(\delta_{i,j}\mathbb{1}_{x_i})_{i=1}^n \in \prod_{i=1}^n \mathscr{C}(x_j, x_i)$ obtained by the universality of the projection family $(\pi_i)_{i\in I}$ (see Remark 1.1.26). Then $(x_i \xrightarrow{\sigma_i} x \xrightarrow{\pi_i} x_i)_{i=1}^n$ turns out to be a direct sum system. Indeed, by definition of σ_j, we have $\pi_i \circ \sigma_j = \delta_{i,j}\mathbb{1}_{x_i}$ for all $1 \le i, j \le n$. Therefore, it only remains to show that

$$\sum_{i=1}^n \sigma_i \circ \pi_i = \mathbb{1}_x$$

holds. This is verified by presenting both sides as column vectors as follows:

$$\begin{aligned}(\text{LHS}) &= \left[\pi_j \circ \sum_{i=1}^n \sigma_i \circ \pi_i\right]_{j=1}^n = \left[\sum_{i=1}^n \delta_{i,j}\pi_i\right]_{j=1}^n = [\pi_j]_{j=1}^n \\ &= [\pi_j \circ \mathbb{1}_x]_{j=1}^n = (\text{RHS}).\end{aligned}$$

(3) $\Rightarrow$ (1). Assume that there exists a direct sum system $(x_i \xrightarrow{\sigma_i} x \xrightarrow{\pi_i} x_i)_{i=1}^n$. Then we show that $(x, (\pi_i)_{i=1}^n)$ turns out to be a product of $(x_i)_{i=1}^n$. To this end it is enough to show that for each $y \in \mathscr{C}_0$ the map

$$\alpha_y \colon \mathscr{C}(y, x) \to \prod_{i=1}^n \mathscr{C}(y, x_i), \quad f \mapsto (\pi_i \circ f)_{i=1}^n$$

is bijective. We construct its inverse by

$$\beta_y \colon \prod_{i=1}^n \mathscr{C}(y, x_i) \to \mathscr{C}(y, x), \quad (f_i)_{i=1}^n \mapsto \sum_{i=1}^n \sigma_i \circ f_i.$$

It immediately follows from the two axioms of a direct sum system that this is actually the inverse of α_y.

(2) $\Leftrightarrow$ (3). This is the dual of the equivalence (1) $\Leftrightarrow$ (3), and hence it is proved by the same argument as in $\mathscr{C}^{\text{op}}$. □

From the argument in the proof of the above proposition, the following holds.

COROLLARY 1.5.15.

(1) *A product $(x, (\pi_i)_{i=1}^n)$ of $(x_i)_{i=1}^n$ is uniquely extended to a direct sum system $(x_i \xrightarrow{\sigma_i} x \xrightarrow{\pi_i} x_i)_{i=1}^n$.*

(2) *A coproduct $(x, (\sigma_i)_{i=1}^n)$ of $(x_i)_{i=1}^n$ is uniquely extended to a direct sum system $(x_i \xrightarrow{\sigma_i} x \xrightarrow{\pi_i} x_i)_{i=1}^n$.*

(3) *For a direct sum system $(x_i \xrightarrow{\sigma_i} x \xrightarrow{\pi_i} x_i)_{i=1}^n$, $(x, (\pi_i)_{i=1}^n)$ is a product of $(x_i)_{i=1}^n$, and $(x, (\sigma_i)_{i=1}^n)$ is a coproduct of $(x_i)_{i=1}^n$. Furthermore, for each $y \in \mathscr{C}_0$, the inverse of the isomorphism*

$$\mathscr{C}(y, \coprod_{i=1}^n x_i) \to \prod_{i=1}^n \mathscr{C}(y, x_i), \ f \mapsto (\pi_i \circ f)_{i=1}^n$$

is given by the correspondence

$$\sum_{i=1}^n \sigma_i \circ f_i \leftarrow\!\shortmid (f_i)_{i=1}^n.$$

PROOF. It remains to show that the extensions in (1) and (2) are unique. Since both statements are shown similarly, we only show the uniqueness in (1). Let $(x_i \xrightarrow{\sigma_i'} x \xrightarrow{\pi_i} x_i)_{i=1}^n$ be another direct sum system. Then for each $1 \le j \le n$, apply

σ_j from the right to both sides of the equality $\sum_{i=1}^n \sigma_i \circ \pi_i = \sum_{i=1}^n \sigma_i' \circ \pi_i$ to have $\sigma_j = \sigma_j'$. □

DEFINITION 1.5.16. Let x, x_1 and x_2 be objects in $\mathscr{C}$. If $x \cong x_1 \amalg x_2$ in $\mathscr{C}$, then each x_i $(i = 1, 2)$ is called a *direct summand* of x. When we have $x \cong \coprod_{i=1}^n x_i$ for objects x, x_i $(1 \le i \le n)$ in $\mathscr{C}$, note that x_i is a direct summand of x for all $1 \le i \le n$ because we have $x \cong x_i \amalg \left(\coprod_{j \ne i} x_j\right)$.

1.5.c. Coproducts in a linear category. Throughout this subsection, again we let $\mathscr{C}$ be a linear category, and I a small set, $x, x_i \in \mathscr{C}_0$ $(i \in I)$, and $(\coprod_{i\in I} x_i, (\sigma_i)_{i\in I})$ a coproduct of $(x_i)_{i\in I}$. We investigate the relationship between $\mathscr{C}(x, \coprod_{i\in I} x_i)$ and $\coprod_{i\in I} \mathscr{C}(x, x_i)$.

LEMMA 1.5.17. *Let $(\coprod_{i\in I} \mathscr{C}(x, x_i), (\overline{\sigma_i})_{i\in I})$ be the standard direct sum in* $\operatorname{Mod} \Bbbk$*, and consider the following diagram:*

$$\begin{array}{ccc} \mathscr{C}(x, x_i) & \xrightarrow{\mathscr{C}(x,\sigma_i)} & \mathscr{C}(x, \coprod_{i\in I} x_i). \\ {\scriptstyle \overline{\sigma_i}}\downarrow & \nearrow{\scriptstyle \phi} & \\ \coprod_{i\in I} \mathscr{C}(x, x_i) & & \end{array} \tag{1.7}$$

By the universality of $\coprod_{i\in I} \mathscr{C}(x, x_i)$, there exists a unique ϕ making this diagram commutative, i.e., $\phi = {}^t[\mathscr{C}(x, \sigma_i)]_{i\in I}$. Then the following hold:

(1) *For each $(f_i)_{i\in I} \in \coprod_{i\in I} \mathscr{C}(x, x_i)$ we have*

$$\phi((f_i)_{i\in I}) = \sum_{i\in I} \sigma_i \circ f_i.$$

(2) *For each $j \in I$ there exists $\pi_j \in \mathscr{C}(\coprod_{i\in I} x_i, x_j)$ such that $\pi_j \circ \sigma_i = \delta_{i,j} \mathbb{1}_j$ $(i, j \in I)$.*

(3) *ϕ is a monomorphism.*

PROOF.

(1) Since for each $(f_i)_{i\in I} \in \coprod_{i\in I} \mathscr{C}(x, x_i)$, $\{i \in I \mid f_i \ne 0\}$ is a finite set, we can define a linear map $\psi \colon \coprod_{i\in I} \mathscr{C}(x, x_i) \to \mathscr{C}(x, \coprod_{i\in I} x_i)$ by setting $\psi((f_i)_{i\in I}) = \sum_{i\in I} \sigma_i \circ f_i$. If we verify that this makes the diagram (1.7) commutative, then by the universality of the coproduct we have $\phi = \psi$ and statement (1) is proved. This is checked by the following computation for each $i \in I$ and $f \in \mathscr{C}(x, x_i)$:

$$\psi(\overline{\sigma_i}(f)) = \psi((\delta_{i,j} f)_{j\in I}) = \sum_{j\in I} \sigma_j \circ \delta_{i,j} f = \sigma_i \circ f = \mathscr{C}(x, \sigma_i)(f).$$

(2) This is proved by a similar argument used in the proof of the implication (1) $\Rightarrow$ (3) in Proposition 1.5.14. Namely, for each $j \in I$, take an element $(\delta_{i,j} \mathbb{1}_{x_j})_{i\in I} \in \coprod_{i\in I} \mathscr{C}(x_i, x_j)$ and set $\pi_j := {}^t[\delta_{i,j} \mathbb{1}_{x_j}]$ to be the canonical morphism corresponding to it obtained by the universality of the injection family $(\sigma_i)_{i\in I}$. Then by definition we have $\pi_j \circ \sigma_i = \delta_{i,j} \mathbb{1}_{x_j}$.

(3) To show that ϕ is a monomorphism, take $f := (f_i)_{i\in I} \in \coprod_{i\in I} \mathscr{C}(x, x_i)$ and assume that $\phi(f) = 0$. Then we have $\sum_{i\in I} \sigma_i \circ f_i = 0$. For each $j \in I$, apply π_j from the left to the both sides to have $\sum_{i\in I} \pi_j \circ \sigma_i \circ f_i = 0$, which shows $f_j = 0$. Therefore $f = 0$, and hence ϕ is a monomorphism.

□

EXERCISE 1.5.18. The linear map ϕ above does not need to be an isomorphism. For instance, verify this fact for the case that $\mathscr{C} := \operatorname{Mod} \Bbbk$, $I = \mathbb{N}$, $x = \coprod_{i \in \mathbb{N}} \Bbbk$, $x_i = \Bbbk$ $(i \in I)$.

An object x such that ϕ above is an isomorphism, is said to be compact. More explicitly we make the following definition.

DEFINITION 1.5.19. Let $\mathscr{C}$ be a linear category. An object $x \in \mathscr{C}_0$ is said to be *compact* if for each small set I, each $(x_i)_{i \in I} \in \mathscr{C}_0^I$ and each coproduct $(\coprod_{i \in I} x_i, (\sigma_i)_{i \in I})$ of $(x_i)_{i \in I}$, the canonical morphism

$$^t[\mathscr{C}(x, \sigma_i)]_{i \in I} \colon \coprod_{i \in I} \mathscr{C}(x, x_i) \to \mathscr{C}(x, \coprod_{i \in I} x_i)$$

obtained by the universality of the standard injection family $(\overline{\sigma_i} \colon \mathscr{C}(x, x_i) \to \coprod_{i \in I} \mathscr{C}(x, x_i))_{i \in I}$ is an isomorphism.

1.6. Algebra homomorphisms and linear functors

DEFINITION 1.6.1. Let A and A' be algebras. A monoid homomorphism $f \colon A \to A'$ is called an (algebra) *homomorphism* from A to A' if it is linear.

EXAMPLE 1.6.2. Let A be an algebra and take $e \in A$. Then both $Ae := \{ae \mid a \in A\}$ and $eA := \{ea \mid a \in A\}$ are subspace of A. The following are algebra homomorphisms

$$\lambda \colon A \to \operatorname{End}_{\Bbbk}(Ae), \lambda(a)(x) := ax \ (a \in A, x \in Ae),$$
$$\rho \colon A^{\mathrm{op}} \to \operatorname{End}_{\Bbbk}(eA), \rho(a)(x) := xa \ (a \in A^{\mathrm{op}}, x \in eA).$$

DEFINITION 1.6.3. Let $\mathscr{C}$ and $\mathscr{C}'$ be linear categories. A functor $F \colon \mathscr{C} \to \mathscr{C}'$ is called a *linear functor* from $\mathscr{C}$ to $\mathscr{C}'$ if the map $F_{y,x} \colon \mathscr{C}(x, y) \to \mathscr{C}'(F(x), F(y))$ is linear for all $x, y \in \mathscr{C}_0$.

EXERCISE 1.6.4. Let $\mathscr{C}$ be a locally small linear category and take $x \in \mathscr{C}_0$. Then show that the representable functor (see Exercise 1.2.5) $\mathscr{C}(x, \text{-})$ is a linear functor $\mathscr{C} \to \operatorname{Mod} \Bbbk$, and that $\mathscr{C}(\text{-}, x)$ is a linear functor $\mathscr{C}^{\mathrm{op}} \to \operatorname{Mod} \Bbbk$.

EXAMPLE 1.6.5. Let $\mathscr{C}$ be a linear category. We can define a linear functor $F \colon \mathscr{C} \to \operatorname{Mod} \Bbbk$ by setting $F(x) := 0$ $(x \in \mathscr{C}_0)$, $F(f) := 0 \colon 0 \to 0$ $(f \in \mathscr{C}(x, y), x, y \in \mathscr{C}_0)$. We denote this functor simply by 0. Similarly we define a linear functor $0 \colon \mathscr{C}^{\mathrm{op}} \to \operatorname{Mod} \Bbbk$.

REMARK 1.6.6. Linear functors between linear singleton categories are nothing but homomorphisms between algebras.

DEFINITION 1.6.7. By $\mathrm{Alg}_{\Bbbk}$ we denote the category whose objects are the small algebras and whose morphisms are the algebra homomorphisms between objects.

CHAPTER 2

Representations

2.1. Representations and modules

In general, to relate an unfamiliar one X to a familiar (well-understood) one Y is said "to represent X by Y." The thing relates them is called a representation of X by Y. Now the multiplication of an algebra A is given by an operation table among various tables and is not well understood, but the multiplication of the endomorphism algebra $\mathrm{End}_{\Bbbk}(M)$ of a vector space M is given by the composition of maps, which is well understood. Therefore we may consider an algebra A unfamiliar, and $\mathrm{End}_{\Bbbk}(M)$ familiar, and what relates them is a map that preserves the structure of an algebra, i.e., an algebra homomorphism

$$\lambda\colon A \to \mathrm{End}_{\Bbbk}(M).$$

Then according to the general statement above we define as follows.

DEFINITION 2.1.1. In the setting above, λ is called a *representation* of A by $\mathrm{End}_{\Bbbk}(M)$ or a *representation* of A over M (or (M, λ) is called a representation of A for short).

Recall now that in general for arbitrary sets X, Y, Z, there exists a bijection

$$Z^{X\times Y} \xrightarrow{\sim} (Z^Y)^X$$

given by the adjoint between the direct product construction $? \times Y$ and the power set construction $?^Y$, where V^U denotes the set of all maps from U to V for all sets U, V. This is obtained by defining μ and λ from each other so that the equality

$$\mu(x, y) = [\lambda(x)](y) \ (x \in X, y \in Y) \tag{2.1}$$

holds for all $\mu \in Z^{X\times Y}$ and $\lambda \in (Z^Y)^X$. By applying this fact to $(X, Y, Z) := (A, M, M)$ in the setting above, we obtain the bijection

$$M^{A\times M} \xrightarrow{\sim} (M^M)^A. \tag{2.2}$$

Suppose that $\mu \in M^{A\times M}$ corresponds to $\lambda \in (M^M)^A$ in this correspondence. Then the necessary and sufficient condition for $\lambda \in (M^M)^A$ to satisfy $\lambda \in \mathrm{Alg}_{\Bbbk}(A, \mathrm{End}_{\Bbbk}(M))$ (Definition 1.6.7) can be written down as follows:

(R1) $\lambda(1_A) = \mathbb{1}_M$,
(R2) $\lambda(ab) = \lambda(a) \circ \lambda(b)$,
(R3) $\lambda(a + b) = \lambda(a) + \lambda(b)$,
(R4) $\lambda(k \cdot a) = k \cdot \lambda(a)$,
(R5) $\lambda(a)(v + w) = \lambda(a)(v) + \lambda(a)(w)$, and
(R6) $\lambda(a)(k \cdot v) = k \cdot \lambda(a)(v)$

$(a, b \in A, v, w \in M, k \in \Bbbk)$, where in order to distinguish from the multiplication of A the scalar multiplication from $\Bbbk$ to M was denoted by $k \cdot v$ $(k \in \Bbbk, v \in M)$.

Using the relationship (2.1), these conditions for λ can be rewritten as conditions for μ as follows, where we set $\mu(a, v) := av$ $(a \in A, v \in M)$ for simplicity.

(M1) $1_A v = v$,
(M2) $(ab)v = a(bv)$,
(M3) $(a + b)v = av + bv$,
(M4) $(k \cdot a)v = k \cdot (av)$,
(M5) $a(v + w) = av + aw$, and
(M6) $a(k \cdot v) = k \cdot (av)$

$(a, b \in A, v, w \in M, k \in \Bbbk)$. Therefore, we define the following concepts.

DEFINITION 2.1.2. Let A be an algebra. A *left A-module* is a pair of the following data that satisfies the axioms below:

Data:

- a vector space M,
- a map $\mu\colon A \times M \to M$, $\mu(a, v) := av \quad (a \in A, v \in M)$.

Axioms: The conditions (M1) to (M6) above hold.

This immediately gives us the following:

PROPOSITION 2.1.3. *Let M be a vector space. Then the bijection* (2.2) *induces a bijection from the set of all representations λ of A over M to the set of all left A-module structures μ of M.*

Therefore, by identifying modules and representations via the bijection above, we may make the following definition.

DEFINITION 2.1.4. Let A be an algebra. A *left A-module* is a pair of the data:

- a vector space M,
- an algebra homomorphism $\lambda\colon A \to \mathrm{End}_{\Bbbk}(M)$.

Similarly, the conditions for a map $\rho \in (M^M)^{A^{\mathrm{op}}}$ to be a representation $\rho \in \mathrm{Alg}(A^{\mathrm{op}}, \mathrm{End}_{\Bbbk}(M))$ of A^{op} can be written down as the conditions obtained by changing λ to ρ in the conditions (R1) to (R6) and exchanging only (R2) to the following

(R2′) $\rho(ab) = \rho(b) \circ \rho(a)$.

If we write $va := [\rho(a)](v)$ $(a \in A, v \in M)$ so that A acts from the right, then this condition can be written in the same form as the associativity:

(M2′) $v(ab) = (va)b \quad (a, b \in A, v \in M)$.

Therefore, we define the concept corresponding to the representation of A^{op} as follows:

DEFINITION 2.1.5. Let A be an algebra. A *right A-module* is a pair of the following data that satisfies the axioms below:

Data:

- a vector space M,
- a map $\mu\colon M \times A \to M$, $\mu(v, a) := va \quad (a \in A, v \in M)$.

Axioms: The following conditions (M1′) to (M6′) hold.

(M1′) $v1_A = v$,
(M2′) $v(ab) = (va)b$,

(M3′) $v(a+b) = va + vb$,
(M4′) $v(k \cdot a) = k \cdot (va)$,
(M5′) $(v+w)a = va + wa$,
(M6′) $(k \cdot v)a = k \cdot (va)$, and

$(a, b \in A, v, w \in M, k \in \Bbbk)$.

As in the case of the left module, by identifying the representation with the module, we can define it as follows:

DEFINITION 2.1.6. Let A be an algebra. A *right A-module* is a pair of the data:

- a vector space M,
- an algebra homomorphism $\rho\colon A^{\mathrm{op}} \to \mathrm{End}_{\Bbbk}(M)$.

In the following, we will adopt the latter definition as the formal definition of left (and right) modules, and we will proceed by identifying it with the former definition. When we say simply modules, we mean right modules.

EXAMPLE 2.1.7.

(1) Let A be an algebra, $e \in A$, and $\lambda\colon A \to \mathrm{End}_{\Bbbk}(Ae)$, $\rho\colon A^{\mathrm{op}} \to \mathrm{End}_{\Bbbk}(eA)$ the algebra homomorphisms in Example 1.6.2. Then $Ae := (Ae, \lambda)$ (resp. $eA := (eA, \rho)$) turns out to be a left A-module (resp. a right A-module).

(2) Let A be an algebra, I a small set, and M_i an A-module for all $i \in I$. Then the direct product

$$\prod_{i \in I} M_i := \{(m_i)_{i\in I} \mid m_i \in M_i \ (i \in I)\}$$

of sets becomes an A-module by the addition, the scalar multiplication, and the A-action defined componentwise as follows (if $I \neq \emptyset$) [1]:

$$(m_i)_{i\in I} + (m'_i)_{i\in I} := (m_i + m'_i)_{i\in I},$$
$$c(m_i)_{i\in I} := (cm_i)_{i\in I},$$
$$(m_i)_{i\in I}a := (m_i a)_{i\in I}, \quad ((m_i)_{i\in I}, (m'_i)_{i\in I} \in \prod_{i\in I} M_i, c \in \Bbbk, a \in A).$$

This is called the *direct product* of the A-modules M_i $(i \in I)$.

When $M_i = M$ for all $i \in I$, we write $M^I := \prod_{i\in I} M_i$. Moreover, if $I = \{1, \dots, n\}$ for some positive integer n, then we write $M^n := M^I$, $M^0 := M^{\emptyset} := 0$.

(3) With the same setup as above, the following will be a *submodule* of the direct product of A-modules (see Definition 2.2.3 for the definition of a submodule):

$$\coprod_{i\in I} M_i := \{(m_i)_{i\in I} \mid \{i \in I \mid m_i \neq 0\} \text{ is a finite set}\}.$$

This is called the *direct sum* of the A-modules M_i $(i \in I)$. If I is a finite set, of course we have $\coprod_{i\in I} M_i = \prod_{i\in I} M_i$.

When $M_i = M$ for all $i \in I$, we write $M^{(I)} := \coprod_{i\in I} M_i$. Moreover, if $I = \{1, \dots, n\}$ for a positive integer n, then we write $M^{(n)} := M^{(I)}$, $M^{(0)} := M^{\emptyset} := 0$.

[1] If $I = \emptyset$, then $\prod_{i\in I} M_i := 0$, the trivial module.

PROPOSITION 2.1.8. *Let A be an algebra, M a right A-module, I a small set, and M_i a* submodule *of M for all $i \in I$* (see Definition 2.2.3). *Since $\{i \in I \mid m_i \neq 0\}$ is a finite set for all $m = (m_i)_{i\in I} \in \coprod_{i\in I} M_i$, it is possible to define a map*

$$\Sigma\colon \coprod_{i\in I} M_i \to M,\ (m_i)_{i\in I} \mapsto \sum_{i\in I} m_i.$$

Then Σ is a homomorphism, and the following are equivalent:

(1) *Σ is an isomorphism.*
(2) (a) *$M = \sum_{i\in I} M_i$;*
 (b) *For each $(m_i)_{i\in I} \in \coprod_{i\in I} M_i$, the condition $\sum_{i\in I} m_i = 0$ implies $m_i = 0$ for all $i \in I$.*

When this is the case, M is called the (inner) direct sum *of $(M_i)_{i\in I}$, and this fact is denoted by $M = \bigoplus_{i\in I} M_i$. In contrast, $\coprod_{i\in I} M_i$ is also called the* outer direct sum *of $(M_i)_{i\in I}$.*

PROOF. It is obvious that Σ is a homomorphism. Since $\operatorname{Im}\Sigma = \sum_{i\in I} M_i$, the condition (a) is equivalent to the fact that Σ is an epimorphism. Moreover, the condition (b) is equivalent to $\operatorname{Ker}\Sigma = 0$, and hence to the fact that Σ is a monomorphism. Therefore, both (a) and (b) hold if and only if Σ is an isomorphism. □

Referring to the fact that for an algebra A, the left (resp. right) A-modules are viewed as representations of A (resp. A^{op}), we define the left (resp. right) modules over a linear category in general as follows.

DEFINITION 2.1.9. Let $\mathscr{C}$ be a linear category. A linear functor $M\colon \mathscr{C} \to \operatorname{Mod}\Bbbk$ (resp. $M\colon \mathscr{C}^{\mathrm{op}} \to \operatorname{Mod}\Bbbk$) is called a *left $\mathscr{C}$-module* (resp. a *right $\mathscr{C}$-module*).

EXAMPLE 2.1.10. Let $\mathscr{C}$ be a linear category and $x \in \mathscr{C}_0$. Then $\mathscr{C}(x, \text{-})$ (resp. $\mathscr{C}(\text{-}, x)$) turns out to be a left $\mathscr{C}$-module (resp. a right $\mathscr{C}$-module) (cf. Excercise 1.2.5).

REMARK 2.1.11. Modules over a linear singleton is nothing but modules over an algebra.

2.2. Module categories of an algebra and of a linear category

DEFINITION 2.2.1. Let A be an algebra, and $M = (M, \rho)$ and $M' = (M', \rho')$ A-modules.

(1) An *A-homomorphism* from M to M' is a linear map $f\colon M \to M'$ satisfying **Axiom:** $f(ma) = f(m)a$ $(x \in M, a \in A)$.

In other words, the following diagram commutes for all $a \in A$:

$$\begin{array}{ccc} M & \xrightarrow{f} & M' \\ {\scriptstyle \rho(a)}\downarrow & & \downarrow{\scriptstyle \rho'(a)} \\ M & \xrightarrow[f]{} & M' \end{array}.$$

A homomorphism from M to M itself is called an *endomorphism* of M.

(2) By $\mathrm{Hom}_A(M, M')$ we denote the set of all A-homomorphisms from M to M', which is a subspace of $\mathrm{Hom}_{\Bbbk}(M, M')$. In particular we set $\mathrm{End}_A(M) := \mathrm{Hom}_A(M, M)$, which becomes a k-algebra by taking the composition of A-homomorphisms as its multiplication. This is called the *endomorphism algebra* of M.
(3) An A-homomorphism is called a *monomorphism* (resp. an *epimorphism*) if it is injective (resp. surjective).

EXAMPLE 2.2.2. Let A be an algebra, M a left A-module, I a small set, and M_i $(i \in I)$ left A-modules.

(1) The identity map $\mathbb{1}_M : M \to M$ is an A-homomorphism.
(2) For each $j \in I$, the map $\pi_j : \prod_{i\in I} M_i \to M_j$, $(m_i)_{i\in I} \mapsto m_j$ is an A-homomorphism, which is called the j-th *projection*.
(3) For each $j \in I$, the map $\sigma_j : M_j \to \coprod_{i\in I} M_i$, $m \mapsto (\delta_{ij}\, m)_{i\in I}$ is an A-homomorphism, which is called the j-th *injection*.

DEFINITION 2.2.3. Let A be an algebra, and $L = (L, \lambda_L), M = (M, \lambda_M)$ left A-modules. We say that L is a *submodule* of M, and denote it by $L \le M$ if the following hold:

(1) $L \le M$ as $\Bbbk$-vector spaces (i.e., L is a subspace of M); and
(2) The inclusion map $\sigma : L \hookrightarrow M$ is an A-homomorphism.

By definition of A-homomorphisms, condition (2) above is equivalent to the following:

(a) $aL := \lambda_M(a)(L) \subseteq L$ for all $a \in A$, and
(b) $\lambda_L(a) = \lambda_M(a)|_L$ for all $a \in A$.

Therefore if a subspace L of M satisfies condition (a), then we can define the submodule L of M by using (b). In this case, we say that the subspace L *defines the submodule* of M, or simply that L is a *submodule* of M.

EXAMPLE 2.2.4. Let A be an algebra, M a left A-module, and $m \in M$. Then $Am := \{am \mid a \in A\}$ is a submodule of M.

As is well-known, for an abelian group M and a family $(M_i)_{i\in I}$ of its subgroups, both $\bigcap_{i\in I} M_i$ and $\sum_{i\in I} M_i$ become also subgroups of M. By the same proof, the following holds:

LEMMA 2.2.5. *Let A be an algebra, M a left A-module, I a small set, and $M_i \le M$ for all $i \in I$. Then both subspaces $\bigcap_{i\in I} M_i$ and $\sum_{i\in I} M_i$ define submodules of M. These submodules are denoted by $\bigcap_{i\in I} M_i$ and $\sum_{i\in I} M_i$, respectively.*

EXERCISE 2.2.6. Prove the lemma above.

DEFINITION 2.2.7. Let A be an algebra, M a left A-module, and $S \subseteq M$.

(1) The smallest submodule of M that contains S is called the submodule of M *generated by* S, and is denoted by $\langle S\rangle$, i.e., we set

$$\langle S\rangle := \bigcap_{S\subseteq L\le M} L \quad \text{(in particular, } \langle\emptyset\rangle = 0).$$

(2) M is said to be *finitely generated* if there exists a finite subset S of M such that $M = \langle S\rangle$.

EXERCISE 2.2.8. Show that the following statements hold for an algebra A and a left A-module M. We define ($\sum_{m\in\emptyset} Am := 0$.)

(1) If S is a subset of M, then we have $\langle S\rangle = \sum_{m\in S} Am$.
(2) Show that the following are equivalent:
  (a) M is finitely generated.
  (b) $M = Am_1 + \cdots + Am_n$ for some $m_1, \ldots, m_n \in M$ and $n \in \mathbb{N}$.
  (c) There exists an epimorphism $A^{(n)} \to M$ for some $n \in \mathbb{N}$.
(3) Assume that A is finite-dimensional. Then M is finitely generated if and only if M is finite-dimensional.

DEFINITION 2.2.9. Let A be an algebra, M a left A-module, and L a submodule of M. By regarding M, L as $\Bbbk$-vector spaces, consider the factor space M/L and let

$$\pi\colon M \to M/L$$

be the canonical epimorphism (i.e, $\pi(x) := x + L \in M/L$ ($x \in M$)). Then there exists a unique way to define a left A-module structure on M/L so that π becomes an A-homomorphism.

Namely, $a\pi(x) := \pi(ax)$ ($a \in A, x \in M$). (Indeed, if $\pi(x) = \pi(y)$ ($x, y \in M$), then $x - y \in \operatorname{Ker}\pi = L$ and since $L \le M$, we have $ax - ay = a(x - y) \in L$, which shows that $\pi(ax) = \pi(ay)$. Thus $\pi(ax)$ does not depend on the choice of representative $x \in M$ of the element $\pi(x)$ in M/L.) This left A-module M/L is called the *factor module* of M by L, and the π above is called the *canonical epimorphism*.

EXAMPLE 2.2.10. Let A be an algebra.

(0) The category

$$\operatorname{MOD} A = ((\operatorname{MOD} A)_0, (\operatorname{MOD} A)_1, \operatorname{dom}, \operatorname{cod}, \circ, \mathbb{1})$$

formed by the (right) A-modules (whose base sets do not need to be small sets) and the A-homomorphisms between them is a linear category, where

$$(\operatorname{MOD} A)_0 := \begin{pmatrix} \text{the set of all right } A\text{-modules } M \\ \text{(the base set of } M \text{ may not be small)} \end{pmatrix},$$

$$(\operatorname{MOD} A)_1 := (\text{the set of all } A\text{-homomorphisms between right } A\text{-modules}),$$

for each element $f\colon X \to Y$ in $(\operatorname{MOD} A)_1$, $\operatorname{dom}(f) := X$, $\operatorname{cod}(f) := Y$,

for each pair $X \xrightarrow{f} Y \xrightarrow{g} Z$ of elements in $(\operatorname{MOD} A)_1$, $(g \circ f)(x) := g(f(x))$ ($x \in X$),

for each $X \in (\operatorname{MOD} A)_0$, $\mathbb{1}_X :=$ the identity map $X \to X$.

Indeed, each $(\operatorname{MOD} A)(X, Y) = \operatorname{Hom}_A(X, Y)$ is a vector space and the compositions are bilinear. The category $\operatorname{MOD} A$ is called the (right) *module category* of A. Moreover, $\operatorname{MOD} A^{\mathrm{op}}$ is called the left *module category* of A, and is denoted by A-MOD.

(1) The full subcategory $\operatorname{Mod}^{\mathfrak{U}} A$ of $\operatorname{MOD} A$ consisting of (right) A-modules M such that the base set of M is small forms a linear category. We usually abbreviate $\mathfrak{U}$ and denote it by $\operatorname{Mod} A$. $\operatorname{Mod} A$ is called the (right) *module category* of A. Moreover, $\operatorname{Mod} A^{\mathrm{op}}$ is called the left *module category* of A, and is denoted by A-Mod.
(2) The full subcategory $\operatorname{mod} A$ of $\operatorname{Mod} A$ consisting of the finitely generated A-modules also forms a linear category.

REMARK 2.2.11. Let A be an algebra.

(1) Similar to the remark for $\operatorname{Mod}\Bbbk$, note that $(\operatorname{Mod} A)_0 \subsetneq (\operatorname{MOD} A)_0$ and that for universes $\mathfrak{U}, \mathfrak{U}'$, if $\mathfrak{U} \subsetneq \mathfrak{U}'$, then we have $(\operatorname{Mod}^{\mathfrak{U}} A)_0 \subsetneq (\operatorname{Mod}^{\mathfrak{U}'} A)_0$.
(2) By Proposition 2.1.3 we can identify the left A-module (M, μ) and the representation (M, λ) of A. This identification can be also formulated as follows: If we define the category $\operatorname{REP} A$ of the whole representations of A suitably, then this category turns out to be isomorphic to the category A-MOD.

EXERCISE 2.2.12. Let A be an algebra. In the category $\operatorname{Mod} A$ (or in $\operatorname{MOD} A$) show that a morphism f is a monomorphism (resp. an epimorphism) as defined in category theory (see Definition 1.1.19) if and only if it also meets the criteria of Definition 2.2.1, i.e., it is an injective (resp. surjective) homomorphism.

EXERCISE 2.2.13. Let A be an algebra, I a small set, and $M, M_i \in (\operatorname{Mod} A)_0$ for all $i \in I$. Further define the morphisms π_i, σ_i as in Example 2.2.2. Then show that $(\prod_{i\in I} M_i, (\pi_i)_{i\in I})$ and $(\coprod_{i\in I} M_i, (\sigma_i)_{i\in I})$ are a product and a coproduct of $(M_i)_{i\in I}$, respectively.

DEFINITION 2.2.14. Let A be an algebra and $0 \neq M \in (\operatorname{Mod} A)_0$. Then it is easy to see that the following statements are equivalent in the category $\operatorname{Mod} A$:

(1) If $M \cong M_1 \amalg M_2$, then $M_1 = 0$ or $M_2 = 0$.
(2) If $M_1, M_2 \leq M$ and $M = M_1 \oplus M_2$, then $M_1 = 0$ or $M_2 = 0$.

If one of them holds (thus if both hold), then M is said to be *indecomposable*.

EXERCISE 2.2.15. Prove the equivalence above.

DEFINITION 2.2.16. Let $\mathscr{C}$ and $\mathscr{D}$ be linear categories. As in Definition 1.3.8 we define the *functor category* $\mathscr{F} := \operatorname{Fun}_{\Bbbk}(\mathscr{C}, \mathscr{D})$ from $\mathscr{C}$ to $\mathscr{D}$ as follows:

- $\mathscr{F}_0 := \{F \mid F\colon \mathscr{C} \to \mathscr{D} \text{ is a linear functor}\}$.
- For each $F, F' \in \mathscr{F}_0$, we set
$$\mathscr{F}(F, F') := \{\alpha \mid \alpha\colon F \Rightarrow F' \text{ is a natural transformation}\}.$$
- The composition in $\mathscr{F}$ is given by the vertical composition of natural transformations.
- For each $F \in \mathscr{F}_0$, $\mathbb{1}_F$ is the identity natural transformation of F defined in Example 1.3.2.

REMARK 2.2.17. In the above in particular, let $\mathscr{C}$ be the linear singleton, and $\mathscr{D} := \operatorname{Mod}\Bbbk$. Then both F and F' become left modules over the algebra $A := G_{\mathscr{C}}$ (see Remark 1.1.15)), and α becomes an A-homomorphism between them. Therefore, $\operatorname{Fun}_{\Bbbk}(\mathscr{C}, \operatorname{Mod}\Bbbk)$ can be identified with the left module category A-Mod, and $\operatorname{Fun}_{\Bbbk}(\mathscr{C}^{\mathrm{op}}, \operatorname{Mod}\Bbbk)$ with the right module category $\operatorname{Mod} A$.

DEFINITION 2.2.18. For a linear category $\mathscr{C}$, we set $\mathscr{C}\text{-}\operatorname{Mod} := \operatorname{Fun}_{\Bbbk}(\mathscr{C}, \operatorname{Mod}\Bbbk)$ (resp. $\operatorname{Mod}\mathscr{C} := \operatorname{Fun}_{\Bbbk}(\mathscr{C}^{\mathrm{op}}, \operatorname{Mod}\Bbbk)$) and call it the *left module category* (resp. *right module category*) of $\mathscr{C}$.

EXERCISE 2.2.19. Let $\mathscr{C}$ be a linear category. In the category $\operatorname{Mod}\mathscr{C}$ (or in $\operatorname{MOD}\mathscr{C}$), show that a morphism $\alpha\colon L \to M$ is a monomorphism (resp. an epimorphism) (see Definition 1.1.19) if and only if $\alpha_x\colon L(x) \to M(x)$ is injective (resp. surjective) for all $x \in \mathscr{C}_0$. (It is recommended to use the concepts of submodules and factor modules defined in Definitions 2.2.21 and 2.2.25 below.)

REMARK 2.2.20. The module categories $\operatorname{Mod}\mathscr{C}$ and $\mathscr{C}\text{-}\operatorname{Mod}$ become light categories (resp. 2-moderate categories that are not light) if $\mathscr{C}$ is a small category (resp. a light category that is not small). For instance, even if $\mathscr{C}$ is a small category, $\operatorname{Mod}(\operatorname{Mod}\mathscr{C})$ becomes a 2-moderate category that is not light because $\operatorname{Mod}\mathscr{C}$ is a light category that is not small. (see Definition A.3.1). In the sequel, we assume that all categories are light categories unless otherwise stated. When we do not assume that the given category is small, we call it "a general category" (see Appendix in Chapter A).

DEFINITION 2.2.21. Let $\mathscr{C}$ be a linear category and L, M left $\mathscr{C}$-modules. We say that L is a *submodule* of M, and denote it by $L \le M$ if the following two conditions are satisfied:

(1) $L(x) \le M(x)$ for all $x \in \mathscr{C}_0$; and
(2) the family $\sigma := (\sigma_x \colon L(x) \hookrightarrow M(x))_{x\in\mathscr{C}_0}$ of inclusion maps becomes a morphism in $\mathscr{C}\text{-}\operatorname{Mod}$.

By definition of morphisms in $\mathscr{C}\text{-}\operatorname{Mod}$ the condition (2) above is equivalent to the following.

(a) $M(f)(L(x)) \le L(y)$ $(x, y \in \mathscr{C}_0, f \in \mathscr{C}(x,y))$, and
(b) $L(f) = M(f)|_{L(x)}$ $(x, y \in \mathscr{C}_0, f \in \mathscr{C}(x,y))$.

Therefore if the family $L(x) \le M(x)$, $(x \in \mathscr{C}_0)$ of subspaces satisfies the condition (a), then it is possible to define the submodule L of M by (b). In this case, we say that the family $(L(x))_{x\in\mathscr{C}_0}$ of subspaces *defines a submodule* of M.

By the same proof as Lemma 2.2.5, the following holds.

LEMMA 2.2.22. *Let $\mathscr{C}$ be a linear category, M a left $\mathscr{C}$-module, I a set, and $M_i \le M$ for all $i \in I$. Then both the families*

$$(\bigcap_{i\in I} M_i(x))_{x\in\mathscr{C}_0} \quad \text{and} \quad (\sum_{i\in I} M_i(x))_{x\in\mathscr{C}_0}$$

of subspaces define submodules of M, which are denoted by $\bigcap_{i\in I} M_i$ and $\sum_{i\in I} M_i$, respectively.

EXERCISE 2.2.23. Prove the lemma above.

DEFINITION 2.2.24. Let $\mathscr{C}$ be a linear category, M a left $\mathscr{C}$-module, and $S \subseteq \bigsqcup_{x\in\mathscr{C}_0} M(x)$.

(1) If a submodule L of M satisfies $S \subseteq \bigsqcup_{x\in\mathscr{C}_0} L(x)$, then we say that L *includes* S, and denote it by $S \subseteq L$. Then the smallest submodule of M that includes S is called the submodule of M *generated by* S, and is denoted by $\langle S\rangle$. Namely, we set
$$\langle S\rangle := \bigcap_{S\subseteq L\le M} L.$$
(2) M is said to be *finitely generated* if there exists a finite set S $(\subseteq M)$ such that $M = \langle S\rangle$.
(3) The full subcategory of $\mathscr{C}$-Mod consisting of the finitely generated left $\mathscr{C}$-modules is denoted by $\mathscr{C}$-mod.

DEFINITION 2.2.25. Let $\mathscr{C}$ be a linear category, M a left $\mathscr{C}$-module, and L a submodule of M. Then a left $\mathscr{C}$-module M/L is defined as follows.

Set $(M/L)(x) := M(x)/L(x)$ $(x \in \mathscr{C}_0)$, and for each morphism $f\colon x \to y$ in $\mathscr{C}$, define $(M/L)(f)$ by the following commutative diagram with exact rows:

$$\begin{array}{ccccccccc} 0 & \longrightarrow & L(x) & \xrightarrow{\sigma_x} & M(x) & \xrightarrow{\pi_x} & M(x)/L(x) & \longrightarrow & 0 \\ & & \downarrow{\scriptstyle L(f)} & & \downarrow{\scriptstyle M(f)} & & \downarrow{\scriptstyle (M/L)(f)} & & \\ 0 & \longrightarrow & L(y) & \xrightarrow[\sigma_y]{} & M(y) & \xrightarrow[\pi_y]{} & M(y)/L(y) & \longrightarrow & 0, \end{array}$$

where π_x, π_y are the canonical epimorphisms. (This diagram is obtained by applying the fundamental homomorphism theorem for $\Bbbk$-vector spaces to the left commutative square.)

Then by this commutative diagram, $\pi := (\pi_x)_{x\in\mathscr{C}_0}$ turns out to be an epimorphism $\pi\colon M \to M/L$ in the category $\mathscr{C}$-Mod. This left $\mathscr{C}$-module M/L (resp. the morphism π) is called the *factor module* of M by L (resp. the *canonical epimorphism* of this factor module construction).

2.3. Finer correspondence between algebras and linear categories

DEFINITION 2.3.1. Let A be an algebra.

(1) An element $e \in A$ is called an *idempotent* if $e^2 = e$.
(2) Idempotents e, f in A are said to be *orthogonal* if $ef = 0 = fe$.
(3) A nonzero idempotent e in A is said to be *primitive* if $e = e_1 + e_2$ (e_1, e_2: orthogonal idempotents in A) implies $e_1 = 0$ or $e_2 = 0$.
(4) A set $\{e_1, \dots, e_n\}$ of idempotents in A is said to be *complete* if $e_1 + \cdots + e_n = 1$.
(5) Let $\{e_1, \dots, e_n\}$ be a complete set of idempotents in A. This is called a *complete set of orthogonal idempotents* if e_i and e_j are orthogonal for all $i \neq j$ with $1 \leq i, j \leq n$, and is called a *complete set of orthogonal primitive idempotents* if further e_i are primitive for all $i = 1, \dots, n$.

The following is easy to show.

EXERCISE 2.3.2. Prove the following statements for an algebra A and an idempotent e in A.

(1) The element $1 - e$ is also an idempotent.
(2) If e has a left inverse or a right inverse in A, then $e = 1$.
(3) Let S be a subset of A. If $eS \subseteq S$, then we have $eS = \{a \in S \mid a = ea\}$.

LEMMA 2.3.3. *The following are equivalent for an algebra A and a nonzero idempotent e in A.*

(1) *e is primitive.*
(2) *eA is an indecomposable right A-module* (see Definition 2.2.14).
(3) *Ae is an indecomposable left A-module.*

PROOF. (2) $\Rightarrow$ (1). Assume that e is not primitive. Then there exist nonzero orthogonal idempotents e_1, e_2 such that $e = e_1 + e_2$. This implies that $eA = e_1A \oplus e_2A$. Indeed, for each $a \in A$, we have $ea = (e_1+e_2)a = e_1a + e_2a \in e_1A + e_2A$, which shows that $eA \leq e_1A + e_2A$. Moreover, $ee_1 = e_1e_1 + e_2e_1 = e_1e_1 = e_1$ shows that $e_1A \leq eA$. Similarly we have $e_2A \leq eA$, which shows that $e_1A + e_2A \leq eA$, and hence $eA = e_1A + e_2A$. Further, for each $a \in e_1A \cap e_2A$, it follows from Exercise

2.3.2(3) that $e_1 a = a = e_2 a$ and hence that $a = e_1(e_2 a) = 0$. Therefore we have $eA = e_1 A \oplus e_2 A$. Here since $e_1 A \neq 0$ and $e_2 A \neq 0$, eA is not indecomposable.

(1) $\Rightarrow$ (2). If eA is not indecomposable, then there exist nonzero submodules P_1, P_2 such that $eA = P_1 \oplus P_2$. Therefore there exists a unique pair $(e_1, e_2) \in P_1 \times P_2$ such that

$$e = e_1 + e_2. \tag{2.3}$$

Since $e_1 \in P_1 \leq eA$, it follows from Exercise 2.3.2(3) that $e_1 = ee_1 = e_1 e_1 + e_2 e_1$. Thus we have

$$e_1 = e_1 e_1 + e_2 e_1. \tag{2.4}$$

On the other hand, since $ee = e$, we have $e_1 e_1 + e_1 e_2 + e_2 e_1 + e_2 e_2 = e_1 + e_2$. By taking $e_1 e_1 + e_1 e_2 \in P_1$ and $e_2 e_1 + e_2 e_2 \in P_2$ into account, we have

$$e_1 = e_1 e_1 + e_1 e_2. \tag{2.5}$$

By the equalities (2.4) and (2.5), we have $e_2 e_1 = e_1 e_2 \in P_2 \cap P_1 = 0$. Thus $e_2 e_1 = e_1 e_2 = 0$. This together with (2.4) shows that $e_1 e_1 = e_1$. Similarly we have $e_2 e_2 = e_2$. Therefore the equality (2.3) gives a decomposition of e into the sum of orthogonal idempotents. Here $e_1 A = P_1 \neq 0$ shows $e_1 \neq 0$. Similarly we have $e_2 \neq 0$. Hence e is not primitive. Here note that $e_1 A = P_1$, $e_2 A = P_2$. Indeed, this follows from the facts that $e_1 A \leq P_1, e_2 A \leq P_2$ and that $eA \leq e_1 A + e_2 A \leq P_1 \oplus P_2 = eA$.

(1) $\Leftrightarrow$ (3). This is shown similarly as above. $\square$

The following is a generalization of the one-to-one correspondence between the set of proportionalities and the set of proportional constants:

PROPOSITION 2.3.4. *Let A be an algebra, $e \in A$ an idempotent, and M an A-module. Then the map*

$$\phi\colon \operatorname{Hom}_A(eA, M) \to Me, f \mapsto f(e)$$

turns out to be an isomorphism of vector spaces, where we set $Me := \{me \mid m \in M\}$. The inverse is given by

$$\psi\colon Me \to \operatorname{Hom}_A(eA, M), me \mapsto \lambda_{me}, (\lambda_{me}(x) := mex \ (x \in eA)).$$

In particular, we have $\operatorname{Hom}_A(eA, A) \cong Ae$, $\operatorname{End}_A(eA) \cong eAe$.

EXERCISE 2.3.5. Verify that the maps ϕ and ψ given above are inverses of each other, and that the isomorphisms $\operatorname{Hom}_A(eA, A) \cong Ae$, $\operatorname{End}_A(eA) \cong eAe$ are also isomorphisms as left A-modules and as $\Bbbk$-algebras, respectively.

REMARK 2.3.6. The proposition above for algebras is generalized to a statement for linear categories, which is well-known as the Yoneda lemma (for linear categories) (see Lemma 2.5.8).

There is a more detailed relationship between the algebras and the linear categories as explained below.

DEFINITION 2.3.7. We denote by $\Bbbk$-$\mathbf{Cat}_{\mathrm{f}}$ the category whose objects are the small linear categories with only finitely many objects, and whose morphisms are the functors F such that the map $F_0\colon \mathscr{C}_0 \to \mathscr{C}'_0$ is injective. (This forms a subcategory of $\Bbbk$-**Cat** defined in Example 4.1.8.) We also define a category $\Bbbk\text{-}\mathbf{Alg}_{\mathrm{coi}}$ as follows:

(1) $(\Bbbk\text{-}\mathbf{Alg}_{\mathrm{coi}})_0$ is the set of pairs (A, E), where A is an algebra, and E is a complete set of orthogonal idempotents in A.
(2) For each $(A, E), (A', E') \in (\Bbbk\text{-}\mathbf{Alg}_{\mathrm{coi}})_0$, $\Bbbk\text{-}\mathbf{Alg}_{\mathrm{coi}}((A, E), (A', E'))$ is the set of linear maps $f\colon A \to A'$ such that f preserves the multiplication[2], and satisfies $f(E) \subseteq E'$.
(3) The composition of morphisms in $\Bbbk\text{-}\mathbf{Alg}_{\mathrm{coi}}$ is defined as the usual composition of maps.
(4) For each $(A, E) \in (\Bbbk\text{-}\mathbf{Alg}_{\mathrm{coi}})_0$, we set $\mathbb{1}_{(A,E)} := \mathbb{1}_A$.

Then by definition of being isomorphic for two objects of a category, (A, E), $(A', E') \in \Bbbk\text{-}\mathbf{Alg}_{\mathrm{coi}}$ are isomorphic if and only if there exists an isomorphism $f\colon A \to A'$ of algebras satisfying that $f(E) = E'$.

PROPOSITION 2.3.8. *The categories $\Bbbk\text{-}\mathbf{Alg}_{\mathrm{coi}}$ and $\Bbbk\text{-}\mathbf{Cat}_{\mathrm{f}}$ are equivalent. Roughly speaking, a pair (A, E) of an algebra A and a complete set E of orthogonal idempotents in A is identified with a linear category with only finitely many objects.*

PROOF. We first define a functor $\mathrm{Cat}\colon \Bbbk\text{-}\mathbf{Alg}_{\mathrm{coi}} \to \Bbbk\text{-}\mathbf{Cat}_{\mathrm{f}}$ as follows:
For each $(A, E) \in (\Bbbk\text{-}\mathbf{Alg}_{\mathrm{coi}})_0$, define $\mathscr{C} = \mathscr{C}_{A,E} = \mathrm{Cat}(A, E)$ by

$$\mathscr{C}_0 := E,\ \mathscr{C}(x, y) := yAx\ (x, y \in \mathscr{C}_0),$$

where the composition is given by the multiplication of A, and for each $x \in \mathscr{C}_0$ $\mathbb{1}_x = x\,(= x1x)$. Conversely, we define a functor $\mathrm{Mat}\colon \Bbbk\text{-}\mathbf{Cat}_{\mathrm{f}} \to \Bbbk\text{-}\mathbf{Alg}_{\mathrm{coi}}$ as follows:

For each $\mathscr{C} \in \Bbbk\text{-}\mathbf{Cat}_{\mathrm{f}}$, define $(A_{\mathscr{C}}, E_{\mathscr{C}}) = \mathrm{Mat}(\mathscr{C})$ by

$$A_{\mathscr{C}} := \mathrm{Mat}_{\mathscr{C}} := \coprod_{x,y\in\mathscr{C}_0} \mathscr{C}(x, y),$$

$$E_{\mathscr{C}} := \{e_{x,x} := (\mathbb{1}_x \delta_{(j,i),(x,x)})_{j,i\in\mathscr{C}_0} \mid x \in \mathscr{C}_0\},$$

where the multiplication is given by the usual multiplication of matrices by regarding each element $(a_{yx})_{y,x} \in A_{\mathscr{C}}$, $(a_{yx} \in \mathscr{C}(x, y), x, y \in \mathscr{C}_0)$ as a matrix, and the identity of $A_{\mathscr{C}}$ is given by $(\mathbb{1}_x\delta_{y,x})_{y,x\in\mathscr{C}_0}$.

Now we check that they are quasi-inverses of each other. Let $(A, E) \in (\Bbbk\text{-}\mathbf{Alg}_{\mathrm{coi}})_0$, and set $\mathscr{C} := \mathrm{Cat}(A, E)$. Then $\mathrm{Mat}(\mathrm{Cat}(A, E)) = (A_{\mathscr{C}}, E_{\mathscr{C}})$, and the natural isomorphism $\Sigma\colon A_{\mathscr{C}} = \coprod_{x,y\in E} yAx \to \bigoplus_{x,y\in E} yAx = A$ (from the outer direct sum to the inner direct sum) turns out to be an algebra isomorphism from $A_{\mathscr{C}}$ to A. Moreover, we see that $\Sigma(E_{\mathscr{C}}) = E$.

Conversely let $\mathscr{C} \in (\Bbbk\text{-}\mathbf{Cat}_{\mathrm{f}})_0$. Then we have $\mathrm{Cat}(\mathrm{Mat}(\mathscr{C})) = \mathscr{C}_{A_{\mathscr{C}},E_{\mathscr{C}}}$. There exists a bijection $\mathscr{C}_0 \to E_{\mathscr{C}} = \mathrm{Cat}(\mathrm{Mat}(\mathscr{C}))_0$, $x \mapsto e_{x,x}$ between the object sets. Moreover, the corresponding local morphism set $\mathscr{C}(x, y)$ and $e_{y,y}A_{\mathscr{C}}e_{x,x} = \mathrm{CatMat}(\mathscr{C})(\mathrm{CatMat}(x), \mathrm{CatMat}(y))$ are naturally isomorphic. □

EXERCISE 2.3.9.[3] In the proof of Proposition 2.3.8 above, give precise definitions of natural transformations $\mathrm{Mat} \circ \mathrm{Cat} \Rightarrow \mathbb{1}_{\Bbbk\text{-}\mathbf{Alg}_{\mathrm{coi}}}$ and $\mathbb{1}_{\Bbbk\text{-}\mathbf{Cat}_{\mathrm{f}}} \Rightarrow \mathrm{Cat} \circ \mathrm{Mat}$, and show that they are natural isomorphisms.

---

[2]Namely, $f(ab) = f(a)f(b)$ for all $a, b \in A$. Here we do not assume $f(1) = 1$.
[3]A precise explanation is given in Sect. B.1.

REMARK 2.3.10. For an algebra A if we set $E := \{1_A\}$, then $(A, E) \in (\Bbbk\text{-}\mathbf{Alg}_{\mathrm{coi}})_0$, and then $\mathrm{Cat}(A, E)$ is isomorphic to A that is regarded as a singleton.

PROPOSITION 2.3.11. *Let $(A, E) \in (\Bbbk\text{-}\mathbf{Alg}_{\mathrm{coi}})_0$, and set $\mathscr{C}_E$ (resp. $\mathscr{C}'_E$) to be the full subcategory of* $\mathrm{Mod}\,A$ *(resp. A-Mod) with the object set $\{eA \mid e \in E\}$ (resp. $\{Ae \mid e \in E\}$). Then the three linear categories $\mathrm{Cat}(A, E)$, $\mathscr{C}_E$, ${\mathscr{C}'_E}^{\mathrm{op}}$ are isomorphic to each other.*

PROOF. We define a linear functor $F\colon \mathrm{Cat}(A, E) \to \mathscr{C}_E$ as follows:

$$F(e) := eA, F(fae) := fae \cdot (\text{-})\colon eA \to fA \quad (e, f \in E, a \in A).$$

We also define a linear functor $F'\colon \mathscr{C}_E \to \mathrm{Cat}(A, E)$ as follows:

$$F'(eA) := e, F'(u) := u(e)\colon e \to f \quad (e, f \in E, u \in \mathscr{C}_E(eA, fA)).$$

Then as is easily seen, F and F' are inverses of each other. Therefore F is an isomorphism.

Next, we define a linear functor $G\colon : \mathscr{C}_E \to {\mathscr{C}'_E}^{\mathrm{op}}$ as follows:

$$G(eA) := Ae, G(u) := (\text{-}) \cdot u(e)\colon Af \to Ae \quad (e, f \in E, u \in \mathscr{C}_E(eA, fA)).$$

We also define a linear functor $G'\colon {\mathscr{C}'_E}^{\mathrm{op}} \to \mathscr{C}_E$ as follows:

$$G'(Ae) := eA, G'(u) := u(f) \cdot (\text{-})\colon eA \to fA \quad (e, f \in E, u \in \mathscr{C}'_E(Af, Ae)).$$

Then again as is easily seen, G and G' are inverses of each other, and hence G is an isomorphism. □

REMARK 2.3.12. For an algebra A and for a set E of idempotents in A (not necessarily a complete set), it is possible to define the category $\mathrm{Cat}(A, E)$ in the same way as above, and the assertion above holds also for this (A, E).

COROLLARY 2.3.13. *Let $(A, E) \in (\Bbbk\text{-}\mathbf{Alg}_{\mathrm{coi}})_0$ and $e, f \in E$. Then the following are equivalent:*

(1) *e and f are isomorphic in the linear category $\mathrm{Cat}(A, E)$ ($e \cong f$), that is, we have $(fae)(ebf) = f$ and $(ebf)(fae) = e$ for some $a, b \in A$.*
(2) *Right modules eA and fA are isomorphic in the category* $\mathrm{Mod}\,A$.
(3) *Left modules Ae and Af are isomorphic in the category A-Mod.*

PROOF. We keep the notation used in the proof of Proposition 2.3.11.
(1) ⇔ (2). This follows by the isomorphisms F, F'.
(2) ⇔ (3). This follows by the isomorphisms G, G'. □

Later the corollary above is generalized to Corollary 2.5.11.

PROPOSITION 2.3.14. *Every finite-dimensional algebra has a complete set of orthogonal primitive idempotents.*

PROOF. We only state how to make a complete set E of orthogonal primitive idempotents.

- Let A be a finite-dimensional algebra. Then since the right A-module A itself is finite-dimensional, it can be decomposed as $A = P_1 \oplus \cdots \oplus P_n$ for some indecomposable right A-modules $P_1, \dots, P_n$.
- This shows that $1 \in A$ is written in the form $1 = e_1 + \cdots + e_n$ ($e_1 \in P_1, \dots, e_n \in P_n$).
- Then we can take $E := \{e_1, \dots, e_n\}$. □

EXERCISE 2.3.15.[4] Complete the proof of the above proposition, referring to the proof of Lemma 2.3.3. Show also that $P_i = e_iA$ for all $i = 1, \dots, n$.

We quote the following theorem without proof. For details, refer to [**1**, 12.9 The Krull–Schmidt Theorem] for instance.[5]

THEOREM 2.3.16 (The Krull–Schmidt Theorem). *Every finite-dimensional module M over an algebra A can be expressed as the direct sum $M = \bigoplus_{i=1}^{m} M_i$ of finitely many indecomposable modules M_i. Moreover, if we have another decomposition $M = \bigoplus_{i=1}^{n} N_i$ of M into indecomposable modules, then we have $m = n$ and there exists a permutation σ of the set $\{1, \dots, n\}$ such that $N_i \cong M_{\sigma(i)}$ for all $i = 1, \dots, n$.*

REMARK 2.3.17. We add the following remarks to Proposition 2.3.14:

- Assume that we have another complete set $E' = \{e'_1, \dots, e'_m\}$ of orthogonal primitive idempotents. Then it yields another decomposition $A = \bigoplus_{i=1}^{m} e'_iA$ of A into indecomposable modules e'_iA, and hence by the Krull–Schmidt theorem, we have $m = n$, and there exists a permutation σ of $\{1, \dots, n\}$ such that $e_iA \cong e'_{\sigma(i)}A$ for all $i = 1, \dots, n$.
- From this fact we see the following:
  If we rewrite E as $E = \{e_{11}, \dots, e_{1a_1}, \dots, e_{n1}, \dots, e_{na_n}\}$, so that $e_{ij} \cong e_{pq}$ in $\mathrm{Cat}(A, E)$ if and only if $i = p$, then as the multiplicities of the isomorphic direct summands, the numbers $a_1, \dots, a_n$ do not depend on the choice of E.

DEFINITION 2.3.18. Let A be a finite-dimensional algebra and $E = \{e_{11}, \dots, e_{1a_1}, \dots, e_{n1}, \dots, e_{na_n}\}$ a complete set of orthogonal primitive idempotents in A, where as above $e_{ij} \cong e_{pq}$ in $\mathrm{Cat}(A, E)$ if and only if $i = p$.

(1) If $a_1 = \cdots = a_n = 1$, or equivalently, if the elements in E are pairwise non-isomorphic in $\mathrm{Cat}(A, E)$, then A is called a *basic algebra*.
(2) If we set $e := e_{11} + e_{21} + \cdots + e_{n1}$, then eAe is a basic algebra with the identity e, which is called a *basic algebra* of A, and this e is called a *basic idempotent* of A.

We quote the Morita theorem below without proof[6]. See e.g., [**1**, 22.4 Corollary] for details.

THEOREM 2.3.19 (Morita). *Let A be a finite-dimensional algebra and e a basic idempotent of A. Then the functor*

$$\mathrm{Hom}_A(eA, \text{-})\colon \operatorname{Mod} A \to \operatorname{Mod} eAe, M \mapsto \mathrm{Hom}_A(eA, M) \cong Me$$

is en equivalence.

From this theorem, the study of $\operatorname{Mod} A$ is reduced to that of $\operatorname{Mod} eAe$ for the basic algebra eAe instead of A. Therefore, without loss of generality, we may assume that A is a basic algebra.

DEFINITION 2.3.20. Let A be an algebra. Then an element $a \in A$ is called a *unit* of A if a has its inverse in A, and is called a *non-unit* otherwise. The algebra A is called *local* if $a + b$ is a non-unit for all non-units $a, b \in A$.

---

[4]See Sect. B.2.a for an answer.
[5]See Sect. B.3 for a proof.
[6]See Sect. B.4 for proofs.

EXERCISE 2.3.21.[7] Show that the following are equivalent for an algebra $A(\neq 0)$:

(1) A is local.
(2) For each $a \in A$, a or $1 - a$ is a unit.

REMARK 2.3.22. Let A be an algebra, and M a right A-module.

(1) By Excercise 2.3.2(1), (2), it is immediate that the set of idempotents in a local algebra is equal to $\{0, 1\}$. Therefore, if the endomorphism algebra $\mathrm{End}_A(M)$ of M is local, then M becomes indecomposable (see Exercise 2.3.23(3) below).
(2) The converse holds if M is finite-dimensional. Namely, if M is finite-dimensional and indecomposable, then $\mathrm{End}_A(M)$ turns out to be local (see Exercise 2.3.24 below). In particular, if A is a finite-dimensional algebra and e is a primitive idempotent of A, then eA is a finite-dimensional indecomposable right A-module, and hence $\mathrm{End}_A(eA) \cong eAe$ becomes local.

EXERCISE 2.3.23.[8] Let A be an algebra, and $0 \neq M \in (\mathrm{Mod}\, A)_0$.

(1) Assume that $M_1, M_2 \leq M, M = M_1 \oplus M_2$. Then for each $i = 1, 2$ show that $e_i \colon M \xrightarrow{\pi_i} M_i \hookrightarrow M$ is an idempotent of $\mathrm{End}_A(M)$ and that $\mathbb{1}_M = e_1 + e_2$, where π_i is the i-th projection $m_1 + m_2 \mapsto m_i$ $(m_1 \in M_1, m_2 \in M_2)$.
(2) Let $e \in \mathrm{End}_A(M)$ be an idempotent. Then show that $M = e(M) \oplus (\mathbb{1}_M - e)(M)$ as right A-modules.
(3) Prove that the following are equivalent:
    (a) M is indecomposable.
    (b) The set of idempotents in $\mathrm{End}_A(M)$ is equal to $\{0, \mathbb{1}_M\}$.

EXERCISE 2.3.24.[9] Prove the following statements for an algebra A and a finite-dimensional right A-module M:

(1) If $f \in \mathrm{End}_A(M)$, then there exists some $n \geq 1$ such that $M = \mathrm{Ker}\, f^n \oplus \mathrm{Im}\, f^n$ (cf. [**1**, Fitting's Lemma]).
(2) If M is indecomposable, then every $f \in \mathrm{End}_A(M)$ is either an isomorphism or *nilpotent* (i.e., $f^n = 0$ for some $n \geq 1$).
(3) If M is indecomposable, then $\mathrm{End}_A(M)$ is a local algebra (cf. Exercise 2.3.21).

DEFINITION 2.3.25. Let $\mathscr{C}$ be a linear category.

(1) $\mathscr{C}$ is called a *spectroid*[10] if the following are satisfied:
    (a) $\mathscr{C}$ is *basic*, i.e., distinct objects are not isomorphic.
    (b) $\mathscr{C}(x, x)$ is a local algebra for all $x \in \mathscr{C}_0$.
    (c) $\mathscr{C}$ is Hom-*finite*, i.e., $\mathscr{C}(x, y)$ is finite-dimensional for all $x, y \in \mathscr{C}_0$.
(2) A spectroid $\mathscr{C}$ is said to be *locally bounded* if $\{y \in \mathscr{C}_0 \mid \mathscr{C}(x, y) \neq 0 \text{ or } \mathscr{C}(y, x) \neq 0\}$ is a finite set for all $x \in \mathscr{C}_0$.
(3) A spectroid is said to be *finite* if it has only a finite number of objects.

---

[7] See Sect. B.2.b for an answer.
[8] See Sect. B.2.c for an answer.
[9] See Sect. B.2.d for an answer.
[10] This terminology first appeared in [**24**]. Before that it was called a *locally finite-dimensional category* in the literature.

EXAMPLE 2.3.26. For an algebra A, choose a complete set $\mathscr{L}$ of representatives of isoclasses of finitely generated indecomposable right A-modules. Then the full subcategory $\operatorname{ind} A$ of $\operatorname{mod} A$ with the object set $\mathscr{L}$ turns out to be a spectroid.

Proposition 2.3.8 gives us the following proposition.

DEFINITION 2.3.27. We set $\Bbbk\text{-}\mathbf{alg}_{\text{copi}}$ as the full subcategory of $\Bbbk\text{-}\mathbf{Alg}_{\text{coi}}$ consisting of the pairs (A, E) of a finite-dimensional algebra A and a complete set E of orthogonal primitive idempotents of A, and $\Bbbk\text{-}\mathbf{Spec}_{\text{f}}$ to be the full subcategory of $\Bbbk\text{-}\mathbf{Cat}_{\text{f}}$ consisting of finite spectroids.

PROPOSITION 2.3.28. *Two categories* $\Bbbk\text{-}\mathbf{alg}_{\text{copi}}$ *and* $\Bbbk\text{-}\mathbf{Spec}_{\text{f}}$ *are equivalent. Thus roughly speaking, we can identify the pair (A, E) of a finite-dimensional algebra A and its complete set E of orthogonal primitive idempotents with the corresponding spectroid* $\operatorname{Cat}(A, E)$.

PROOF. We keep the notations used in the proof of Proposition 2.3.8. The claim follows by Proposition 2.3.8 if we show the following two statements:

(1) $\operatorname{Cat}(A, E) \in (\Bbbk\text{-}\mathbf{Spec}_{\text{f}})_0$, and
(2) $\operatorname{Mat}(\mathscr{C}) \in (\Bbbk\text{-}\mathbf{alg}_{\text{copi}})_0$

for all $(A, E) \in (\Bbbk\text{-}\mathbf{alg}_{\text{copi}})_0$ and $\mathscr{C} \in \Bbbk\text{-}\mathbf{Spec}_{\text{f}}$.

(1) (a) Since A is basic, so is $\operatorname{Cat}(A, E)$.

(b) Let $e \in E$. Then by Remark 2.3.22 (2), $\operatorname{Cat}(A, E)(e, e) = eAe$ is local.

(c) Let $e, f \in E$. Then since A is finite-dimensional, so is $\operatorname{Cat}(A, E)(e, f) = fAe$. Finally, it is obvious that $\operatorname{Cat}(A, E)_0 = E$ is a finite set.

(2) Since $\mathscr{C}_0$ is a finite set, and $\mathscr{C}$ is Hom-finite, we see that $A := \operatorname{Mat}_{\mathscr{C}}$ is finite-dimensional. It is obvious that $E_{\mathscr{C}} = \{e_{x,x} \mid x \in \mathscr{C}_0\}$ is a complete set of orthogonal idempotents. It remains to show that each $e_{x,x}$ is a primitive idempotent, or equivalently, that $e_{x,x}A$ is indecomposable. By definition of A, we have an isomorphism $e_{x,x}Ae_{x,x} \cong \mathscr{C}(x, x)$ as algebras, which is local by definition of $\mathscr{C}$. Therefore so is $\operatorname{End}_A(e_{x,x}A) \cong e_{x,x}Ae_{x,x}$, and hence $e_{x,x}A$ is indecomposable by Remark 2.3.22 (1). □

EXAMPLE 2.3.29. Let $C := \Bbbk[x, y]/\langle x^2, y^2\rangle$, and set $\bar{a} := a + \langle x^2, y^2\rangle \in C$ for all $a \in \Bbbk[x, y]$. Then C is a 4-dimensional algebra with a basis consisting of $\bar{1}, \bar{x}, \bar{y}, \overline{xy}$. Moreover, $B := \Bbbk\bar{1} \oplus \Bbbk\overline{xy}$ is a *subalgebra* [11] of C, and $\Bbbk\overline{xy}$ is the set of all non-units in B, which is closed under sums. Thus B is a local algebra. Here we set

$$A := \begin{pmatrix} B & \Bbbk\bar{y} \\ \Bbbk\bar{x} & B \end{pmatrix} := \left\{ \begin{pmatrix} a_{11} & a_{12} \\ a_{21} & a_{22} \end{pmatrix} \middle| \, a_{11}, a_{22} \in B, a_{21} \in \Bbbk\bar{x}, a_{12} \in \Bbbk\bar{y} \right\}.$$

Then it is a 6-dimensional subalgebra of the algebra $M_2(C)$ (cf. Example 1.5.4 (4′)). Moreover, if we set $e_{ij} := (\bar{1}\delta_{(x,y),(i,j)})_{x,y} \in M_2(C)$ $(i, j \in \{1, 2\})$, then $E := \{e_{11}, e_{22}\}$ is a complete set of orthogonal idempotents in A, and since $e_{ii}Ae_{ii} \cong B$ $(i = 1, 2)$ is local, e_{ii} $(i = 1, 2)$ are primitive. Thus $(A, E) \in (\Bbbk\text{-}\mathbf{alg}_{\text{copi}})_0$.

[11] Thus it is a subspace of C, and forms an algebra with the addition and multiplication that are restrictions of those of C, having the same identity element as C.

In general, for a linear category $\mathscr{C}$, let $\mathscr{B}(x,y)$ be a basis of $\mathscr{C}(x,y)$ for all $x,y \in \mathscr{C}_0$, and consider the quiver $\mathscr{B} := (\mathscr{C}_0, \bigsqcup_{x,y\in\mathscr{C}_0} \mathscr{B}(x,y), \mathrm{dom}, \mathrm{cod})$. Then $\mathscr{C}$ is given by $\mathscr{C}_0 = \mathscr{B}_0$, $\mathscr{C}(x,y) = \Bbbk\mathscr{B}(x,y)$ for all $x,y \in \mathscr{B}_0$, and the composition table $T\colon \mathscr{B}(y,z) \times \mathscr{B}(x,y) \to \mathscr{C}(x,z)$ for all $x,y,z \in \mathscr{B}_0$ that is the restriction of the composition $\mathscr{C}(y,z) \times \mathscr{C}(x,y) \to \mathscr{C}(x,z)$ of $\mathscr{C}$. Therefore we can present $\mathscr{C}$ by $\mathscr{B}$ (and the composition table T that can be implicitly given). Thus $\mathscr{C}$ can be graphically presented by expressing each element b of $\mathscr{B}(x,y)$ by an arrow $x \xrightarrow{b} y$.

In this way, $\mathrm{Cat}(A,E)$ ($\in (\Bbbk\text{-}\mathbf{Spec}_{\mathrm{f}})_0$) is presented by the following:

$$\overline{xy}e_{11} \circlearrowright \underset{\circlearrowleft\ \bar{1}e_{11}}{e_{11}} \underset{\bar{y}e_{12}}{\overset{\bar{x}e_{21}}{\rightleftarrows}} \underset{\circlearrowleft\ \bar{1}e_{22}}{e_{22}} \circlearrowleft \overline{xy}e_{22} \cong \mathscr{C} := \overline{xy} \circlearrowright \underset{\circlearrowleft\ \bar{1}}{e_{11}} \underset{\bar{y}}{\overset{\bar{x}}{\rightleftarrows}} \underset{\circlearrowleft\ \bar{1}}{e_{22}} \circlearrowleft \overline{xy}\,.$$

Moreover, we conversely have $(A,E) = \mathrm{Mat}(\mathscr{C}) = (\mathrm{Mat}_{\mathscr{C}}, E_{\mathscr{C}})$. In particular, $A = \mathrm{Mat}_{\mathscr{C}}$.

THEOREM 2.3.30. *Let $(A,E) \in (\Bbbk\text{-}\mathbf{alg}_{\mathrm{copi}})_0$, $\mathscr{C} \in (\Bbbk\text{-}\mathbf{Spec}_{\mathrm{f}})_0$, and assume that $(A,E) = \mathrm{Mat}(\mathscr{C})$. Then the linear categories A-Mod and $\mathscr{C}$-Mod are equivalent.*

PROOF. We define a linear functor $\Phi\colon A\text{-Mod} \to \mathscr{C}\text{-Mod}$ as follows:

On objects: Let $M \in A\text{-Mod}_0$. Then define a functor $\Phi(M)\colon \mathscr{C} \to \mathrm{Mod}\,\Bbbk$ by

$$\Phi(M)(x) := e_{x,x}M, \Phi(M)(f) := f\cdot(\text{-})\colon e_{x,x}M \to e_{y,y}M$$

$(x,y \in \mathscr{C}_0, f \in \mathscr{C}(x,y))$.

On morphisms: Let $u\colon M \to N$ be a morphism in A-Mod. Then define $\Phi(u)\colon \Phi(M) \to \Phi(N)$ by

$$\Phi(u)_x := u|_{e_{x,x}M}\colon e_{x,x}M \to e_{x,x}N \quad (x \in \mathscr{C}_0).$$

Since u is a homomorphism of A-modules, we see that $\Phi(u)$ is a natural transformation, i.e., a morphism in $\mathrm{Mod}\,\mathscr{C}$.

Further, we define a linear functor $\Psi\colon \mathscr{C}\text{-Mod} \to A\text{-Mod}$ as follows:

On objects: Let $V \in \mathscr{C}\text{-Mod}_0$. Then define $\Psi(V) \in A\text{-Mod}_0$ by setting $\Psi(V) := \coprod_{x\in\mathscr{C}_0} M(x)$ as a vector space and by giving a left A-action by

$$a\cdot v := \left(\sum_{x\in\mathscr{C}_0} V(a_{y,x})v_x\right)_{y\in\mathscr{C}_0}$$

for all $a = (a_{y,x})_{y,x\in\mathscr{C}_0} \in A$ $(a_{y,x} \in \mathscr{C}(x,y))$ and $v = (v_x)_{x\in\mathscr{C}_0} \in \Psi(V)$. Note that the right-hand side is the usual matrix multiplication by regarding elements of $\Psi(V)$ as column vectors.

On morphisms: Let $\alpha\colon V \to W$ be a morphism in $\mathscr{C}$-Mod. Then define $\Psi(\alpha)\colon \Psi(V) \to \Psi(W)$ by

$$\coprod_{x\in\mathscr{C}_0} \alpha_x\colon \coprod_{x\in\mathscr{C}_0} V(x) \to \coprod_{x\in\mathscr{C}_0} W(x).$$

Since α is a natural transformation, $\Psi(\alpha)$ becomes a homomorphism of A-modules.

Finally it is easily verified that Φ and Ψ are quasi-inverses of each other. □

EXERCISE 2.3.31. In the proof above, show that Φ, Ψ are linear functors, and that they are mutually quasi-inverses.

REMARK 2.3.32. The theorem above reduces problems on representations of a finite-dimensional algebra A to those on representations of the finite spectroid $\mathscr{C} \in (\Bbbk\text{-}\mathbf{Spec}_{\mathrm{f}})_0$ corresponding to the basic algebra of A.

2.4. Products and direct sums of modules over a linear category

DEFINITION 2.4.1. Let $\mathscr{C}$ be a linear category, I a small set, and $M_i \in (\operatorname{Mod}\mathscr{C})_0$ for all $i \in I$. Then

(1) We define a right $\mathscr{C}$-module $\prod_{i\in I} M_i$ (resp. $\coprod_{i\in I} M_i$) as follows, and call it the *direct product* (resp. *direct sum*) of $(M_i)_{i\in I}$: For each morphism $f\colon x \to y$ in $\mathscr{C}$

$$\left(\prod_{i\in I} M_i\right)(x) := \prod_{i\in I} M_i(x),$$

$$\left(\prod_{i\in I} M_i\right)(f) := \prod_{i\in I} M_i(f)\colon \prod_{i\in I} M_i(y) \to \prod_{i\in I} M_i(x),$$
$$(m_i)_{i\in I} \mapsto (M_i(f)(m_i))_{i\in I}$$

$$\left(\coprod_{i\in I} M_i\right)(x) := \coprod_{i\in I} M_i(x),$$

$$\left(\coprod_{i\in I} M_i\right)(f) := \coprod_{i\in I} M_i(f)\colon \coprod_{i\in I} M_i(y) \to \coprod_{i\in I} M_i(x).$$
$$(m_i)_{i\in I} \mapsto (M_i(f)(m_i))_{i\in I}$$

In the above, the direct products (resp. direct sums) on the right-hand sides stand for those in $\operatorname{Mod}\Bbbk$. Note that by construction $\coprod_{i\in I} M_i$ is a submodule of $\prod_{i\in I} M_i$.

(2) Let $j \in I$ and $\pi_{j,x}\colon \prod_{i\in I} M_i(x) \to M_j(x)$ be the j-th projection in $\operatorname{Mod}\Bbbk$ for all $x \in \mathscr{C}_0$. Then $\pi_j := (\pi_{j,x})_{x\in\mathscr{C}_0}$ turns out to be a morphism $\prod_{i\in I} M_i \to M_j$ in $\operatorname{Mod}\mathscr{C}$, and is called the j-th *projection*. We call $(\pi_i)_{i\in I}$ the *family of canonical projections* of the direct product $\prod_{i\in I} M_i$. Here we set $\pi'_j = (\pi'_{j,x})_{x\in\mathscr{C}_0}\colon \coprod_{i\in I} M_i \to M_j$ to be the composite of the inclusion $\coprod_{i\in I} M_i \hookrightarrow \prod_{i\in I} M_i$ and π_j.

(3) Let $j \in I$ and $\sigma_{j,x}\colon M_j(x) \to \coprod_{i\in I} M_i(x)$ be the j-th injection in $\operatorname{Mod}\Bbbk$ for all $x \in \mathscr{C}_0$. Then $\sigma_j := (\sigma_{j,x})_{x\in\mathscr{C}_0}$ turns out to be a morphism $M_j \to \coprod_{i\in I} M_i$ in $\operatorname{Mod}\mathscr{C}$, and is called the j-th *injection*. We call $(\sigma_i)_{i\in I}$ the *family of canonical injections* of the direct sum $\coprod_{i\in I} M_i$.

EXERCISE 2.4.2. In Definition 2.4.1, show that $(\prod_{i\in I} M_i, (\pi_j)_{j\in I})$ is a product in the category $\operatorname{Mod}\mathscr{C}$, and $(\coprod_{i\in I} M_i, (\sigma_j)_{j\in I})$ is a coproduct in the category $\operatorname{Mod}\mathscr{C}$ (cf. Definition 1.1.23).

DEFINITION 2.4.3. Let $\mathscr{C}$ be a linear category, I a small set, $M, N, M_i \in (\operatorname{Mod}\mathscr{C})_0$ for all $i \in I$, $\sigma\colon \coprod_{i\in I} M_i \hookrightarrow \prod_{i\in I} M_i$ the inclusion, and $(\pi_i\colon \prod_{j\in I} M_j \to M_i)_{i\in I}$ the family of canonical projections.

(1) A family $(f_i)_{i\in I}$ of elements in $(\operatorname{Mod}\mathscr{C})(M, N)$ is said to be *summable* if for any $x \in \mathscr{C}_0$ and $m \in M(x)$, there exists a finite subset $I_{x,m}$ of I such

that $f_{i,x}(m) = 0$ for all $i \in I \setminus I_{x,m}$, where we set $f_i := (f_{i,x} \colon M(x) \to N(x))_{x \in \mathscr{C}_0}$.

When this is the case, it is possible to define a linear map

$$\sum_{i \in I} f_{i,x} \colon M(x) \to N(x)$$

for all $x \in \mathscr{C}_0$ by $\left(\sum_{i \in I} f_{i,x}\right)(m) := \sum_{i \in I} f_{i,x}(m)$ for all $m \in M(x)$ because the right-hand side is a finite sum. This defines a morphism $\sum_{i \in I} f_i \in (\operatorname{Mod}\mathscr{C})(M, N)$ by $\sum_{i \in I} f_i := \left(\sum_{i \in I} f_{i,x}\right)_{x \in \mathscr{C}_0}$.

(2) A morphism $p \colon M \to \prod_{i \in I} M_i$ to a direct product is said to be *locally column-finite* if it factors through the inclusion σ, i.e., if there exists a morphism $p' \colon M \to \coprod_{i \in I} M_i$ such that $p = \sigma \circ p'$. Further, a family $(p_i)_{i \in I} \in \prod_{i \in I} (\operatorname{Mod}\mathscr{C})(M, M_i)$ of morphisms is said to be *locally column-finite* if the canonical morphism $[p_i]_{i \in I} \colon M \to \prod_{i \in I} M_i$ defined by the universality of the family $(\pi_i)_{i \in I}$ is locally column-finite.

(3) A family $(M_i \xrightarrow{q_i} M \xrightarrow{p_i} M_i)_{i \in I}$ of morphisms in $\operatorname{Mod}\mathscr{C}$ is called a *direct sum system* if

(a) $p_i \circ q_j = \delta_{i,j} \mathbb{1}_{M_i}$ $(i, j \in I)$;

(b) The family $(p_i)_{i \in I}$ is locally column-finite;

(c) $\sum_{i \in I} q_i \circ p_i = \mathbb{1}_M$.

In the above, note that the condition (b) implies the condition

(d) The family $(q_i \circ p_i)_{i \in I}$ of elements in $(\operatorname{Mod}\mathscr{C})(M, M)$ is summable,

which makes it possible to define the left-hand side of (c). Indeed, by the condition (b), we have $p_x = \sigma_x \circ p'_x$ for all $x \in \mathscr{C}_0$, and hence $p_x(m) = (p_{i,x}(m))_{i \in I} \in \coprod_{i \in I} M_i(x)$ for all $m \in M(x)$. Thus there exists a finite subset $I_{x,m}$ of I such that for each $i \in I \setminus I_{x,m}$ we have $p_{i,x}(m) = 0$, and thus $(q_{i,x} \circ p_{i,x})(m) = 0$. Therefore (d) holds.

Note that if I is a finite set, then the the condition (b) holds trivially. The concepts above for an infinite set I are mainly used in Section 7.4.

LEMMA 2.4.4. *Let $\mathscr{C}$ be a linear category, I a small set, and $M, M_i \in (\operatorname{Mod}\mathscr{C})_0$ for all $i \in I$. Then the following hold:*

(1) *The family $(M_i \xrightarrow{\sigma_i} \coprod_{i \in I} M_i \xrightarrow{\pi'_i} M_i)_{i \in I}$ of morphisms in $\operatorname{Mod}\mathscr{C}$ is a direct sum system, which is called the* canonical direct sum system *of the family $(M_i)_{i \in I}$.*

(2) *If $\phi \colon M \to \coprod_{i \in I} M_i$ is an isomorphism in $\operatorname{Mod}\mathscr{C}$, then $(M_i \xrightarrow{\phi^{-1} \circ \sigma_i} M \xrightarrow{\pi'_i \circ \phi} M_i)_{i \in I}$ is a direct sum system.*

(3) *Any direct sum system $(M_i \xrightarrow{q_i} M \xrightarrow{p_i} M_i)_{i \in I}$ has this form, i.e., there exists an isomorphism $\phi \colon M \to \coprod_{i \in I} M_i$ such that for each $i \in I$ the diagram*

$$\begin{array}{ccccc}
M_i & \xrightarrow{q_i} & M & \xrightarrow{p_i} & M_i \\
\| & & \downarrow{\scriptstyle \phi} & & \| \\
M_i & \xrightarrow{\sigma_i} & \coprod_{i \in I} M_i & \xrightarrow{\pi'_i} & M_i
\end{array}$$

commutes. Note that this ϕ is uniquely determined because the commutativity of the left square above shows that ϕ^{-1} is the canonical morphism $\phi^{-1} = {}^t[q_i]_{i \in I}$ determined by the universality of the injection family

$(\sigma_i)_{i\in I}$. *This* ϕ *is called the* canonical isomorphism *determined by this direct sum system.*

PROOF. (1) The condition (b) of a direct sum system clearly follows from the construction of π'_i. By definition it is clear that $\pi'_{j,x} \circ \sigma_{i,x} = \delta_{i,j}\mathbb{1}_{M_i(x)}$ and $\sum_{i\in I} \sigma_{i,x} \circ \pi'_{i,x} = \mathbb{1}_{M(x)}$ for all $x \in \mathscr{C}_0$, which shows the conditions (a) and (c). Statement (2) is also obvious.

(3) By the universality of the direct sum (coproduct), there exists a unique $q\colon \coprod_{i\in I} M_i \to M$ such that $q_i = q \circ \sigma_i$ for all $i \in I$. By the universality of the direct product, there also exists a unique $p\colon M \to \prod_{i\in I} M_i$ such that $p_i = \pi_i \circ p$ for all $i \in I$, and this p is written as $p = \sigma\circ\phi$ for some $\phi\colon M \to \coprod_{i\in I} M_i$ because $(p_i)_{i\in I}$ is locally column-finite, where $\sigma\colon \coprod_{i\in I} M_i \hookrightarrow \prod_{i\in I} M_i$ is the inclusion. Then it suffices to show that $\phi\circ q = \mathbb{1}_{\coprod_{i\in I} M_i}$ and that $q\circ\phi = \mathbb{1}_M$. Indeed, if these are true, then ϕ is an isomorphism with $q = \phi^{-1}$, which shows that $q_i = q\circ\sigma_i = \phi^{-1}\circ\sigma_i$ for all $i \in I$ and that $p_i = \pi_i \circ p = \pi_i \circ \sigma \circ \phi = \pi'_i \circ \phi$ for all $i \in I$, as required.

(i) Verification of $\phi\circ q = \mathbb{1}_{\coprod_{i\in I} M_i}$. Consider the following diagram for all $i \in I$:

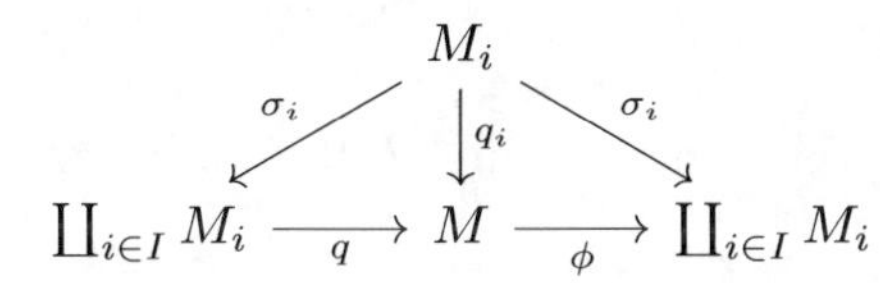

By construction of q the left triangle is commutative. If we show the commutativity of the right triangle, then we have $\sigma_i = (\phi\circ q)\circ\sigma_i$ for all $i \in I$. On the other hand we clearly have $\sigma_i = \mathbb{1}_{\coprod_{i\in I} M_i} \circ \sigma_i$ for all $i \in I$. Hence the universality of $\coprod_{i\in I} M_i$ implies that $\phi\circ q = \mathbb{1}_{\coprod_{i\in I} M_i}$. Therefore it is enough to show that $\sigma_i = \phi\circ q_i$ for all $i \in I$. Let $i \in I$. Then for each $j \in I$ we have

$$\pi_j \circ \sigma \circ \phi \circ q_i = \begin{Bmatrix} \pi_i \circ p \circ q_i = p_i \circ q_i = \mathbb{1}_{M_i} & (j = i) \\ \pi_j \circ p \circ q_i = p_j \circ q_i = 0 & (j \neq i) \end{Bmatrix} = \pi_j \circ \sigma \circ \sigma_i.$$

Therefore the universality of $\prod_{j\in I} M_j$ together with the fact that σ is a monomorphism shows that $\phi\circ q_i = \sigma_i$.

(ii) Verification of $q\circ\phi = \mathbb{1}_M$. Since $(\sigma_i\circ\pi'_i)_{i\in I}$ is summable, we have

$$\begin{aligned} q\circ\phi &= q\circ(\mathbb{1}_{\coprod_{i\in I} M_i})\circ\phi = q\circ\left(\sum_{i\in I}\sigma_i\circ\pi'_i\right)\circ\phi \\ &= \left(\sum_{i\in I} q\circ\sigma_i\circ\pi'_i\circ\phi\right) = \sum_{i\in I} q_i\circ p_i = \mathbb{1}_M. \end{aligned}$$

□

The following is immediate from Lemma 2.4.4.

PROPOSITION 2.4.5. *Let* $\mathscr{C}$ *be a linear category,* I *a small set, and* $M, M_i \in (\operatorname{Mod}\mathscr{C})_0$ *for all* $i \in I$. *The the following are equivalent in* $\operatorname{Mod}\mathscr{C}$*:*

(1) $M \cong \coprod_{i\in I} M_i$*;*

(2) *There exists a direct sum system* $(M_i \xrightarrow{q_i} M \xrightarrow{p_i} M_i)_{i\in I}$.

DEFINITION 2.4.6. Let I, J be small sets, $(L_i)_{i\in I}, (M_j)_{j\in J}$ families of elements of $(\operatorname{Mod}\mathscr{C})_0$, and $(f_{j,i})_{(j,i)\in J\times I} \in \prod_{j\in J}\prod_{i\in I}(\operatorname{Mod}\mathscr{C})(L_i, M_j)$. Then $(f_{j,i})_{(j,i)\in J\times I}$

is called *locally column-finite* if $(f_{j,i})_{j\in J}$ is locally column-finite for all $i \in I$. The set of those $(f_{j,i})_{j\in J}$ is denoted by $\prod'_{j\in J}\prod_{i\in I}(\operatorname{Mod}\mathscr{C})(L_i, M_j)$.

The following gives the matrix presentation of a morphism between modules with direct sum decompositions.

PROPOSITION 2.4.7. *Let I, J be small sets and $(L_i \xrightarrow{q_i} L \xrightarrow{p_i} L_i)_{i\in I}, (M_j \xrightarrow{s_j} M \xrightarrow{r_j} M_j)_{j\in J}$ direct sum systems in $\operatorname{Mod}\mathscr{C}$. Then the following is an isomorphism in $\operatorname{Mod}\Bbbk$:*

$$(\operatorname{Mod}\mathscr{C})(L, M) \to \prod_{j\in J}{}' \prod_{i\in I}(\operatorname{Mod}\mathscr{C})(L_i, M_j).$$
$$f \mapsto (r_j \circ f \circ q_i)_{(j,i)\in J\times I}$$

We denote its inverse by

$$\prod_{j\in J}{}' \prod_{i\in I}(\operatorname{Mod}\mathscr{C})(L_i, M_j) \to (\operatorname{Mod}\mathscr{C})(L, M)$$
$$(f_{j,i})_{(j,i)\in J\times I} \mapsto [f_{j,i}]_{(j,i)\in J\times I}.$$

Therefore if we set $f_{j,i} := r_j \circ f \circ q_i$ $(i \in I, j \in J)$, then we have $f = [f_{j,i}]_{(j,i)\in J\times I}$, which is called the matrix presentation *of f with respect to these direct sum systems* (cf. Proposition 1.1.27).

PROOF. Let $(L_i \xrightarrow{\sigma_i} \coprod_{i\in I} L_i \xrightarrow{\pi'_i} L_i)_{i\in I}$, $(M_j \xrightarrow{\tau_j} \coprod_{j\in J} M_j \xrightarrow{\rho'_j} M_j)_{j\in J}$ be the canonical direct sum systems, and set $\phi\colon L \to \coprod_{i\in I} L_i$, $\psi\colon M \to \coprod_{j\in J} M_j$ to be the canonical isomorphisms determined by them, respectively. Moreover, let $(\rho_j\colon \prod_{j\in J} M_j \to M_j)_{j\in J}$ be the canonical projection family, and $\sigma\colon \coprod_{j\in J} M_j \to \prod_{j\in J} M_j$ the inclusion (Note here that $\rho'_j = \rho_j \circ \sigma$ $(j \in J)$.), and set

$$\alpha\colon (\operatorname{Mod}\mathscr{C})(\coprod_{i\in I} L_i, \prod_{j\in J} M_j) \to \prod_{j\in J}\prod_{i\in I}(\operatorname{Mod}\mathscr{C})(L_i, M_j), g \mapsto (\rho_j \circ g \circ \sigma_i)_{j,i}$$

to be the bijection in Proposition 1.1.27 obtained by the universalities of products and coproducts. By Remark 1.1.20, $\sigma_* := (\operatorname{Mod}\mathscr{C})(\coprod_{i\in I} L_i, \sigma)$:

$$(\operatorname{Mod}\mathscr{C})(\coprod_{i\in I} L_i, \coprod_{j\in J} M_j) \to (\operatorname{Mod}\mathscr{C})(\coprod_{i\in I} L_i, \prod_{j\in J} M_j)$$

is a monomorphism, and hence so is the composite $\beta := \alpha \circ \sigma_*$:

$$\beta\colon (\operatorname{Mod}\mathscr{C})(\coprod_{i\in I} L_i, \coprod_{j\in J} M_j) \to \prod_{j\in J}\prod_{i\in I}(\operatorname{Mod}\mathscr{C})(L_i, M_j),\ g \mapsto (\rho'_j \circ g \circ \sigma_i)_{j,i}.$$

Since by definition we have $\operatorname{Im}\beta = \prod'_{j\in J}\prod_{i\in I}(\operatorname{Mod}\mathscr{C})(L_i, M_j)$, β gives the isomorphism

$$(\operatorname{Mod}\mathscr{C})(\coprod_{i\in I} L_i, \coprod_{j\in J} M_j) \to \prod_{j\in J}{}'\prod_{i\in I}(\operatorname{Mod}\mathscr{C})(L_i, M_j),\ g \mapsto (\rho'_j \circ g \circ \sigma_i)_{j,i}.$$

Therefore by composing this and the isomorphism

$$(\operatorname{Mod}\mathscr{C})(L, M) \xrightarrow{(\operatorname{Mod}\mathscr{C})(\phi^{-1},\psi)} (\operatorname{Mod}\mathscr{C})(\coprod_{i\in I} L_i, \coprod_{j\in J} M_j),\ f \mapsto \psi \circ f \circ \phi^{-1},$$

we obtain the desired isomorphism $f \mapsto (\rho'_j \circ \psi \circ f \circ \phi^{-1} \circ \sigma_i)_{j,i} = (r_j \circ f \circ q_i)_{j,i}$. $\square$

REMARK 2.4.8 (Meaning of local column-finiteness). Under the setting of Proposition 2.4.7, take an arbitrary $g \in (\operatorname{Mod}\mathscr{C})(\coprod_{i\in I} L_i, \prod_{j\in J} M_j)$. Then g is locally column-finite if and only if for each $x \in \mathscr{C}_0$ and $l \in \coprod_{i\in I} L_i(x)$, we have $g_x(l) \in \coprod_{j\in J} M_j(x)$, thus if and only if there exists a finite subset $J_{x,l}$ of J such that $\pi_j(g_x(l)) = 0$ for all $j \in J \setminus J_{x,l}$.

LEMMA 2.4.9. *Let I, J, K be small sets and $(L_i \xrightarrow{q_i} \coprod_{i\in I} L_i \xrightarrow{p_i} L_i)_{i\in I}, (M_j \xrightarrow{s_j} \coprod_{j\in J} M_j \xrightarrow{r_j} M_j)_{j\in J}, (N_k \xrightarrow{u_k} \coprod_{k\in K} N_k \xrightarrow{t_k} N_k)_{k\in K}$ the canonical direct sum systems in $\operatorname{Mod}\mathscr{C}$. Then the following hold:*

(1) *Let $x \in \mathscr{C}_0$. Then we have*

$$f_x(l) = [f_{j,i,x}]_{(j,i)\in J\times I}(l_i)_{i\in I} = \left(\sum_{i\in I} f_{j,i,x}(l_i)\right)_{j\in J}$$

for all $f = [f_{j,i}]_{(j,i)\in J\times I} \in (\operatorname{Mod}\mathscr{C})(\coprod_{i\in I} L_i, \coprod_{j\in J} M_j)$ and $l = (l_i)_{i\in I} \in \coprod_{i\in I} L_i(x)$. Namely, it has the same form as the usual product of a matrix and a column vector if we regard $(l_i)_{i\in I}$ as a column vector.

(2) *Let $f \in (\operatorname{Mod}\mathscr{C})(\coprod_{i\in I} L_i, \coprod_{j\in J} M_j)$, $g \in (\operatorname{Mod}\mathscr{C})(\coprod_{j\in J} M_j, \coprod_{k\in K} N_k)$, and $f = [f_{j,i}]_{(j,i)\in J\times I}, g = [g_{k,j}]_{(k,j)\in K\times J}$ be their matrix presentations with respect to the given direct sum systems. Then for each $i \in I, k \in K$, the family $(g_{k,j} \circ f_{j,i})_{j\in J} \in (\operatorname{Mod}\mathscr{C})(L_i, N_k)^J$ is summable, and we have*

$$g \circ f = \left[\sum_{j\in J} g_{k,j} \circ f_{j,i}\right]_{(k,i)\in K\times I}.$$

Namely, their composition corresponds to the products of matrices.

PROOF. Set $L := \coprod_{i\in I} L_i, M := \coprod_{j\in J} M_j, N := \coprod_{k\in K} N_k$.

(1) Since $l = \mathbb{1}_{L(x)}(l) = \sum_{i\in I} q_{i,x} \circ p_{i,x}(l)$, $f_x = \mathbb{1}_{M(x)} \circ f_x = \sum_{j\in J} s_{j,x} \circ r_{j,x} \circ f_x$, we have

$$\begin{aligned}
f_x(l) &= \left(\sum_{j\in J} s_{j,x} \circ r_{j,x} \circ f_x\right)\left(\sum_{i\in I} q_{i,x} \circ p_{i,x}(l)\right) \\
&= \sum_{j\in J}\sum_{i\in I} s_{j,x} \circ r_{j,x} \circ f_x \circ q_{i,x} \circ p_{i,x}(l) \\
&= \sum_{j\in J}\sum_{i\in I} s_{j,x}(f_{j,i,x}(l_i)) \\
&= \sum_{j\in J} s_{j,x}\left(\sum_{i\in I} f_{j,i}(l_i)\right) = \left(\sum_{i\in I} f_{j,i,x}(l_i)\right)_{j\in J}.
\end{aligned}$$

(2) This follows from the computation below:

$$
\begin{aligned}
g \circ f = g \circ \mathbb{1}_M \circ f = g \circ \left(\sum_{j\in J} s_j \circ r_j\right) \circ f &= \sum_{j\in J} g \circ s_j \circ r_j \circ f \\
&= \sum_{j\in J} \mathbb{1}_N \circ g \circ s_j \circ r_j \circ f \circ \mathbb{1}_L \\
&= \sum_{j\in J} \left(\sum_{k\in K} u_k \circ t_k\right) \circ g \circ s_j \circ r_j \circ f \circ \left(\sum_{i\in I} q_i \circ p_i\right) \\
&= \sum_{k\in K}\sum_{i\in I} u_k \circ \left(\sum_{j\in J} t_k \circ g \circ s_j \circ r_j \circ f \circ q_i\right) \circ p_i \\
&= \sum_{k\in K}\sum_{i\in I} u_k \circ \left(\sum_{j\in J} g_{k,j} \circ f_{j,i}\right) \circ p_i \\
&= \left[t_{k'} \circ \sum_{k\in K}\sum_{i\in I} u_k \circ \left(\sum_{j\in J} g_{k,j} \circ f_{j,i}\right) \circ p_i \circ q_{i'}\right]_{(k',i')\in K\times I} \\
&= \left[\sum_{j\in J} g_{k',j} \circ f_{j,i'}\right]_{(k',i')\in K\times I} = \left[\sum_{j\in J} g_{k,j} \circ f_{j,i}\right]_{(k,i)\in K\times I}.
\end{aligned}
$$

□

2.5. Finitely generated projective modules over a linear category

Throughout this section we let $\mathscr{C}$ be a linear category.

DEFINITION 2.5.1. Let $f\colon L \to M$ be a morphism of $\mathscr{C}$-modules. A submodule $\operatorname{Ker} f$ of L is defined by $(\operatorname{Ker} f)(x) := \operatorname{Ker}(f_x\colon L(x) \to M(x))$, $(x \in \mathscr{C}_0)$, and a submodule $\operatorname{Im} f$ of M is defined by $(\operatorname{Im} f)(x) := \operatorname{Im}(f_x\colon L(x) \to M(x))$, $(x \in \mathscr{C}_0)$, which are called the *kernel* and the *image* of f, respectively.

REMARK 2.5.2. Let $f\colon L \to M$ be a morphism of $\mathscr{C}$-modules.

(1) f is an epimorphism if and only if $\operatorname{Im} f = M$.
(2) f is a monomorphism if and only if $\operatorname{Ker} f = 0$.

EXERCISE 2.5.3. Let $f\colon L \to M$ be a morphism in $\operatorname{Mod}\mathscr{C}$. Then prove the following.

(1) Let $\sigma\colon \operatorname{Ker} f \to L$ be the inclusion. Then this turns out to be the kernel of f in the category $\operatorname{Mod}\mathscr{C}$ (see Definition 1.5.9).
(2) Let $\pi\colon M \to M/\operatorname{Im} f$ be the canonical epimorphism. Then this turn out to be the cokernel of f in the category $\operatorname{Mod}\mathscr{C}$. Therefore we set $\operatorname{Coker} f := M/\operatorname{Im} f$ and call it the *cokernel* of f.

DEFINITION 2.5.4.

(1) A sequence $M_1 \xrightarrow{f_1} M_2 \xrightarrow{f_2} \cdots \xrightarrow{f_{n-1}} M_n$ of morphisms of $\mathscr{C}$-modules is called *exact* if $\operatorname{Im} f_i = \operatorname{Ker} f_{i+1}$ $(1 \le i \le n-2)$.

(2) An exact sequence of morphisms of $\mathscr{C}$-modules of the form $0 \to L \xrightarrow{f} M \xrightarrow{g} N \to 0$ is called a *short exact sequence*.

REMARK 2.5.5. In the above, by Definition 2.5.1, a sequence of morphisms of $\mathscr{C}$-modules

$$0 \to L \xrightarrow{f} M \xrightarrow{g} N \to 0$$

is exact if and only if the sequence

$$0 \to L(x) \xrightarrow{f_x} M(x) \xrightarrow{g_x} N(x) \to 0$$

of $\Bbbk$-linear maps is exact for all $x \in \mathscr{C}_0$.

DEFINITION 2.5.6. A $\mathscr{C}$-module X is called *projective* if for each short exact sequence $0 \to L \xrightarrow{f} M \xrightarrow{g} N \to 0$ of morphisms of $\mathscr{C}$-modules, the sequence

$$0 \to (\operatorname{Mod}\mathscr{C})(X, L) \xrightarrow{(\operatorname{Mod}\mathscr{C})(X,f)} (\operatorname{Mod}\mathscr{C})(X, M) \xrightarrow{(\operatorname{Mod}\mathscr{C})(X,g)} (\operatorname{Mod}\mathscr{C})(X, N) \to 0$$

is an exact sequence of linear maps of vector spaces. The full subcategory of $\operatorname{Mod}\mathscr{C}$ consisting of all projective $\mathscr{C}$-modules is denoted by $\operatorname{Prj}\mathscr{C}$.

LEMMA 2.5.7.

(1) *Let I be a small set and $(X_i)_{i\in I}$ a family of elements of* $(\operatorname{Mod}\mathscr{C})_0$. *If X_i is projective for all $i \in I$, then so is $\coprod_{i\in I} X_i$.*

(2) *Let $X, X_1 \in (\operatorname{Mod}\mathscr{C})_0$, and assume that X_1 is a direct summand of X. If X is projective, then so is X_1* (see Definition 1.5.16).

PROOF. Take an arbitrary short exact sequence

$$0 \to L \xrightarrow{f} M \xrightarrow{g} N \to 0 \tag{2.6}$$

in $\operatorname{Mod}\mathscr{C}$.

(1) Set $X := \coprod_{i\in I} X_i$. By the formula (1.6) in Exercise 1.3.4 we have a natural isomorphism

$$\alpha\colon (\operatorname{Mod}\mathscr{C})(X, \text{-}) \to \prod_{i\in I} (\operatorname{Mod}\mathscr{C})(X_i, \text{-}).$$

Using this we obtain a commutative diagram

$$\begin{array}{ccccccccc}
0 \longrightarrow & (\operatorname{Mod}\mathscr{C})(X, L) & \longrightarrow & (\operatorname{Mod}\mathscr{C})(X, M) & \longrightarrow & (\operatorname{Mod}\mathscr{C})(X, N) & \longrightarrow 0 \\
& \downarrow{\scriptstyle \alpha_L} & & \downarrow{\scriptstyle \alpha_M} & & \downarrow{\scriptstyle \alpha_N} & \\
0 \to & \prod_{i\in I}(\operatorname{Mod}\mathscr{C})(X_i, L) & \to & \prod_{i\in I}(\operatorname{Mod}\mathscr{C})(X_i, M) & \to & \prod_{i\in I}(\operatorname{Mod}\mathscr{C})(X_i, N) & \to 0
\end{array}$$

in $\operatorname{Mod}\Bbbk$ from the short exact sequence (2.6), which gives an isomorphism between the upper and the lower sequences. Since each X_i is projective, and the direct product of exact sequences is an exact sequence in $\operatorname{Mod}\Bbbk$, the lower sequence is exact. Therefore the upper sequence is exact as well.

(2) Let $(X_i \xrightarrow{\sigma_i} X \xrightarrow{\pi_i} X_i)_{i=1}^2$ be a direct sum system. For the sequence

$$0 \to (\operatorname{Mod}\mathscr{C})(X_1, L) \to (\operatorname{Mod}\mathscr{C})(X_1, M) \xrightarrow{(\operatorname{Mod}\mathscr{C})(X_1,g)} (\operatorname{Mod}\mathscr{C})(X_1, N) \to 0$$

obtained from the short exact sequence (2.6) the exactness conditions at $(\operatorname{Mod}\mathscr{C})(X_1, L)$ and at $(\operatorname{Mod}\mathscr{C})(X_1, M)$ hold in general by Remark 1.5.10 because f is a kernel of g (in the sense of Definition 1.5.9). It remains to verify that $(\operatorname{Mod}\mathscr{C})(X_1, g)$ is an epimorphism. In the commutative diagram

$$\begin{array}{ccc} (\operatorname{Mod}\mathscr{C})(X, M) & \xrightarrow{(\operatorname{Mod}\mathscr{C})(X,g)} & (\operatorname{Mod}\mathscr{C})(X, N) \\ {\scriptstyle (\operatorname{Mod}\mathscr{C})(\sigma_1,M)}\downarrow & & \downarrow{\scriptstyle (\operatorname{Mod}\mathscr{C})(\sigma_1,N)} \\ (\operatorname{Mod}\mathscr{C})(X_1, M) & \xrightarrow{(\operatorname{Mod}\mathscr{C})(X_1,g)} & (\operatorname{Mod}\mathscr{C})(X_1, N) \end{array}$$

the projectivity of X implies that $(\operatorname{Mod}\mathscr{C})(X, g)$ is an epimorphism. Since $\pi_1\sigma_1 = \mathbb{1}_{X_1}$, we have $(\operatorname{Mod}\mathscr{C})(\sigma_1, N)(\operatorname{Mod}\mathscr{C})(\pi_1, N) = \mathbb{1}_{(\operatorname{Mod}\mathscr{C})(X,N)}$, which shows that $(\operatorname{Mod}\mathscr{C})(\sigma_1, N)$ is also an epimorphism. Therefore the clockwise composite is an epimorphism, and hence so is the counter-clockwise composite by the commutativity. This shows that $(\operatorname{Mod}\mathscr{C})(X_1, g)$ is an epimorphism as well.

□

The Yoneda lemma, which we will discuss next, is obtained by transferring the proof of Proposition 2.3.4 to linear categories as algebras with multiple objects. After all, it is a generalization of the one-to-one correspondence between the set of proportionalities and the set of proportional constants. This will be used immediately afterwards to prove the projectivity of representable functors.

LEMMA 2.5.8 (The Yoneda lemma). *There exists an isomorphism*

$$\phi_{x,F} \colon (\operatorname{Mod}\mathscr{C})(\mathscr{C}(\text{-}, x), F) \xrightarrow{\sim} F(x)$$

that is natural in $x \in \mathscr{C}_0$ *and in* $F \in (\operatorname{Mod}\mathscr{C})_0$.

To accurately state this naturality, we first define the functor

$$Y \colon \mathscr{C} \to \operatorname{Mod}\mathscr{C}$$

by $Y(x) := \mathscr{C}(\text{-}, x), Y(f) := \mathscr{C}(\text{-}, f),\ (x \in \mathscr{C}_0, f \in \mathscr{C}_1)$, *which is called the* Yoneda embedding. *We also define a functor*

$$\operatorname{ev} \colon \mathscr{C}^{\mathrm{op}} \times \operatorname{Mod}\mathscr{C} \to \operatorname{Mod}\Bbbk$$

by $\operatorname{ev}(x, F) := F(x)$ *and* $\operatorname{ev}(f, \alpha) := \alpha_{x'} \circ F(f) \colon F(x) \to F(x') \to F'(x')\ \ (f \in \mathscr{C}(x', x), \alpha \in (\operatorname{Mod}\mathscr{C})(F, F'), x, x' \in \mathscr{C}_0, F, F' \in (\operatorname{Mod}\mathscr{C})_0)$.

Then the precise assertion of the lemma is that there exists a natural isomorphism:

$$\begin{array}{ccc} \mathscr{C}^{\mathrm{op}} \times \operatorname{Mod}\mathscr{C} & \xrightarrow{Y^{\mathrm{op}} \times \operatorname{Mod}\mathscr{C}} & (\operatorname{Mod}\mathscr{C})^{\mathrm{op}} \times \operatorname{Mod}\mathscr{C} \\ {\scriptstyle \operatorname{ev}}\downarrow & \overset{\phi}{\underset{\sim}{\Longleftarrow}} & \downarrow{\scriptstyle (\operatorname{Mod}\mathscr{C})(?,\text{-})} \\ \operatorname{Mod}\Bbbk & \hookrightarrow & \operatorname{Mod}^2\Bbbk \end{array}$$

(see Exercise 1.2.5 for representable functors, Definition 1.2.6 (resp. 1.2.8) for the direct product of categories (resp. functors), and Remark 1.5.7(2) for $\operatorname{Mod}^2\Bbbk$). *Here, if* $\mathscr{C}$ *is a small category, then* $\operatorname{Mod}^2\Bbbk$ *above can be replaced by* $\operatorname{Mod}\Bbbk$ (cf. [**33**, Proposition 11]).

PROOF. Let $x \in \mathscr{C}_0$ and $F \in (\operatorname{Mod}\mathscr{C})_0$. First, we define an isomorphism

$$\phi_{x,F} \colon (\operatorname{Mod}\mathscr{C})(\mathscr{C}(\text{-}, x), F) \to F(x).$$

For each $\alpha \in (\operatorname{Mod}\mathscr{C})(\mathscr{C}(\text{-},x),F)$, by noting that $\alpha_x\colon \mathscr{C}(x,x) \to F(x)$ is a morphism in $\operatorname{Mod}\Bbbk$, we set

$$\phi_{x,F}(\alpha) := \alpha_x(\mathbb{1}_x)\ (\in F(x))$$

and define $\phi := (\phi_{x,F})_{(x,F)\in(\mathscr{C}^{\mathrm{op}}\times\operatorname{Mod}\mathscr{C})_0}$.

Next, we define the inverse

$$\psi_{x,F}\colon F(x) \to (\operatorname{Mod}\mathscr{C})(\mathscr{C}(\text{-},x),F)$$

of $\phi_{x,F}$. For each $v \in F(x)$ define a natural transformation $\psi_{x,F}(v) = (\psi_{x,F}(v)_y\colon \mathscr{C}(y,x) \to F(y))_{y\in\mathscr{C}_0}$ that is a morphism in $\operatorname{Mod}\mathscr{C}$, as follows: For each $y \in \mathscr{C}_0$ and $f \in \mathscr{C}(y,x)$, by noting that $F(f)\colon F(x) \to F(y)$ is a morphism in $\operatorname{Mod}\Bbbk$, set

$$\psi_{x,F}(v)_y(f) := F(f)(v)\ (\in F(y))$$

and define $\psi := (\psi_{x,F})_{(x,F)\in(\mathscr{C}^{\mathrm{op}}\times\operatorname{Mod}\mathscr{C})_0}$.

It is easy to verify that $\phi_{x,F}$ and $\psi_{x,F}$ defined above are linear and inverses of each other, and that ϕ is in fact a natural transformation. Finally, we refer the reader to Example A.3.5 for the fact that $(\operatorname{Mod}\mathscr{C})(\mathscr{C}(\text{-},x),F)$ is a 2-class, i.e., that $(\operatorname{Mod}\mathscr{C})(\mathscr{C}(\text{-},x),F) \in (\operatorname{Mod}^2\Bbbk)_0$. □

EXERCISE 2.5.9. In the proof above, verify that $\phi_{x,F}$ and $\psi_{x,F}$ are linear and inverses of each other, and that ϕ is a natural transformation.

LEMMA 2.5.10. *In the proof of the Yoneda lemma, in particular, if F has the form $F = \mathscr{C}(\text{-},y)$ for some $y \in \mathscr{C}_0$, then the isomorphism $\psi_{x,\mathscr{C}(\text{-},y)}\colon \mathscr{C}(x,y) \to \operatorname{Mod}\mathscr{C}(\mathscr{C}(\text{-},x),\mathscr{C}(\text{-},y))$ is given by*

$$\psi_{x,\mathscr{C}(\text{-},y)}(f) = \mathscr{C}(\text{-},f)\ \ (f \in \mathscr{C}(x,y)).$$

Thus the map $\mathscr{C}(x,y) \to \operatorname{Mod}\mathscr{C}(\mathscr{C}(\text{-},x),\mathscr{C}(\text{-},y))$, $f \mapsto \mathscr{C}(\text{-},f)$ is a bijection.

PROOF. For simplicity we set $\alpha := \psi_{x,\mathscr{C}(\text{-},y)}$. Then it is enough to show that $\alpha(f)_z(g) = \mathscr{C}(z,f)(g)(= fg)$ for all $z \in \mathscr{C}_0$ and $g \in \mathscr{C}(z,x)$. By the naturality of $\alpha(f)$ the diagram

$$\begin{CD}
\mathscr{C}(x,x) @>{\alpha(f)_x}>> \mathscr{C}(x,y)\\
@V{\mathscr{C}(g,x)}VV @VV{\mathscr{C}(g,y)}V\\
\mathscr{C}(z,x) @>>{\alpha(f)_z}> \mathscr{C}(z,y)
\end{CD}$$

is commutative. Hence

$$\begin{aligned}
\alpha(f)_z(g) &= (\alpha(f)_z \circ \mathscr{C}(g,x))(\mathbb{1}_x) = (\mathscr{C}(g,y) \circ \alpha(f)_x)(\mathbb{1}_x)\\
&= \mathscr{C}(g,y)(f) = fg.
\end{aligned}$$

□

From the lemma above, it follows that two objects in $\mathscr{C}$ being isomorphic is equivalent to the corresponding representable functors being isomorphic in the functor category, which is a generalization of Corollary 2.3.13. Namely we have the following.

COROLLARY 2.5.11. *The following are equivalent for all $x, y \in \mathscr{C}_0$:*

(1) *$x \cong y$ in the category $\mathscr{C}$.*
(2) *$\mathscr{C}(\text{-},x) \cong \mathscr{C}(\text{-},y)$ in the category $\operatorname{Mod}\mathscr{C}$.*

PROOF. (1) $\Rightarrow$ (2). Assume that $f\colon x \to y$ and $g\colon y \to x$ are inverses of each other in the category $\mathscr{C}$. Then $\alpha := \mathscr{C}(\text{-}, f)\colon \mathscr{C}(\text{-}, x) \to \mathscr{C}(\text{-}, y)$ and $\beta := \mathscr{C}(\text{-}, g)\colon \mathscr{C}(\text{-}, y) \to \mathscr{C}(\text{-}, x)$ are inverses of each other in the category $\operatorname{Mod}\mathscr{C}$. Indeed, $\beta \circ \alpha = \mathscr{C}(\text{-}, g \circ f) = \mathscr{C}(\text{-}, \mathbb{1}_x) = \mathbb{1}_{\mathscr{C}(\text{-},x)}$, and similarly we have $\alpha \circ \beta = \mathbb{1}_{\mathscr{C}(\text{-},y)}$.

(2) $\Rightarrow$ (1). Assume that $\alpha\colon \mathscr{C}(\text{-}, x) \to \mathscr{C}(\text{-}, y)$ and $\beta\colon \mathscr{C}(\text{-}, y) \to \mathscr{C}(\text{-}, x)$ are inverses of each other in $\operatorname{Mod}\mathscr{C}$. Then by Lemma 2.5.10, $\alpha = \mathscr{C}(\text{-}, f), \beta = \mathscr{C}(\text{-}, g)$ for some morphisms $f\colon x \to y$ and $g\colon y \to x$ in $\mathscr{C}$. Here since $\beta \circ \alpha = \mathbb{1}_{\mathscr{C}(\text{-},x)}$, we have $\mathscr{C}(\text{-}, g \circ f) = \mathscr{C}(\text{-}, \mathbb{1}_x)$, which implies that $g \circ f = \mathbb{1}_x$ by Lemma 2.5.10. Similarly, from the fact that $\alpha \circ \beta = \mathbb{1}_{\mathscr{C}(\text{-},y)}$, it follows that $f \circ g = \mathbb{1}_y$. □

By the same proof, the following holds for left modules (covariant functors):

LEMMA 2.5.12. *There exists an isomorphism*

$$\phi'_{x,F}\colon (\mathscr{C}\text{-Mod})(\mathscr{C}(x, \text{-}), F) \xrightarrow{\sim} F(x)$$

that is natural in $x \in \mathscr{C}_0$ and in $F \in (\mathscr{C}\text{-Mod})_0$. In particular, if F has the form $F = \mathscr{C}(y, \text{-})$ for some $y \in \mathscr{C}_0$, then the isomorphism ${\phi'}^{-1}_{x,\mathscr{C}(y,\text{-})}\colon \mathscr{C}(y, x) \to \mathscr{C}\text{-Mod}(\mathscr{C}(x, \text{-}), \mathscr{C}(y, \text{-}))$ is given by

$$f \mapsto \mathscr{C}(f, \text{-})\ (f \in \mathscr{C}(x, y)).$$

Moreover, the following are equivalent for all $x, y \in \mathscr{C}_0$:

(1) *$x \cong y$ in the category $\mathscr{C}$.*
(2) *$\mathscr{C}(x, \text{-}) \cong \mathscr{C}(y, \text{-})$ in the category $\mathscr{C}$-Mod.*

REMARK 2.5.13. In the above, we have explained the Yoneda lemma for linear categories. This lemma holds not only for those categories, but also for general categories. Namely, when $\mathscr{C}$ is a light category, there exists a bijection

$$\phi_{x,F}\colon \operatorname{Fun}(\mathscr{C}^{\mathrm{op}}, \mathbf{Set})(\mathscr{C}(\text{-}, x), F) \xrightarrow{\sim} F(x)$$

that is natural in $x \in \mathscr{C}_0$ and in $F \in \operatorname{Fun}(\mathscr{C}^{\mathrm{op}}, \mathbf{Set})_0$. In particular, if F has the form $F = \mathscr{C}(\text{-}, y)$ for some $y \in \mathscr{C}_0$, then the bijection $\phi^{-1}_{x,\mathscr{C}(\text{-},y)}\colon \mathscr{C}(x, y) \to \operatorname{Fun}(\mathscr{C}^{\mathrm{op}}, \mathbf{Set})(\mathscr{C}(\text{-}, x), \mathscr{C}(\text{-}, y))$ is given by

$$f \mapsto \mathscr{C}(\text{-}, f)\ (f \in \mathscr{C}(x, y)).$$

Moreover, the following are equivalent for all $x, y \in \mathscr{C}_0$:

(1) $x \cong y$ in the category $\mathscr{C}$.
(2) $\mathscr{C}(\text{-}, x) \cong \mathscr{C}(\text{-}, y)$ in the category $\operatorname{Fun}(\mathscr{C}^{\mathrm{op}}, \mathbf{Set})$.

The corresponding assertion holds for covariant functors.

PROPOSITION 2.5.14. *The representable functor $\mathscr{C}(\text{-}, x)$ (resp. $\mathscr{C}(x, \text{-})$) is projective in $\operatorname{Mod}\mathscr{C}$ (resp. $\mathscr{C}$-Mod) for all $x \in \mathscr{C}_0$.*

PROOF. Since both can be proved in the same way, we only prove the statement for $\mathscr{C}(\text{-}, x)$. Let

$$(E) : 0 \to L \xrightarrow{f} M \xrightarrow{g} N \to 0$$

be an exact sequence of morphisms of $\mathscr{C}$-modules. Then by the Yoneda lemma, the sequence

$$(E') : 0 \to (\operatorname{Mod}\mathscr{C})(\mathscr{C}(\text{-}, x), L) \xrightarrow{(\operatorname{Mod}\mathscr{C})(\mathscr{C}(\text{-},x),f)} (\operatorname{Mod}\mathscr{C})(\mathscr{C}(\text{-}, x), M)$$
$$\xrightarrow{(\operatorname{Mod}\mathscr{C})(\mathscr{C}(\text{-},x),g)} (\operatorname{Mod}\mathscr{C})(\mathscr{C}(\text{-}, x), N) \to 0$$

is isomorphic to the sequence

$$0 \to L(x) \xrightarrow{f_x} M(x) \xrightarrow{g_x} N(x) \to 0.$$

By Remark 2.5.5, this sequence is exact. Hence so is (E'). $\square$

LEMMA 2.5.15. *The following are equivalent for a right $\mathscr{C}$-module M:*

(1) *M is finitely generated.*
(2) *There exists an epimorphism $\alpha\colon \coprod_{i=1}^n \mathscr{C}(\text{-}, x_i) \to M$ for some finite number of objects $x_1, \dots, x_n$ in $\mathscr{C}$.*

PROOF. (1) $\Rightarrow$ (2). Assume that $M = \langle v_1, \dots, v_n\rangle$ for some $\{v_1, \dots, v_n\} \subseteq \bigsqcup_{x\in\mathscr{C}_0} M(x)$. Then for each $i = 1, \dots, n$ there exists an $x_i \in \mathscr{C}_0$ such that $v_i \in M(x_i)$. Take here

$$\alpha := {}^t[\psi_{x_i,M}(v_i)]_{i=1}^n\colon \coprod_{i=1}^n \mathscr{C}(\text{-}, x_i) \to M.$$

Then since $\psi_{x_i,M}(v_i)(\mathbb{1}_{x_i}) = v_i$ $(i = 1, \dots, n)$, we have $M = \langle v_1, \dots, v_n\rangle \le \operatorname{Im}\alpha \le M$, and hence we see that α is an epimorphism.

(2) $\Rightarrow$ (1). Assume that (2) holds, and set $v_i := \alpha_{x_i}(\mathbb{1}_{x_i})$ $(i = 1, \dots, n)$. Take any L such that $\{v_1, \dots, v_n\} \subseteq L \le M$. Then for each $x \in \mathscr{C}_0$, we have $(\operatorname{Im}\alpha)(x) = \sum_{i=1}^n\{M(f)v_i \mid f \in \mathscr{C}(x, x_i)\} = \sum_{i=1}^n\{L(f)v_i \mid f \in \mathscr{C}(x, x_i)\} \le L(x)$, which shows that $\operatorname{Im}\alpha \le L$. Therefore $M = \operatorname{Im}\alpha \le \langle v_1, \dots, v_n\rangle \le M$, and hence $M = \langle v_1, \dots, v_n\rangle$. $\square$

DEFINITION 2.5.16. We denote by $\operatorname{prj}\mathscr{C}$ the full subcategory of $\operatorname{Mod}\mathscr{C}$ consisting of finitely generated projective $\mathscr{C}$-modules.

DEFINITION 2.5.17. Let $x \in \mathscr{C}_0$ and $e \in \mathscr{C}(x, x)$.

(1) e is called an *idempotent* in $\mathscr{C}$ if $e^2 = e$.
(2) An idempotent e is said to *split* if we have $e = \sigma\pi$ for some pair $y \xrightarrow{\sigma} x \xrightarrow{\pi} y$ of morphisms in $\mathscr{C}$ with $y \in \mathscr{C}_0$ satisfying $\pi\sigma = \mathbb{1}_y$.
(3) $\mathscr{C}$ is said to be *idempotent complete* if every idempotent in $\mathscr{C}$ splits.

EXAMPLE 2.5.18. Suppose that $x \cong x_1 \amalg x_2$ in $\mathscr{C}$, and let $(x_i \xrightarrow{\sigma_i} x \xrightarrow{\pi_i} x_i)_{i=1}^2$ be the corresponding direct sum system in $\mathscr{C}$. Then $e_i := \sigma_i\pi_i$ turns out to be idempotent for all $i = 1, 2$. Indeed, $e_i^2 = \sigma_i(\pi_i\sigma_i)\pi_i = \sigma_i\pi_i = e_i$.

LEMMA 2.5.19. *Let $x, x_1 \in \mathscr{C}_0$. If $\mathscr{C}$ is idempotent complete, then the following are equivalent:*

(1) *x_1 is a direct summand of x.*
(2) *We have $\pi\sigma = \mathbb{1}_{x_1}$ for some pair $x_1 \xrightarrow{\sigma} x \xrightarrow{\pi} x_1$ of morphisms. (In particular by this equality, σ is a section and π is a retraction.)*

PROOF. (1) $\Rightarrow$ (2). This is clear from definition.

(2) $\Rightarrow$ (1). Assume that (2) holds. Then it is immediate that $\mathbb{1}_x - \sigma\pi \in \mathscr{C}(x, x)$ is also an idempotent. Since $\mathscr{C}$ is idempotent complete, we have $\mathbb{1}_x - \sigma\pi = \sigma'\pi'$ for some pair $x_2 \xrightarrow{\sigma'} x \xrightarrow{\pi'} x_2$ of morphisms in $\mathscr{C}$ with $x_2 \in \mathscr{C}_0$ satisfying $\pi'\sigma' = \mathbb{1}_{x_2}$. Therefore we have

$$\mathbb{1}_x = \sigma\pi + \sigma'\pi'. \tag{2.7}$$

It remains to show that $\pi'\sigma = 0, \pi\sigma' = 0$ because in that case we have $x \cong x_1 \amalg x_2$ by taking σ, σ' (resp. π, π') as the injections (resp. projections). Now by the equality (2.7), we have $\sigma = \mathbb{1}_x\sigma = \sigma\pi\sigma + \sigma'\pi'\sigma = \sigma + \sigma'\pi'\sigma$. which shows that $0 = \sigma'\pi'\sigma$. By composing π' from the left to both sides, we obtain $0 = \pi'\sigma$. The remaining equality is obtained similarly. □

REMARK 2.5.20. By the lemma above, $\mathscr{C}$ is idempotent complete if and only if all idempotents in $\mathscr{C}$ are given in the form of Example 2.5.18.

PROPOSITION 2.5.21. *The following are equivalent:*

(1) *All idempotents in $\mathscr{C}$ have their kernels.*
(2) *$\mathscr{C}$ is idempotent complete.*

PROOF. (1) $\Rightarrow$ (2). Let $e\colon X \to X$ be an idempotent in $\mathscr{C}$. Then since $\mathbb{1}_X - e$ is also an idempotent in $\mathscr{C}$, there exists a kernel $K := \operatorname{Ker}(\mathbb{1}_X - e) \xrightarrow{\sigma} X$ of $\mathbb{1}_X - e$ by (1) (see Definition 1.5.9). Since we have $(\mathbb{1}_x - e)e = e - e^2 = 0$, $e = \sigma\pi$ for a unique $\pi\colon X \to K$ by the universality of the kernel. Then the computation

$$\sigma\pi\sigma = e\sigma = (\mathbb{1}_X - (\mathbb{1}_X - e))\sigma = \sigma - (\mathbb{1}_X - e)\sigma = \sigma = \sigma\mathbb{1}_K$$

shows that $\pi\sigma = \mathbb{1}_K$ because σ is a monomorphism as a kernel morphism. Hence e splits.

(2) $\Rightarrow$ (1). Let $e\colon X \to X$ be an idempotent in $\mathscr{C}$. Then since $\mathbb{1}_X - e$ is also an idempotent in $\mathscr{C}$, there exists a pair $X \xrightarrow{\pi} K \xrightarrow{\sigma} X$ of morphisms such that $\sigma\pi = \mathbb{1}_X - e, \pi\sigma = \mathbb{1}_K$ by (2). It is enough to show that σ is a kernel of e.

(i) $e\sigma = 0$. Indeed, since π is an epimorphism as a retraction, from the fact that $e\sigma\pi = e(\mathbb{1}_X - e) = e - e = 0$, it follows that $e\sigma = 0$.

(ii) σ has the universality, i.e., for any morphism $Y \xrightarrow{f} X$ in $\mathscr{C}$ satisfying $ef = 0$, there exists a unique $g\colon Y \to K$ such that $f = \sigma g$. Indeed, since $f = f - ef = (\mathbb{1}_X - e)f = \sigma\pi f$, we can take $g = \pi f\colon Y \to K$. Since σ is a monomorphism, the uniqueness of g follows.

□

EXAMPLE 2.5.22. Let $\mathscr{A}$ be a category. A *zero* object in $\mathscr{A}$ is an object x in $\mathscr{A}$ such that $\#\mathscr{A}(x, y) = 1 = \#\mathscr{A}(y, x)$ for all $y \in \mathscr{A}_0$. A $\mathbb{Z}$-linear category $\mathscr{A}$ is called *additive* if it has a zero object, and $x \oplus y$ exists in $\mathscr{A}$ for all $x, y \in \mathscr{A}_0$. An additive category $\mathscr{A}$ is called *abelian* if it has kernels and cokernels for all morphisms, and each monomorphism (resp. epimorphism) is a kernel (resp. cokernel) morphism of some morphism. Therefore a linear abelian category (e.g., $\operatorname{Mod}\mathscr{C}$) is an idempotent complete additive category.

LEMMA 2.5.23. *If X is a projective $\mathscr{C}$-module, and $Y \xrightarrow{\pi} X$ is an epimorphism, then π is a retraction. Hence in particular, X becomes a direct summand of Y.*

PROOF. Let $Z := \operatorname{Ker}\pi$. Then we have an exact sequence $0 \to Z \hookrightarrow Y \xrightarrow{\pi} X \to 0$, which yields an exact sequence $(\operatorname{Mod}\mathscr{C})(X, Y) \xrightarrow{(\operatorname{Mod}\mathscr{C})(X,\pi)} (\operatorname{Mod}\mathscr{C})(X, X) \to 0$, and hence $\mathbb{1}_X = (\operatorname{Mod}\mathscr{C})(X, \pi)(\sigma) = \pi\sigma$ for some $\sigma \in (\operatorname{Mod}\mathscr{C})(X, Y)$. □

LEMMA 2.5.24. *The following are equivalent for a $\mathscr{C}$-module M:*

(1) *M is finitely generated projective.*
(2) *There exists a finite number of objects $x_1, \dots, x_n$ in $\mathscr{C}$ such that M is a direct summand of $\coprod_{i=1}^n \mathscr{C}(\text{-}, x_i)$.*

PROOF. (1) $\Rightarrow$ (2). Assume that (1) holds. Then since M is finitely generated, there exists an epimorphism $\alpha\colon \coprod_{i=1}^n \mathscr{C}(\text{-}, x_i) \to M$ for some $x_1, \dots, x_n \in \mathscr{C}_0$ by Lemma 2.5.15. Since M is projective, α is a retraction by Lemma 2.5.23, and hence M becomes a direct summand of $\coprod_{i=1}^n \mathscr{C}(\text{-}, x_i)$.

(2) $\Rightarrow$ (1). Assume that (2) holds. Then since each $\mathscr{C}(\text{-}, x_i)$ is projective, so is M by Lemma 2.5.7. M is also finitely generated because by (2) there exists a retraction $\coprod_{i=1}^n \mathscr{C}(\text{-}, x_i) \to M$, which is an epimorphism. $\square$

If the category $\mathscr{C}$ above is an idempotent complete additive category, (e.g., taking $\operatorname{Mod}\mathscr{C}$ instead of $\mathscr{C}$), then finitely generated projective modules have much simpler forms.

LEMMA 2.5.25. *Let $\mathscr{D}$ be an idempotent complete additive linear category, and M a $\mathscr{D}$-module. Then the following are equivalent:*

(1) *M is finitely generated projective.*
(2) *$M \cong \mathscr{D}(\text{-}, x)$ for some $x \in \mathscr{D}_0$.*

PROOF. (2) $\Rightarrow$ (1). This is clear from Lemma 2.5.24.

(1) $\Rightarrow$ (2). Assume that (1) holds. Then by Lemma 2.5.24 M is a direct summand of $\coprod_{i=1}^n \mathscr{D}(\text{-}, x_i)$ for some $x_1, \dots, x_n \in \mathscr{D}_0$. Since $\mathscr{D}$ is additive, there exists $y := \coprod_{i=1}^n x_i$ in $\mathscr{D}$. Then since $\mathscr{D}(\text{-}, y) \cong \coprod_{i=1}^n \mathscr{D}(\text{-}, x_i)$, M becomes a direct summand of $\mathscr{D}(\text{-}, y)$. Thus there exists a pair $M \underset{\pi}{\overset{\sigma}{\rightleftarrows}} \mathscr{D}(\text{-}, y)$ satisfying $\pi\sigma = \mathbb{1}_M$. Here, $\varepsilon := \sigma\pi \in \mathscr{D}(\text{-}, y)$ is an idempotent. Set $e := \phi_{y,\mathscr{D}(\text{-},y)}(\varepsilon) \in \mathscr{D}(y, y)$. By Lemma 2.5.10 $\psi_{y,\mathscr{D}(\text{-},y)}\colon \mathscr{D}(y, y) \to (\operatorname{Mod}\mathscr{D})(\mathscr{D}(\text{-}, y), \mathscr{D}(\text{-}, y))$ is given by $\psi_{y,\mathscr{D}(\text{-},y)}(f) = \mathscr{D}(\text{-}, f)$ $(f \in \mathscr{D}(y, y))$. Hence we have $\mathscr{D}(\text{-}, e) = \varepsilon$. Therefore $\mathscr{D}(\text{-}, e^2) = \varepsilon^2 = \varepsilon = \mathscr{D}(\text{-}, e)$. By applying $\phi_{y,\mathscr{D}(\text{-},y)}$ to both sides, we get $e^2 = e$. Thus e is also an idempotent in $\mathscr{D}$. Since $\mathscr{D}$ is idempotent complete, we have $e = sr$ for some pair $y \underset{s}{\overset{r}{\rightleftarrows}} x$ in $\mathscr{D}$ with $x \in \mathscr{D}_0$ satisfying $rs = \mathbb{1}_x$. Therefore we obtain the diagram

$$M \underset{\pi}{\overset{\sigma}{\rightleftarrows}} \mathscr{D}(\text{-}, y) \underset{\mathscr{D}(\text{-},s)}{\overset{\mathscr{D}(\text{-},r)}{\rightleftarrows}} \mathscr{D}(\text{-}, x).$$

It can be seen from the following calculations that the leftward and rightward composites are the inverses of each other:

$$\begin{aligned}(\pi\mathscr{D}(\text{-}, s))(\mathscr{D}(\text{-}, r)\sigma) &= \pi\mathscr{D}(\text{-}, sr)\sigma = \pi\mathscr{D}(\text{-}, e)\sigma = \pi\varepsilon\sigma \\ &= \pi\sigma\pi\sigma = \mathbb{1}_M, \\ (\mathscr{D}(\text{-}, r)\sigma)(\pi\mathscr{D}(\text{-}, s)) &= \mathscr{D}(\text{-}, r)\varepsilon\mathscr{D}(\text{-}, s) = \mathscr{D}(\text{-}, r)\mathscr{D}(\text{-}, e)\mathscr{D}(\text{-}, s) = \mathscr{D}(\text{-}, res) \\ &= \mathscr{D}(\text{-}, rsrs) = \mathscr{D}(\text{-}, \mathbb{1}_x) = \mathbb{1}_{\mathscr{D}(\text{-},x)}.\end{aligned}$$

As a consequence, we have $M \cong \mathscr{D}(\text{-}, x)$. $\square$

2.6. Ideals and factor categories of a linear category

The definition of an ideal I of a ring R and the construction of the factor ring R/I are generalized to the setting of linear categories as follows:

DEFINITION 2.6.1. Let $\mathscr{C}$ be a linear category. A submodule of the left $(\mathscr{C}^{\mathrm{op}} \times \mathscr{C})$-module $\mathscr{C}(\text{-}, ?)\colon \mathscr{C}^{\mathrm{op}} \times \mathscr{C} \to \operatorname{Mod}\Bbbk$, $(x, y) \mapsto \mathscr{C}(x, y)$ $((x, y) \in (\mathscr{C}_0^{\mathrm{op}} \times \mathscr{C}_0) \cup$

$(\mathscr{C}_1^{\mathrm{op}} \times \mathscr{C}_1))$ is called an *ideal* of $\mathscr{C}$[12]. Namely, by Definition 2.2.21, an ideal I of $\mathscr{C}$ is a left $(\mathscr{C}^{\mathrm{op}} \times \mathscr{C})$-module defined by a family $I(x,y) \le \mathscr{C}(x,y)$, $((x,y) \in \mathscr{C}_0^{\mathrm{op}} \times \mathscr{C}_0)$ of subspaces satisfying the condition that $g \circ I(x,y) \circ f \le I(x',y')$, $(x,y,x',y' \in \mathscr{C}_0, f \in \mathscr{C}(x',x), g \in \mathscr{C}(y,y'))$.

DEFINITION 2.6.2. Let $\mathscr{C}$ be a linear category, and I its ideal. Then a linear category $\mathscr{C}/I$ which is called the *factor category* of $\mathscr{C}$ by I is defined as follows:

- $(\mathscr{C}/I)_0 := \mathscr{C}_0$.
- $(\mathscr{C}/I)(x,y) := \mathscr{C}(x,y)/I(x,y)$, $(x,y \in (\mathscr{C}/I)_0)$. Set the canonical epimorphism $\mathscr{C}(x,y) \to \mathscr{C}(x,y)/I(x,y)$ to be $\pi_{y,x}$.
- We define the composite of morphisms $a := \pi_{y,x}(f)\colon x \to y$ and $b := \pi_{z,y}(g)\colon y \to z$ in $\mathscr{C}/I$ $(f \in \mathscr{C}(x,y), g \in \mathscr{C}(y,z))$ as

$$b \circ a = \pi_{z,y}(g) \circ \pi_{y,x}(f) := \pi_{z,x}(g \circ f). \tag{2.8}$$

The right-hand side does not depend on the choice of representables f, g of a, b. Indeed, if $f' \in \mathscr{C}(x,y), g' \in \mathscr{C}(y,z)$ satisfy $\pi_{y,x}(f) = \pi_{y,x}(f'), \pi_{z,y}(g) = \pi_{z,y}(g')$, then $u := f' - f \in I(x,y), v := g' - g \in I(y,z)$, which shows that $g' \circ f' = (g+v) \circ (f+u) = g \circ f + g \circ u + v \circ f + v \circ u \in g \circ f + I(x,z)$. Therefore $\pi_{z,x}(g' \circ f') = \pi_{z,x}(g \circ f)$, as desired.

We can define a quiver morphism $\pi\colon \mathscr{C} \to \mathscr{C}/I$ by $\pi_0 := \mathbb{1}_{\mathscr{C}_0}$, $\pi(f) := \pi_{y,x}(f)$ for $f \in \mathscr{C}(x,y)$ with $x, y \in \mathscr{C}_0$, which turns out to be a linear functor by the formula (2.8). The functor π is called the *canonical functor* to the factor category.

The fundamental homomorphism theorem for rings is generalized as follows:

THEOREM 2.6.3. *Let $F\colon \mathscr{C} \to \mathscr{D}$ be a linear functor between linear categories satisfying the following conditions:*

1. *$F_0\colon \mathscr{C}_0 \to \mathscr{D}_0$ is a bijection; and*
2. *F induces an epimorphism $\mathscr{C}(x,y) \to \mathscr{D}(Fx,Fy)$ for all $x, y \in \mathscr{C}_0$.*

Then an ideal $\operatorname{Ker} F$ *of $\mathscr{C}$ is defined by*

$$(\operatorname{Ker} F)(x,y) := \{f \in \mathscr{C}(x,y) \mid F(f) = 0\} \quad (x,y \in \mathscr{C}_0),$$

and there exists a unique isomorphism $\bar{F}\colon \mathscr{C}/\operatorname{Ker} F \to \mathscr{D}$ such that $F = \bar{F} \circ \pi$, where $\pi\colon \mathscr{C} \to \mathscr{C}/\operatorname{Ker} F$ is the canonical functor.

EXERCISE 2.6.4. Prove the theorem above.

2.7. Constructions of algebras and of linear categories by quivers

In the sequel, we assume that all quivers are locally finite unless otherwise stated. In this section, we will briefly explain how to construct algebras and linear categories using quivers so that the examples can be computed.

EXAMPLE 2.7.1.

(1) Let $Q_0 = \{1,2\}, Q_1 = \{\alpha, \beta\}, s(\alpha) = t(\beta) = 1, t(\alpha) = s(\beta) = 2$. Then the quiver $Q = (Q_0, Q_1, s, t)$ is displayed as $1 \underset{\beta}{\overset{\alpha}{\rightleftarrows}} 2$.

[12]Since $\mathscr{C}$ is a linear category, $\mathscr{C}(\text{-},?)$ is, in fact, a $\mathscr{C}$-$\mathscr{C}$-bimodule, and the I above is automatically its $\mathscr{C}$-$\mathscr{C}$-subbimodule (see Definition 7.2.1).

(2) Define a quiver $\tilde{Q} = (\tilde{Q}_0, \tilde{Q}_1, s, t)$ by

$$\tilde{Q}_0 := \mathbb{Z},\ \tilde{Q}_1 := \{\alpha_i \mid i \in \mathbb{Z}\},\ s(\alpha_i) := i,\ t(\alpha_i) := i+1\ (i \in \mathbb{Z}).$$

Then this is displayed as

$$\tilde{Q} = (\cdots -1 \xrightarrow{\alpha_{-1}} 0 \xrightarrow{\alpha_0} 1 \xrightarrow{\alpha_1} \cdots).$$

Define a quiver morphism $f\colon \tilde{Q} \to Q$ as follows: If $i \in \mathbb{Z}$ is odd, then set $f(i) := 1, f(\alpha_i) := \alpha$, otherwise set $f(i) := 2, f(\alpha_i) := \beta$. This quiver morphism corresponds to a covering $\tilde{Q} \to Q$ of figures.

DEFINITION 2.7.2. Let Q be a quiver.

(1) A sequence $\mu := (\alpha_n, \dots, \alpha_1)$ $(n \geq 1)$ of arrows in Q is called a *path* of Q if the arrows are connected from the right to the left, more precisely if $s(\alpha_{i+1}) = t(\alpha_i)$ for all $i = 1, \dots, n-1$. The number $|\mu| := n$ is called the *length* of μ, the vertices $s(\mu) := s(\alpha_1)$, $t(\mu) := t(\alpha_n)$ are called the *source* and the *target* of μ, respectively. Moreover, for each vertex $x \in Q_0$, the symbol e_x is called a path of length 0, and we set $s(e_x) := x =: t(e_x)$.

(2) We define a category $\mathbb{P}Q$ from Q as follows:
First, set $(\mathbb{P}Q)_0 := Q_0$, and for any $x, y \in (\mathbb{P}Q)_0$, set

$$(\mathbb{P}Q)(x,y) := \{\mu \mid \mu \text{ is a path of } Q, s(\mu) = x, t(\mu) = y\}.$$

Define the composition by the concatenation of paths. Namely, for each $x \in Q_0$, let e_x be the identity morphism at x, and for paths $(\alpha_m, \dots, \alpha_1) \in \mathbb{P}Q(x,y)$ and $(\beta_n, \dots, \beta_1) \in \mathbb{P}Q(y,z)$ with $x, y, z \in (\mathbb{P}Q)_0$, define their composite to be $(\beta_n, \dots, \beta_1) \cdot (\alpha_m, \dots, \alpha_1) := (\beta_n, \dots, \beta_1, \alpha_m, \dots, \alpha_1)$. This category is called the *free* category of Q.

(3) The $\mathbb{k}$-linearlization of the category $\mathbb{P}Q$ is denoted by $\mathbb{k}[Q]$, and is called the *path category* of Q, i.e., we set $\mathbb{k}[Q]_0 := Q_0$, and for any $x, y \in (\mathbb{k}[Q])_0$,

$$\mathbb{k}[Q](x,y) := (\text{the vector space with the basis } (\mathbb{P}Q)(x,y)).$$

The composition is given by the $\mathbb{k}$-bilinearization of that of $\mathbb{P}Q$. Then $\mathbb{k}[Q]$ turns out to be a linear category. The pair (Q, I) of a quiver Q and an ideal I of $\mathbb{k}[Q]$ is called a *quiver with relations*. Without loss of generality, we assume that $I \cap (Q_1 \cup \{e_x \mid x \in Q_0\}) = \emptyset$.

(4) The ideal of $\mathbb{k}[Q]$ generated by the paths of length at least 1 is denoted by $\mathbb{k}[Q]^+$, and an ideal $I \leq (\mathbb{k}[Q]^+)^2$ is called *admissible* if for any $x, y \in Q_0$, there exists some $h_{x,y} \geq 2$ such that $(\mathbb{k}[Q]^+)^{h_{x,y}}(x,y) \subseteq I(x,y)$. A pair (Q, I) of a quiver Q and an admissible ideal I of $\mathbb{k}Q$ is called a *bound quiver*.

DEFINITION 2.7.3. Let Q be a finite quiver.

(1) $\mathbb{k}Q := \mathrm{Mat}_{\mathbb{k}[Q]}$ is called the *path algebra* of Q (see the notation in the proof of Proposition 2.3.8). This is identified with the algebra A defined as the vector space with basis the set of all paths $(\mathbb{P}Q)_1$ of Q, having the multiplication defined as follows: for any paths μ and λ, their product is given by the composite $\mu \cdot \lambda$ if $s(\mu) = t(\lambda)$ (i.e., they are connected) and 0 otherwise. Thus there exists an isomorphism $A \to \mathbb{k}Q$, $a \mapsto e_{y,y}ae_{x,x}$ $(a \in (\mathbb{P}Q)_1, x := s(a), y := t(a))$. Under this isomorphism, the complete set $E_{\mathbb{k}[Q]} = \{e_{x,x} \mid x \in Q_0\}$ of orthogonal idempotents of $\mathbb{k}[Q]$ corresponds to $\{e_x \mid x \in Q_0\}$. We usually identify $\mathbb{k}Q$ with $\mathbb{k}[Q]$.

(2) As in the case of the category $k[Q]$, the ideal of $\Bbbk Q$ generated by the paths of length at least 1 is denoted by $\Bbbk Q^+$, and an ideal I is called an *admissible* ideal if $(\Bbbk Q^+)^h \subseteq I \subseteq (\Bbbk Q^+)^2$ for some $h \geq 2$, and a pair (Q, I) of a quiver Q and an admissible ideal I of $\Bbbk Q$ is called a *bound quiver*.

REMARK 2.7.4.

(1) If Q is a quiver (resp. finite quiver), then a quiver with relations (Q, I) defines a linear category (resp. an algebra) $\Bbbk[Q, I] := \Bbbk[Q]/I$ (resp. $\Bbbk(Q, I) := \Bbbk Q/I$). In this way it is possible to construct so many linear categories (resp. algebras) that all basic algebras can be obtained in this way when $\Bbbk$ is an algebraically closed field. Note that we can regard $\Bbbk(Q, I) = \mathrm{Mat}_{k[Q,I]}$ when Q is a finite quiver.

(2) Let Q be a finite quiver, and consider a bound quiver (Q, I). Then $\Bbbk(Q, I)$ turns out to be a finite-dimensional basic algebra[13], and its Jacobson radical is given by $\Bbbk Q^+/I$.

(3) If $(Q, I), (Q', I')$ are bound quivers with Q, Q' finite quivers, then $\Bbbk(Q, I) \cong \Bbbk(Q', I')$ implies $Q \cong Q'$.

EXAMPLE 2.7.5.

(1) Let $Q = (1 \circlearrowleft \alpha)$ (1 vertex and 1 arrow). Then $\Bbbk Q$ is an algebra with identity $e_1 =: \alpha^0$ that has a basis $\{\alpha^n \mid n \geq 0\}$. The correspondence $\alpha^n \mapsto x^n$ $(n \in \mathbb{N})$ defines an isomorphism $\Bbbk Q \to \Bbbk[x]$, where $\Bbbk[x]$ is the polynomial ring with one variable x over $\Bbbk$.

(2) Let $Q = (1 \xrightarrow{\alpha_1} 2 \xrightarrow{\alpha_2} \cdots \xrightarrow{\alpha_{n-1}} n)$. Then $\Bbbk Q$ is an algebra with basis $\{\mu_{ji} \mid 1 \leq i \leq j \leq n\}$, where $\mu_{ji} := \alpha_{j-1} \cdots \alpha_i$ (if $i < j$), $\mu_{ji} := e_i$ (if $i = j$). This is isomorphic to the algebra $T_n(\Bbbk)$ of lower triangular $n \times n$ matrices over $\Bbbk$. An isomorphism is defined by the correspondence $\mu_{ji} \mapsto e_{ji}$, where e_{ji} is the $n \times n$ matrix $(\delta_{(q,p),(j,i)})_{q,p}$ for all $i, j = 1, \ldots, n$.

(3) Consider the quiver Q in Example 2.7.1. Then $\Bbbk Q$ is an infinite-dimensional algebra. The ideal

$$I := \langle \beta\alpha - e_1, \alpha\beta - e_2 \rangle$$

is not an admissible ideal, but the factor algebra $A := \Bbbk Q/I$ is 4-dimensional and isomorphic to the matrix algebra $M_2(\Bbbk)$ of 2×2 matrices over $\Bbbk$. An isomorphism is defined by $e_i \mapsto e_{ii}$ $(i = 1, 2)$, $\alpha \mapsto e_{21}$, $\beta \mapsto e_{12}$.

(4) Consider again the quiver Q in Example 2.7.1. Then $I := \langle \alpha\beta\alpha, \beta\alpha\beta \rangle$ is an admissible ideal of $\Bbbk Q$, and the factor algebra $A := \Bbbk Q/I$ is 6-dimensional, and is a *self-injective* algebra (i.e., A is an injective right A-module).

(5) Consider the quiver morphism $f \colon \tilde{Q} \to Q$ in Example 2.7.1. A $\Bbbk$-linearlization of f gives a functor $\Bbbk f \colon \Bbbk[\tilde{Q}] \to \Bbbk[Q]$. For the ideal I in (4) above, define an ideal of $\Bbbk[\tilde{Q}]$ by $\tilde{I} := \langle \alpha_{i+2}\alpha_{i+1}\alpha_i \mid i \in \mathbb{Z} \rangle$. Then, since $(\Bbbk f)(\tilde{I}) = I$, the functor $\Bbbk f$ induces a functor $\Bbbk[\tilde{Q}]/\tilde{I} \to \Bbbk Q/I$. This gives a model of a covering of the algebra $A := \Bbbk Q/I$. We refer the reader to Exercise 6.2.22 for a method to compute this as a smash product.

EXAMPLE 2.7.6. Let Q be a (locally finite) quiver, and I an admissible ideal of $\Bbbk[Q]$. Then $\Bbbk[Q]/I$ is a locally bounded spectroid.

[13] For instance, $M_n(\Bbbk)$ $(n \geq 2)$ is not a basic algebra, but its module category is equivalent to that of $\Bbbk$.

Finally, we cite the following theorem without proof (see for example **[6]**, **[14]** for details).

THEOREM 2.7.7 (Gabriel's theorem). *Assume that* $\Bbbk$ *is an algebraically closed field, and let* A *be a finite-dimensional algebra. Then the basic algebra* A' *of* A *is expressed as* $A' \cong \Bbbk(Q, I)$ *for some finite quiver* Q *and some admissible ideal* I *of* $\Bbbk Q$*. Therefore* $\operatorname{Mod} A$ *and* $\operatorname{Mod} \Bbbk(Q, I)$ *turn out to be equivalent. Moreover, this* Q *is uniquely determined by* A *up to isomorphisms (in the category of quivers and quiver morphisms, cf.* Remark 2.7.4(3)*).*

2.8. The representation category of a quiver and the module category of a linear category

Throughout this section, we let Q be a locally finite quiver, and I an ideal of $\Bbbk[Q]$.

DEFINITION 2.8.1. In this definition, we regard the category $\operatorname{Mod} \Bbbk$ as a quiver by forgetting its composition and identities[14].

(1) A quiver morphism V from a Q to the quiver $\operatorname{Mod} \Bbbk$ is called a *representation* of Q. Namely, V maps each vertex $x \in Q_0$ to a vector space $V(x) \in (\operatorname{Mod} \Bbbk)_0$, and each arrow $a\colon x \to y$ in Q to a linear map $V(a)\colon V(x) \to V(y)$.

(2) Let V be a representation of Q. For each morphism $\mu\colon x \to y$ in $\Bbbk[Q]$, we define a linear map $V(\mu)\colon V(x) \to V(y)$ as follows:
 (a) In the case that $\mu \in \mathbb{P}Q(x, y), |\mu| = 0$, i.e., when $\mu = e_x$ for some $x \in Q_0$, we set $V(\mu) := \mathbb{1}_{V(x)}$;
 (b) In the case that $\mu \in \mathbb{P}Q(x, y), |\mu| = 1$, i.e., when $\mu = a$ for some $a \in Q(x, y)$, we set $V(\mu) := V(a)$;
 (c) In the case that $\mu \in \mathbb{P}Q(x, y), |\mu| \geq 2$, i.e., when $\mu = a_n \cdots a_2 a_1$ for some $a_1, a_2, \ldots, a_n \in Q_1$, and $n \geq 2$, we set
$$V(\mu) := V(a_n) \circ \cdots \circ V(a_2) \circ V(a_1).$$
 (d) For a general μ, since it is uniquely written in the form $\mu = \sum_{i=1}^{n} k_i \mu_i$ for some $k_1, \ldots, k_n \in \Bbbk, \mu_1, \ldots, \mu_n \in \mathbb{P}Q(x, y)$, and $n \geq 1$, we set
$$V(\mu) := \sum_{i=1}^{n} k_i V(\mu_i).$$

(3) A representation V of Q is called a *representation* of (Q, I) if $V(\mu) = 0$ for all $\mu \in I(x, y), x, y \in Q_0$.

(4) Let V, W be representations of Q. A family $(f_x\colon V(x) \to W(x))_{x \in Q_0}$ of linear maps is called a *morphism* from V to W if it satisfies the following axiom:

 Axiom: For each arrow $a\colon x \to y$ in Q, the diagram
$$\begin{array}{ccc} V(x) & \xrightarrow{f_x} & W(x) \\ {\scriptstyle V(a)}\downarrow & & \downarrow{\scriptstyle W(a)} \\ V(y) & \xrightarrow[f_y]{} & W(y) \end{array}$$
 is commutative.

[14] This does not mean that we delete the identities, but that we just forget the map $\mathbb{1}\colon (\operatorname{Mod} \Bbbk)_0 \to (\operatorname{Mod} \Bbbk)_1$, $x \mapsto \mathbb{1}_x$.

(5) Let $U \xrightarrow{f} V \xrightarrow{g} W$ be morphisms of representations of Q. Then the composite $g \circ f : U \to W$ is defined as

$$g \circ f := (g_x \circ f_x)_{x \in Q_0}.$$

Here, as easily seen, $\mathbb{1}_V := (\mathbb{1}_{V(x)})_{x \in Q_0}$ is a morphism $V \to V$, which turns out to be the identity of V under the composition defined above.

(6) By the composition and identities above, the set of all representations of Q and the set of all morphisms between them form a category, which is denoted by $\operatorname{Rep}_k Q$. Moreover, for each pair of representations V, W, the set $(\operatorname{Rep}_k Q)(V, W)$ turns out to be a vector space by the following operations, and hence $\operatorname{Rep}_k Q$ becomes a linear category.

$$\begin{aligned} f + g &:= (f_x + g_x)_{x \in Q_0} \ (f, g \in (\operatorname{Rep}_k Q)(V, W)), \\ kf &:= (kf_x)_{x \in Q_0} \ (f \in (\operatorname{Rep}_k Q)(V, W), k \in \mathbb{k}). \end{aligned}$$

(7) The full subcategory of $\operatorname{Rep}_k Q$ consisting of the representations of (Q, I) is denoted by $\operatorname{Rep}_k(Q, I)$.

The following theorem enables us to present the category of left $\mathbb{k}[Q, I]$-modules by the category of representations of (Q, I). It is simpler to present objects in the latter than in the former because there are fewer arrows in Q than in $\mathbb{P}Q$. In general, the fewer arrows there are, the simpler the display will be.

THEOREM 2.8.2. *We have an isomorphism*

$$\mathbb{k}[Q, I]\text{-Mod} \cong \operatorname{Rep}_{\mathbb{k}}(Q, I)$$

of linear categories.

PROOF. We define a linear functor $\phi : \mathbb{k}[Q, I]\text{-Mod} \to \operatorname{Rep}_{\mathbb{k}}(Q, I)$ as follows:

On objects: Let $M \in \mathbb{k}[Q, I]\text{-Mod}_0$. Then set the composite $\mathbb{k}[Q] \xrightarrow{\pi} \mathbb{k}[Q, I] \xrightarrow{M} \operatorname{Mod} \mathbb{k}$ to be M', where π is the canonical functor. Using this we define $\phi(M) \in (\operatorname{Rep}_{\mathbb{k}} Q)_0$ as

$$\phi(M)(x) := M'(x) = M(x) \ (x \in Q_0), \quad \phi(M)(a) := M'(a) \ (a \in Q_1).$$

Here, for each path μ in Q, we have $\phi(M)(\mu) = M'(\mu)$ by Definition 2.8.1(2), and it also holds that $M'(\mu) = 0$ if $\mu \in I$ by construction of M'. Therefore we have $\phi(M)(\mu) = 0$ for all $\mu \in I$, which shows that $\phi(M) \in \operatorname{Rep}_{\mathbb{k}}(Q, I)_0$.

On morphisms: Let $f : M \to N$ be a morphism in $k[Q, I]$-Mod. Then $f = (f_x : M(x) \to N(x))_{x \in Q_0}$ is a family of linear maps, and is a natural transformation from M to N, from which we see that $f : \phi(M) \to \phi(N)$ satisfies the axiom in Definition 2.8.1(4) by considering morphisms in $\mathbb{k}[Q, I]$ of the form $a + I$ for some $a \in Q_1$. Therefore we may set $\phi(f) := f$.

We next define a linear functor $\psi : \operatorname{Rep}_{\mathbb{k}}(Q, I) \to \mathbb{k}[Q, I]\text{-Mod}$ as follows:

On objects: Let $V \in \operatorname{Rep}_{\mathbb{k}}(Q, I)$. By Definition 2.8.1(2), V is extended to a representation of the category $\mathbb{k}[Q]$. Then since $V(\mu) = 0$ for all $\mu \in I$, this V induces a representation of $k[Q, I]$, which we set to be $\psi(V)$: $\psi(V)(\rho + I) := V(\rho)$ for all $\rho \in \mathbb{k}[Q](x, y)$, $(x, y \in Q_0)$.

On morphisms: Let $f\colon V \to W$ be a morphism in $\mathrm{Rep}_{\Bbbk}(Q,I)$. It is easy to see that $\psi(f) := f$ is a natural transformation from $\psi(V)$ to $\psi(W)$ by using several times the fact that f satisfies the commutativity of the diagram in Definition 2.8.1(4). It is clear that ϕ and ψ are inverses of each other. $\square$

THEOREM 2.8.3. *Let Q be a finite quiver, and I an admissible ideal. Then we have an equivalence*

$$\Bbbk(Q,I)\text{-Mod} \simeq \Bbbk[Q,I]\text{-Mod}$$

of linear categories.

PROOF. In this setting, $\Bbbk(Q,I) \in (\Bbbk\text{-}\mathbf{alg}_{\mathrm{copi}})_0$ corresponds to $k[Q,I] \in (\Bbbk\text{-}\mathbf{Spec}_{\mathrm{f}})_0$ under the equivalence given in Proposition 2.3.8, thus $\Bbbk(Q,I) = \mathrm{Mat}_{\Bbbk[Q,I]}$. Therefore the assertion follows from Theorem 2.3.30. $\square$

REMARK 2.8.4. By combining Theorems 2.8.2 and 2.8.3, we obtain an equivalence

$$\Bbbk(Q,I)\text{-Mod} \simeq \mathrm{Rep}_{\Bbbk}(Q,I).$$

Therefore, when an algebra A is presented in the form $\Bbbk(Q,I)$ as above, this equivalence enables us to present left A-modules (resp. homomorphisms of A-modules) as representations of (Q,I) (resp. morphisms of representations).

EXAMPLE 2.8.5. Consider the algebra A in Example 2.3.29. Namely, let

$$A := \begin{pmatrix} B & \Bbbk\bar{y} \\ \Bbbk\bar{x} & B \end{pmatrix},$$

where $C := \Bbbk[x,y]/\langle x^2, y^2\rangle$, and $B := \Bbbk\bar{1} \oplus \Bbbk\overline{xy} \subseteq C$. Then, by taking $E := \{e_{11}, e_{22}\}$, we have $(A,E) \in (\Bbbk\text{-}\mathbf{alg}_{\mathrm{copi}})_0$, and

$$\mathrm{Cat}(A,E) \cong \mathscr{C} := \overline{xy} \circlearrowright \overset{\bar{1}\,\circlearrowright}{e_{11}} \underset{\bar{y}}{\overset{\bar{x}}{\rightleftarrows}} \overset{\bar{1}\,\circlearrowright}{e_{22}} \circlearrowleft \overline{xy} \quad \in (\Bbbk\text{-}\mathbf{Spec}_{\mathrm{f}})_0.$$

Here define a bound quiver (Q,I) by $Q := 1 \underset{\beta}{\overset{\alpha}{\rightleftarrows}} 2$, $I := \langle \alpha\beta\alpha, \beta\alpha\beta\rangle$. Then we have isomorphisms $\Bbbk[Q,I] \cong \mathscr{C}$ and $k(Q,I) \cong A$. Thus A is presented by the bound quiver (Q,I).

CHAPTER 3

Classical Covering Theory

In this chapter, we review the classical covering theory by Gabriel. Throughout this chapter, we assume that all functors $F\colon \mathscr{C} \to \mathscr{B}$ of linear categories are *linear functors* (see Definition 1.6.3).

3.1. Galois coverings

Since the propositions below will be stated in general and proved in Chapter 5, the proofs are omitted in this chapter.

DEFINITION 3.1.1. A pair $(\mathscr{C}, X)$ of a linear category $\mathscr{C}$ and a group homomorphism $X\colon G \to \mathrm{Aut}(\mathscr{C})$ is called a linear category with a *G-action* or simply a *G-category*. A spectoroid with a G-action is called a *G-spectoroid*. When there is no risk of confusion, we abbreviate as $\mathscr{C} = (\mathscr{C}, X)$. In particular, we denote the trivial G-action $G \to \mathrm{Aut}(\mathscr{C}), a \mapsto \mathbb{1}_{\mathscr{C}}$ $(a \in G)$ by 1, and set $\Delta(\mathscr{C}) := (\mathscr{C}, 1)$.

DEFINITION 3.1.2. Let $\mathscr{C} = (\mathscr{C}, X)$ be a G-spectoroid.

(1) If we have $X(a)x \neq x$ for all $1 \neq a \in G$ and $x \in \mathscr{C}_0$, thus if for each $x \in \mathscr{C}_0$, the map

$$\rho_x\colon G \to Gx, \quad \rho_x(a) := X(a)x \ (a \in G)$$

is bijective, then the G-action X is said to be *free*, where we set $Gx := \{X(a)x \mid a \in G\}$.

(2) The G-action X is said to be *locally bounded* if the set $\{a \in G \,|\, \mathscr{C}(X(a)x, y) \neq 0\}$ is finite for all $x, y \in \mathscr{C}_0$.

REMARK 3.1.3. In statement (2) above, since we have

$$\mathscr{C}(X(a)x, y) \cong \mathscr{C}(x, X(a^{-1})y) \quad (a \in G),$$

the condition in (2) is equivalent to saying that $\{a \in G \mid \mathscr{C}(x, X(a)y) \neq 0\}$ is a finite set.

DEFINITION 3.1.4. Let $\mathscr{C}, \mathscr{B}$ be spectoroids, and $F\colon \mathscr{C} \to \mathscr{B}$ a functor. Assume that $\mathscr{C} = (\mathscr{C}, X)$ has a free and locally bounded G-action X.

(1) If $F = F \circ X(a)$ for all $a \in G$, then F is said to be *strictly G-invariant*.

(2) If F is strictly G-invariant, and if the conditions (a), (b) (resp. (a), (b), (c)) below hold, then F is called a *Galois precovering* (resp. *Galois covering*) with group G.

   (a) $F^{-1}(Fx) = Gx$ for all $x \in \mathscr{C}_0$, i.e., the map $G \to F^{-1}(Fx)$ $(a \mapsto X(a)x)$ is an epimorphism (since X is free, it becomes a bijection).

(b) F induces the following isomorphisms of vector spaces for all $x, y \in \mathscr{C}_0$

$$\coprod_{a\in G} \mathscr{C}(X(a)x, y) \to \mathscr{B}(Fx, Fy),$$
$$(f_a)_{a\in G} \mapsto \sum_{a\in G} F(f_a)$$
$$\coprod_{b\in G} \mathscr{C}(x, X(b)y) \to \mathscr{B}(Fx, Fy),$$
$$(f_b)_{b\in G} \mapsto \sum_{b\in G} F(f_b)$$

(c) $F\colon \mathscr{C}_0 \to \mathscr{B}_0$ is a surjection.

Next, given a free and locally bounded spectoroid $\mathscr{C}$ with a G-action we construct a Galois covering from $\mathscr{C}$. "The classical orbit category" defined below forms a skeleton (see Definition 1.4.7) of "the orbit category defined in Chapter 5" (this follows from [**5**, Proposition 2.11] and Proposition 6.1.3). To distinguish these we add a small "c" (standing for the word "classical") to the classical orbit category as $\mathscr{C}/_{\!c}G$, and simply denote the general orbit cagegory as "$\mathscr{C}/G$."

Definition 3.1.5. Let $\mathscr{C} = (\mathscr{C}, X)$ be a spectoroid with a free and locally bounded G-action. Then we define a linear category $\mathscr{C}/_{\!c}G$ as follows, and call it the (classical) *orbit category* of $\mathscr{C}$ by G.

- $(\mathscr{C}/_{\!c}G)_0 := \{Gx \mid x \in \mathscr{C}_0\}$.
- For any $u, v \in (\mathscr{C}/_{\!c}G)_0$, $(\mathscr{C}/_{\!c}G)(u, v)$ is given by the set of all elements $(f_{yx})_{\substack{x\in u\\ y\in v}}$ in $\prod_{\substack{x\in u\\ y\in v}} \mathscr{C}(x, y)$ satisfying the condition
$$X(a)(f_{yx}) = f_{X(a)y, X(a)x} \ (a \in G, x, y \in \mathscr{C}_0).$$
Here we note that in the expression $(f_{yx})_{\substack{x\in u\\ y\in v}}$ of each element in $(\mathscr{C}/_{\!c}G)(u, v)$, u, v are denoted without using their representatives.
- For each $u \xrightarrow{f} v \xrightarrow{g} w$ with $f = (f_{yx})_{\substack{x\in u\\ y\in v}}$, $g = (g_{zy})_{\substack{y\in v\\ z\in w}}$, we set
$$gf := \left(\sum_{y\in v} g_{zy} \cdot f_{yx}\right)_{\substack{x\in u\\ z\in w}}.$$
Note that since the action X is locally bounded, each entry of gf is a finite sum.

Definition 3.1.6. In the setting above, we define the canonical functor $P\colon \mathscr{C} \to \mathscr{C}/_{\!c}G$ as follows:

For each $x \in \mathscr{C}_0$, set $Px := Gx$, and for each $h \in \mathscr{C}(x, y)$,
$$Ph := (\delta_{\rho_x^{-1}(p), \rho_y^{-1}(q)} X(\rho_x^{-1}(p))h)_{\substack{p\in Gx\\ q\in Gy}} : Px \to Py.$$
Here, note that Ph can be defined because X is a free action (i.e., each ρ_x is bijective). Without using the inverse maps, this can be written as follows:
$$Ph := (\delta_{a,b} X(a)h)_{\substack{X(a)x\in Gx\\ X(b)y\in Gy}} : Px \to Py,$$
which is easier to understand.

Proposition 3.1.7. *$P\colon \mathscr{C} \to \mathscr{C}/_{\!c}G$ is a Galois covering with group G.*

By the following proposition, we see that an arbitrary Galois covering $\mathscr{C} \to \mathscr{B}$ from a spectoroid $\mathscr{C}$ with a free and locally bounded G-action can be uniquely expressed as the composite of the canonical functor above and an isomorphism $\mathscr{C}/_{\mathrm{c}}G \to \mathscr{B}$.

PROPOSITION 3.1.8. *In the setting as above, $P\colon \mathscr{C} \to \mathscr{C}/_{\mathrm{c}}G$ is a universal strictly G-invariant functor. Namely, it is strictly G-invariant, and for any strictly G-invariant functor $E\colon \mathscr{C} \to \mathscr{C}'$, there exists a unique functor $H\colon \mathscr{C}/_{\mathrm{c}}G \to \mathscr{C}'$ such that $E = HP$.*

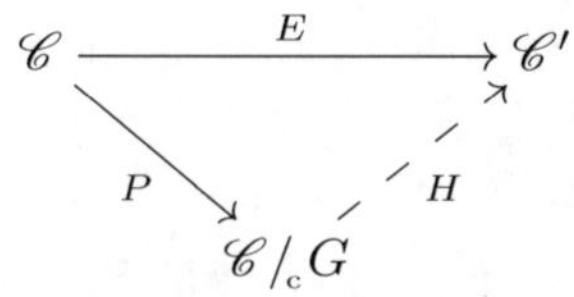

Moreover here, E is a Galois covering with group G if and only if H is an isomorphism.

3.2. Module categories of $\mathscr{C}$ and of $\mathscr{C}/_{\mathrm{c}}G$

Next, we describe the relationships between the module categories of $\mathscr{C}$ and that of $\mathscr{C}/_{\mathrm{c}}G$. The contents of this section will also be discussed in detail in Chapter 7, therefore we will only give an outline here.

DEFINITION 3.2.1.

(1) For a linear functor $F\colon \mathscr{A} \to \mathscr{B}$ of linear categories, we define a functor $F^{\bullet}\colon \operatorname{Mod}\mathscr{B} \to \operatorname{Mod}\mathscr{C}$ by $F^{\bullet}M := M \circ F^{\mathrm{op}}$. This is called the *pullup* of F.
(2) $F^{\bullet}$ has a left adjoint $F_{\bullet}$, which is called the *pushdown* of F.
(3) For a G-category $\mathscr{C} = (\mathscr{C}, X)$, $\operatorname{Mod}\mathscr{C}$ turns out to be a G-category by defining a G-action $\operatorname{Mod}X\colon G \to \operatorname{Aut}(\operatorname{Mod}\mathscr{C})$ on it by

$$(\operatorname{Mod}X)(a)M := {}^{a}M := M \circ X(a^{-1})$$
$$(a \in G, M \in (\operatorname{Mod}\mathscr{C})_0).$$

REMARK 3.2.2.

(1) In statement (3) above, the following holds (cf. (8.1)):

$$ {}^{a}\mathscr{C}(\text{-}, x) = \mathscr{C}(X(a^{-1})(\text{-}), x) \cong \mathscr{C}(\text{-}, X(a)x) \quad (a \in G, x \in \mathscr{C}_0). \tag{3.1}$$

(2) Let $\mathscr{C} = (\mathscr{C}, X)$ be a spectoroid with a free and locally bounded G-action, and $P\colon \mathscr{C} \to \mathscr{C}/_{\mathrm{c}}G$ the canonical functor. Then we have $P_{\bullet}\,\mathscr{C}(\text{-}, x) \cong (\mathscr{C}/_{\mathrm{c}}G)(\text{-}, Gx)$, and in statement (3), since $P_{\bullet}$ is right exact, $P_{\bullet}$ induces a functor $\operatorname{mod}\mathscr{C} \to \operatorname{mod}\mathscr{C}/_{\mathrm{c}}G$.

The following is a theorem that is fundamental to the classical covering theory (**[23]**).

THEOREM 3.2.3. *Let $\mathscr{C} = (\mathscr{C}, X)$ be a locally bounded spectoroid with a free and locally bounded G-action, and $P\colon \mathscr{C} \to \mathscr{C}/_{\mathrm{c}}G$ the canonical functor. Then the following hold.*

(1) *If M is an idecomposable $\mathscr{C}$-module, then $P_{\bullet}M$ is an indecomposable $\mathscr{C}/_{\mathrm{c}}G$-module.*
(2) *$P_{\bullet}\colon \operatorname{ind}\mathscr{C} \to \operatorname{ind}(\mathscr{C}/_{\mathrm{c}}G)$ is a Galois precovering with group G.*

(3) *If* $\mathscr{C}$ *is* locally representation-finite, *then* $P_{\bullet}$ *above is a Galois covering with group* G.

Here, in statement (2), $\operatorname{ind}\mathscr{C}$ *(resp.* $\operatorname{ind}(\mathscr{C}/_{\mathrm{c}}G)$ *) is the full subcategory of* $\operatorname{mod}\mathscr{C}$ *(resp.* $\operatorname{mod}(\mathscr{C}/_{\mathrm{c}}G)$ *) defined by a complete set of representatives of isoclasses of indecomposable* $\mathscr{C}$*-modules (resp.* $\mathscr{C}/_{\mathrm{c}}G$*-modules), where we assume that* $\operatorname{ind}\mathscr{C}$ *satisfies the condition that* ${}^{a}M \in \operatorname{ind}\mathscr{C}$ *for all* $a \in G, M \in \operatorname{ind}\mathscr{C}$. *In statement* (3), $\mathscr{C}$ *is said to be* locally representation-finite *if for each* $x \in \mathscr{C}_0$*, there exists only a finite number of indecomposable* $\mathscr{C}$*-modules* M *with* $M(x) \neq 0$ *up to isomorphisms.*

The following is an example using quivers. For the Auslander-Reiten quiver of a bound quiver, see, for example, reference **[14]**.

EXAMPLE 3.2.4. Using the quiver $\tilde{Q}$ and the ideal $\tilde{I}$ in Example 2.7.5(5), consider the linear category $\mathscr{C} := \Bbbk[\tilde{Q}]/\tilde{I}$. Let G be an infinite cyclic group with a generator a. Define a G-action X on $\mathscr{C}$ by

$$X(a)(i) := i + 2 \ (i \in \mathbb{Z}).$$

Then $\mathscr{C}/_{\mathrm{c}}G$ can be written as $\mathscr{C}/_{\mathrm{c}}G = \Bbbk Q/I$ by using the quiver Q and the ideal I in Example 2.7.5(4). The Galois covering $P\colon \mathscr{C} \to \mathscr{C}/_{\mathrm{c}}G$ given by the canonical functor is written as $\Bbbk f$ in Example 2.7.5(5) (see Example 6.1.4 for this computation).

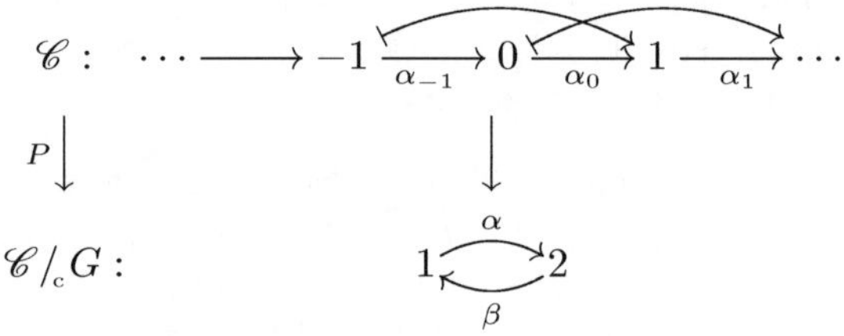

Then $\mathscr{C}$ is locally representation-finite, and the source (resp. target) category of $P_{\bullet}\colon \operatorname{ind}\mathscr{C} \to \operatorname{ind}(\mathscr{C}/_{\mathrm{c}}G)$ is given as the factor category of the path category of the following quiver by the *mesh ideal*[1].

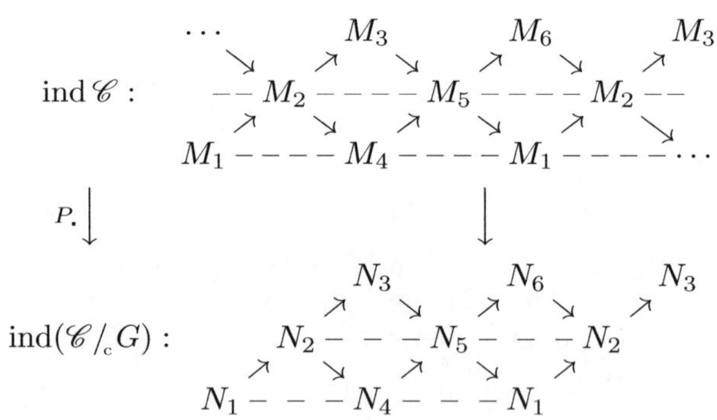

Here, the quiver of $\operatorname{ind}\mathscr{C}$ continues infinitely to the left and right; the generator a of G acts to move two vertices to the right (to the vertex with the same symbol); and in the quiver of $\operatorname{ind}(\mathscr{C}/_{\mathrm{c}}G)$, modules with the same symbol are identical. Moreover, we have $P_{\bullet}M_i = N_i$ $(i = 1, \ldots, 6)$. In addition, if we define a G-grading on $A := \mathscr{C}/_{\mathrm{c}}G$ by $\deg(\alpha) := 1, \deg(\beta) := a^{-1}$, then we have $\mathscr{C} \cong A\#G$. See Definition 5.6.1 and Section 6.2 for the definition of $A\#G$ and for the way of computation, respectively.

[1]The ideal generated by the sum of the paths of length 2 from the left-end vertex of each dashed line to the right end vertex.

CHAPTER 4

Basics of 2-Categories

In this chapter we summarize necessary facts on 2-categories that will be used later. We refer the reader to **[17]** and **[31]**.

4.1. 2-categories

A category is defined to be consisting of a set of objects, a set of morphisms, a composition of morphisms, and a family of identities that satisfies associativity and unitality. Here, the morphisms show the relations between objects. A 2-category is a notion obtained from a category by adding 2-morphisms giving relations between morphisms, vertical and horizontal compositions of 2-morphisms, and 2-identities that still satisfy the associativity and unitality for each of the two kinds of compositions, and in addition, a law between these compositions, called an *interchange law*, and a law between 2-identities and the horizontal composition. Or, another way to look at it is to say that the notion of a 2-category is obtained from that of a category $\mathscr{C}$ by replacing the word "set" in the defining condition of the category that $\mathscr{C}(x, y)$ be a set for all $x, y \in \mathscr{C}_0$ by the word "category." The definition below is based on the latter view. For the time being, we will deal with the general categories not only with light categories. First, we define general 2-categories.

DEFINITION 4.1.1. A 2-*category* is a sequence of the following data that satisfies the axioms below.

Data:

(1) A non-empty set $\mathbf{C}_0$,
(2) A family of categories $(\mathbf{C}(x, y))_{x,y\in\mathbf{C}_0}$,
(3) A family of functors $\circ := (\circ_{x,y,z} \colon \mathbf{C}(y, z) \times \mathbf{C}(x, y) \to \mathbf{C}(x, z))_{x,y,z\in\mathbf{C}_0}$,
(4) A family of functors $(u_x \colon \mathbf{1} \to \mathbf{C}(x, x))_{x\in\mathbf{C}_0}$

Axioms:

(associativity) The following diagram is commutative for all $x, y, z, w \in \mathbf{C}_0$:

$$\begin{array}{ccc}
\mathbf{C}(z,w) \times \mathbf{C}(y,z) \times \mathbf{C}(x,y) & \xrightarrow{\circ\times\mathbf{1}} & \mathbf{C}(y,w) \times \mathbf{C}(x,y) \\
{\scriptstyle \mathbf{1}\times\circ}\big\downarrow & & \big\downarrow{\scriptstyle \circ} \\
\mathbf{C}(z,w) \times \mathbf{C}(x,z) & \xrightarrow[\circ]{} & \mathbf{C}(x,w)
\end{array}$$

(unitality) The following diagram is commutative for all $x, y \in \mathbf{C}_0$:

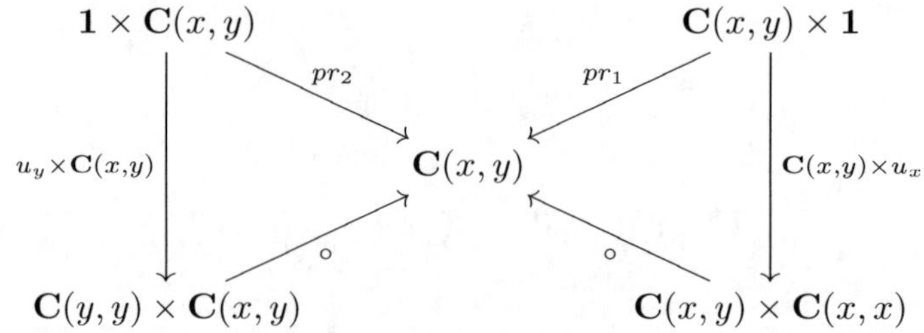

Elements of $\mathbf{C}_0$ are called *objects* of $\mathbf{C}$, objects (resp. morphisms, compositions) of $\mathbf{C}(x, y)$ (with $x, y \in \mathbf{C}_0$) are called 1-*morphisms* (resp. 2-*morphisms, vertical compositions*) of $\mathbf{C}$. Furthermore, $\circ_{x,y,z}$ (with $x, y, z \in \mathbf{C}_0$) are called *horizontal compositions* of $\mathbf{C}$, and we set $\mathbb{1}_x := u_x(*)$, $\mathbb{1}_{\mathbb{1}_x} := u_x(\mathbb{1}_*)$ for all $x \in \mathbf{C}_0$.

We denote vertical compositions (i.e., compositions of categories $\mathbf{C}(x, y)$) by the symbol $\bullet$ to distinguish it from horizontal compositions $\circ$. Sometimes we omit the symbol for horizontal composition to write gf for $g \circ f$.

In a 2-category $\mathbf{C}$ we display objects $x, y \in \mathbf{C}_0$, 1-morphisms $f, g \in \mathbf{C}(x, y)_0$, and a 2-morphism $\alpha \in \mathbf{C}(x, y)(f, g)$ as follows:

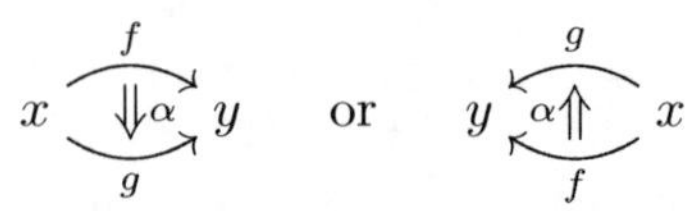

For the diagram

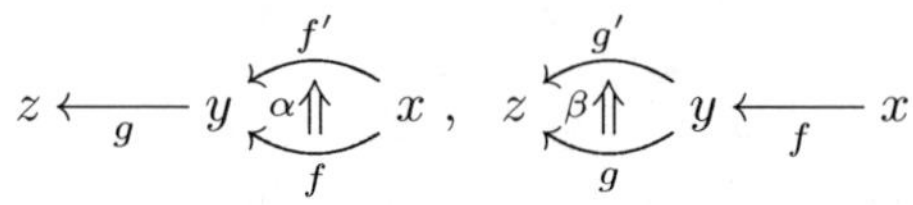

consisting of objects, 1-morphisms, and 2-morphisms, we sometimes use the abbreviations $g \circ \alpha := \mathbb{1}_g \circ \alpha$, and $\beta \circ f := \beta \circ \mathbb{1}_f$.

REMARK 4.1.2. The axioms above are expressed by equalities of 1-morphisms and 2-morphisms as follows:

Associativity: For a diagram

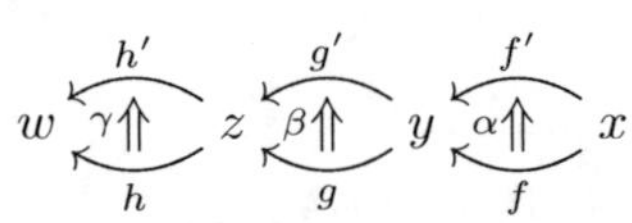

in a 2-category $\mathbf{C}$ we have $(h \circ g) \circ f = h \circ (g \circ f)$, $(\gamma \circ \beta) \circ \alpha = \gamma \circ (\beta \circ \alpha)$.

Unitality: For a diagram

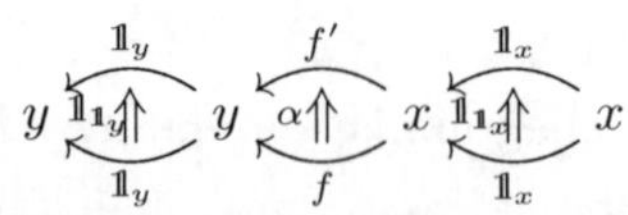

in a 2-category $\mathbf{C}$ we have $\mathbb{1}_y \circ f = f = f \circ \mathbb{1}_x$, $\mathbb{1}_{\mathbb{1}_y} \circ \alpha = \alpha = \alpha \circ \mathbb{1}_{\mathbb{1}_x}$.

NOTATION 4.1.3. For a 2-category $\mathbf{C}$ we set

$$\mathbf{C}_1 := \bigcup_{x,y \in \mathbf{C}_0} \mathbf{C}(x, y)_0, \quad \mathbf{C}_2 := \bigcup_{x,y \in \mathbf{C}_0} \mathbf{C}(x, y)_1.$$

These are disjoint unions.

REMARK 4.1.4. Let x, y, z be objects in a 2-category $\mathbf{C}$. Then the condition that $\circ_{x,y,z}$ is a functor is stated as follows:

(1) Preservation of compositions is equivalent to saying that for each diagram

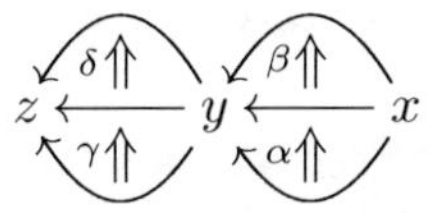

we have $(\delta \bullet \gamma) \circ (\beta \bullet \alpha) = (\delta \circ \beta) \bullet (\gamma \circ \alpha)$,

which is called the *interchange law*.

(2) Preservation of identities is equivalent to saying that for each diagram

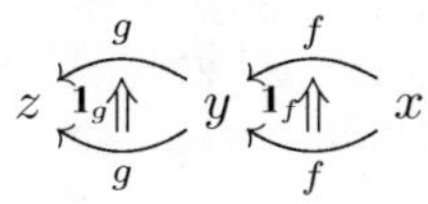

we have $\mathbb{1}_g \circ \mathbb{1}_f = \mathbb{1}_{g \circ f}$.

Indeed, (1) (LHS) $= \circ_{x,y,z}(\delta \bullet \gamma, \beta \bullet \alpha) = \circ_{x,y,z}((\delta, \beta) \bullet (\gamma, \alpha)) = (\circ_{x,y,z}(\delta, \beta)) \bullet (\circ_{x,y,z}(\gamma, \alpha)) =$ (RHS).

(2) (LHS) $= \circ_{x,y,z}(\mathbb{1}_g, \mathbb{1}_f) = \circ_{x,y,z}(\mathbb{1}_{(g,f)}) =$ (RHS).

EXAMPLE 4.1.5. Any category I can be seen as a 2-category by letting the set of all identity 2-morphisms be the set of 2-morphisms $I_2 := \{\mathbb{1}_a \mid a \in I_1\}$. More precisely, I can be identified with the 2-category $\mathbf{C}$ defined as follows:

(1) $\mathbf{C}_0 := I_0$.
(2) Regard $\mathbf{C}(x, y) := I(x, y)$ as a discrete category for all $x, y \in I_0$.
(3) The functor $\circ_{x,y,z}$ is defined as $(\mathbb{1}_x, \mathbb{1}_x) \mapsto \mathbb{1}_x$ if $x = y = z$ and the trivial map otherwise.
(4) We define a functor $u_x \colon \mathbf{1} \to \mathbf{C}(x, x)$ by $u_x(*) := \mathbb{1}_x, u_x(\mathbb{1}_*) := \mathbb{1}_{\mathbb{1}_x}$ for all $x \in I_0$.

DEFINITION 4.1.6. Let $\mathbf{C}$ be a 2-category, and $k \geq 1$.

(1) The set $\mathbf{C}_0$ is called the *object set* of $\mathbf{C}$, the set $\mathbf{C}(x, y)_0$ is called a *local 1-morphism set* of $\mathbf{C}$ for each $x, y \in \mathbf{C}_0$, and the set $\mathbf{C}(x, y)(f, g)$ is called a *local 2-morphism set* for each $f, g \in \mathbf{C}(x, y)_0$.
(2) $\mathbf{C}$ is called a *small* 2-category if $\mathbf{C}_0$ and all local r-morphism sets are small for each $r = 1, 2$.
(3) $\mathbf{C}$ is called a *light* 2-category if $\mathbf{C}_0$ is a class and all local r-morphism sets are small for each $r = 1, 2$.
(4) $\mathbf{C}$ is called a *k-moderate* 2-category if $\mathbf{C}_0$ and all local r-morphism sets are k-classes for each $r = 1, 2$. (1-moderate 2-categories are simply called *moderate 2-categories*.)

EXAMPLE 4.1.7. Fix a set $\mathscr{U}$ of (light) categories (e.g., the set of all small categories). Then we define a 2-category $\mathbf{C}$ as follows:

(1) Set $\mathbf{C}_0 := \mathscr{U}$.
(2) For each $x, y \in \mathbf{C}_0$ let $\mathbf{C}(x, y)$ be the functor category. Namely, its object set $\mathbf{C}(x, y)_0$ is the set of all functors $x \to y$, and the morphisms are the natural transformations between them and the compositions are given by the vertical compositions of natural transformations $((\beta, \alpha) \mapsto \beta \bullet \alpha)$.

(3) To determine family $\circ := (\circ_{x,y,z} \colon \mathbf{C}(y,z) \times \mathbf{C}(x,y) \to \mathbf{C}(x,z))_{x,y,z \in \mathbf{C}_0}$ of functors with two variables take functors and natural transformations

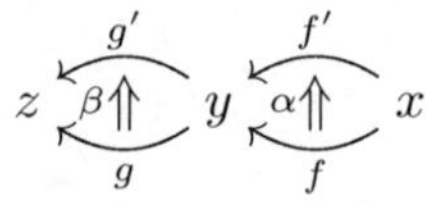

$(x, y, z \in \mathbf{C}_0)$

in $\mathbf{C}$. Then we define $g \circ f \colon x \to z$ to be the usual composite of functors, and $\beta \circ \alpha$ to be the horizontal composite of natural transformations.

(4) For each $x \in \mathbf{C}_0$ we set $u_x(*) := \mathbb{1}_x,\ u_x(\mathbb{1}_*) := \mathbb{1}_{\mathbb{1}_x}$.

It is easily verified that the data given above satisfy associativity and unitality. We denote this 2-category $\mathbf{C}$ by $\mathbf{Cat}(\mathscr{U})$. If $\mathscr{U}$ contains a light category $\mathscr{C}$ that is not small, then the local 1-morphism set $\mathrm{Fun}(\mathscr{C}, \mathscr{C})_0$ is not a class because $\mathbb{1}_{\mathscr{C}}$ is not a small set. Hence $\mathbf{Cat}(\mathscr{U})$ cannot be even a moderate 2-category (it becomes a 2-moderate 2-category). If $\mathscr{U}$ is the class of all small categories, then we denote $\mathbf{Cat}(\mathscr{U})$ just by $\mathbf{Cat}$, which is a light 2-category. If $\mathscr{U}$ is the whole of light categories, then we denote $\mathbf{Cat}(\mathscr{U})$ by $\mathbf{CAT}$, which turns out to be a 2-moderate 2-category.

EXAMPLE 4.1.8. Fix a set $\mathscr{S}$ of linear categories, and define a 2-category $\mathbb{k}\text{-}\mathbf{Cat}(\mathscr{S})$ in a similar way as above, i.e., by taking $\mathscr{S}$ as the object set, the linear functors between the objects as the 1-morphisms, the natural transformations between linear functors as the 2-morphisms, and so on. When $\mathscr{S}$ is the set of all linear small categories (resp. all linear light categories) $\mathbb{k}\text{-}\mathbf{Cat}(\mathscr{S})$ is denoted by $\mathbb{k}\text{-}\mathbf{Cat}$ (resp. $\mathbb{k}\text{-}\mathbf{CAT}$).

DEFINITION 4.1.9. From a 2-category $\mathbf{C}$ the following three new 2-categories are defined:

$\mathbf{C}^{\mathrm{op}}$:=(the 2-category obtained from $\mathbf{C}$ by reversing the 1-morphisms of $\mathbf{C}$),

$\mathbf{C}^{\mathrm{co}}$:=(the 2-category obtained from $\mathbf{C}$ by reversing the 2-morphisms of $\mathbf{C}$),

and

$\mathbf{C}^{\mathrm{coop}} := (\mathbf{C}^{\mathrm{co}})^{\mathrm{op}} = (\mathbf{C}^{\mathrm{op}})^{\mathrm{co}}$.

In the sequel unless otherwise stated we again assume that all categories are light and all 2-categories are light.

DEFINITION 4.1.10. Let $\mathbf{C}$ be a 2-category, $x, y \in \mathbf{C}_0$, $f, g \in \mathbf{C}(x,y)_0$, and $\alpha \in \mathbf{C}(x,y)(f,g)$.

(1) If we have $\beta \bullet \alpha = \mathbb{1}_f, \alpha \bullet \beta = \mathbb{1}_g$ for some $\beta \in \mathbf{C}(x,y)(g,f)$, then α is said to be an *isomorphism* (*2-isomorphism*). Such a β is called the *inverse* (*2-inverse*) of α, which is unique if it exists, and hence we denote it by α^{-1}.

(2) If there exists an isomorphism $\alpha \in \mathbf{C}(x,y)(f,g)$, then we say that f and g are *isomorphic*, and denote it by $f \cong g$.

(3) If we have $f'f = \mathbb{1}_x, ff' = \mathbb{1}_y$ for some $f' \in \mathbf{C}(y,x)$, then f is called an *isomorphism* (*1-isomorphism*). Such an f' is called the *inverse* of f, which is unique if it exists, and hence we denote it by f^{-1}.

(4) If we have $f'f \cong \mathbb{1}_x, ff' \cong \mathbb{1}_y$ for some $f' \in \mathbf{C}(y,x)$, then f is called an *equivalence*. Such an f' is called a *quasi-inverse* of f, which is not uniquely determined in general.

(5) If there exists an isomorphism $f \in \mathbf{C}(x,y)$, then x and y are said to be *isomorphic*, and denote it by $x \cong y$.

(6) If there exists an equivalence $f \in \mathbf{C}(x, y)$, then x and y are said to be *equivalent*, and denote it by $x \simeq y$.

DEFINITION 4.1.11. Let $\mathbf{C}$, $\mathbf{D}$ be 2-categories. A pair of the following data that satisfies the axioms below is called a *2-functor* from $\mathbf{C}$ to $\mathbf{D}$, and denote it by $\mathbf{C} \to \mathbf{D}$

Data:

- A map $X\colon \mathbf{C}_0 \to \mathbf{D}_0$,
- A family of functors $({}_yX_x\colon \mathbf{C}(x,y) \to \mathbf{D}(X(x), X(y)))_{(x,y)\in\mathbf{C}_0\times\mathbf{C}_0}$, where we use the abbreviation $X(f)$ for ${}_yX_x(f)$ for all $f \in \mathbf{C}(x,y)_0$.

Axioms:

(1) For each $x, y, z \in \mathbf{C}_0$ the following is commutative:

$$\begin{array}{ccc} \mathbf{C}(y,z) \times \mathbf{C}(x,y) & \xrightarrow{\quad\circ\quad} & \mathbf{C}(x,z) \\ {\scriptstyle {}_zX_y \times {}_yX_x}\big\downarrow & & \big\downarrow{\scriptstyle {}_zX_x} \\ \mathbf{D}(X(y),X(z)) \times \mathbf{D}(X(x),X(y)) & \xrightarrow[\circ]{} & \mathbf{D}(X(x),X(z)) \end{array}$$

(2) For each $x \in \mathbf{C}_0$ the following is commutative:

$$\begin{array}{ccc} \mathbf{1} & \xrightarrow{u_x} & \mathbf{C}(x,x) \\ & {\scriptstyle u_{X(x)}}\searrow & \big\downarrow{\scriptstyle {}_xX_x} \\ & & \mathbf{D}(X(x),X(x)) \end{array}$$

Thus in forms of equalities of 1-morphisms and 2-morphisms the axioms above are expressed as follows:

(1) For each diagram

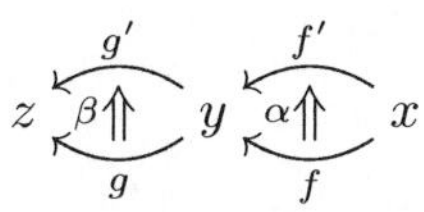

in $\mathbf{C}$ we have $X(g \circ f) = X(g) \circ X(f)$, $X(\beta \circ \alpha) = X(\beta) \circ X(\alpha)$.

(2) $X(\mathbb{1}_x) = \mathbb{1}_{X(x)}$, $X(\mathbb{1}_{\mathbb{1}_x}) = \mathbb{1}_{\mathbb{1}_{X(x)}}$ for all $x \in \mathbf{C}_0$.

REMARK 4.1.12. In the definition above, since each ${}_yX_x$ is a functor, X preserves identities and compositions (i.e., vertical compositions of 2-morphisms) in the category $\mathbf{C}(x,y)$. Thus we also have

$$X(\mathbb{1}_f) = \mathbb{1}_{X(f)} \ (f \in \mathbf{C}(x,y)_0),$$

$$X(\beta \bullet \alpha) = X(\beta) \bullet X(\alpha) \ (f \xrightarrow{\alpha} g \xrightarrow{\beta} h; \alpha, \beta \in \mathbf{C}(x,y)_1).$$

Here, we shall observe that the correspondence of small categories to their left module categories (resp. right module categories) extends to a 2-functor (see Definition 4.1.9 for the notation).

EXAMPLE 4.1.13. We construct the following three 2-functors.

(1) $(\text{-})\text{-Mod} := \operatorname{Fun}_{\Bbbk}(\text{-}, \operatorname{Mod}\Bbbk)\colon \Bbbk\text{-}\mathbf{Cat} \to \Bbbk\text{-}\mathbf{CAT}^{\mathrm{op}}$;

(2) $(\text{-})^{\mathrm{op}}\colon \Bbbk\text{-}\mathbf{Cat} \to \Bbbk\text{-}\mathbf{Cat}^{\mathrm{co}}$;

(3) $\operatorname{Mod}(\text{-})\colon \Bbbk\text{-}\mathbf{Cat} \to \Bbbk\text{-}\mathbf{CAT}^{\mathrm{coop}}$

For (1): Definition of a 2-functor (-)-Mod.

(Objects): For each object $\mathscr{C}$ in $\Bbbk$-**Cat**, we set $\mathscr{C}$-Mod $:= \mathrm{Fun}_{\Bbbk}(\mathscr{C}, \mathrm{Mod}\,\Bbbk) \in \Bbbk\text{-}\mathbf{CAT}_0$. Here note that $\mathrm{Fun}_{\Bbbk}(\mathscr{C}, \mathrm{Mod}\,\Bbbk)$ is a light category by Proposition A.3.4.

(1-morphisms): For each 1-morphism $F\colon \mathscr{C} \to \mathscr{D}$ in $\Bbbk$-**Cat**, we define a 1-morphism

$$F\text{-Mod}\colon \mathscr{D}\text{-Mod} \to \mathscr{C}\text{-Mod}$$

in $\Bbbk$-**CAT** as follows: For each morphism $\alpha\colon M \to N$ in $\mathscr{D}$-Mod, consider the diagram

$$\mathscr{C} \xrightarrow{F} \mathscr{D} \underset{N}{\overset{M}{\Downarrow \alpha}} \mathrm{Mod}\,\Bbbk$$

and define $(F\text{-Mod})(\alpha)\colon (F\text{-Mod})(M) \to (F\text{-Mod})(N)$ as

$$\alpha \circ F\colon M \circ F \to N \circ F$$

(see Definition 1.3.6).

(2-morphisms): For each 2-morphism $\mathscr{C} \underset{F}{\overset{E}{\Downarrow \phi}} \mathscr{D}$ in $\Bbbk$-**Cat**, define ϕ-Mod: E-Mod $\Rightarrow$ F-Mod as follows: For each object M in $\mathscr{D}$-Mod, first consider the diagram

$$\mathscr{C} \underset{F}{\overset{E}{\Downarrow \phi}} \mathscr{D} \xrightarrow{M} \mathrm{Mod}\,\Bbbk$$

and define $(\phi\text{-Mod})_M\colon (E\text{-Mod})(M) \Rightarrow (F\text{-Mod})(M)$ as

$$M \circ \phi\colon M \circ E \Rightarrow M \circ F,$$

and then set $\phi\text{-Mod} := ((\phi\text{-Mod})_M)_{M \in (\mathscr{D}\text{-Mod})_0}$.

(2) Definition of a 2-functor $(\text{-})^{\mathrm{op}}$. We define this by extending the definition in Exercise 1.2.5(3) as follows. For each diagram $\mathscr{C} \underset{F}{\overset{E}{\Downarrow \phi}} \mathscr{D}$ consisting of objects, 1-morphisms, and a 2-morphism in $\Bbbk$-**Cat**, $(\text{-})^{\mathrm{op}}$ maps it to the diagram $\mathscr{C}^{\mathrm{op}} \underset{F^{\mathrm{op}}}{\overset{E^{\mathrm{op}}}{\Uparrow \phi^{\mathrm{op}}}} \mathscr{D}^{\mathrm{op}}$ in $\Bbbk$-**Cat**, where $F^{\mathrm{op}}, \phi^{\mathrm{op}}$ are defined by $F^{\mathrm{op}}(x) := F(x), F^{\mathrm{op}}(f) := F(f), \phi^{\mathrm{op}}_x := \phi_x$ $(x \in \mathscr{C}_0, f \in \mathscr{C}_1)$, respectively. Here note that the directions of 1-morphisms do not change, but those of 2-morphisms are reversed because $\phi^{\mathrm{op}}_x = \phi_x \in \mathscr{D}(E(x), F(x)) = \mathscr{D}^{\mathrm{op}}(F^{\mathrm{op}}(x), E^{\mathrm{op}}(x))$.

(3) Definition of a 2-functor Mod(-). This is defined as the composite of (1) and (2) above: $\mathrm{Mod}(\text{-}) := (\text{-})\text{-Mod} \circ (\text{-})^{\mathrm{op}} = ((\text{-})^{\mathrm{op}})\text{-Mod} = \Bbbk\text{-}\mathbf{Cat}((\text{-})^{\mathrm{op}}, \mathrm{Mod}\,\Bbbk)$. Then since the directions of both 1-morphisms and 2-morphisms are reversed, we see that $\mathrm{Mod}(\text{-})\colon \Bbbk\text{-}\mathbf{Cat} \to \Bbbk\text{-}\mathbf{CAT}^{\mathrm{coop}}$.

DEFINITION 4.1.14. Let $X, Y\colon \mathbf{C} \to \mathbf{D}$ be 2-functors of 2-categories. A family $(\rho_x\colon X(x) \to Y(x))_{x \in \mathbf{C}_0}$ of 1-morphisms of $\mathbf{D}$ that satisfies the axioms below is called a *strict 2-natural transformation* from X to Y, and is denoted by $\rho\colon X \Rightarrow Y$.

Axioms: For each $x, y \in \mathbf{C}_0$ the following is commutative:

$$\begin{CD} \mathbf{C}(x,y) @>{{}_yX_x}>> \mathbf{D}(X(x),X(y)) \\ @V{{}_yY_x}VV @VV{\mathbf{D}(X(x),\rho_y)}V \\ \mathbf{D}(Y(x),Y(y)) @>>{\mathbf{D}(\rho_x,Y(y))}> \mathbf{D}(X(x),Y(y)). \end{CD}$$

In other forms by equalities of 1-morphisms and 2-morphisms, for each diagram

$$x \underset{g}{\overset{f}{\underset{\Downarrow\alpha}{\rightrightarrows}}} y$$

in $\mathbf{C}$ we have $\rho_y \circ X(f) = Y(f) \circ \rho_x$, $\quad \mathbb{1}_{\rho_y} \circ X(\alpha) = Y(\alpha) \circ \mathbb{1}_{\rho_x}$.

If the first (resp. second) equalities hold, then ρ is said to have a *strict* 1-*naturality* (resp. *strict* 2-*naturality*).

Strict 2-natural transformations are sometimes too strong to apply. This notion is weakened by using 2-morphisms as follows:

DEFINITION 4.1.15. Let $X, Y\colon \mathbf{C} \to \mathbf{D}$ be 2-functors of 2-categories. A pair of the following data that satisfies the axioms below is called a 2-*natural transformation* from X to Y, and is denoted by $\rho\colon X \Rightarrow Y$.

Data:

- A family $(\rho_x\colon X(x) \to Y(x))_{x \in \mathbf{C}_0}$ of 1-morphisms of $\mathbf{D}$,
- A family $(\rho_f\colon \rho_y \circ X(f) \Rightarrow Y(f) \circ \rho_x)_{f \in \mathscr{C}_1, f\colon x \to y}$ of 2-morphisms of $\mathbf{D}$.

Axioms:

- For each $x, y \in \mathbf{C}_0$ $\rho_{x,y} := (\rho_f)_{f \in \mathbf{C}(x,y)_0}$ is a natural transformation of functors:

$$\begin{array}{ccc} \mathbf{C}(x,y) & \xrightarrow{\quad {}_yX_x \quad} & \mathbf{D}(X(x),X(y)) \\ {\scriptstyle {}_yY_x}\big\downarrow & \overset{\rho_{x,y}}{\swarrow\!\!\!\swarrow} & \big\downarrow{\scriptstyle \mathbf{D}(X(x),\rho_y)} \\ \mathbf{D}(Y(x),Y(y)) & \xrightarrow[\mathbf{D}(\rho_x,Y(y))]{} & \mathbf{D}(X(x),Y(y)). \end{array}$$

In other forms by diagrams of 1-morphisms and 2-morphisms, for each diagram

$$x \underset{g}{\overset{f}{\underset{\Downarrow\alpha}{\rightrightarrows}}} y$$

in $\mathbf{C}$ there exists a 2-isomorphism

$$\begin{array}{ccc} X(x) & \xrightarrow{X(f)} & X(y) \\ {\scriptstyle \rho_x}\big\downarrow & \overset{\rho_f}{\underset{\sim}{\swarrow\!\!\!\swarrow}} & \big\downarrow{\scriptstyle \rho_y} \\ Y(x) & \xrightarrow[Y(f)]{} & Y(y) \end{array} \quad \text{that makes} \quad \begin{array}{ccc} \rho_y \circ X(f) & \overset{\rho_f}{\underset{\sim}{\Longrightarrow}} & Y(f) \circ \rho_x \\ {\scriptstyle \mathbb{1}_{\rho_y} \circ X(\alpha)}\Big\Downarrow & & \Big\Downarrow{\scriptstyle Y(\alpha) \circ \mathbb{1}_{\rho_x}} \\ \rho_y \circ X(g) & \underset{\rho_g}{\overset{\sim}{\Longrightarrow}} & Y(g) \circ \rho_x \end{array} \quad \text{commutative,}$$

i.e., $\rho_g \bullet (\mathbb{1}_{\rho_y} \circ X(\alpha)) = (Y(\alpha) \circ \mathbb{1}_{\rho_x}) \bullet \rho_f$. If there exists a 2-isomorphism ρ_f in the diagram on the left (resp. if the diagram on the right is commutative), then ρ is said to have the 1-*naturality* (resp. 2-*naturality*). In particular, a strict 2-natural transformation is nothing but a 2-natural transformation ρ for which ρ_f is the identity 2-morphism for all $f \in \mathbf{C}_1$.

EXAMPLE 4.1.16. Let $X\colon \mathbf{C} \to \mathbf{D}$ be a 2-functor of 2-categories. Then

$$\mathbb{1}_X := ((\mathbb{1}_{X(x)})_{x\in\mathbf{C}_0}, (\mathbb{1}_{X(f)})_{f\in\mathbf{C}_1})$$

is a 2-natural transformation $X \Rightarrow X$, which is called the *identity* 2-*natural transformation*.

Vertical compositions and horizontal compositions for 2-natural transformations are defined in a similar way for usual natural transformations as follows:

DEFINITION 4.1.17 (Vertical composition). For a diagram

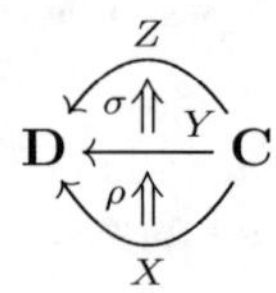

consisting of 2-categories, 2-functors and 2-natural transformations we define the *vertical composite* $\sigma \bullet \rho\colon X \Rightarrow Z$ of ρ and σ as follows:

$$\begin{aligned}
\sigma \bullet \rho &:= (((\sigma \bullet \rho)_x)_{x\in\mathbf{C}_0}, ((\sigma \bullet \rho)_f)_{f\in\mathbf{C}_1}),\\
(\sigma \bullet \rho)_x &:= \sigma_x \circ \rho_x \quad (x \in \mathbf{C}_0),\\
(\sigma \bullet \rho)_f &= (\sigma_f \circ \rho_x) \bullet (\sigma_y \circ \rho_f) \quad (f \in \mathbf{C}_1, f\colon x \to y).
\end{aligned}$$

It immediately follows from the commutative diagram below that this is again a 2-natural transformation:

$$\begin{array}{ccccc}
\sigma_y \circ \rho_y \circ X(f) & \overset{\sigma_y\circ\rho_f}{\Longrightarrow} & \sigma_y \circ Y(f) \circ \rho_x & \overset{\sigma_f\circ\rho_x}{\Longrightarrow} & Z(f) \circ \sigma_x \circ \rho_x\\
{\scriptstyle \sigma_y\circ\rho_y\circ X(\alpha)}\Big\Downarrow & & \Big\Downarrow{\scriptstyle \sigma_y\circ Y(\alpha)\circ\rho_x} & & \Big\Downarrow{\scriptstyle Z(\alpha)\circ\sigma_x\circ\rho_x}\\
\sigma_y \circ \rho_y \circ X(g) & \underset{\sigma_y\circ\rho_g}{\Longrightarrow} & \sigma_y \circ Y(g) \circ \rho_x & \underset{\sigma_g\circ\rho_x}{\Longrightarrow} & Z(g) \circ \sigma_x \circ \rho_x.
\end{array}$$

DEFINITION 4.1.18 (Horizontal composition of a 2-functor and a 2-natural transformation). For the diagram

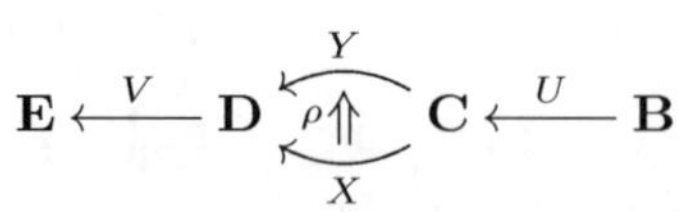

consisting of 2-categories, 2-functors and 2-natural transformation, we define the composite $\rho \circ U$ and $V \circ \rho$ of a 2-functor and a 2-natural transformation as follows:

$$\begin{aligned}
\rho \circ U &:= ((\rho_{U(y)})_{y\in\mathbf{B}_0}, (\rho_{U(g)})_{g\in\mathbf{B}_1})\colon X \circ U \Rightarrow Y \circ U,\\
V \circ \rho &:= ((V(\rho_x))_{x\in\mathbf{C}_0}, (V(\rho_f))_{f\in\mathbf{C}_1})\colon V \circ X \Rightarrow V \circ Y.
\end{aligned}$$

As is easily seen, these are again 2-natural transformations.

DEFINITION 4.1.19 (Horizontal composition). For a diagram

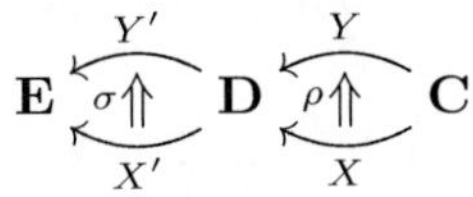

consisting of 2-categories, 2-functors and 2-natural transformation, it follows from Definitions 4.1.17 and 4.1.18 that

$$(\sigma \circ Y) \bullet (X' \circ \rho) = (Y' \circ \rho) \bullet (\sigma \circ X).$$

These are 2-natural transformations $X' \circ X \Rightarrow Y' \circ Y$. We set this common 2-natural transformation to be $\sigma \circ \rho$, and call it the *horizontal composite* of ρ and σ. In particular by definition, we have $\mathbb{1}_{X'} \circ \rho = X' \circ \rho$, $\sigma \circ \mathbb{1}_X = \sigma \circ X$. More explicitly,

$$\begin{aligned}
\sigma \circ \rho &:= (((\sigma \circ \rho)_x)_{x \in \mathbf{C}_0}, ((\sigma \circ \rho)_f)_{f \in \mathbf{C}_1}),\\
(\sigma \circ \rho)_x &:= \sigma_{Y(x)} \circ X'(\rho_x) = Y'(\rho_x) \circ \sigma_{X(x)} \quad (x \in \mathbf{C}_0),\\
(\sigma \circ \rho)_f &= \sigma_{Y(f)} \bullet X'(\rho_f) = Y'(\rho_f) \bullet \sigma_{X(f)} \quad (f \in \mathbf{C}_1).
\end{aligned}$$

Next we introduce a notion that gives relationship between 2-natural transformations.

DEFINITION 4.1.20. Let $X, Y\colon \mathbf{C} \to \mathbf{D}$ be 2-functors of 2-categories, and $\rho, \sigma\colon X \Rightarrow Y$ 2-natural transformations between them. A family $\Phi = (\Phi_x\colon \rho_x \Rightarrow \sigma_x)_{x \in \mathbf{C}_0}$ of 2-morphisms of $\mathbf{D}$ satisfying the axioms below is called a *modification* from ρ to σ, and is denoted by $\Phi\colon \rho \rightsquigarrow \sigma$.

Axioms: For each diagram

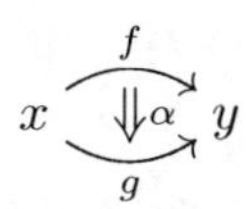

in $\mathbf{C}$, the following diagram is commutative:

$$\begin{array}{ccc}
\rho_y \circ X(f) & \overset{\rho_f}{\underset{\sim}{\Longrightarrow}} & Y(f) \circ \rho_x \\
{\scriptstyle \Phi_y \circ X(\alpha)} \Big\Downarrow & & \Big\Downarrow {\scriptstyle Y(\alpha) \circ \Phi_x} \\
\sigma_y \circ X(g) & \underset{\sigma_g}{\overset{\sim}{\Longrightarrow}} & Y(g) \circ \sigma_x
\end{array}$$

Namely, $\sigma_g \bullet (\Phi_y \circ X(\alpha)) = (Y(\alpha) \circ \Phi_x) \bullet \rho_f$.

DEFINITION 4.1.21. Let $X, Y\colon \mathbf{C} \to \mathbf{D}$ be 2-functors of 2-categories, $\rho, \sigma\colon X \Rightarrow Y$ two 2-natural transformations, and $\Phi\colon \rho \rightsquigarrow \sigma$ a modification.

(1) If Φ_x is a 2-isomorphism for all $x \in \mathbf{C}_0$, then $\Phi = (\Phi_x)_{x \in \mathbf{C}_0}$ is called an *isomorphism* (*isomorphic modification*).
(2) If there exists an isomorphic modification $\Phi\colon \rho \rightsquigarrow \sigma$, then ρ and σ are said to be *isomorphic*, which is denoted by $\rho \cong \sigma$.
(3) If we have $\rho'\rho = \mathbb{1}_X$, $\rho\rho' = \mathbb{1}_Y$ for some 2-natural transformation $\rho'\colon Y \Rightarrow X$, then ρ is called a 2-*natural isomorphism*. Since ρ' is uniquely determined by ρ, we denote it by ρ^{-1}, and call it the *inverse* (the *inverse natural transformation*) of ρ.

(4) If we have $\rho'\rho \cong \mathbb{1}_X, \rho\rho' \cong \mathbb{1}_Y$ for some 2-natural transformation $\rho'\colon Y \Rightarrow X$, then ρ is called a 2-*natural equivalence*, and this ρ' is called a *quasi-inverse* of ρ.

EXAMPLE 4.1.22. We fix a set $\mathbf{U}$ of 2-categories, (e.g., the whole of small 2-categories). Then we define a 2-category $\mathbf{C}$ as follows:

(1) Set $\mathbf{C}_0 := \mathbf{U}$.
(2) For each $x, y \in \mathbf{C}_0$ let $\mathbf{C}(x, y)$ be the 2-functor categories. Namely, it is the category, the object set $\mathbf{C}(x, y)_0$ of which is given by the whole of 2-functors $x \to y$, the morphisms of which are the 2-natural transformations and the compositions of which are the vertical compositions of 2-natural transformations.
(3) To determine a family $\circ := (\circ_{x,y,z}\colon \mathbf{C}(y, z) \times \mathbf{C}(x, y) \to \mathbf{C}(x, z))_{x,y,z\in\mathbf{C}_0}$ of functors with 2-variables take 2-functors and 2-natural transformations

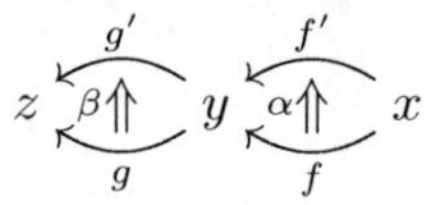

$(x, y, z \in \mathbf{C}_0)$.

Then we define $g \circ f\colon x \to z$ to be the usual composite of 2-functors, and $\beta \circ \alpha$ to be the horizontal composite of 2-natural transformations.
(4) For each $x \in \mathbf{C}_0$ we set $u_x(*) := \mathbb{1}_x,\ u_x(\mathbb{1}_*) := \mathbb{1}_{\mathbb{1}_x}$.

It is easily verified that the data above satisfy associativity and unitality. This 2-category $\mathbf{C}$ is denoted by 2-$\mathbf{Cat}(\mathbf{U})$. The 2-category 2-$\mathbf{Cat}(\mathbf{U})$ has modifications as "3-morphisms" and forms a "3-category".

The 2-category obtained from 2-$\mathbf{Cat}(\mathbf{U})$ by restricting the 2-morphisms to strict 2-natural transformations is denoted by 2-$\mathbf{Cat}^{\mathrm{st}}(\mathbf{U})$.

4.2. String diagrams

String diagrams make computations of 2-morphisms in a 2-category or natural transformations drastically easy in many cases. In this book we usually prove propositions on 2-morphisms or natural transformations using string diagrams. We let $\mathbf{C}$ be a 2-category and we explain how to draw them below. (In this book we compose 1-morphisms (resp. 2-morphisms) from right to left (resp. from bottom to top) according to **[45]**[1]. We refer the reader to **[35**, §2**]** as a reference, but note that it composes 1-morphisms (resp. 2-morphisms) from left to right (resp. from bottom to top).)

(1) An object x of $\mathbf{C}$ is displayed as an area.

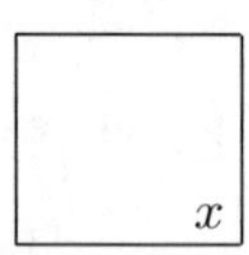

(2) 1-morphism $f\colon x \to y$ is displayed as a string (vertical line) (going from right to left):

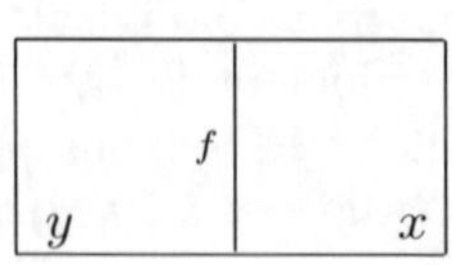

[1] The auther first learned how to draw string diagrams by these movies.

We draw $\mathbb{1}_x\colon x \to x$ by a dotted vertical line or nothing for all $x \in \mathbf{C}_0$:

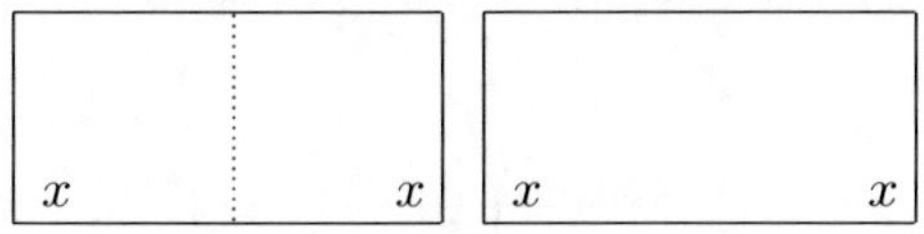

(3) The composite $g \circ f\colon x \to z$ of a 1-morphism $f\colon x \to y$ and $g\colon y \to z$ is displayed as follows (composed from right to left):

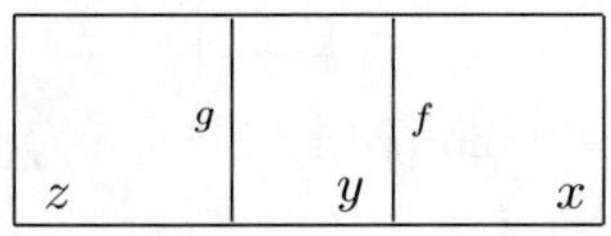

(4) A 2-morphism 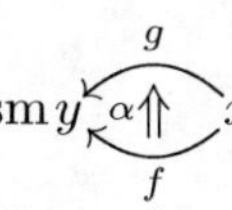 is displayed as a vertex drawn as a white circle with its name inside as follows (draw going from bottom to top):

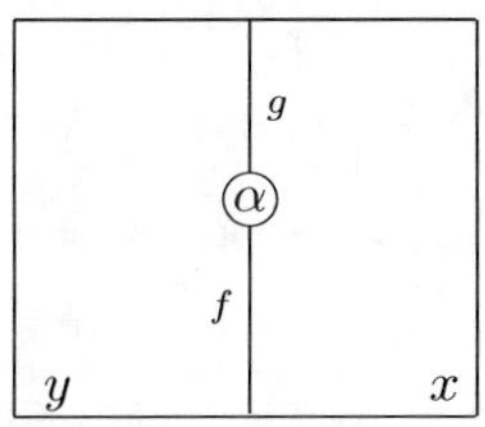

In particular, when $f = f_m \circ \cdots \circ f_1$, $g = g_n \circ \cdots \circ g_1$ are composites of 1-morphisms, a 2-morphism $\alpha\colon f_m \circ \cdots f_1 \Rightarrow g_n \circ \cdots g_1$ from a composite of 1-morphisms to a composite of 1-morphisms is displayed as follows:

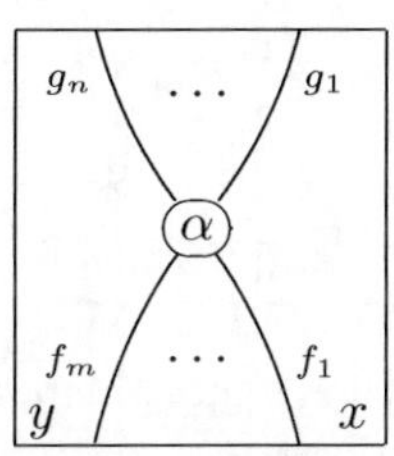

(5) For two 2-morphisms 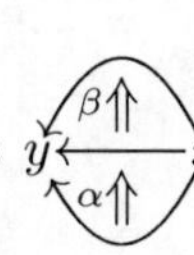 the vertical composite $\beta \bullet \alpha$ is displayed as follows (composed from bottom to top):

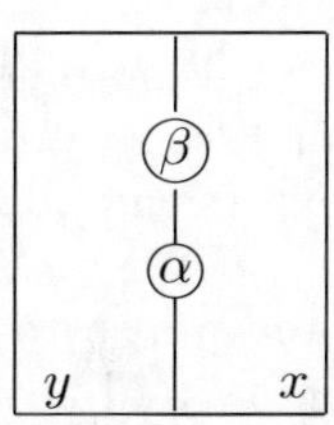

(6) For two 2-morphisms 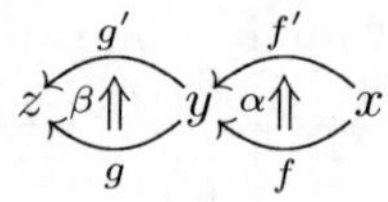 the horizontal composite $\beta \circ \alpha$ is displayed as follows:

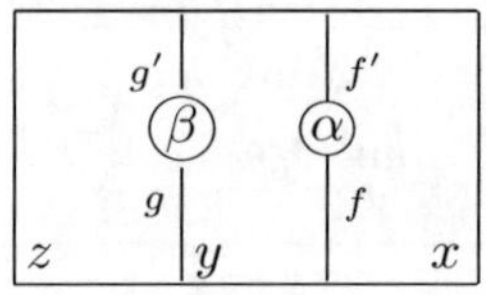

(7) We omit the identity 2-morphism $\mathbb{1}_f \colon f \Rightarrow f$:

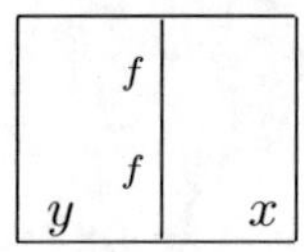

(8) For four 2-morphisms 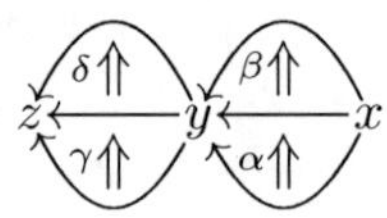 since we have the interchange law $(\delta \bullet \gamma) \circ (\beta \bullet \alpha) = (\delta \circ \beta) \bullet (\gamma \circ \alpha)$, the following displays the common 2-morphism of both sides:

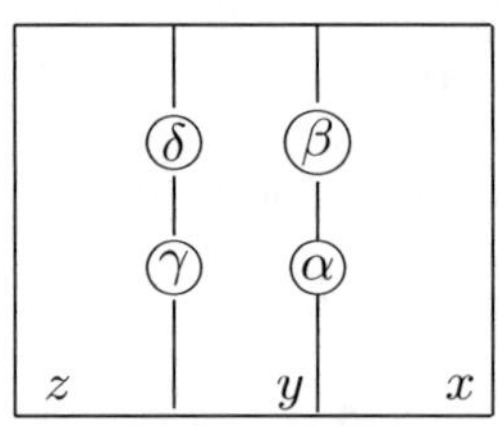

(9) By substituting the identity 2-morphisms for 2-morphisms above suitably we obtain the following *slide equations*:

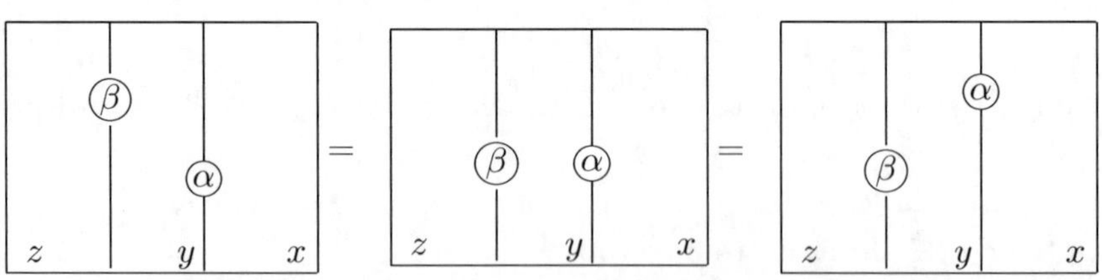

Thanks to computations (8) and (9), 2-morphisms become very simple. We freely make use of string diagrams below.

(10) We usually omit areas expressing objects of $\mathbf{C}$ and draw only lines and vertices.

We apply the drawing rules above also to the 2-category $\mathbf{Cat}(\mathscr{U})$ (Example 4.1.7) consisting of categories, functors and natural transformations.

EXAMPLE 4.2.1. Axioms of 2-natural transformations or modifications are displayed by string diagrams as follows:

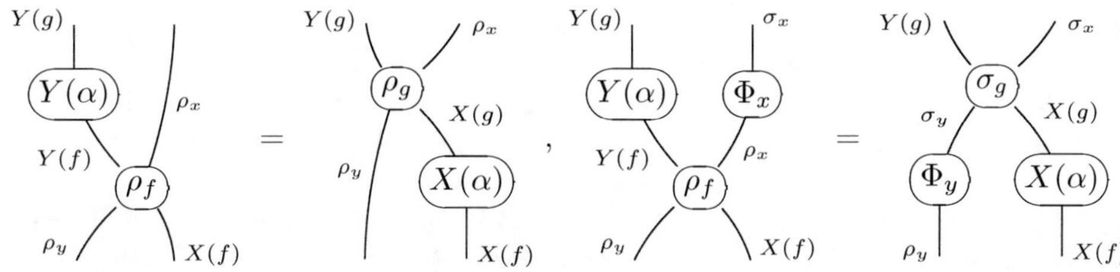

EXAMPLE 4.2.2. Let $\mathscr{C}$ be a category. Then by the following way we can use string diagrams for objects and morphisms of a category (cf. [**35**, §2.1]). Recall that we let $\mathbf{1}$ be a discrete category with a unique object $*$. Then each object x of $\mathscr{C}$ is identified with a functor $\mathbf{1} \to \mathscr{C}$ $(* \mapsto x, \mathbb{1}_* \mapsto \mathbb{1}_x)$. We denote this functor also by x. Thus $x(*) := x, x(\mathbb{1}_*) := \mathbb{1}_x$. Then we can regard each morphism $f\colon x \to y$ of $\mathscr{C}$ as a natural transformation

$$\mathbf{1} \underset{y}{\overset{x}{\Downarrow f}} \mathscr{C}\ , \quad (f_*\colon x(*) \to y(*)) := (f\colon x \to y).$$

Here, consider two functors $F, F'\colon \mathscr{C} \to \mathscr{D}$ between categories and a family $\alpha := (\alpha_x\colon F(x) \to F'(x))_{x \in \mathscr{C}_0}$ of morphisms in $\mathscr{D}$. Then α is a natural transformation if and only if for each morphism $f\colon x \to y$ of $\mathscr{C}$, the diagram

$$\begin{array}{ccc} F(x) & \xrightarrow{\alpha_x} & F'(x) \\ {\scriptstyle F(f)}\downarrow & & \downarrow{\scriptstyle F'(f)} \\ F(y) & \xrightarrow[\alpha_y]{} & F'(y) \end{array}$$

is commutative, i.e., we have $F'(f) \circ \alpha_x = \alpha_y \circ F(f)$. This is expressed by the following slide equation:

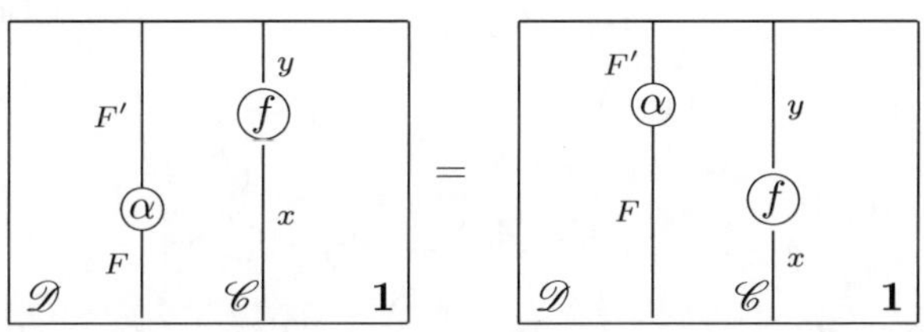

(Note that since $F \circ x = F(x), \alpha \circ x = \alpha_x$, and so on, the equality above is written as $(F' \circ f) \bullet (\alpha \circ x) = (\alpha \circ y) \bullet (F \circ f)$.) Therefore we may display the common morphism of $\mathscr{D}$ by the following diagram:

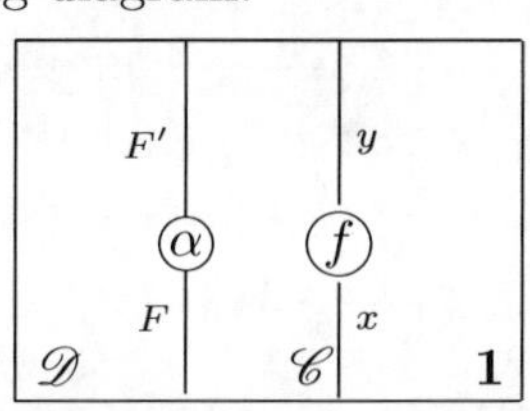

This will be used later in the proof of Theorem 4.4.7 that gives a meaning of adjoint functors.

4.3. Lax functors, colax functors, and pseudofunctors

Definition of 2-functors are sometimes too strong to discuss general theory of 2-categories. We define weaker notions using 2-morphisms in 2-categories below. These are called "lax functors" (functors with relaxed conditions). There is a general definition of lax functors from a 2-category to a 2-category (Definition 8.2.1). However, since at present we do not need this general definition, we instead give a definition of those from a category to a 2-category. First of all we start from colax functors, which are obtained by substituting 2-morphisms for equalities in the definition of functors.

DEFINITION 4.3.1. Let I be a category and $\mathbf{C}$ a 2-category. A sequence of the following data satisfying the axioms below is called a *colax functor* from I to $\mathbf{C}$, and is denoted by $X\colon I \to \mathbf{C}$.

Data:

- A map $X\colon I_0 \to \mathbf{C}_0$,
- A family of maps $({}_jX_i\colon I(i,j) \to \mathbf{C}(X(i), X(j)))_{i,j\in I_0}$, (where ${}_jX_i(a)$ is abbreviated as $X(a)$ for all $a \in I(i,j)$),
- A family $(X_i\colon X(\mathbb{1}_i) \Rightarrow \mathbb{1}_{X(i)})_{i\in I_0}$ of 2-morphisms of $\mathbf{C}$,
- A family $(X_{b,a}\colon X(ba) \Rightarrow X(b)X(a))_{(b,a)\in \mathrm{com}(I_1)}$ of 2-morphisms of $\mathbf{C}$, where $\mathrm{com}(I_1) := \{(b,a) \in I_1 \times I_1 \mid \mathrm{dom}(b) = \mathrm{cod}(a)\}$.

Axioms:

(1) The following are commutative for all morphisms $a\colon i \to j$ of I:

$$\begin{array}{ccc} X(a\mathbb{1}_i) & \xRightarrow{X_{a,\mathbb{1}_i}} & X(a)X(\mathbb{1}_i) \\ & \searrow\!\!\!= & \Big\Downarrow{\scriptstyle X(a)X_i} \\ & & X(a)\mathbb{1}_{X(i)} \end{array} \qquad \begin{array}{ccc} X(\mathbb{1}_j a) & \xRightarrow{X_{\mathbb{1}_j,a}} & X(\mathbb{1}_j)X(a) \\ & \searrow\!\!\!= & \Big\Downarrow{\scriptstyle X_jX(a)} \\ & & \mathbb{1}_{X(j)}X(a) \end{array} .$$

(2) The following are commutative for all paths $i \xrightarrow{a} j \xrightarrow{b} k \xrightarrow{c} l$ in I:

$$\begin{array}{ccc} X(cba) & \xRightarrow{X_{c,ba}} & X(c)X(ba) \\ {\scriptstyle X_{cb,a}}\Big\Downarrow & & \Big\Downarrow{\scriptstyle X(c)X_{b,a}} \\ X(cb)X(a) & \xRightarrow[X_{c,b}X(a)]{} & X(c)X(b)X(a). \end{array}$$

REMARK 4.3.2. In the form of string diagrams the axioms of colax functors $X\colon I \to \mathbf{C}$ is expressed as follows: First, 2-morphisms $X_i, X_{b,a}$ are displayed as follows, respectively:

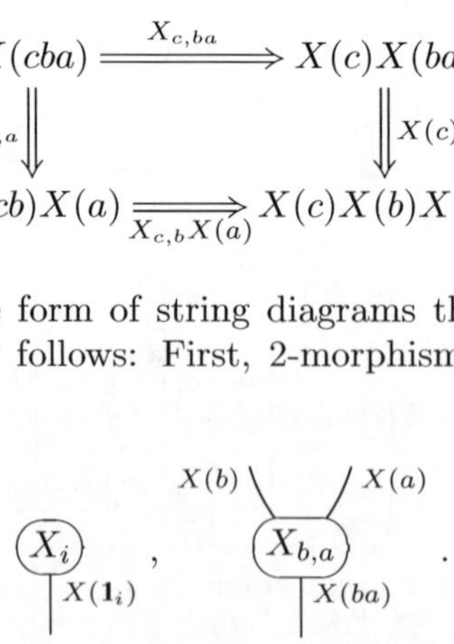

Using these the axioms (a), (b) are displayed as follows by string diagrams, respectively:

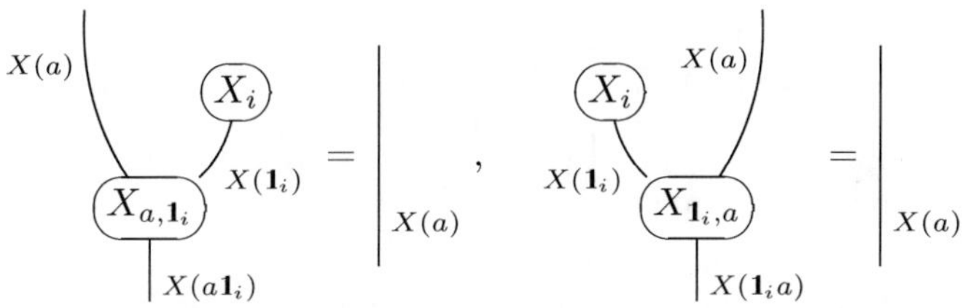

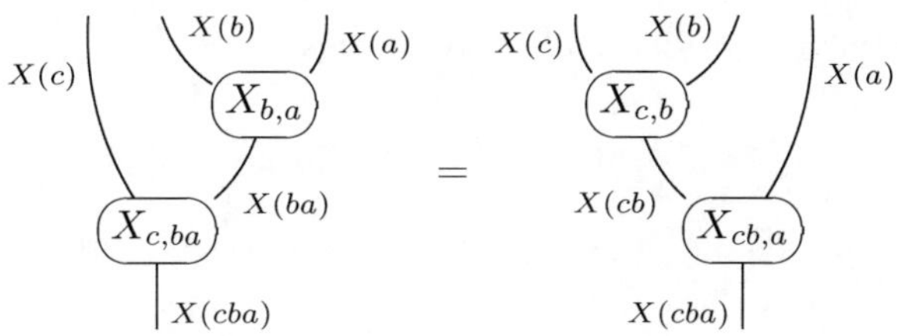

EXAMPLE 4.3.3. A colax functor $X\colon \mathbf{1} \to \mathbf{Cat}$ from $\mathbf{1}$ to the 2-category $\mathbf{Cat}$ of all small categories is nothing but a comonad $T := X(\mathbb{1}_*)$ over a small category $\mathscr{C} := X(*)$.

DEFINITION 4.3.4. Let $\mathbf{C}$ be a 2-category.

- A colax functor $I \to \mathbf{C}^{\mathrm{co}}$ is called a *lax functor* $I \to \mathbf{C}$.
- A colax functor $X\colon I \to \mathbf{C}$ is called a *pseudofunctor* if all $X_i, X_{b,a}$ are 2-isomorphisms.
- A lax functor $X\colon I \to \mathbf{C}$ is called a *lax pseudofunctor* if all $X_i, X_{b,a}$ are 2-isomorphisms. Clearly, this notion is equivalent to that of a pseudofunctor defined above. This terminology is used to distinguish these equivalent notions. To contrast, the latter should be called a *colax pseudofunctor*, but we call it simply a pseudofunctor in this book.
- A colax functor $X\colon I \to \mathbf{C}$ with all $X_i, X_{b,a}$ identity 2-morphisms is nothing but a 2-functor $I \to \mathbf{C}$.

4.4. Adjoints and equivalences

We can consider adjoints for 1-morphisms in 2-categories ("inner" adjoints) and adjoints for 2-functors between 2-categories ("outer" adjoints). We start from inner adjoints.

4.4.a. Adjoints and equivalences in a 2-category.

DEFINITION 4.4.1. Let $\mathbf{C}$ be a 2-category.

(1) A quadruple $(f, g, \eta, \varepsilon)$ of the following data is called a *pre-adjoint system* in $\mathbf{C}$.

Data:

- 1-morphisms of $\mathbf{C}$ $x \underset{g}{\overset{f}{\rightleftarrows}} y$,
- 2-morphisms of $\mathbf{C}$ $\eta\colon \mathbb{1}_x \Rightarrow g \circ f$ and $\varepsilon\colon f \circ g \Rightarrow \mathbb{1}_y$.

Namely, by collecting them in a diagram in $\mathbf{C}$, we have

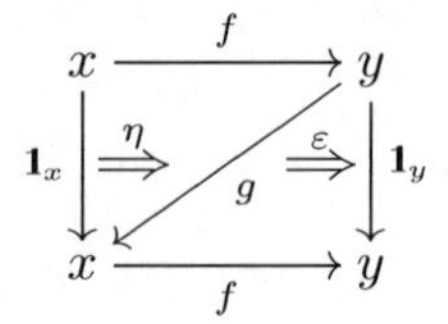

(2) A pre-adjoint system $(f, g, \eta, \varepsilon)$ is called an *equivalence system* if η, ε are 2-isomorphisms.

(3) A pre-adjoint system is called an *adjoint system* if it satisfies the following axioms:

Axioms: The following diagrams are commutative:

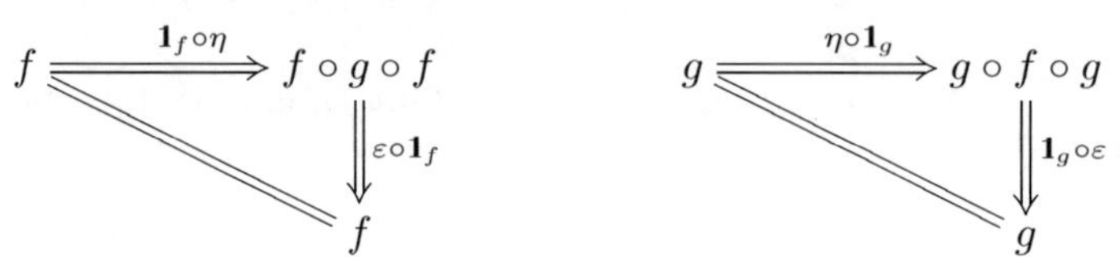

When this is the case, f is called a *left adjoint* to g, g is called a *right adjoint* to f, and (f, g) is called an adjoint pair, and we denote this by $f \dashv g$. The 2-morphism η (resp. ε) is called a *unit* (resp. a *counit*) of this adjoint system.

(4) An adjoint system $(f, g, \eta, \varepsilon)$ is called an *adjoint equivalence system* if both ε and η are 2-isomorphisms, also (f, g) is called an *adjoint equivalence pair*.

REMARK 4.4.2. The axioms of adjoint system $(f, g, \eta, \varepsilon)$ are displayed by string diagrams as follows:

(4.1)

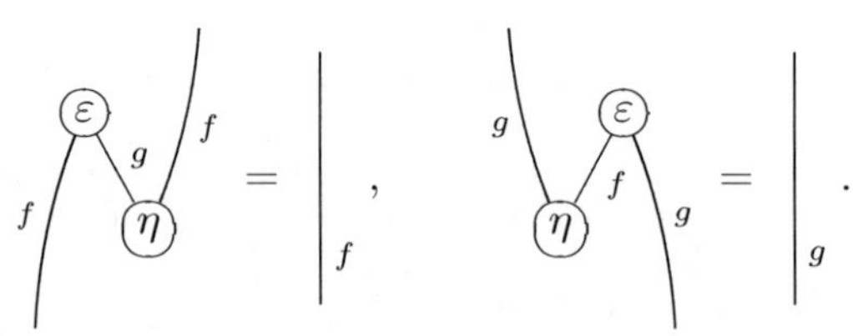

By the forms above these equations are also called *zigzag equations*. This is interpreted as follows: We may fold straight strings in the form of the letter 'N' or 'И', and conversely expand the folded strings to straight ones.

LEMMA 4.4.3. *If an equivalence system $(f, g, \eta, \varepsilon)$ in a 2-category $\mathbf{C}$ satisfies one of the zigzag equations, the so does the other.*

PROOF. Call the left (resp. right) one of the zigzag equation (4.1)(L) (resp. (4.1)(R)). Assume that (4.1)(R) holds. We show that (4.1)(L) holds. Take the inverses of both sides of (4.1)(R) to obtain the following:

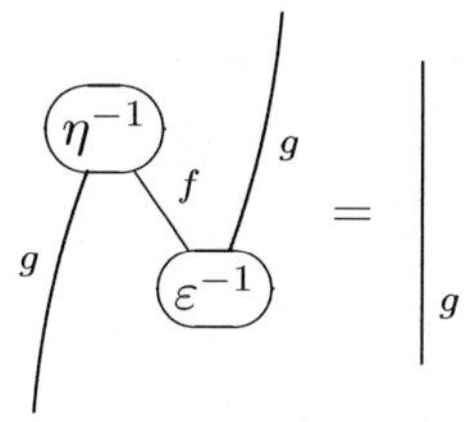

Substitute this left-hand side for g in the left-hand side of (4.1)(L), and continue to transform it. Then finally it is shown to be equal to $\mathbb{1}_f$ as follows:

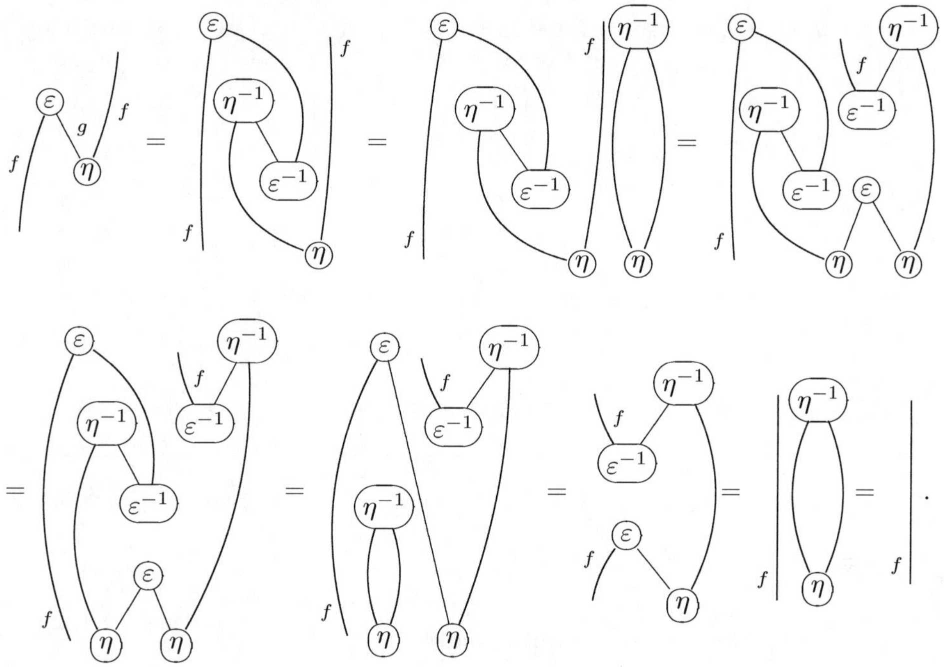

Similarly it is shown that (4.1)(L) implies (4.1)(R). □

THEOREM 4.4.4. *In a 2-category* $\mathbf{C}$ *let* $f\colon x \to y$ *be an equivalence,* $g\colon y \to x$ *a 1-morphism,* $\eta\colon \mathbb{1}_x \Rightarrow gf$ *a 2-isomorphism. Then there exists a unique 2-isomorphism* $\varepsilon\colon fg \Rightarrow \mathbb{1}_y$ *such that* $(f, g, \eta, \varepsilon)$ *is an adjoint equivalence system.*

PROOF. Since $f\colon x \to y$ is an equivalence, it has a quasi-inverse $g'\colon y \to x$. Thus there exist 2-isomorphisms $\alpha\colon fg' \Rightarrow \mathbb{1}_y$ and $\beta\colon \mathbb{1}_x \Rightarrow g'f$. Set $\varepsilon := \alpha \bullet (f\eta^{-1}g') \bullet (fg\alpha^{-1})\colon fg \Rightarrow fgfg' \Rightarrow fg' \Rightarrow \mathbb{1}_y$. Then ε is also a 2-isomorphism, which is displayed by the following string diagram:

$\varepsilon :=$

.

We show that this is the desired one. By Lemma 4.4.3 it is enough to show the right equation of (4.1). This is proved by the following string diagrams:

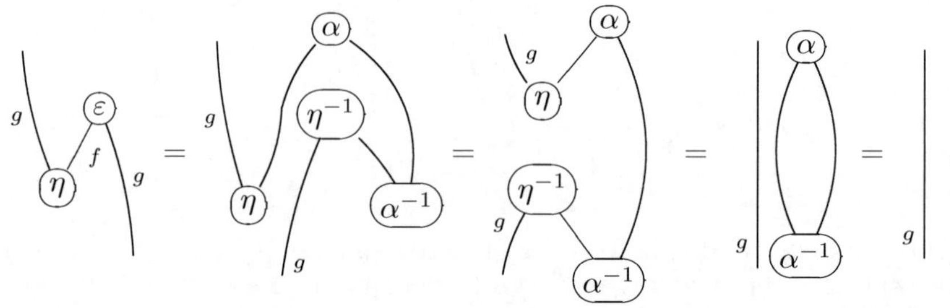

To show the uniqueness of ε let $\varepsilon' \colon fg \Rightarrow \mathbb{1}_y$ be another 2-isomorphism satisfying zigzag equations with η. Then the following string diagrams prove the fact that $\varepsilon = \varepsilon'$:

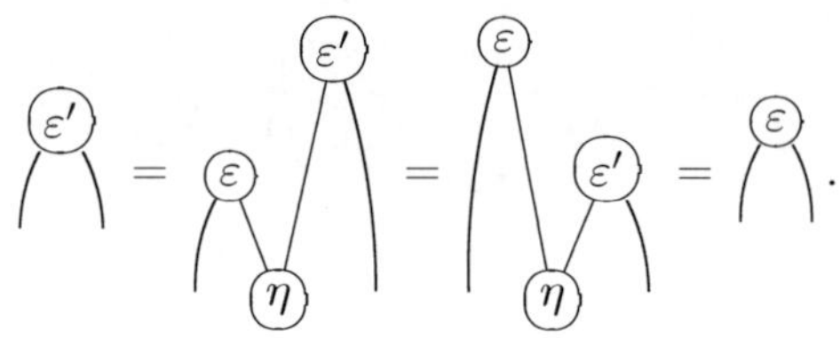

□

We immediately have the following from the theorem above.

COROLLARY 4.4.5. *Let* $x \underset{g}{\overset{f}{\rightleftarrows}} y$ *be* 1*-morphisms between objects in a* 2*-category* $\mathbf{C}$*. Then the following are equivalent:*

(1) *f and g are quasi-inverses to each other;*
(2) *(f, g) is an adjoint equivalence pair.*

REMARK 4.4.6. In the case that a 2-category $\mathbf{C}$ consists of categories, functors and natural transformations, namely when $\mathbf{C} = \mathbf{CAT}$ (cf. Example 4.1.7), the following are equivalent for a pair of 1-morphisms $\mathscr{C} \underset{G}{\overset{F}{\rightleftarrows}} \mathscr{D}$ in $\mathbf{C}$ by Theorem 1.4.4:

(1) F is an equivalence of categories;
(2) (F, G) is an adjoint equivalence pair for some functor $G \colon \mathscr{D} \to \mathscr{C}$; and
(3) F is full, faithful, and dense.

In MacLane's book [**34**, IV. Adjoints, 4. Equivalence of categories] the proof from (1) to (2) was through (3). Here we proved the equivalence of (1) and (2) using a general theory of 2-categories.

When we first give a proof of the equivalence between (1) and (2) by the corollary above, we may prove the implication from (2) to (3) instead of from (1) to (3) although the latter was proved in the proof of Theorem 1.4.4. In that case, the proof of the equality $F_{y,x} \circ F' = \mathbb{1}_{\mathscr{D}(F(x),F(y))}$ can be given without using the faithfulness of G as follows:

For any $g \in \mathscr{D}(F(x), F(y))$, consider the diagram

$$\begin{array}{ccccc} F(x) & \xrightarrow{F(\eta_x)} & F(G(F(x))) & \xrightarrow{\varepsilon_{F(x)}} & F(x) \\ {\scriptstyle F(F'(g))}\downarrow & & \downarrow{\scriptstyle F(G(g))} & & \downarrow{\scriptstyle g} \\ F(y) & \xrightarrow[F(\eta_y)]{} & F(G(F(y))) & \xrightarrow[\varepsilon_{F(y)}]{} & F(y). \end{array}$$

By the definition of $F'(g)$ and by the naturality of ε, this is commutative, and by the zigzag equations, both of the composites of horizontal arrows are identities, which shows that $F(F'(g)) = g$.

Next we give the following equivalence for the case that the 2-category consists of categories, functors, and natural transformations, which gives the origin of the name of adjoint system.

THEOREM 4.4.7. *For functors* $\mathscr{C} \underset{R}{\overset{L}{\rightleftarrows}} \mathscr{D}$ *between categories in the 2-category* **CAT** *the following are equivalent:*

(1) *There exists an adjoint system* $(L, R, \eta, \varepsilon)$*; and*
(2) *Two functors* $\mathscr{D}(L(\text{-}), ?)$ *and* $\mathscr{C}(\text{-}, R(?)) \colon \mathscr{C}^{\mathrm{op}} \times \mathscr{D} \to$ **Set** *of two variables are naturally isomorphic.*

PROOF. (1) $\Rightarrow$ (2). Assume (1) holds. Then we define a natural isomorphism $\omega \colon \mathscr{D}(L(\text{-}), ?) \Rightarrow \mathscr{C}(\text{-}, R(?))$ as follows:

For each $(x, y) \in (\mathscr{C}^{\mathrm{op}} \times \mathscr{D})_0$ we set $\omega_{x,y} \colon \mathscr{D}(L(x), y) \to \mathscr{C}(x, R(y))$ and $\omega'_{x,y} \colon \mathscr{C}(x, R(y)) \to \mathscr{D}(L(x), y)$ by

$$\begin{cases} \omega_{x,y}(f) := R(f) \circ \eta_x \colon x \to RL(x) \to R(y) & (f \in \mathscr{D}(L(x), y)), \\ \omega'_{x,y}(g) := \varepsilon_y \circ L(g) \colon L(x) \to LR(y) \to y & (g \in \mathscr{C}(x, R(y))). \end{cases} \tag{4.2}$$

and we define $\omega := (\omega_{x,y}), \omega' := (\omega'_{x,y})$. Using string diagrams as explained in Example 4.2.2, these are displayed as follows:

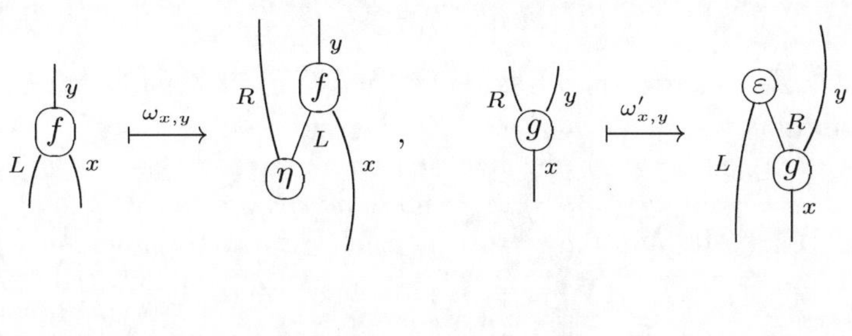

Therefore by zigzag equations we have the following equalities:

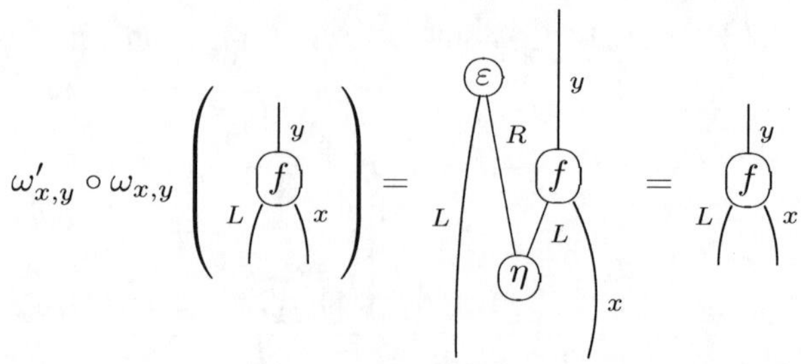

Similarly it is shown that $\omega_{x,y} \circ \omega'_{x,y}(g) = g$. Thus $\omega_{x,y}$ and $\omega'_{x,y}$ are inverse maps to each other, and hence, $\omega_{x,y}$ is a bijection. The naturality of ω is equivalent to saying that for each morphism $a \colon x' \to x$ of $\mathscr{C}$ and each path $L(x) \xrightarrow{f} y \xrightarrow{b} y'$ in $\mathscr{D}$ we have

$$\omega(b \circ f \circ L(a)) = R(b) \circ \omega(f) \circ a,$$

which follows from the fact that η is a natural transformation as follows:

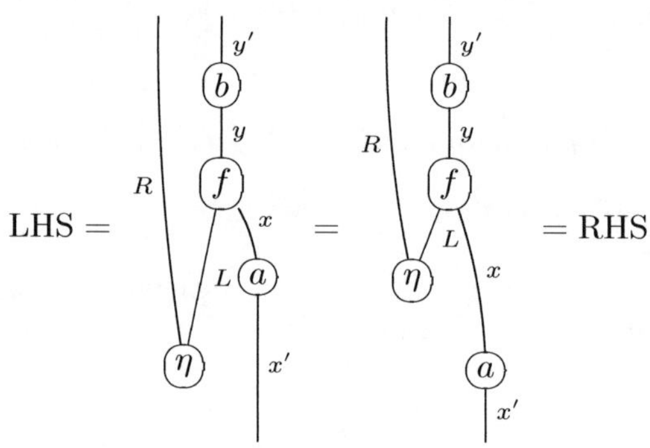

(2) $\Rightarrow$ (1). Assume that there exists a natural isomorphism $\omega \colon \mathscr{D}(L(\text{-}), ?) \Rightarrow \mathscr{C}(\text{-}, R(?))$. Then we define natural transformations $\eta \colon \mathbb{1}_{\mathscr{C}} \Rightarrow RL$ and $\varepsilon \colon LR \Rightarrow \mathbb{1}_{\mathscr{D}}$ by

$$\eta_x := \omega_{x,L(x)}(\mathbb{1}_{L(x)}) \in \mathscr{C}(x, RL(x)) \quad (x \in \mathscr{C}_0^{\mathrm{op}}),$$
$$\varepsilon_y := \omega^{-1}_{R(y),y}(\mathbb{1}_{R(y)}) \in \mathscr{D}(LR(y), y) \quad (y \in \mathscr{D}_0)$$

and we set $\eta := (\eta_x)_{x \in \mathscr{C}_0^{\mathrm{op}}}, \varepsilon := (\varepsilon_y)_{y \in \mathscr{D}_0}$. Then it is easily verified that (i) these are natural transformations; and (ii) these satisfy the zigzag equations. □

EXERCISE 4.4.8. Verify the last two claims (i) and (ii) in the proof above.

Here are some commonly used facts about adjoints in the 2-category **CAT**.

COROLLARY 4.4.9. *In the 2-category* **CAT** *let* $\mathscr{C} \underset{R}{\overset{L}{\rightleftarrows}} \mathscr{D}$ *be functors of categories, and* $(L, R, \eta, \varepsilon)$ *an adjoint system. Then the following hold:*

(1) *Relations between* L *and* η*:*
   (a) L *is full* $\iff$ η_x *is a retraction for all* $x \in \mathscr{C}_0$.
   (b) L *is faithful* $\iff$ η_x *is a monomorphism for all* $x \in \mathscr{C}_0$.
   (c) L *is fully faithful* $\iff$ η *is a natural isomorphism.*

(2) *Relations between* R *and* ε:
   (a) R *is full* $\iff \varepsilon_y$ *is a section for all* $y \in \mathscr{D}_0$.
   (b) R *is faithful* $\iff \varepsilon_y$ *is an epimorphism for all* $y \in \mathscr{D}_0$.
   (c) R *is fully faithful* $\iff \varepsilon$ *is a natural isomorphism.*

(3) *On denseness:*
   (a) *If* η *is a natural isomorphism, then* R *is dense.*
   (b) *If* ε *is a natural isomorphism, then* L *is dense.*

PROOF. Statement (3) is obvious. Since (2) is shown in the same way as (1), we only show statement (1). Let $x', x \in \mathscr{C}_0$. Then the defining formula (4.2) in the proof of Theorem 4.4.7 yields the following diagram:

$$\begin{array}{ccc} & \mathscr{C}(x',x) & \\ {}^{L(\text{-})}\swarrow & & \searrow^{\mathscr{C}(x',\eta_x)} \\ \mathscr{D}(Lx',Lx) & \xrightarrow[\cong]{\omega_{x',Lx}} & \mathscr{C}(x',RLx) \end{array}$$

The commutativity of this diagram follows from the fact that each $f \in \mathscr{C}(x',x)$ is mapped as follows:

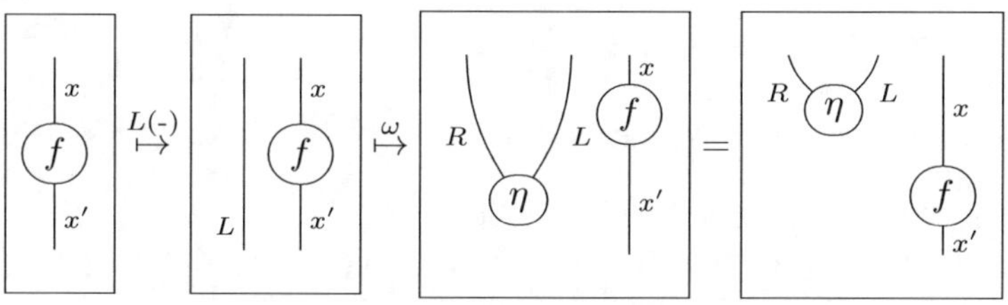

The right end of this is $\mathscr{C}(x',\eta_x)(f)$. Then statement (a) follows from the following equivalences:

L is full $\iff \forall x', x \in \mathscr{C}_0, L(\text{-})$ is an epimorphism (by definition)
$\iff \forall x', x \in \mathscr{C}_0, \mathscr{C}(x',\eta_x)$ is an epimorphism (since ω is an isomorphism)
$\iff \forall x \in \mathscr{C}_0, \eta_x$ is a retraction (by Proposition 1.1.22(2))

Statement (b) is shown by the following equivalences:

L is faithful $\iff \forall x', x \subset \mathscr{C}_0, L(\text{-})$ is a monomorphism (by definition)
$\iff \forall x', x \in \mathscr{C}_0, \mathscr{C}(x',\eta_x)$ is a monomorphism (since ω is an isom.)
$\iff \forall x \in \mathscr{C}_0, \eta_x$ is a monomorphism (by Remark 1.1.20(1))

Statement (c) follows from (a) and (b) because as easily seen a morphism is an isomorphism if it is both a monomorphism and a retraction. □

Again, we return to general adjoints, and state their properties. First of all, we show that left (resp. right) adjoints exist at most one up to isomorphisms, and that a 1-morphism that is isomorphic to a left (resp. right) adjoint is also a left (resp. right) adjoint.

PROPOSITION 4.4.10. *Let* $f, f' \colon x \to y$, $g, g' \colon y \to x$ *be* 1*-morphisms between objects in a* 2*-category* $\mathbf{C}$ *and assume that* $f \dashv g$. *Then the following hold:*

(1) $f \dashv g' \iff g \cong g'$.
(2) $f' \dashv g \iff f \cong f'$.

PROOF. Let $(f, g, \eta, \varepsilon)$ be an adjoint system. We only show statement (1) because (2) is shown similarly.

($\Rightarrow$). Let $(f, g', \eta', \varepsilon')$ be another adjoint system. Then we define $\alpha\colon g \Rightarrow g'$, $\beta\colon g' \Rightarrow g$ by the following string diagrams:

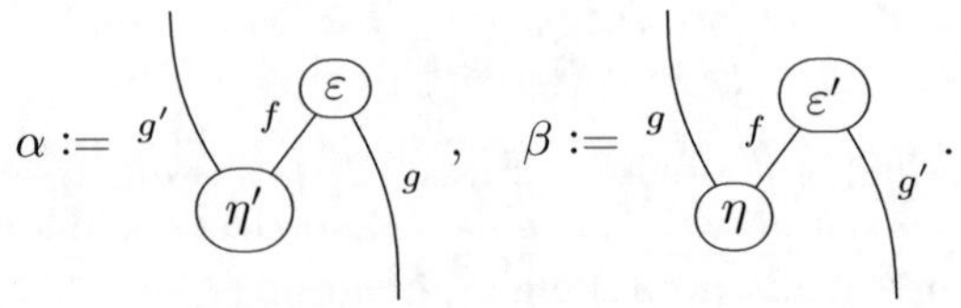

Then we see that $\alpha \bullet \beta = \mathbb{1}_{g'}$ holds by the following computation:

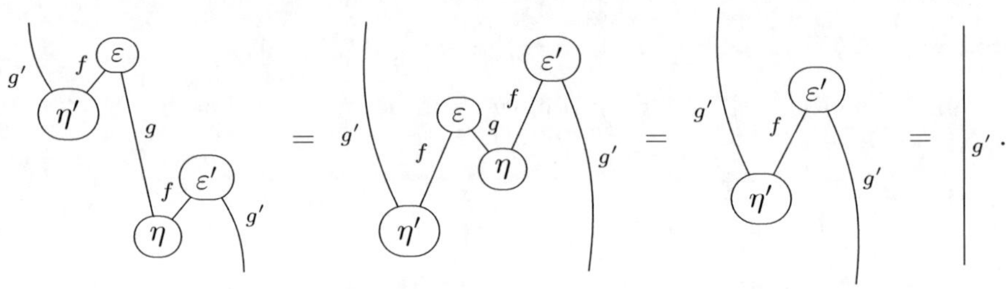

Similarly it is shown that $\beta \bullet \alpha = \mathbb{1}_g$. Hence $g \cong g'$.

($\Leftarrow$). Let $\alpha\colon g \Rightarrow g'$ be a natural isomorphism. We define $\eta'\colon \mathbb{1}_x \Rightarrow g' \circ f$, $\varepsilon'\colon f \circ g' \Rightarrow \mathbb{1}_y$ by the following string diagrams:

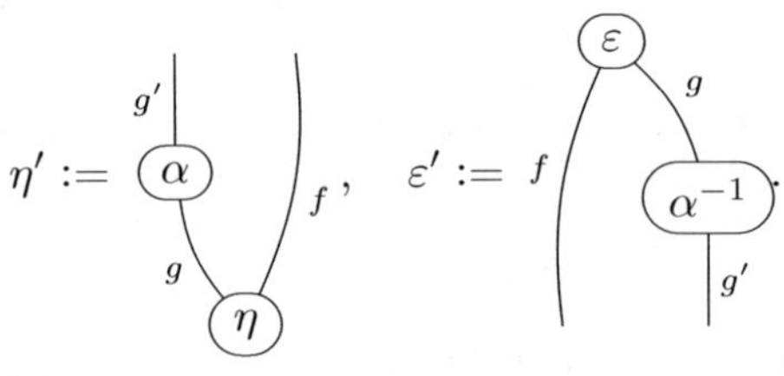

Then it is enough to show that η', ε' satisfy the zigzag equations. One is shown by the following:

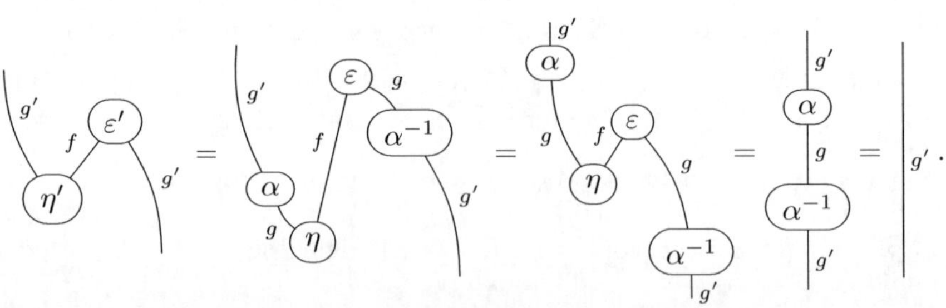

The remaining equation is shown similarly. Hence $(f, g', \eta', \varepsilon')$ turns out to be an adjoint system and we have $f \dashv g'$. $\square$

The second one is as follows.

PROPOSITION 4.4.11. *Let* $x \underset{g}{\overset{f}{\rightleftarrows}} y \underset{g'}{\overset{f'}{\rightleftarrows}} z$ *(with* $\perp$ *in each) be two adjoints in a 2-category* $\mathbf{C}$.

Then we have an adjoint $x \underset{g \circ g'}{\overset{f' \circ f}{\rightleftarrows}} z$ *(with* $\perp$*).*

PROOF. Let $(f, g, \eta, \varepsilon), (f', g', \eta', \varepsilon')$ be adjoint systems. We define $\eta'' \colon \mathbb{1}_x \Rightarrow (g \circ g') \circ (f' \circ f)$ and $\varepsilon'' \colon (f' \circ f) \circ (g \circ g') \Rightarrow \mathbb{1}_z$ by the following string diagrams:

$\eta'' :=$ 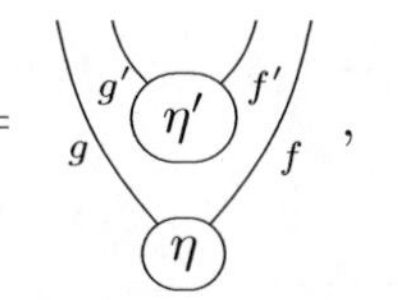 , $\varepsilon'' :=$ 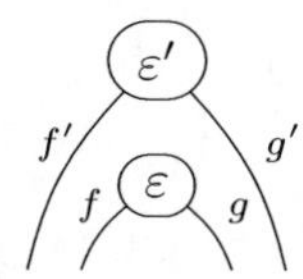

Then by the zigzag equations for η, ε and for η', ε' and by slide equations the zigzag equations for η'', ε'' hold as follows:

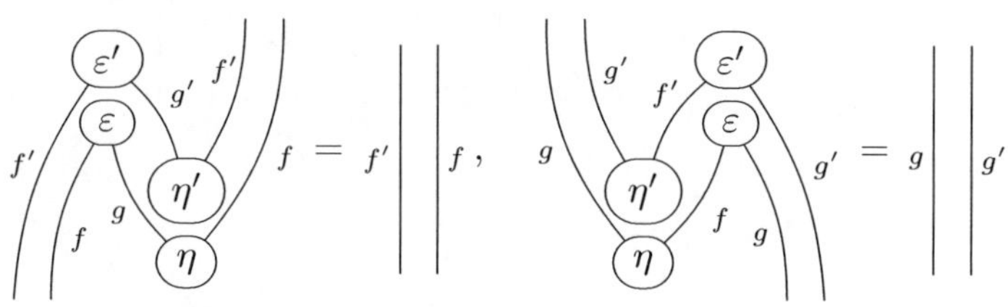

Therefore we have $f' \circ f \dashv g \circ g'$. □

The following is the last one, which gives an alternative proof of continuity (resp. cocontinuity) of right (resp. left) adjoints (see the next subsection).

PROPOSITION 4.4.12. *Consider the following four adjoint pairs in a 2-category* $\mathbf{C}$*:*

(4.3) 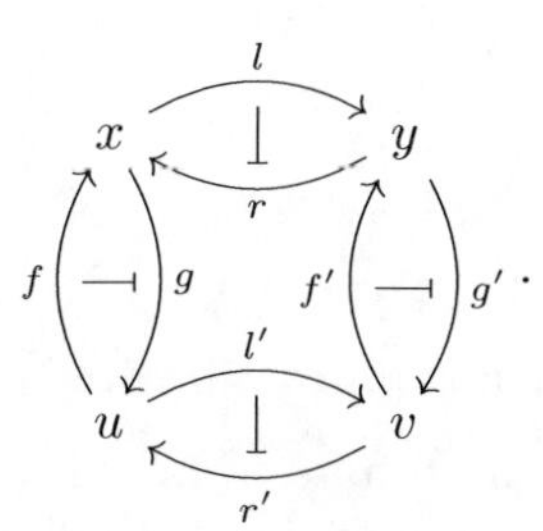

Then we have $l \circ f \cong f' \circ l' \iff r' \circ g' \cong g \circ r$.

PROOF. First, from the four pairs (4.3) of adjoint we obtain the following two adjoint pairs by using Proposition 4.4.11:

$$l \circ f \dashv g \circ r, \quad f' \circ l' \dashv r' \circ g'.$$

Therefore by Proposition 4.4.10 if $l \circ f \cong f' \circ l'$, then we have $l \circ f \dashv r' \circ g'$. Again by Proposition 4.4.10 it follows from this together with $l \circ f \dashv g \circ r$ that $r' \circ g' \cong g \circ r$. The converse is shown similarly. □

REMARK 4.4.13 (An alternative proof of Proposition 4.4.12). In this setting, the proof can also be done by constructing the isomorphism directly using string diagrams.

Let $(f, g, \alpha, \beta), (f', g', \alpha', \beta'), (l, r, \eta, \varepsilon), (l', r', \eta', \varepsilon')$ be an adjoint system. Since the necessity and the sufficiency can be shown in the same way, we only show the necessity here. Now assume the left-hand side, that is, suppose that there exists a natural isomorphism $\gamma\colon l \circ f \Rightarrow f' \circ l'$. Here, define natural transformations $\delta\colon r' \circ g' \Rightarrow g \circ r$ and $\delta'\colon g \circ r \Rightarrow r' \circ g'$ by the following string diagrams, respectively:

$\delta :=$ 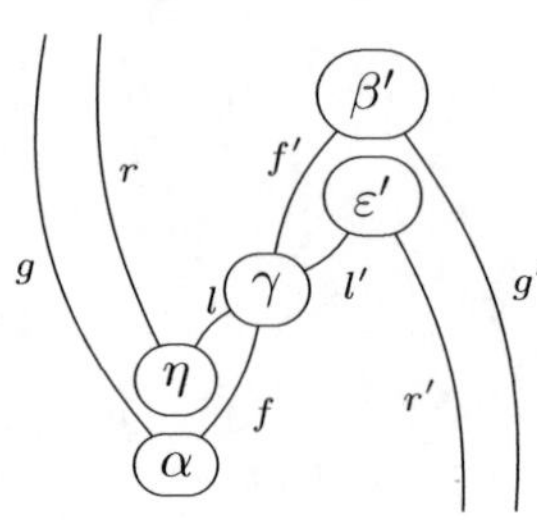

, $\delta' :=$ 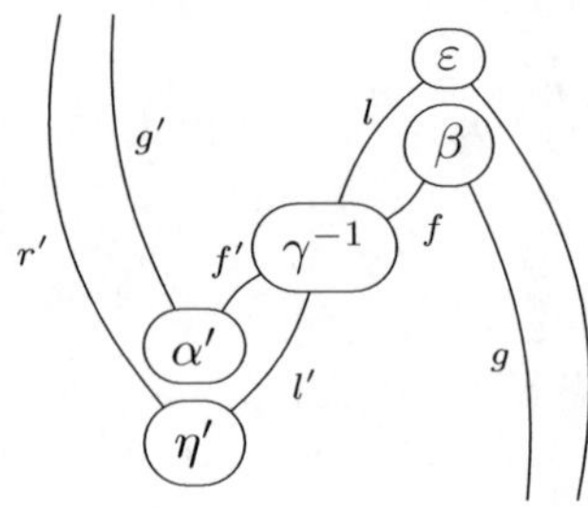

Then it is enough to show that $\delta \bullet \delta' = \mathbb{1}_{g \circ r}$ and $\delta' \bullet \delta = \mathbb{1}_{r' \circ g'}$. The left-hand side of the first equation is given by the following string diagram, by using the slide equation:

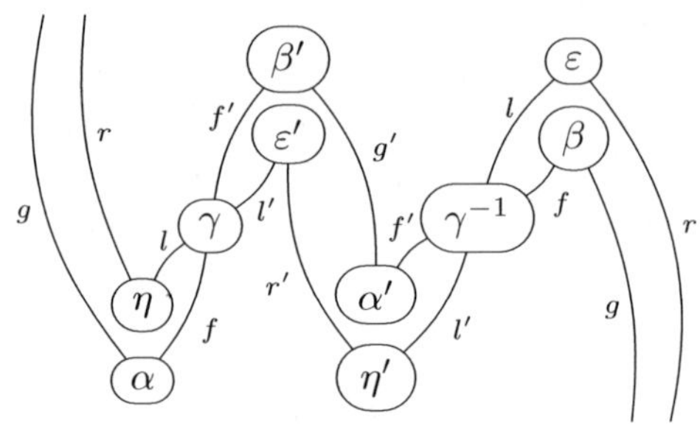

Therefore by the slide equation, the zigzag equations of four adjoints, and the equality $\gamma^{-1} \bullet \gamma = \mathbb{1}_{l \circ f}$, this string diagram is equal to $\mathbb{1}_{g \circ r}$, the right-hand side of the first equation. The second equation also follows by the similar computation. As a consequence, δ turns out to be a natural isomorphism, and hence we obtain $g \circ r \cong r' \circ g'$. □

4.5. Adjoints and limits, colimits

In this section, we apply the results of the previous section to the adjoints in **CAT**. This needs knowledge of a limit $\varprojlim F$, a colimit $\varinjlim F$, and the functors defined by them such as the limit functor $\varprojlim\colon \mathscr{C}^I \to \mathscr{C}$, the colimit functor $\varinjlim\colon \mathscr{C}^I \to \mathscr{C}$, and the diagonal functor $\Delta\colon \mathscr{C} \to \mathscr{C}^I$ for a functor $F \in \mathrm{Fun}(I, \mathscr{C})_0 = (\mathscr{C}^I)_0$, where I is a small category and $\mathscr{C} \in \mathbf{CAT}_0$. The explanation of these is left to introductory books on category theory (e.g., [**38**, 2.3.2 Limits and colimits]), but we give the definitions of these and a brief description of the adjoints $\varinjlim \dashv \Delta$, $\Delta \dashv \varprojlim$ below[2].

[2]The essential points are covered by them. The reader is encouraged to mimic the explanation of products and coproducts, and to draw diagrams of their universality.

DEFINITION 4.5.1. Let I be a small category, and $\mathscr{C} \in \mathbf{CAT}_0$. The *diagonal functor* $\Delta\colon \mathscr{C} \to \mathscr{C}^I$ is defined as follows:

- For each $x \in \mathscr{C}_0$ we define a functor $\Delta(x)\colon I \to \mathscr{C}$ by $(i \xrightarrow{a} j) \mapsto (x \xrightarrow{\mathbf{1}_x} x)$. Namely, $\Delta(x)(i) := x, \Delta(x)(a) := \mathbb{1}_x$ $(i \in I_0, a \in I_1)$.
- For each morphism $f\colon x \to y$ in $\mathscr{C}$ we define a natural transformation $\Delta(f)\colon \Delta(x) \Rightarrow \Delta(y)$ by $\Delta(f)_i := f\colon \Delta(x)(i)(= x) \to \Delta(y)(i)(= y)$ $(i \in I_0)$.

DEFINITION 4.5.2. Let $I, \mathscr{C}$ be as above, and $F\colon I \to \mathscr{C}$ a functor. An element (x, π) of $\bigsqcup_{x \in \mathscr{C}_0} \mathscr{C}^I(\Delta(x), F)$ is called a *cone* of F, and if it has the following universality, then it is called a *limit* of F:

Universality of a cone. For each $z \in \mathscr{C}_0$ the following is a bijection:

$$\mathscr{C}^I(z, \pi) \circ \Delta\colon \mathscr{C}(z, x) \to \mathscr{C}^I(\Delta(z), F), f \mapsto \pi \circ \Delta(f). \tag{4.4}$$

This pair (x, π) is uniquely determined up to isomorphisms[3]. Therefore we denote it by $(\varprojlim F, \pi_F)$. This is naturally extended to a functor $\varprojlim\colon \mathscr{C}^I \to \mathscr{C}$.

Dually a *cocone*, a *colimit* $(\varinjlim F, \sigma_F)$ of F, and a functor $\varinjlim\colon \mathscr{C}^I \to \mathscr{C}$ are defined.

EXERCISE 4.5.3. In the setting above, show that the bijection (4.4) induces an isomorphism

$$\mathscr{C}(z, \varprojlim F) \to \mathscr{C}^I(\Delta(z), F)$$

that is natural in $z \in \mathscr{C}_0$ and in $F \in (\mathscr{C}^I)_0$. From this we obtain an adjoint $\Delta \dashv \varprojlim$. Dually we have an adjoint $\varinjlim \dashv \Delta$.

REMARK 4.5.4. When I is a small set, we can regard it as a discrete category, and a functor $F \in (\mathscr{C}^I)_0$ as a sequence $(F(x))_{x\in I}$. Then the following holds:

$$\varprojlim F = \prod_{x \in I} F(x), \quad \varinjlim F = \coprod_{x \in I} F(x).$$

DEFINITION 4.5.5 (Continuity and cocontinuity of a functor). Let I be a small category, $R\colon \mathscr{C} \to \mathscr{D}$ a functor of categories. Then R induces the following functor between functor categories: $R^I := \mathrm{Fun}(I, R)\colon \mathscr{C}^I \to \mathscr{D}^I$, $(\alpha\colon F \Rightarrow F') \mapsto (R \circ \alpha\colon R \circ F \Rightarrow R \circ F')$.

(1) Assume that there exist both limits $\varprojlim F$ and $\varprojlim R^I(F)$ for all $F \in (\mathscr{C}^I)_0$. Then for each $F \in (\mathscr{C}^I)_0$ we have a unique morphism

$$\varprojlim R^I(\pi_F) := \varprojlim R(\pi_{F,i})_{i \in I_0}\colon R(\varprojlim F) \to \varprojlim R^I(F)$$

between cones by the universality of limits, where $(\varprojlim F, \pi_F)$ is a universal cone of F. This defines a *canonical natural transformation* $\varprojlim R^I(\pi) := (\varprojlim R^I(\pi_F))_{F \in (\mathscr{C}^I)_0}\colon R \circ \varprojlim \Rightarrow \varprojlim \circ R^I$. If this is a natural isomorphism, then R is said to be *continuous* with respect to I (or to *preserve limits*). (When I is the discrete category of a small set, R is said to *preserve products*.) In particular, if R is continuous for all small (resp. k-moderate) categories I, then R is said to be *small-continuous* (resp. *k-moderate-continuous*) $(k \in \mathbb{N})$.

---

[3]cones (x, π) and (x', π') are called *isomorphic* if there exists an isomorphism $f\colon x' \to x$ in $\mathscr{C}$ such that $\pi' = \pi \circ \Delta(f)$.

(2) Assume that there exist both colimits $\varinjlim F$ and $\varinjlim R^I(F)$ for all $F \in (\mathscr{C}^I)_0$. Then for each $F \in (\mathscr{C}^I)_0$ we have a unique morphism

$$\varinjlim R^I(\sigma_F) := \varinjlim R(\sigma_{F,i})_{i\in I_0} \colon \varinjlim R^I(F) \to R(\varinjlim F)$$

between cocones by the universality of colimits, where $(\varinjlim F, \sigma_F)$ is a universal cocone of F. This defines a *canonical natural transformation* $\varinjlim R^I(\sigma) := (\varinjlim R^I(\sigma_F))_{F\in(\mathscr{C}^I)_0} \colon \varinjlim \circ R^I \Rightarrow R \circ \varinjlim$. If this is a natural isomorphism, then R is said to be *cocontinuous* with respect to I (or to *preserve colimits*). (When I is the discrete category of a small set, R is said to *preserve coproducts*.) In particular, if R is cocontinuous for all small (resp. k-moderate) categories I, then R is said to be *small-cocontinuous* (resp. *k-moderate-cocontinuous*) ($k \in \mathbb{N}$).

LEMMA 4.5.6. *Let I be a small category, and $R\colon \mathscr{C} \to \mathscr{D}$ a functor of categories. Then we have $R^I \circ \Delta = \Delta \circ R$.*

PROOF. For each $x \in \mathscr{C}_0, f \in \mathscr{C}_1, i \in I_0, a \in I_1$ we have

$$\begin{aligned}(R^I \circ \Delta)(x)(i) &= R(\Delta(x)(i)) = R(x) = \Delta(R(x))(i),\\ (R^I \circ \Delta)(x)(a) &= R(\Delta(x)(a)) = R(\mathbb{1}_x) = \mathbb{1}_{R(x)} = \Delta(R(x))(a),\\ (R^I \circ \Delta)(f)_i &= R(\Delta(f)_i) = R(f) = \Delta(R(f))_i.\end{aligned}$$

□

LEMMA 4.5.7. *Let I be a small category and $R\colon \mathscr{C} \to \mathscr{D}$ a functor of categories.*

(1) *Assume that both limits $\varprojlim F, \varprojlim R^I(F)$ exist for all $F \in (\mathscr{C}^I)_0$. Then for R to be continuous with respect to I if and only if $R \circ \varprojlim \cong \varprojlim \circ R^I$ in the category* $\mathrm{Fun}(\mathscr{C}^I, \mathscr{D})$.
(2) *Assume that both colimits $\varinjlim F, \varinjlim R^I(F)$ exist for all $F \in (\mathscr{C}^I)_0$. Then for R to be continuous with respect to I if and only if $\varinjlim \circ R^I \cong R \circ \varinjlim$ in the category* $\mathrm{Fun}(\mathscr{C}^I, \mathscr{D})$.

PROOF. We only show statement (1) because (2) is shown similarly. Necessity is trivial by the definition of continuity. To show the sufficiency assume that there exists a natural isomorphism $\delta\colon R \circ \varprojlim \Rightarrow \varprojlim \circ R^I$ in the category $\mathrm{Fun}(\mathscr{C}^I, \mathscr{D})$. Then it is enough to show that for each $F \in (\mathscr{C}^I)_0$, $\varprojlim R^I(\pi_F)$ is an isomorphism. If δ_F were a morphism from the cone $(R(\varprojlim F), R(\pi_F))$ to the cone $(\varprojlim R^I(F), \pi_{R^I(F)})$ for all $F \in (\mathscr{C}^I)_0$, then we would have $\delta_F = \varprojlim R^I(\pi_F)$ by the universality of limits, but note that this is not the case in general [4]. Now take an $F \in (\mathscr{C}^I)_0$. Then $\pi_F = (\pi_{F,i}\colon \varprojlim F \to F(i))_{i\in I_0}$ is regarded as a morphism $\pi_F = (\pi_{F,i})_{i\in I_0}\colon \Delta(\varprojlim F) \to F$ in $\mathscr{C}^I$. Here by the naturality of δ the following

[4] For example, if $R \circ \varprojlim$ has an automorphism γ that is not equal to $\mathbb{1}_{R\circ\varprojlim}$, then $\beta := \varprojlim R^I(\pi) \bullet \gamma$ is a natural isomorphism $R \circ \varprojlim \Rightarrow \varprojlim \circ R^I$, but β_F is not a morphism from the cone $(R(\varprojlim F), R(\pi_F))$ to the cone $(\varprojlim R^I(F), \pi_{R^I(F)})$ for some F.

diagram is commutative:

$$\begin{array}{ccc}
(R\circ \varprojlim)(\Delta(\varprojlim F)) & \xrightarrow[\cong]{\delta_{\Delta(\varprojlim F)}} & (\varprojlim \circ R^I)(\Delta(\varprojlim F)) \\
{\scriptstyle (R\circ\varprojlim)(\pi_F)}\downarrow & & \downarrow{\scriptstyle (\varprojlim \circ R^I)(\pi_F)} \\
(R\circ \varprojlim)(F) & \xrightarrow[\delta_F]{\cong} & (\varprojlim \circ R^I)(F)
\end{array}.$$

But it immediately follows by the universality of limits that $(R \circ \varprojlim)(\pi_F) = \mathbb{1}_{R(\varprojlim F)}$. Therefore by the commutativity of the diagram above, $\varprojlim R^I(\pi_F)$ turns out to be an isomorphism. $\square$

PROPOSITION 4.5.8. *Let I be a small category, and consider an adjoint pair* $\mathscr{C} \underset{R}{\overset{L}{\rightleftarrows}} \mathscr{D}$ ($L \dashv R$) *in a 2-category* **CAT**.

(1) *Assume that there exist both limits $\varprojlim F, \varprojlim R^I(F)$ for all $F \in (\mathscr{D}^I)_0$. Then R is continuous with respect to I.*
(2) *Assume that there exist both colimits $\varinjlim F, \varinjlim L^I(F)$ for all $F \in (\mathscr{C}^I)_0$. Then L is cocontinuous with respect to I.*

PROOF.

(1) We have the following adjoint pairs in the 2-category **CAT**:

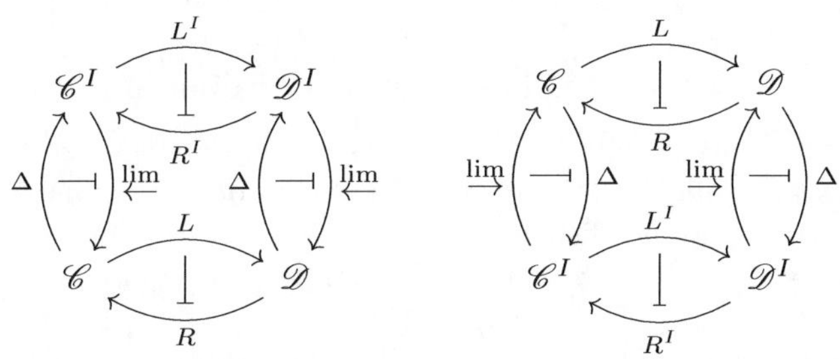

Apply Proposition 4.4.12 to the four adjoint pairs on the left to obtain

$$L^I \circ \Delta \simeq \Delta \circ L \iff R \circ \varprojlim \simeq \varprojlim \circ R^I.$$

But by Lemma 4.5.6 we have $L^I \circ \Delta = \Delta \circ L$. Therefore we have $R \circ \varprojlim \cong \varprojlim \circ R^I$. Thus R is continuous with respect to I by Lemma 4.5.7.

(2) This is similarly shown from the four adjoint pairs on the right and the equality $R^I \circ \Delta = \Delta \circ R$.

$\square$

REMARK 4.5.9. Since in the 2-category **CAT**, Theorem 4.4.7 that gives a characterization of adjoints holds, by using condition (2) of this theorem, we can show the following:

(1) Proposition 4.4.12 for the case that $\mathbf{C} = \mathbf{CAT}$. To prove this we can use the Yoneda lemma (Remark 2.5.13). By the natural isomorphisms

$$u(\text{-}, (r' \circ g')(?)) \cong v(l'(\text{-}), g'(?)) \cong y((f' \circ l')(\text{-}), ?),$$
$$u(\text{-}, (g \circ r)(?)) \cong x(f(\text{-}), r(?)) \cong y((l \circ f)(\text{-}), ?)$$

of functors of two variables, $f' \circ l' \cong l \circ f$ implies $u(\text{-}, (r' \circ g')(?)) \cong u(\text{-}, (g \circ r)(?))$. Hence $r' \circ g' \cong g \circ r$.

(2) In Proposition 4.5.8 (1) if $\varprojlim F$ exists, then automatically $(R(\varprojlim F), R^I(\pi_F))$ is a universal cone of $R^I(F)$. Hence we do not need to assume the existence of $\varprojlim R^I(F)$.

(3) In Proposition 4.5.8 (2) if $\varinjlim F$ exists, then automatically $(L(\varinjlim F), L^I(\sigma_F))$ is a universal cocone of $L^I(F)$. Hence we do not need to assume the existence of $\varinjlim L^I(F)$.

EXERCISE 4.5.10. Prove statement (2) in Remark 4.5.9 above.

In abelian categories, any morphism f has its kernel (resp. cokernel) by definition, and the kernel (resp. cokernel) of f is given as the equalizer (resp. coequalizer) of f and 0, which is a special type of limit (resp. colimit). Moreover, for a linear functor $F\colon \mathscr{C} \to \mathscr{D}$ of abelian categories, F is left exact (resp. right exact) if and only if F preserves kernels (resp. cokernels) ([**38**, Corollary 4.2.31, 4.2.32]). Therefore the following holds by Proposition 4.5.8:

COROLLARY 4.5.11. *Let* $\mathscr{C} \underset{R}{\overset{L}{\rightleftarrows}} \mathscr{D}$ ($L \dashv R$) *be an adjoint pair of abelian categories. Then L is right exact, and R is left exact.*

REMARK 4.5.12. In the above we assumed that the category I is a small category, but the same assertions hold even if we assume that I is a k-moderate category for some $k \geq 1$.

4.6. 2-adjoints and 2-equivalences

Next we define "outer adjoints." To investigate relationships between 2-categories we need to consider a "3-category" consisting of 2-categories having modifications as 3-morphisms, and discuss inside this 3-category. But we do not explain 3-categories because it is too complicated. Definitions of adjoints in a strong sense will be given by applying definitions from the preceding section to the setting of 2-category 2-$\mathbf{Cat}(\mathbf{U})$, where $\mathbf{U}$ is a suitable set of 2-categories containing 2-categories $\mathbf{C}, \mathbf{D}$ between which a 2-adjoint system is defined. (When we consider inside the "3-category" 2-$\mathbf{Cat}(\mathbf{U})$ we remove the word "strong".) Explicitly speaking, we make the following definitions.

DEFINITION 4.6.1. Let $\mathbf{C}, \mathbf{D}$ be 2-categories.

(1) In the diagram consisting of 2-functors and 2-natural morphisms

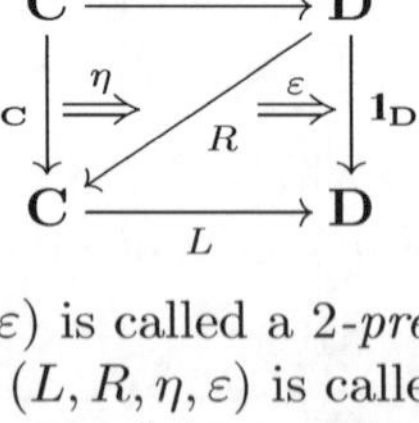

the quadruple $(L, R, \eta, \varepsilon)$ is called a 2-*pre-adjoint system*.

(2) A 2-pre-adjoint system $(L, R, \eta, \varepsilon)$ is called a *strong* 2-*equivalence system* if both η and ε are 2-natural isomorphisms. If this is the case, we say that L and R are *strong* 2-*equivalences*, and that they are *strong* 2-*quasi-inverses* of each other. $\mathbf{C}$ and $\mathbf{D}$ are said to be *strongly* 2-*equivalent* if there exists a strong 2-equivalence $L\colon \mathbf{C} \to \mathbf{D}$.

(3) A 2-pre-adjoint system $(L, R, \eta, \varepsilon)$ is called a 2-*equivalence system* if both η and ε are 2-natural equivalences (cf. Definition 4.1.21), i.e., if there exists 2-natural transformations $\eta' \colon R \circ L \Rightarrow \mathbb{1}_{\mathbf{C}}$ and $\varepsilon' \colon \mathbb{1}_{\mathbf{D}} \Rightarrow L \circ R$ such that $\eta' \bullet \eta \cong \mathbb{1}_{\mathbf{1_C}}, \eta \bullet \eta' \cong \mathbb{1}_{R \circ L}, \varepsilon' \bullet \varepsilon \cong \mathbb{1}_{L \circ R}, \varepsilon \bullet \varepsilon' \cong \mathbb{1}_{\mathbf{1_D}}$. If this is the case, we say that L and R are 2-*equivalences* and that they are 2-*quasi-inverses* of each other. Further $\mathbf{C}$ and $\mathbf{D}$ are said to be 2-*equivalent* if there exists a 2-equivalence $L \colon \mathbf{C} \to \mathbf{D}$.

(4) A 2-pre-adjoint system $(L, R, \eta, \varepsilon)$ is called a *strict* 2-*adjoint system* if it satisfies the following axiom:

Axioms: The following diagrams are commutative:

$$\begin{array}{ccc} L & \xRightarrow{\mathbf{1}_L \circ \eta} & L \circ R \circ L \\ & \diagdown\!\!\diagdown & \Downarrow{\scriptstyle \varepsilon \circ \mathbf{1}_L} \\ & & L \end{array} \qquad \begin{array}{ccc} R & \xRightarrow{\eta \circ \mathbf{1}_R} & R \circ L \circ R \\ & \diagdown\!\!\diagdown & \Downarrow{\scriptstyle \mathbf{1}_R \circ \varepsilon} \\ & & R \end{array}$$

in other words, the satsify the zigzag equations (4.1). Then L is called a *left* 2-*adjoint* to R, R is called a *right* 2-*adjoint* to L, and we denote it by $L \dashv R$. 2-natural transformations η, ε are called *2-unit*, *2-counit* of this strict 2-adjoint system, respectively.

(5) A strict 2-adjoint system $(L, R, \eta, \varepsilon)$ is called a *strict* 2-*adjoint strong equivalence system*, and (L, R) is called a *strict* 2-*adjoint strong equivalence pair* if both η and ε are 2-natural isomorphisms. By Corollary 4.4.5 L, R are strong 2-quasi-inverse to each other if and only if (L, R) is a strict 2-adjoint strong equivalence pair.

(6) A 2-pre-adjoint system $(L, R, \eta, \varepsilon)$ is called a 2-*adjoint system* if it satisfies

Axioms:

$$(\varepsilon \circ \mathbb{1}_L) \bullet (\mathbb{1}_L \circ \eta) \cong \mathbb{1}_L, \quad (\mathbb{1}_R \circ \varepsilon) \bullet (\eta \circ \mathbb{1}_R) \cong \mathbb{1}_R.$$

The following is obvious from the definition of 2-functors.

PROPOSITION 4.6.2. *A 2-functor $L \colon \mathbf{C} \to \mathbf{D}$ between 2-categories sends 2-isomorphisms, (resp. strict) 2-adjoint systems, 2-equivalences in $\mathbf{C}$ to 2-isomorphisms, (resp. strict) 2-adjoint systems, 2-equivalences in $\mathbf{D}$, respectively.*

THEOREM 4.6.3. *Fix a set $\mathbf{U}$ of 2-categories, and let $\mathbf{C} \underset{R}{\overset{L}{\rightleftarrows}} \mathbf{D}$ be 2-functors between 2-categories in the 2-category* 2-$\mathbf{Cat}(\mathbf{U})$ (cf. Example 4.1.22). *Then the following are equivalent:*

(1) *There exists a strict 2-adjoint system $(L, R, \eta, \varepsilon)$* (cf. Definition 4.6.1 (4)).

(2) *Two 2-functors $\mathbf{D}(L(\text{-}), ?), \mathbf{C}(\text{-}, R(?)) \colon \mathbf{C}^{\mathrm{op}} \times \mathbf{D} \to \mathbf{Cat}$ are 2-naturally equivalent (i.e., there exists a 2-natural equivalence between them,* cf. Definition 4.1.21 (4)*).*

PROOF. (1) $\Rightarrow$ (2). Let $x \in \mathbf{C}_0, y \in \mathbf{D}_0$. By extending the proof of the implication (1) $\Rightarrow$ (2) in Theorem 4.4.7 we define a functors $\omega_{x,y} \colon \mathbf{D}(Lx, y) \to \mathbf{C}(x, Ry)$ and $\omega'_{x,y} \colon \mathbf{C}(x, Ry) \to \mathbf{D}(Lx, y)$ by

$$\omega_{x,y} \colon (\alpha \colon f \Rightarrow g) \mapsto (R(\alpha) \circ \eta_x \colon R(f) \circ \eta_x \Rightarrow R(g) \circ \eta_x),$$
$$\omega'_{x,y} \colon (\alpha' \colon f' \Rightarrow g') \mapsto (\varepsilon_y \circ L(\alpha) \colon \varepsilon_y \circ L(f) \Rightarrow \varepsilon_y \circ L(g)).$$

for each morphism $\alpha\colon f \Rightarrow g$ of $\mathbf{D}(Lx, y)$ and each morphism $\alpha'\colon f' \Rightarrow g'$ of $\mathbf{C}(x, Ry)$. Then both

$$\Phi_{x,y} := (\varepsilon_f \circ L\eta_x)_{f\in\mathbf{D}(Lx,y)_0}\colon \omega'_{x,y} \circ \omega_{x,y} \Rightarrow \mathbb{1}_{\mathbf{D}(Lx,y)} \text{ and}$$
$$\Psi_{x,y} := (R\varepsilon_y \circ \eta_{f'})_{f'\in\mathbf{C}(x,Ry)_0}\colon \omega_{x,y} \circ \omega'_{x,y} \Rightarrow \mathbb{1}_{\mathbf{C}(x,Ry)}$$

are natural isomorphisms, and both

$$\omega := (\omega_{x,y})_{x,y}\colon \mathbf{D}(L(\text{-}), ?) \Rightarrow \mathbf{C}(\text{-}, R(?)) \text{ and}$$
$$\omega' := (\omega'_{x,y})_{x,y}\colon \mathbf{C}(\text{-}, R(?)) \Rightarrow \mathbf{D}(L(\text{-}), ?)$$

are 2-natural transformations, and finally both

$$\Phi := (\Phi_{x,y})_{x,y}\colon \omega' \circ \omega \rightsquigarrow \mathbb{1}_{\mathbf{D}(L(\text{-}),?)} \text{ and}$$
$$\Psi := (\Psi_{x,y})_{x,y}\colon \omega \circ \omega' \rightsquigarrow \mathbb{1}_{\mathbf{C}(\text{-},R(?))}$$

are isomorphic modifications. Hence ω is a 2-natural equivalence and ω' is its quasi-inverse.

(2) $\Rightarrow$ (1). This is verified by similarly extending the construction in the proof of the implication (2) $\Rightarrow$ (1) in Theorem 4.4.7, which is left to the reader as an exercise. □

EXERCISE 4.6.4. Complete the proof by filling in the arguments omitted in the proof above.

REMARK 4.6.5. By Theorem 4.6.3 (2), in particular, we have an equivalence of categories:

$$\mathbf{D}(L(x), y) \simeq \mathbf{C}(x, R(y)) \tag{4.5}$$

for all $x \in \mathbf{C}_0, y \in \mathbf{D}_0$. Further if we take 2-$\mathbf{Cat}^{\mathrm{st}}(\mathbf{U})$ instead of 2-$\mathbf{Cat}(\mathbf{U})$ in Theorem 4.6.3, namely if both η and ε are strict 2-natural transformations, then the 2-natural equivalence becomes a 2-natural isomorphism, and the equivalence (4.5) above becomes an isomorphism.

CHAPTER 5

2-Categorical Covering Theory under Pseudo-actions of a Group

Throughout this chapter, we assume that $\Bbbk$ is a commutative ring[1]. We denote by $\Bbbk$-**Cat** the 2-category consisting of the small linear categories, the linear functors between them, and the natural transformations between those functors. In the setting of the classical covering theory, the following conditions are assumed on the linear category $\mathscr{C}$:

(1) $\mathscr{C}$ is *basic* (i.e., distinct objects are not isomorphic);
(2) $\mathscr{C}(x, x)$ are local algebras for all $x \in \mathscr{C}_0$;
(3) The G-action is *free* (i.e., $1 \neq a \in G, x \in \mathscr{C}_0$ implies $ax \neq x$);
(4) The G-action is *locally bounded* (i.e., $\{a \in G \mid \mathscr{C}(ax, y) \neq 0\}$ is a finite set for all $x, y \in \mathscr{C}_0$).

These conditions are considerable obstacles in applications. Here, we remove all these assumptions and introduce a general covering theory. In the sequel, we generalize the arguments in the setting of strict group actions in [**7**] to the setting of pseudo group actions. In the original paper, very detailed proofs were omitted, especially explanations and proofs on modifications, which were added to this book. Concepts needed for generalization are taken from [**9**], and the precise arguments for the generalization to pseudo-actions of a group were first published in [**12**].

5.1. 2-categories of G-categories and of pseudo-G-categories

Let $\mathscr{C}$ be a $\Bbbk$-category. Recall that an *action* of G on $\mathscr{C}$ is a group homomorphism $X \colon G \to \mathrm{Aut}(\mathscr{C})$, and a pair $(\mathscr{C}, X)$ is called a category with G-action, or simply a G-category. This notion is generalized as follows:

First, regard the 2-category $\Bbbk$-**CAT** defined in Example 4.1.8 as a category by forgetting 2-morphisms, the vertical and horizontal compositions. Regard also the group $G = (G, \cdot)$ as a category $(\{*\}, G, \mathrm{dom}, \mathrm{cod}, \cdot)$ with a unique object. Then a G-category $(\mathscr{C}, X)$ is nothing but a functor $X' \colon G \to \Bbbk\text{-}\mathbf{CAT}$ satisfying the conditions

$$X'(*) := \mathscr{C}, X'(a) := X(a)\ (a \in G).$$

Here, we generalize the notion of G-category as follows by utilizing the 2-categorical structure of $\Bbbk$-**CAT** that was not used above:

Definition 5.1.1. A *pseudofunctor* (resp. *lax functor*, *colax functor*, *lax pseudofunctor*) $X \colon G \to \Bbbk\text{-}\mathbf{CAT}$ is called a category with a pseudo-G-action (resp. a lax G-action, a colax G-action, a lax pseudo-G-action) or shortly a *pseudo-G-category*

[1] Then we consider $\Bbbk$-modules instead of vector spaces. If the reader is not familiar with modules over a commutative ring, he/she may assume that $\Bbbk$ is a field as before.

(resp. a *lax G-category*, a *colax G-category*, a *lax pseudo-G-category@lax pseudo-G-category*). If $G(*) = \mathscr{C}$, then it is denoted by the pair $(\mathscr{C}, X)$. For instance, a precise definition of a colax G-category $(\mathscr{C}, X)$ is given as follows. This consists of the following data that satisfies the axioms below:

Data:

(1) a $\Bbbk$-category $\mathscr{C}$,
(2) a family $(X(a)\colon \mathscr{C} \to \mathscr{C})_{a\in G}$ of functors,
(3) a natural transformation $X_*\colon X(1) \Rightarrow \mathbb{1}_{\mathscr{C}}$,
(4) a family $(X_{b,a}\colon X(ba) \Rightarrow X(b)\circ X(a))_{a,b\in G}$ of natural transformations.

Axioms:

(a$_1$) 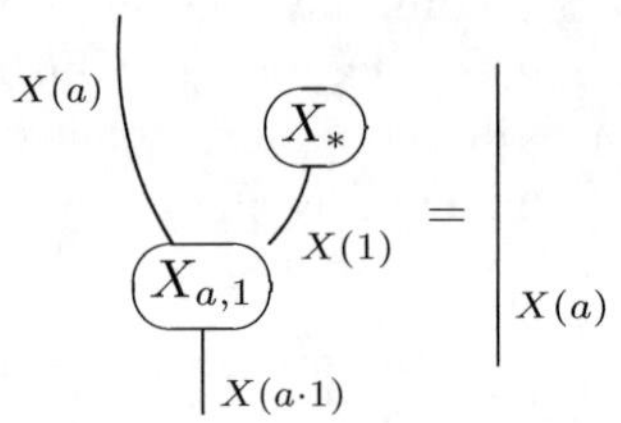

, (a$_2$) 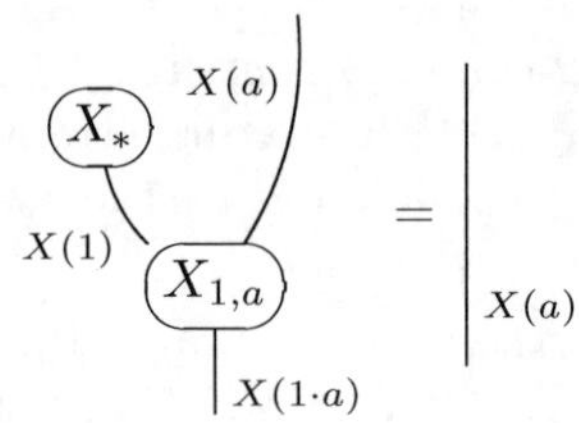

$(a \in G)$

(b) 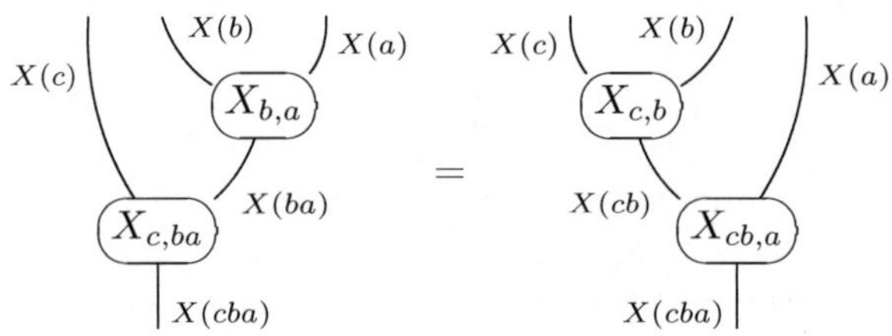

$(a, b, c \in G)$

In particular, when $\mathscr{C}$ is a small category, $(\mathscr{C}, X)$ is a colax functor $X\colon G \to \Bbbk$-**Cat** satisfying $X(*) = \mathscr{C}$.

A *pseudo-G-category* is a colax G-category $(\mathscr{C}, X)$ such that X_* and all $X_{b,a}$ $(a, b \in G)$ are natural isomorphisms.

REMARK 5.1.2. Let $(\mathscr{C}, X)$ be a pseudo-G-category. Then $X(a)$ is an autoequivalence for each $a \in G$. Indeed, $X(a)\circ X(a^{-1}) \cong X(aa^{-1}) = X(1) \cong \mathbb{1}_{\mathscr{C}}$. Similarly, $X(a^{-1})\circ X(a) \cong \mathbb{1}_{\mathscr{C}}$.

LEMMA 5.1.3. *Assume that a triple* (1), (2), (4) *of the data of a pseudo-G-category satisfies the axiom* (b) *and that $X(1)\colon \mathscr{C} \to \mathscr{C}$ is an equivalence having an adjoint system $(X(1), Y, \eta, \varepsilon)$ with both η, ε natural isomorphisms. Then the following six conditions are equivalent:*

(a$_1$), (a$_2$),
(a$'_1$)*:* (a$_1$) *for* $a = 1$, (a$'_2$)*:* (a$_2$) *for* $a = 1$,

(a$''_1$) : $X_* =$ 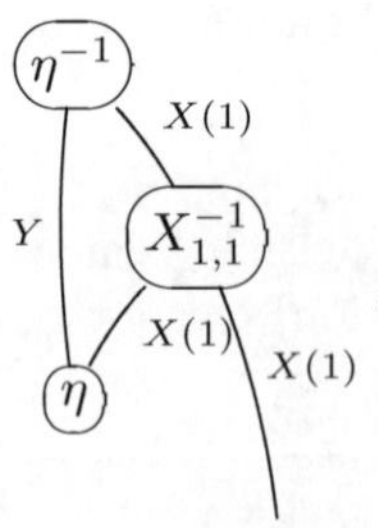

, *and* (a$''_2$) : $X_* =$ 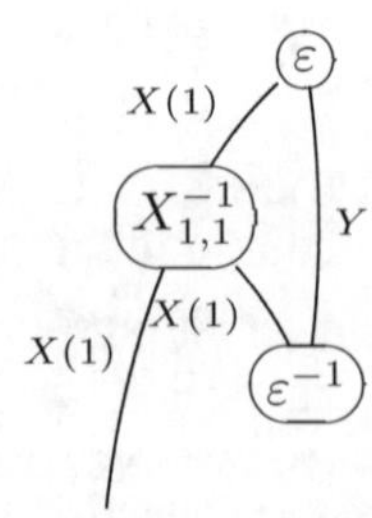

.

Therefore in particular, if either (a_1') *or* (a_2') *holds, then the following holds:*

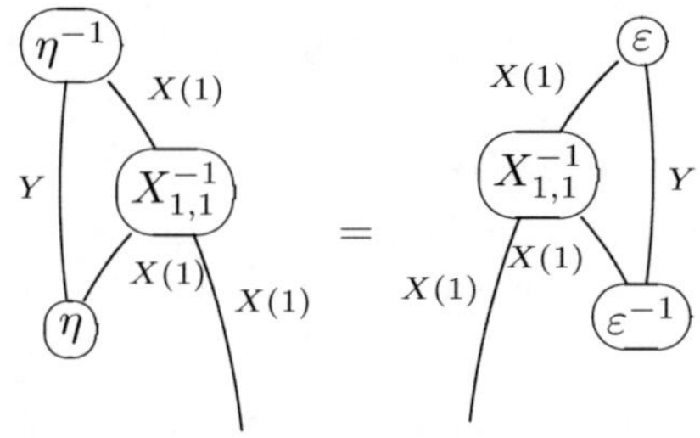

PROOF. $(a_1) \Rightarrow (a_1')$. Trivial.

$(a_1') \Rightarrow (a_1'')$. Assume that (a_1') holds. Vertically compose $X_{1,1}^{-1}$ to it from the below to have

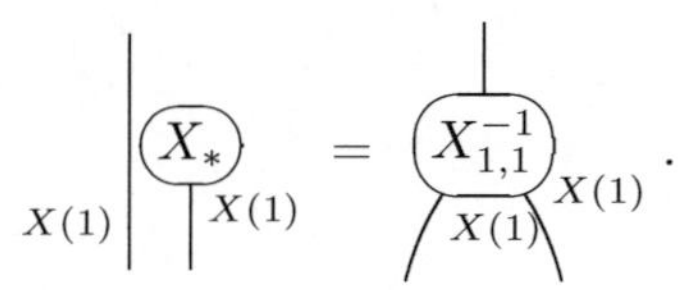

Using this (a_1'') is shown as follows:

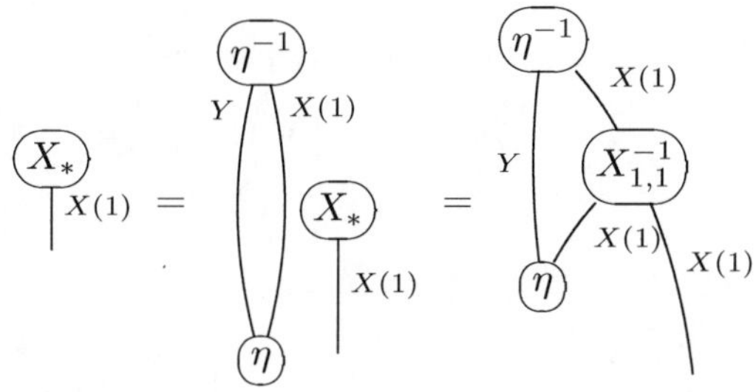

$(a_1'') \Rightarrow (a_2)$. Consider the condition (b) with the value $b = 1$. Then compose $X_{c,a}^{-1}$ to it from the below to obtain the following:

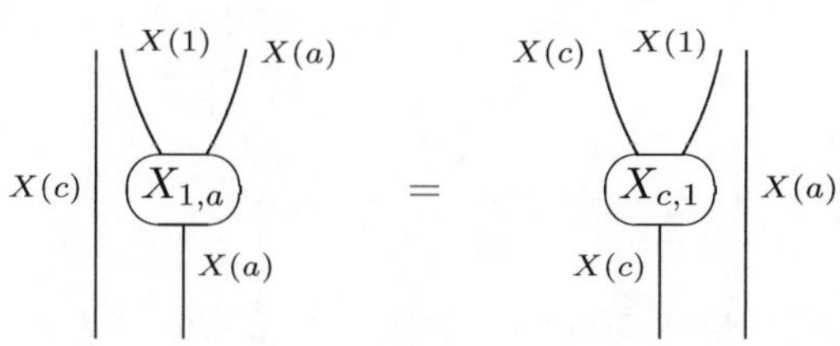

$(a, c \in G)$.

Assume here that (a_1'') holds. Then by applying the formula above for $c := 1$ we can show (a_2) as follows:

LHS = 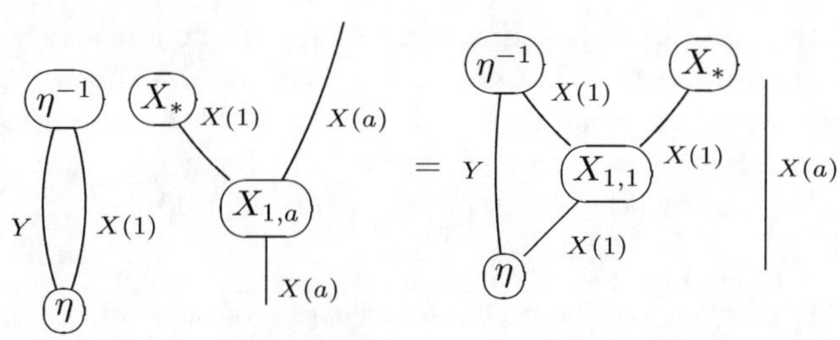 = RHS.

Similarly we can show the following implications: (a$_2$) $\Rightarrow$ (a$_2'$) $\Rightarrow$ (a$_2''$) $\Rightarrow$ (a$_1$). □

REMARK 5.1.4. Using the lemma above we see the following: Consider the condition

(a′) $X(1)$ is an auto-equivalence.

Then a pseudo-G-category is determined by the data (1), (2), (4) with all $X_{b,a}$ natural isomorphisms ($a, b \in G$) in Definition 5.1.1 and the axioms (a′), (b). Namely, the definition above is equivalent to that of weak action of G to a category given by Deligne et. al. **[21]**, **[22]**, **[18]**[2].

DEFINITION 5.1.5. An *equivariant* functor from a pseudo-G-category $(\mathscr{C}, X)$ to a pseudo-G-category $(\mathscr{C}', X')$ is a pair (E, ρ) of the following data that satisfies the axioms below.

Data:

- a functor $E\colon \mathscr{C} \to \mathscr{C}'$
- a family $\rho = (\rho_a\colon X'(a)E \Rightarrow EX(a))_{a\in G}$ of natural isomorphisms.

Axioms: For any $a, b \in G$ the following equalities hold:

(5.1) 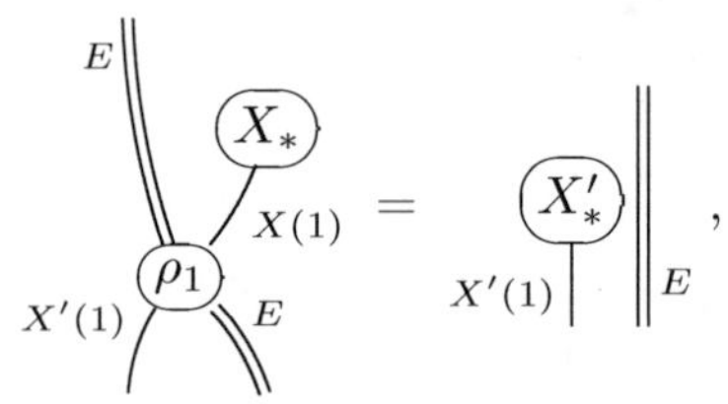

(5.2) 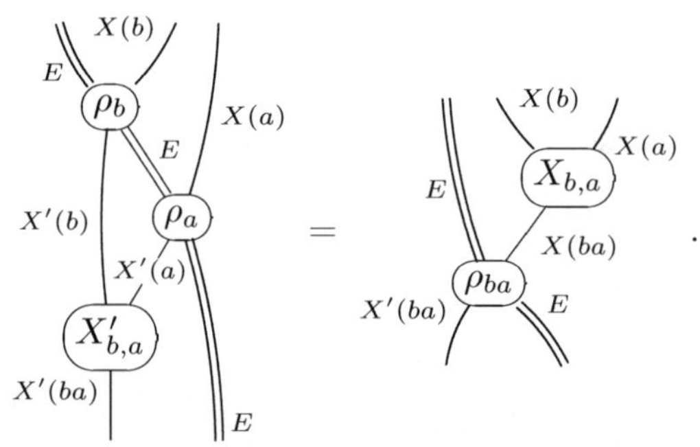

If in addition, all ρ_a above can be taken as identities, i.e., $X'(a)E = EX(a)$ for all $a \in G$, then E is said to be *strictly G-equivariant*.

A G-equivariant functor (E, ρ) is called *G-equivariantly equivalent* if E is an equivalence.

REMARK 5.1.6. In the above diagram we draw the lines representing E as double lines to emphasize it. By focusing on the double lines we see that the formula (5.1) allows us to cross the string of E with the string of $X'(1)$ or to separate the former from the latter by sliding the former on the latter; and the formula (5.2) allows us to slide the string of E on the Y-line of $X_{b,a}$ or $X'_{b,a}$.

[2]They assume that all $X(a)$ ($a \in G$) are autoequivalences, which is stronger than (a′), but follows by Remark 5.1.2. Note that they define a lax functor version not a colax version.

DEFINITION 5.1.7. Let $(E, \rho), (E', \rho') \colon (\mathscr{C}, X) \to (\mathscr{C}', X')$ be G-equivariant functors between pseudo-G-categories. Then a *morphism* from (E, ρ) to (E', ρ') is a natural transformation $\eta \colon E \Rightarrow E'$ that satisfies the following equation for all $a \in G$:

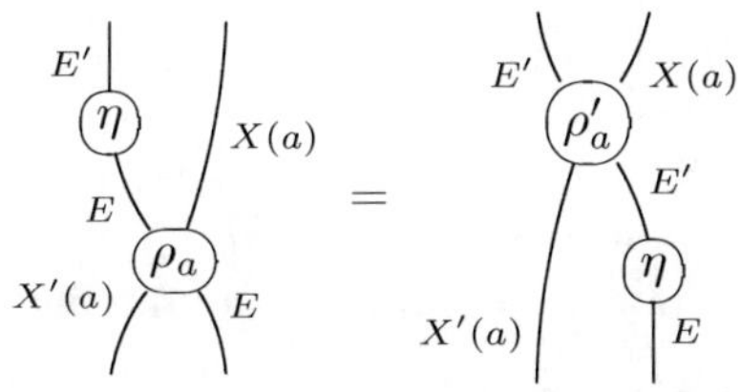

The following is easy to show and the proof is left to the reader.

LEMMA 5.1.8. *Let* $(\mathscr{C}, X) \xrightarrow{(E,\rho)} (\mathscr{C}', X') \xrightarrow{(E',\rho')} (\mathscr{C}'', X'')$ *be G-equivariant functors between pseudo-G-categories. Then*

$$(E', \rho') \circ (E, \rho) := (E' \circ E, \rho' \circ \rho) \colon \mathscr{C} \to \mathscr{C}''$$

is also a G-equivariant functor, which is called the composite *of (E, ρ) and (E', ρ'), where we set $\rho' \circ \rho := ((E' \circ \rho_a) \bullet (\rho'_a \circ E))_{a \in G}$, i.e., $\rho' \circ \rho$ is a family of the following natural transformations:*

$$(\rho' \circ \rho)_a :=$$

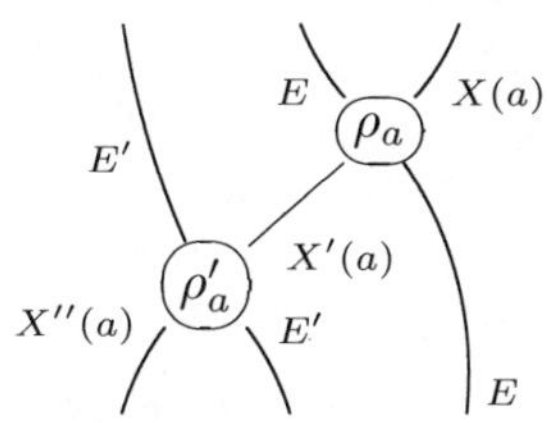

Here each $(\rho' \circ \rho)_a$ turns out also to be a natural transformation because so is each ρ_a, ρ'_a.

EXERCISE 5.1.9. Prove the lemma above.

DEFINITION 5.1.10. We define a 2-category G-**Cat** (resp. G-$\mathbf{Cat}^{\mathrm{s}}$) as follows:

- objects are the small categories with pseudo-G-actions (resp. with G-actions) $(\mathscr{C}, X) \colon G \to \Bbbk\text{-}\mathbf{Cat}$.
- 1-morphisms are the G-equivariant functors between these objects.
- The identity 1-morphism of the object $(\mathscr{C}, X)$ is given by $(\mathbb{1}_{\mathscr{C}}, (\mathbb{1}_{\mathbf{1}_{\mathscr{C}}})_{a \in G})$.
- 2-morphisms are the morphisms between G-equivariant functors.
- The identity 2-morphism of the 1-morphism $(E, \rho) \colon (\mathscr{C}, X) \to (\mathscr{C}', X')$ is the identity natural transformation $\mathbb{1}_E$ of E.
- The composition of 1-morphisms is the one given in the lemma above.
- The vertical and the horizontal compositions of 2-morphisms are the usual ones defined for natural transformations.

REMARK 5.1.11. There is an inclusion of the full 2-subcategory:

$$G\text{-}\mathbf{Cat}^{\mathrm{s}} \subseteq G\text{-}\mathbf{Cat}.$$

5.2. The 2-category of G-graded categories

In this section we define G-graded categories and a 2-category of those. The 1-morphisms defined here are weaker than the usual ones. This notion is weakened because the functor $\omega'_{\mathscr{B}}\colon (\mathscr{B}\#G)/G \to \mathscr{B}$ defined for a G-graded category $\mathscr{B}$ would not preserve degrees in a usual sense. Throughout this section all categories considered as G-graded are always assumed to be $\Bbbk$-linear.

DEFINITION 5.2.1. (1) A *G-graded* category consists of the following data that satisfies the axioms below:

Data:

- A linear category $\mathscr{B}$,
- A family $D = (D^a(x,y))_{x,y\in\mathscr{B}_0, a\in G}$ of subspaces $D^a(x,y)$ $(x, y \in \mathscr{B}_0)$ of $\mathscr{B}(x,y)$ that gives an (inner) direct sum decomposition $\mathscr{B}(x,y) = \bigoplus_{a\in G} D^a(x,y)$ $(x,y \in \mathscr{B}_0)$ of $\Bbbk$-spaces . This D is called a *grading* of $\mathscr{B}$.

Axioms: $D^b(y,z)\circ D^a(x,y) \subseteq D^{ba}(x,z)$ $(x, y \in \mathscr{B};\ a, b \in G)$.

When we deal with multiple gradings (e.g., see Remark 5.10.5), we use this precise notation, but usually we fix a grading D for $\mathscr{B}$ and simply write $\mathscr{B}$ for $(\mathscr{B}, D)$. In this case we also simply write $\mathscr{B}^a(x,y)$ for $D^a(x,y)$.

For each $a \in G$, an element f in $\mathscr{B}^a(x,y)$ is called a *homogeneous* element of $\mathscr{B}(x,y)$ (or a *homogeneous morphism* of $\mathscr{B}$), and a is called the *degree* of f and we set $\deg f := a$ for all $x, y \in \mathscr{B}$.

(2) A *degree-preserving functor* is a pair (H, r) of a linear functor $H\colon \mathscr{B} \to \mathscr{A}$ of G-graded categories and a map $r\colon \mathscr{B}_0 \to G$ that satisfies the condition

$$H(\mathscr{B}^{r_y a}(x,y)) \subseteq \mathscr{A}^{a r_x}(Hx, Hy) \quad (x, y \in \mathscr{B}, a \in G).$$

(The condition above is equivalent to $H(\mathscr{B}^a(x,y)) \subseteq \mathscr{A}^{r_y^{-1} a r_x}(Hx, Hy)$). This r is called a *degree adjuster* of H.

(3) A linear functor $H\colon \mathscr{B} \to \mathscr{A}$ of G-graded categories is called a *strictly degree-preserving functor* if it satisfies $H(\mathscr{B}^a(x,y)) \subseteq \mathscr{A}^a(Hx, Hy)$ $(x, y \in \mathscr{B};\ a \in G)$. Note that this condition is equivalent to saying that $(H, 1)$ is a degree-preserving functor, where 1 is the constant map $\mathscr{B}_0 \to G, x \mapsto 1 \in G$ $(x \in \mathscr{B}_0)$.

(4) Let $(H, r), (I, s)\colon \mathscr{B} \to \mathscr{A}$ be degree-preserving functors. Then a natural transformation $\theta\colon H \Rightarrow I$ is called a *morphism* of degree-preserving functors if $\theta x \in \mathscr{A}^{s_x^{-1} r_x}(Hx, Ix)$ $(x \in \mathscr{B})$.

The following is easily verified.

LEMMA 5.2.2. *Let* $\mathscr{B} \xrightarrow{(H,r)} \mathscr{B}' \xrightarrow{(H',r')} \mathscr{B}''$ *be degree-preserving functors. Then*

$$(H' \circ H, (r_x r'_{Hx})_{x\in\mathscr{B}})\colon \mathscr{B} \to \mathscr{B}''$$

is also a degree-preserving functor, which is called a composite *of* (H, r) *and* (H', r') *and is denoted by* $(H', r') \circ (H, r)$.

DEFINITION 5.2.3. A 2-category G-**GrCat** (resp. G-$\mathbf{GrCat}^{\mathrm{s}}$) is defined as follows:

- The objects are the G-graded small categories.
- The 1-morphisms are the degree-preserving functors (resp. strictly degree-preserving functors) between them.
- The identity 1-morphism of an object $\mathscr{B}$ is given by $(\mathbb{1}_{\mathscr{B}}, 1)$.
- The 2-morphisms are the morphisms between degree-preserving functors.
- The identity 2-morphism of a 1-morphism $(H, r)\colon \mathscr{B} \to \mathscr{A}$ is given by the identity $\mathbb{1}_H$ of H, which is a 2-morphism because $(\mathbb{1}_H)x = \mathbb{1}_{Hx} \in \mathscr{A}^1(Hx, Hx) = \mathscr{A}^{r_x^{-1} r_x}(Hx, Hx)$ $(x \in \mathscr{B}_0)$.
- The composition of 1-morphisms is given as in the lemma above.
- The vertical composition and the horizontal composition of 2-morphisms are those usually defined for natural transformations.

5.3. G-covering functors

In this section, we generalize the Galois coverings introduced in Chapter 3.

DEFINITION 5.3.1. We define a 2-functor $\Delta\colon \Bbbk\text{-}\mathbf{Cat} \to G\text{-}\mathbf{Cat}$, which is called a *diagonal 2-functor* as follows:

- For each object $\mathscr{C}$ of $\Bbbk$-**Cat** we set $\Delta(\mathscr{C})(*) := \mathscr{C}, \Delta(\mathscr{C})(a) := \mathbb{1}_{\mathscr{C}}$ $(a \in G)$. Thus, it is the G-category $\mathscr{C}$ with the trivial G-action.
- For each 1-morphism $F\colon \mathscr{C} \to \mathscr{C}'$ of $\Bbbk$-**Cat**, $\Delta(F)\colon \Delta(\mathscr{C}) \to \Delta(\mathscr{C}')$ is given by the 1-morphism $(F, (\mathbb{1}_F\colon \mathbb{1}_{\mathscr{C}} \circ F \Rightarrow F \circ \mathbb{1}_{\mathscr{C}})_{a \in G})$ in G-**Cat**.
- Let $F, F'\colon \mathscr{C} \to \mathscr{C}'$, $\alpha\colon F \Rightarrow F'$ be 1-morphisms and a 2-morphism in $\Bbbk$-**Cat**. Then we define a 2-morphism $\Delta(\alpha)\colon \Delta(F) \Rightarrow \Delta(F')$ in G-**Cat** by $\Delta(\alpha) := \alpha$.

REMARK 5.3.2. If $F\colon \mathscr{C} \to \mathscr{C}'$ is a 1-morphism in $\Bbbk$-**Cat**, then $\Delta F\colon \Delta(\mathscr{C}) \to \Delta(\mathscr{C}')$ is a strictly G-equivariant functor in G-**Cat**.

As a special case of Lemma 5.1.8 we obtain the following:

LEMMA 5.3.3. *If $H\colon \mathscr{B} \to \mathscr{A}$ is a 1-morphism of $\Bbbk$-**Cat**, and $(F, \psi)\colon \mathscr{C} \to \Delta\mathscr{B}$ is a 1-morphism of G-**Cat**, then*

$$\Delta(H) \circ (F, \psi) = (H \circ F, (H \circ \psi_a)_{a \in G})\colon \mathscr{C} \to \Delta(\mathscr{A}).$$

PROPOSITION 5.3.4. *Let $\mathscr{B}$ be a linear category and $(F, \psi)\colon \mathscr{C} \to \Delta(\mathscr{B})$ a 1-morphism of G-**Cat**. We define the following two linear maps for all $x, y \in \mathscr{C}_0$*

$$(F, \psi)^{(1)}_{x,y}\colon \coprod_{a \in G} \mathscr{C}(X(a)x, y) \to \mathscr{B}(Fx, Fy),\ (f_a)_{a \in G} \mapsto \sum_{a \in G} F(f_a) \circ \psi_a x,$$

$$(F, \psi)^{(2)}_{x,y}\colon \coprod_{b \in G} \mathscr{C}(x, X(b)y) \to \mathscr{B}(Fx, Fy),\ (f_b)_{b \in G} \mapsto \sum_{b \in G} (\psi_b y)^{-1} \circ F(f_b).$$

Then $(F, \psi)^{(1)}_{x,y}$ is an isomorphism if and only if $(F, \psi)^{(2)}_{x,y}$ is.

PROOF. Let $x, y \in \mathscr{C}_0$ and consider the following diagram:

$$\begin{array}{ccc}
\coprod_{a\in G}\mathscr{C}(X(a)x, y) & \xrightarrow{(F,\psi)^{(1)}_{x,y}} & \mathscr{B}(Fx, Fy) \\
{\scriptstyle X(a^{-1})}\downarrow & & \| \\
\coprod_{a\in G}\mathscr{C}(X(a^{-1})(X(a)x), X(a^{-1})y) & & \| \\
{\scriptstyle \coprod_{a\in G}\mathscr{C}(X_{a^{-1},a}x\circ(X_*x)^{-1}, X(a^{-1})y)}\downarrow & & \| \\
\coprod_{a\in G}\mathscr{C}(x, X(a^{-1})y) & & \| \\
{\scriptstyle t}\downarrow & & \| \\
\coprod_{a^{-1}\in G}\mathscr{C}(x, X(a^{-1})y) & \xrightarrow[(F,\psi)^{(2)}_{x,y}]{} & \mathscr{B}(Fx, Fy),
\end{array}$$

where t is an isomorphism define by $t((g_{a^{-1}})_{a\in G}) := (g_{a^{-1}})_{a^{-1}\in G}$. Since X is a pseudo-G-action, all the vertical morphisms are isomorphisms. Therefore to prove the assertion it is enough to show the commutativity of the diagram above. Now take any $(f_a)_{a\in G} \in \coprod_{a\in G}\mathscr{C}(X(a)x, y)$. Then it suffices to show the following equality:

$$\sum_{a^{-1}\in G}(\psi_{a^{-1}}y)^{-1}\circ F(X(a^{-1})f_a\circ X_{a^{-1},a}x\circ(X_*x)^{-1}) = \sum_{a\in G}Ff_a\circ\psi_a x$$

To this end it is enough to show the equality

$$F(X(a^{-1})f_a\circ X_{a^{-1},a}x) = \psi_{a^{-1}}y\circ Ff_a\circ\psi_a x\circ F(X_*x) \tag{5.3}$$

for all $a \in G$. Here set $\Delta(\mathscr{B}) = (\mathscr{B}, X')$. Thus $X'(a) = \mathbb{1}_{\mathscr{B}}$ for all $a \in G$ and $X'_* = \mathbb{1}_{\mathbf{1}_{\mathscr{B}}}: X'(1) \Rightarrow \mathbb{1}_{\mathscr{B}}$, $X'_{b,a} = \mathbb{1}_{\mathbf{1}_{\mathscr{B}}}: X(ba) \Rightarrow X(b)\circ X(a)$ for all $a, b \in G$. The equality (5.3) is verified by the following string diagram:

(RHS) = 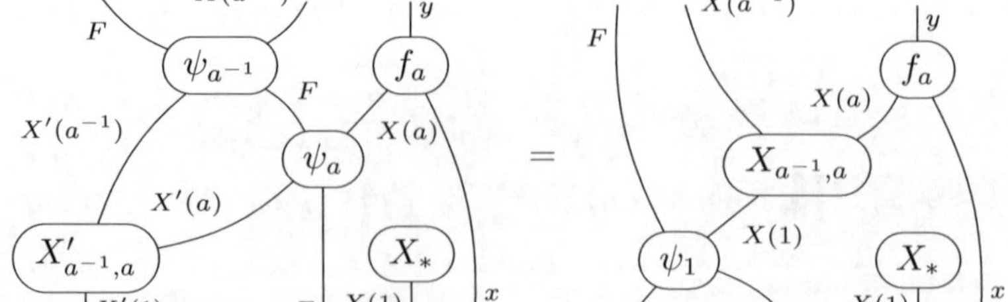

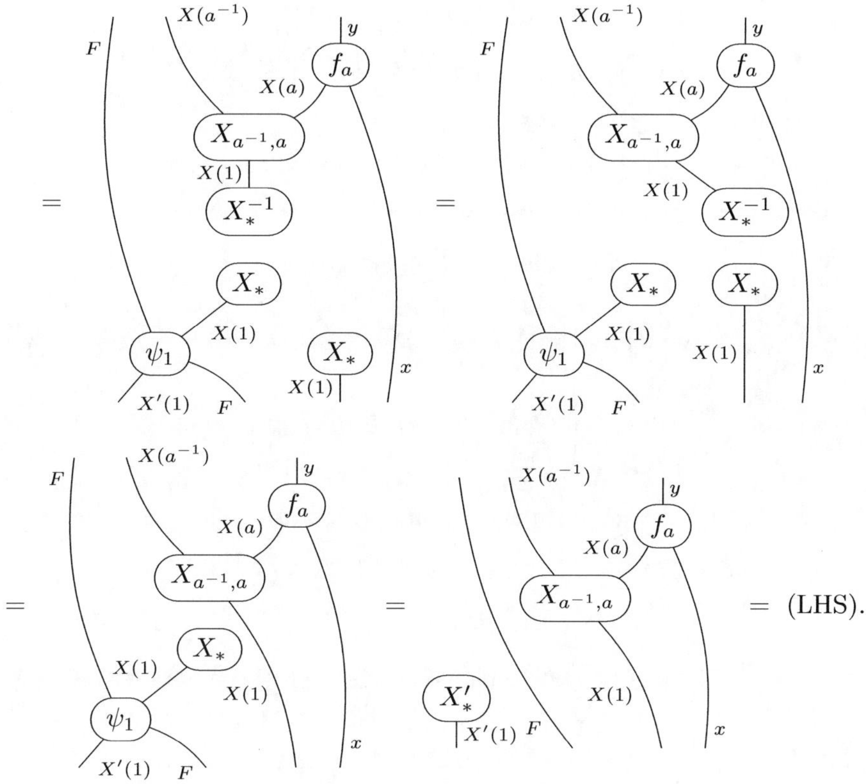

□

DEFINITION 5.3.5. Let $\mathscr{B}$ be a linear category and $(F, \psi)\colon \mathscr{C} \to \Delta(\mathscr{B})$ a 1-morphism of G-**Cat**.

(1) (F, ψ) is called a *G-precovering* if $(F, \psi)^{(1)}_{x,y}$ is an isomorphism for all $x, y \in \mathscr{C}_0$. (Note that by Proposition 5.3.4 this condition is equivalent to saying that $(F, \psi)^{(2)}_{x,y}$ is an isomorphism.)

(2) If (F, ψ) is a G-precovering and the functor F is *dense* (i.e., for each $y \in \mathscr{B}_0$ there exists an $x \in \mathscr{C}_0$ such that $Fx \cong y$), then (F, ψ) is called a *G-covering*.

5.4. Orbit categories

In the sequel, for a sequence $(M_i)_{i\in I}$ of modules we need to distinguish its inner direct sum $\bigoplus_{i\in I} M_i$ and its outer direct sum $\coprod_{i\in I} M_i$. In other places we usually identify them by the canonical isomorphism

$$\Sigma\colon \coprod_{i\in I} M_i \to \bigoplus_{i\in I} M_i, \quad (m_i)_{i\in I} \mapsto \sum_{i\in I} m_i.$$

DEFINITION 5.4.1. We define a 2-functor $?/G\colon G\text{-}\mathbf{Cat} \to \Bbbk\text{-}\mathbf{Cat}$, $x \mapsto x/G$ $(x \in (G\text{-}\mathbf{Cat})_i, i = 0, 1, 2)$ as follows, which is called the *orbit 2-functor*.

(0) On objects: For a pseudo-G-category $\mathscr{C}$ we define a linear category $\mathscr{C}/G$ as follows and call it the *orbit category* of $\mathscr{C}$ by G:

- $(\mathscr{C}/G)_0 := \mathscr{C}_0$.

- For any $x, y \in (\mathscr{C}/G)_0$, set $(\mathscr{C}/G)(x,y) := \coprod_{a\in G} \mathscr{C}(X(a)x, y)$. Here, let
$$\sigma_a^{\mathscr{C}/G} \colon \mathscr{C}(X(a)x,y) \to (\mathscr{C}/G)(x,y), \quad f \mapsto (\delta_{b,a} f)_{b\in G}$$
be the canonical injection, and set
$$(\mathscr{C}/G)^a(x,y) := \sigma_a^{\mathscr{C}/G}(\mathscr{C}(X(a)x,y)), (x,y \in \mathscr{C}_0 = (\mathscr{C}/G)_0, a \in G). \tag{5.4}$$
Then $(\mathscr{C}/G)(x,y)$ is written as an inner direct sum:
$$(\mathscr{C}/G)(x,y) = \bigoplus_{a\in G} (\mathscr{C}/G)^a(x,y). \tag{5.5}$$
- For any $f = (f_a)_{a\in G} \in (\mathscr{C}/G)(x,y), g = (g_b)_{b\in G} \in (\mathscr{C}/G)(y,z)$, set
$$g \circ f := \left(\sum_{\substack{a,b\in G\\ ba=c}} g_b \circ X(b) f_a \circ X_{b,a} x \right)_{c\in G}. \tag{5.6}$$
In the above, each term is the composite of the following:
$$X(ba)x \xrightarrow{X_{b,a}x} X(b)X(a)x \xrightarrow{X(b)f_a} X(b)y \xrightarrow{g_b} z.$$
- For each $x \in (\mathscr{C}/G)_0$, the identity $\mathbb{1}_x^{\mathscr{C}/G}$ in $\mathscr{C}/G$ is given as follows:
$$\mathbb{1}_x^{\mathscr{C}/G} = \sigma_1^{\mathscr{C}/G}(X_* x) = (\delta_{a,1} X_* x)_{a\in G} \in \coprod_{a\in G} \mathscr{C}(X(a)x, x).$$
Note by this form that we have $\deg \mathbb{1}_x^{\mathscr{C}/G} = 1$.

(1) On 1-morphisms: Let $\mathscr{C} = (\mathscr{C}, X), \mathscr{C}' = (\mathscr{C}', X')$ be pseudo-G-categories and $(F, \psi) \colon \mathscr{C} \to \mathscr{C}'$ a 1-morphism in G-**Cat**. Then we define a 1-morphism
$$(F,\psi)/G \colon \mathscr{C}/G \to \mathscr{C}'/G$$
in $\mathbb{k}$-**Cat** as follows:

- For each $x \in (\mathscr{C}/G)_0 = \mathscr{C}_0$, we set
$$((F,\psi)/G)(x) := F(x). \tag{5.7}$$
- For each $x, y \in (\mathscr{C}/G)_0$ and each $f = (f_a)_{a\in G} \in (\mathscr{C}/G)(x,y)$, we set $((F,\psi)/G)(f) := (F(f_a) \circ \psi_a x)_{a\in G}$. Here, each component is the composite of the following:
$$X'(a)Fx \xrightarrow{\psi_a x} FX(a)x \xrightarrow{F(f_a)} Fy.$$

(2) On 2-morphisms: Let $\mathscr{C} = (\mathscr{C}, X), \mathscr{C}' = (\mathscr{C}', X')$ be objects of G-**Cat**, $(F,\psi), (F',\psi') \colon \mathscr{C} \to \mathscr{C}'$ 1-morphisms in G-**Cat**, and $\zeta \colon (F,\psi) \Rightarrow (F',\psi')$ a 2-morphism in G-**Cat**. Then we define a 2-morphism
$$\zeta/G \colon (F,\psi)/G \Rightarrow (F',\psi')/G$$
in $\mathbb{k}$-**Cat** by
$$(\zeta/G)x := \sigma_1^{\mathscr{C}'/G}(\zeta x \circ X_* F(x)) \in (\mathscr{C}'/G)^1(F(x), F'(x)) \quad (x \in \mathscr{C}_0), \tag{5.8}$$
which is possible because we have
$$\zeta x \circ X_* F(x) \in \mathscr{C}'(X(1)F(x), F'(x)).$$

REMARK 5.4.2.

(1) We can express the composite in the orbit category above by the following string diagrams:

(5.9)
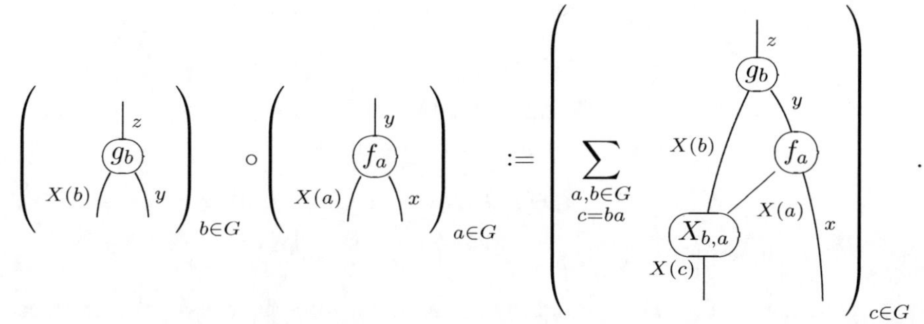

(2) If $\mathscr{C}$ is a light category with a pseudo-G-action, then by definition, $\mathscr{C}/G$ is also a light category.

LEMMA 5.4.3. *The composition of* $\mathscr{C}/G$ *defined by the formula* (5.6) *satisfies the associativity.*

PROOF. Let $x \xrightarrow{f} y \xrightarrow{g} z \xrightarrow{h} w$ be morphisms in $\mathscr{C}/G$, and set $f = (f_a : X(a)x \to y)_{a\in G}, g = (g_b : X(b)y \to z)_{b\in G}, h = (h_c : X(c)z \to w)_{c\in G}$. By the formula (5.9), $h \circ (g \circ f), (h \circ g) \circ f$ are given as follows, respectively:

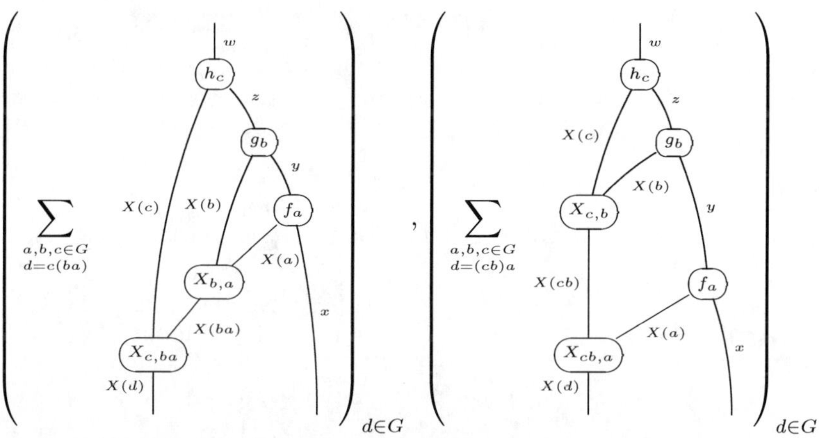

These coincide by the axiom (b) of Definition 5.1.1. □

EXERCISE 5.4.4. By using the formula (5.9), show that for each $x \in (\mathscr{C}/G)_0$, the following holds:

$$\mathbb{1}_x^{\mathscr{C}/G} = (\delta_{a,1}X_*x)_{a\in G} =$$

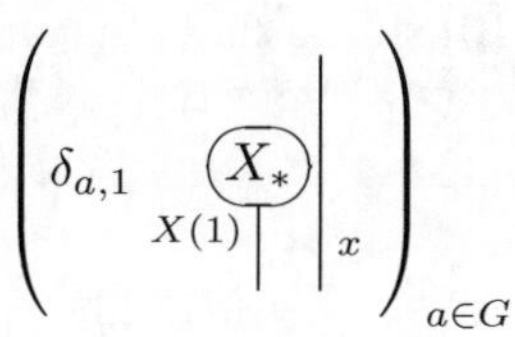

REMARK 5.4.5. We note the following:

(1) The orbit category $\mathscr{C}/G$ is a G-graded category by the direct sum decomposition (5.5) in Definition 5.4.1(0) and by the definition of the composition.

(2) $(F, \psi)/G$ is a strictly degree-preserving functor for all 1-morphism (F, ψ) in G-**Cat**.
(3) ζ/G is a 2-morphism in G-**GrCat** for all 2-morphism ζ in G-**Cat**. Therefore the orbit 2-functor can be seen as a 2-functor

$$?/G\colon G\text{-}\mathbf{Cat} \to G\text{-}\mathbf{GrCat}.$$

Considering the forgetful 2-functor $\mathrm{Fgt}\colon G\text{-}\mathbf{GrCat} \to \Bbbk\text{-}\mathbf{Cat}$ that forgets the G-gradings, we can regard the 2-functor

$$?/G\colon G\text{-}\mathbf{Cat} \to \Bbbk\text{-}\mathbf{Cat}$$

as the composite $\mathrm{Fgt} \circ (?/G)$.

EXAMPLE 5.4.6 (Orbit categories and skew group algebras). We first recall the definition of skew group algebras. Assume that an algebra A has a G-action X. For the free A-module $\bigoplus_{a\in G} A$, we set $f * a := (f\delta_{a,b})_{b\in G}$ $(f \in A, a \in G)$. Then every element $(f_a)_{a\in G}$ is uniquely expressed as a finite sum

$$(f_a)_{a\in G} = \sum_{a\in G} f_a * a. \tag{5.10}$$

The skew group algebra $A*G$ is an A-module $\bigoplus_{a\in G} A$ with a multiplication defined by

$$(g * b) \cdot (f * a) := (g \cdot X(b)f) * (ba) \quad (f, g \in A, a, b \in G).$$

Now consider the case where $\mathscr{C} = A$ is a $\Bbbk$-algebra (= a linear singleton) with a G-action X, and $\mathscr{C}_0 = \{x\}$. Then by regarding $A * G$ as a singleton with a unique object x, we have the equality

$$\mathscr{C}/G = A * G$$

of categories. Thus the orbit category for a G-category is regarded as the skew group algebra with multiple objects. Indeed, we have $(\mathscr{C}/G)_0 = \mathscr{C}_0 = \{x\} = (A*G)_0$, and since $\mathscr{C}_1 = \mathscr{C}(x, x) = A$, we have $(\mathscr{C}/G)_1 = (\mathscr{C}/G)(x, x) = \bigoplus_{g\in G} \mathscr{C}(X(g)x, x) = \bigoplus_{g\in G} \mathscr{C}(x, x) = \bigoplus_{g\in G} A = (A * G)_1$. Now in general, the multiplication of $A * G$ is given by

$$g \cdot f := \sum_{c\in G} \left(\sum_{\substack{a,b\in G \\ ba=c}} g_b \cdot X(b) f_a \right) * c$$

for all elements $f := \sum_{a\in G} f_a * a$ and $g := \sum_{b\in G} g_b * b$ of $A * G$. Then the correspondence (5.10) shows that this multiplication coincides with the composition of $\mathscr{C}/G$ defined by (5.6) (Note that $X_{b,a}x = \mathbb{1}_x$ because X is a strict G-action).

DEFINITION 5.4.7. Let $\mathscr{C} = (\mathscr{C}, X) \in G\text{-}\mathbf{Cat}_0$. Then we define a 1-morphism $(P, \phi)\colon \mathscr{C} \to \Delta(\mathscr{C}/G)$ in G-**Cat** as follows, which is called a *canonical covering functor*. (It is shown that this is actually a G-covering in Proposition 5.4.12.)

(1) For each $x \in (\mathscr{C})_0$, we set $Px := x$.
(2) For each morphism $f\colon x \to y$ in $\mathscr{C}$, we set

$$Pf := \sigma_1^{\mathscr{C}/G}(f \circ X_* x) = (\delta_{a,1} f \circ X_* x)_{a\in G} \in (\mathscr{C}/G)^1(Px, Py).$$

(3) For each $a \in G$, we define a natural transformation $\phi_a \colon P \Rightarrow PX(a)$

$$\begin{array}{ccc} \mathscr{C} & \xrightarrow{P} & \mathscr{C}/G \\ {\scriptstyle X(a)}\downarrow & \overset{\phi_a}{\swarrow\!\!\!\swarrow} & \| \\ \mathscr{C} & \xrightarrow[P]{} & \mathscr{C}/G \end{array}$$

by $\phi_a x := \sigma_a^{\mathscr{C}/G}(\mathbb{1}_{X(a)x}) = (\delta_{b,a}\mathbb{1}_{X(a)x})_{b\in G}$ $(x \in \mathscr{C})$. We set $\phi := (\phi_a)_{a\in G}$. Note by this form that $\deg \phi_a x = a$.

EXERCISE 5.4.8. In the definition above, show that $P \colon \mathscr{C} \to \mathscr{C}/G$ is a linear functor.

The pair (P, ϕ) is shown to be a 1-morphism in G-**Cat** in the next lemma.

LEMMA 5.4.9. *In Definition 5.4.7, $\phi_a := (\phi_a x)_{x\in\mathscr{C}} \colon P \Rightarrow PX(a)$ is a natural isomorphism for all $a \in G$, and $(P, \phi) \colon \mathscr{C} \to \Delta(\mathscr{C}/G)$ turns out to be a G-equivariant functor.*

PROOF. We divide the proof into several steps.

CLAIM 1. *ϕ_a is a natural transformation for all $a \in G$.*

To show this it is enough to show the commutativity of the following diagram for each morphism $f \colon x \to y$ in $\mathscr{C}$.

$$\begin{array}{ccc} Px & \xrightarrow{\phi_a x} & P(X(a)x) \\ {\scriptstyle Pf}\downarrow & & \downarrow{\scriptstyle P(X(a)f)} \\ Py & \xrightarrow[\phi_a y]{} & P(X(a)y) \end{array} \tag{5.11}$$

It is easily verified by using the formula (5.9) that both the clockwise composite and the counter-clockwise composite are equal to $(\delta_{b,a}X(a)f)_{b\in G}$. Thus ϕ_a is a natural transformation.

CLAIM 2. *$\phi_a x$ is an isomorphism in $\mathscr{C}/G$ for all $a \in G, x \in \mathscr{C}_0$.*

It is enough to show that $(\phi_a x)^{-1}$ is given by

(5.12)

$$(\phi_a x)^{-1} = (\delta_{b,a^{-1}}X_* x \circ X^{-1}_{a^{-1},a}x)_{b\in G} =$$

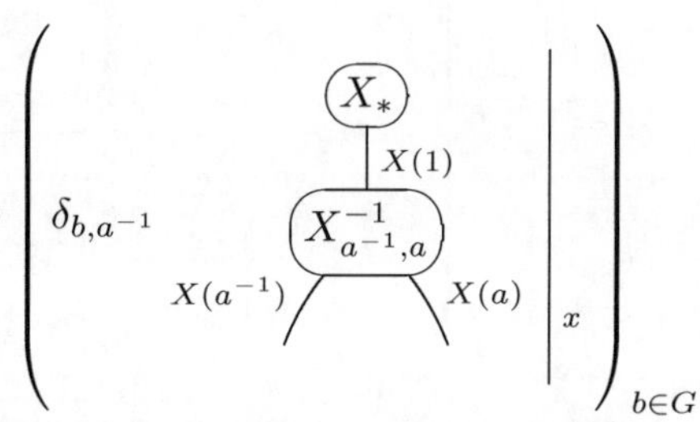

To this end we set $\phi'_a x$ to be the right-hand side. From the definition of composition, it is immediate that $\phi'_a x \circ \phi_a x = \mathbb{1}_x^{\mathscr{C}/G}$. To show that $\phi_a x \circ \phi'_a x = \mathbb{1}_{X(a)x}^{\mathscr{C}/G}$ we first show the following:

(5.13)

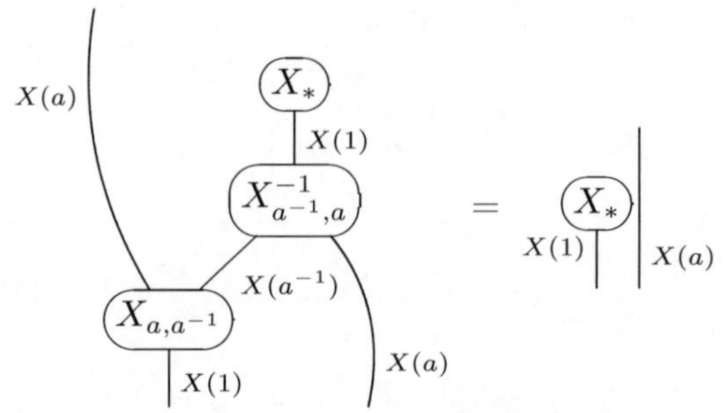

By Definition 5.1.1(b), (a$_1$), (a$_2$), we have

$(\mathrm{LHS}) \bullet X_{1,a} =$

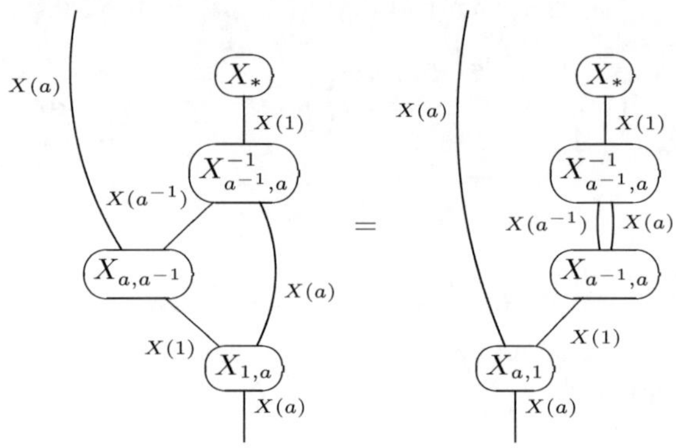

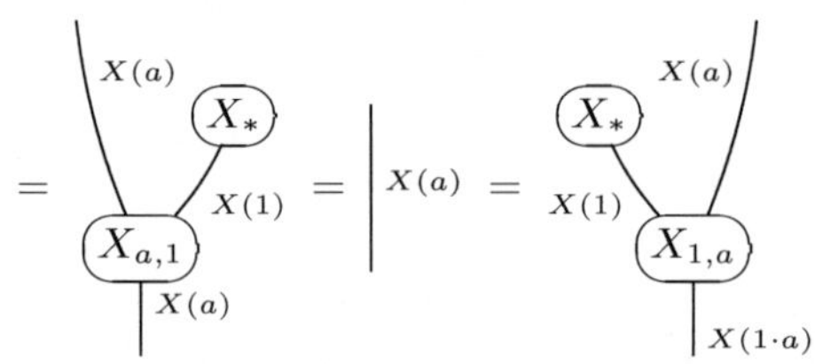

$= (\mathrm{RHS}) \bullet X_{1,a}.$

Here since $X_{1,a}$ is an isomorphism, we have (LHS) = (RHS). Then using (5.13) we have

$\phi_a x \circ \phi'_a x =$

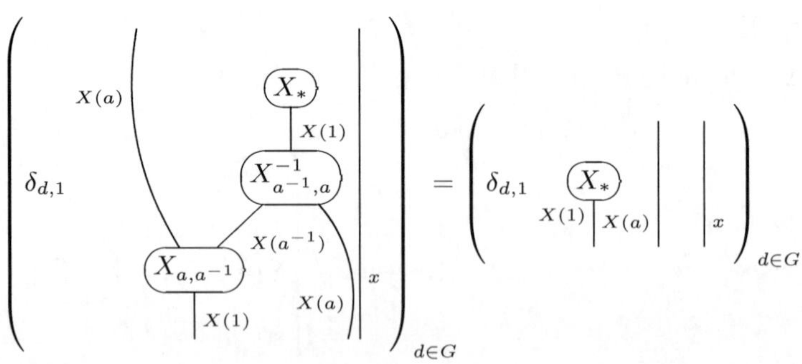

$= \mathbb{1}^{\mathscr{C}/G}_{X(a)x},$

as desired.

CLAIM 3. *(P, ϕ) satisfies* (5.1).

It is enough to show the commutativity of the diagram

$$\begin{array}{ccc} P & \overset{\phi_1}{\Longrightarrow} & PX(1) \\ & \underset{\mathbb{1}_P}{\searrow\!\!\searrow} & \Big\Downarrow {\scriptstyle P\circ X_*} \\ & & P \end{array} \quad . \tag{5.14}$$

Thus, it is enough to show that for each $x \in \mathscr{C}_0$ we have the equality $(PX_*x)\circ\phi_1 x = \mathbb{1}_{Px}$. But this follows from the formula (5.9) and the axiom (a_2) of the Definition 5.1.1.

CLAIM 4. *(P, ϕ) satisfies* (5.2).

It is enough to show the commutativity of the diagram

$$\begin{array}{ccc} P & \overset{\phi_a}{\Longrightarrow} & PX(a) \\ {\scriptstyle \phi_{ba}}\Big\Downarrow & & \Big\Downarrow{\scriptstyle \phi_b X(a)} \\ PX(ba) & \underset{PX_{b,a}}{\Longrightarrow} & PX(b)X(a) \end{array} \quad . \tag{5.15}$$

Thus, it is enough to show that for each $x \in \mathscr{C}_0$ we have the equality $\phi_b X(a)x\circ\phi_a x = PX_{b,a}x \circ \phi_{ba}x$. But it is also easily verified from the formula (5.9) and the axiom (a_2) of Definition 5.1.1 that both sides are equal to $(\delta_{c,ba}X_{b,a}x)_{c\in G}$.

By the two claims above (P, ϕ) turns out to be a G-equivariant functor. $\square$

REMARK 5.4.10. For a colax G-category $(\mathscr{C}, X)$, the orbit category $\mathscr{C}/G$ is defined in the same way, which turns out to be a G-graded category. Because in the proof of Claim 2 in Lemma 5.4.9 we used the fact that $X_{1,a}$ is a natural isomorphism, we have to use a weaker definition gor morphisms of colax G-categories, namely we have to replace natural isomorphisms by natural transformations in Definition 5.1.5.

REMARK 5.4.11. We note the following:

(1) In the definition of orbit categories, the classical definition took the G-orbits Gx of objects $x \in \mathscr{C}_0$ as the objects. But since in our setting we removed the assumption of free action, it is not possible to define the canonical functor P in the same way as in the classical setting. By this reason we set $(\mathscr{C}/G)_0 := \mathscr{C}_0$. But two objects in the same G-orbit turn out to be isomorphic (indeed, if $x \in \mathscr{C}_0, a \in G$, then $\phi_a x\colon x \to X(a)x$ is an isomorphism in $\mathscr{C}/G$). Note that by writing Px for $x \in \mathscr{C}_0$, we can make it explicit that x is an object of $\mathscr{C}/G$. For instance we may write $\mathbb{1}_x^{\mathscr{C}/G} = \mathbb{1}_{Px}$, and equalities (5.4), (5.7), (5.8) can be written as

$$(\mathscr{C}/G)^a(Px, Py) = \sigma_a^{\mathscr{C}/G}(\mathscr{C}(X(a)x, y)) \quad (x, y \in \mathscr{C}_0, a \in G), \tag{5.16}$$

$$((F, \psi)/G)(Px) := P'F(x), \tag{5.17}$$

$$(\zeta/G)Px := \sigma_1^{\mathscr{C}'/G}(\zeta x \circ X_*F(x)) \in (\mathscr{C}'/G)^1(P'F(x), P'F'(x)), \tag{5.18}$$

respectively, where $(P', \phi')\colon \mathscr{C}' \to \mathscr{C}'/G$ is the canonical covering functor.

(2) In the above, our definition of the orbit category is not "left-right symmetric," namely the direct sum is taken over the first variable. There exists, however, a definition that takes the direct sum over the second

variable, and also for G-categories there exists a "left-right symmetric" definition that is close to the original definition by Gabriel. Here, we have chosen this one because it gives a generalization of the skew group algebra construction. In **[5]** we have chosen a left-right symmetric definition for G-categories, and give explicit isomorphisms between three orbit categories.

PROPOSITION 5.4.12. *Let $(\mathscr{C}, X) \in G\text{-}\mathbf{Cat}_0$. Then the canonical covering functor $(P, \phi)\colon \mathscr{C} \to \Delta(\mathscr{C}/G)$ is a G-covering. More precisely, for each $x, y \in \mathscr{C}_0$ the homomorphism*

$$(P, \phi)^{(1)}_{x,y}\colon \coprod_{a \in G} \mathscr{C}(X(a)x, y) \to (\mathscr{C}/G)(Px, Py)$$

turns out to be the identity.

PROOF. By definition of $\mathscr{C}/G$ and of P it is obvious that (P, ϕ) is dense. Let $x, y \in \mathscr{C}_0$. Then it is enough to show that $(P, \phi)^{(1)}_{x,y}$ above is the identity. This is verified as follows: for each $f = (f_a)_{a \in G} \in \coprod_{a \in G} \mathscr{C}(X(a)x, y)$ we have

$$\begin{aligned}
(P, \phi)^{(1)}_{x,y}(f) &= \sum_{a \in G} P(f_a) \circ \phi_a x \\
&= \sum_{a \in G} (\delta_{b,1} f_a \circ X_*(X(a)x))_{b \in G} \circ (\delta_{c,a} \mathbb{1}_{X(a)x})_{c \in G} \\
&= \sum_{a \in G} \left(\sum_{\substack{b,c \in G \\ d = bc}} \delta_{b,1} f_a \circ X_*(X(a)x) \circ \delta_{c,a} X(b)(\mathbb{1}_{X(a)x}) \circ X_{b,c} x \right)_{d \in G} \\
&\overset{*}{=} \sum_{a \in G} \left(\delta_{d,a} f_a \circ X_*(X(a)x) \circ \mathbb{1}_{X(1)X(a)x} \circ X_{1,a} x \right)_{d \in G} \\
&= (f_a \circ X_*(X(a)x) \circ X_{1,a}\, x)_{a \in G} = (f_a)_{a \in G} \\
&= f.
\end{aligned}$$

For the equality $\overset{*}{=}$ above note that $X(b)\colon \mathscr{C} \to \mathscr{C}$ is a functor (cf. Definition 5.1.1(2)). □

EXERCISE 5.4.13. Draw string diagrams for the equalities in the proof above.

LEMMA 5.4.14. *Let $(\mathscr{C}, X) \in G\text{-}\mathbf{Cat}_0$ and $H\colon \mathscr{C}/G \to \mathscr{D}$ a 1-morphism of $\Bbbk\text{-}\mathbf{Cat}$. Then the composite 1-morphism $(F, \psi)\colon \mathscr{C} \xrightarrow{(P,\phi)} \Delta(\mathscr{C}/G) \xrightarrow{\Delta(H)} \Delta(\mathscr{D})$ is a G-covering if and only if H is an equivalence.*

PROOF. We use Theorem 1.4.4 to show that a functor is an equivalence. Since F and H coincide on the objects, (F, ψ) is dense if and only if H is also dense. Moreover for each $x, y \in \mathscr{C}_0$ we have the following commutative diagram by Proposition 5.4.12:

$$\begin{array}{ccc}
\coprod_{a \in G} \mathscr{C}(X(a)x, y) & \xrightarrow{(F,\psi)^{(1)}_{x,y}} & \mathscr{D}(Fx, Fy) \\
{\scriptstyle (P,\phi)^{(1)}_{x,y}} \Big\| & \nearrow_{H_{y,x}} & \\
(\mathscr{C}/G)(x, y) & &
\end{array}$$

by which we see that $(F, \psi)^{(1)}_{x,y}$ is an isomorphism if and only if $H_{y,x}$ is an isomorphism. $\square$

5.4.a. Orbit category by an auto-equivalence. Given an automorphism g of a linear category $\mathscr{C}$, an action $\mathbb{Z} \overset{X}{\twoheadrightarrow} \langle g \rangle \hookrightarrow \mathrm{Aut}(\mathscr{C})$ of the infinite cyclic group $\mathbb{Z} = (\mathbb{Z}, +, 0, -)$ (resp. the cyclic group $\langle g \rangle$) to $\mathscr{C}$ is defined by $X(n) := g^n$ $(n \in \mathbb{Z})$, and $\mathscr{C}/\langle g \rangle = \mathscr{C}/\mathbb{Z}$ is defined as a special case of the orbit category of pseudo-G-category in this section. However, if g is just an auto-equivalence, not an automorphism of $\mathscr{C}$, the action X above is not defined because there does not exist g^{-1}, and even if we take a quasi-inverse $\overline{g}$ of g instead of g^{-1} the X above does not give a G-action to $\mathscr{C}$ as explained in the introduction. Therefore it is not possible to apply the construction above. (Cf. the construction of a cluster category [**15**, §1]. The definition given there is not accurate as g above is not an automorphism.)

In this subsection, as an example of the orbit category of a pseudo-G-category, we explain how to justify the orbit category of a linear category $\mathscr{C}$ by an auto-equivalence g using our construction given in this section. (See [**5**] for other justifications.) Namely, given an auto-equivalence $g \colon \mathscr{C} \to \mathscr{C}$ of a linear category $\mathscr{C}$ and its adjoint equivalence system $\mathbf{A} := (g, \overline{g}, \eta, \varepsilon)$, we will show that it is possible to canonically define a pseudofunctor $X = X_{\mathbf{A}} \colon \mathbb{Z} \to \Bbbk\text{-}\mathbf{CAT}$ satisfying $X(*) = \mathscr{C}$, and thus a pseudo-$\mathbb{Z}$-category $(\mathscr{C}, X)$. By this construction, the *orbit category* $\mathscr{C}/g$ *by an auto-equivalence* g is strictly defined as $\mathscr{C}/g := (\mathscr{C}, X)/\mathbb{Z}$ (Definition 5.4.1). We also show that given a linear functor $E \colon \mathscr{C} \to \mathscr{C}'$, an auto-equivalence g (resp. g') of a linear category $\mathscr{C}$ (resp. $\mathscr{C}'$), and a natural isomorphism $\sigma \colon g' \circ E \Rightarrow E \circ g$, it is possible to canonically define a $\mathbb{Z}$-equivariant functor $(E, \Sigma) \colon (\mathscr{C}, X) \to (\mathscr{C}', X')$ that is an extension of (E, σ), where X (resp. X') is the structure of a pseudo-$\mathbb{Z}$-category on $\mathscr{C}$ (resp. $\mathscr{C}'$) canonically defined as above.

PROPOSITION 5.4.15. *Let $E \colon \mathscr{C} \to \mathscr{C}'$ be a linear functor of linear categories, g (resp. g') an auto-equivalence of $\mathscr{C}$ (resp. $\mathscr{C}'$), $\mathbf{A} := (g, \overline{g}, \eta, \varepsilon)$ (resp. $\mathbf{A}' := (g', \overline{g}', \eta', \varepsilon')$) its adjoint equivalence system, and $\sigma \colon g' \circ E \Rightarrow E \circ g$ a natural isomorphism. Then the following hold:*

(1) *There exists a canonical construction of a pseudofunctor $X = X_{\mathbf{A}} \colon \mathbb{Z} \to \Bbbk\text{-}\mathbf{CAT}$ such that $X(*) = \mathscr{C}, X(1) = g$, and $X(-1) = \overline{g}$.*
(2) *Set $X' := X_{\mathbf{A}'}$. Then there exists a $\mathbb{Z}$-equivariant functor $(E, \Sigma) \colon (\mathscr{C}, X) \to (\mathscr{C}', X')$ that is an extension of (E, σ), i.e., that satisfies $\Sigma_1 = \sigma$.*

PROOF. (1) We first define a family $(X(n) \colon \mathscr{C} \to \mathscr{C})_{n \in \mathbb{Z}}$ of functors by setting

$$X(n) := \begin{cases} g^n & (n > 0) \\ \mathbb{1}_{\mathscr{C}} & (n = 0) \\ (\overline{g})^{|n|} & (n < 0) \end{cases}$$

for all $n \in \mathbb{Z}$. We also define a natural isomorphism $X_* \colon X(0) (= \mathbb{1}_{\mathscr{C}}) \Rightarrow \mathbb{1}_{X(*)} (= \mathbb{1}_{\mathscr{C}})$ by $X_* := \mathbb{1}_{\mathbb{1}_{\mathscr{C}}}$. Finally, we define a family $(X_{m,n} \colon X(m+n) \Rightarrow X(m) \circ X(n))_{m,n \in \mathbb{Z}}$ of natural isomorphisms as follows:

(i) In the case that $mn \geq 0$,

$$X(m+n) = \begin{cases} g^{m+n} = g^m \circ g^n = X(m) \circ X(n) & (m, n > 0) \\ X(n) = \mathbb{1}_{\mathscr{C}} \circ X(n) = X(0) \circ X(n) & (m = 0) \\ X(m) = X(m) \circ \mathbb{1}_{\mathscr{C}} = X(m) \circ X(0) & (n = 0) \\ (\bar{g})^{-m-n} = (\bar{g})^{-m} \circ (\bar{g})^{-n} = X(m) \circ X(n) & (m, n < 0). \end{cases}$$

As is shown above, it always holds that $X(m+n) = X(m) \circ X(n)$ in this case, we set $X_{m,n} := \mathbb{1}_{X(m+n)}$.

(ii) In the case that $m > 0, n < 0$, we divide this case into the following two cases:

(ii-1) When $m \geq \lvert n \rvert$, we have	(ii-2) When $m < \lvert n \rvert$, we have $m+n < 0$, and
$\begin{cases} m+n = m - \lvert n \rvert \geq 0, \\ X(m+n) = g^{m-\lvert n \rvert}, \\ X(m) \circ X(n) = g^m \circ (\bar{g})^{\lvert n \rvert}. \end{cases}$	$\begin{cases} \lvert m+n \rvert = \lvert n \rvert - m, \\ X(m+n) = (\bar{g})^{\lvert n \rvert - m}, \\ X(m) \circ X(n) = g^m \circ (\bar{g})^{\lvert n \rvert}. \end{cases}$

Therefore we define $X_{m,n} \colon X(m+n) \Rightarrow X(m) \circ X(n)$ by the left (resp. right) string diagram below in the case (ii-1) (resp. (ii-2)).

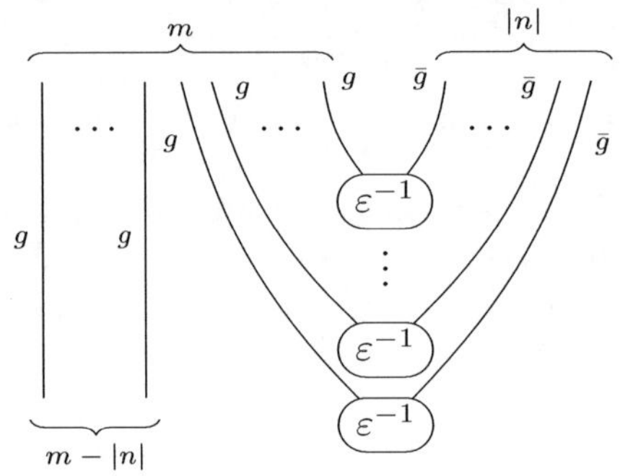

,
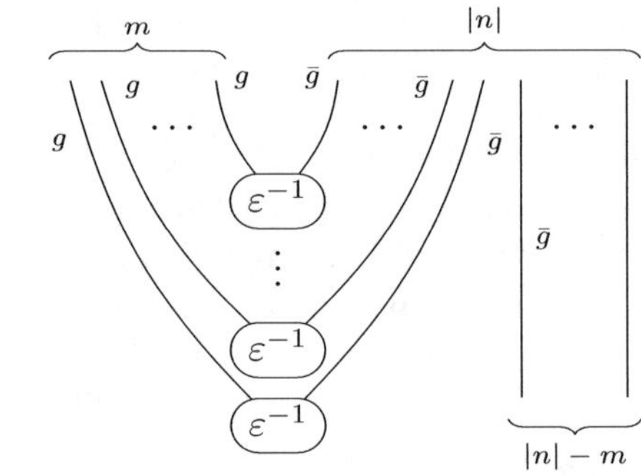

(iii) In the case that $m < 0, n > 0$, we divide this case into the following two cases:

(iii-1) When $\lvert m \rvert \geq n$, we have $m+n \leq 0$, and	(iii-2) When $\lvert m \rvert < n$, we have
$\begin{cases} \lvert m+n \rvert = \lvert m \rvert - n, \\ X(m+n) = (\bar{g})^{\lvert m \rvert - n}, \\ X(m) \circ X(n) = (\bar{g})^{\lvert m \rvert} \circ g^n. \end{cases}$	$\begin{cases} m+n = n - \lvert m \rvert > 0, \\ X(m+n) = g^{n-\lvert m \rvert}, \\ X(m) \circ X(n) = (\bar{g})^{\lvert m \rvert} \circ g^n. \end{cases}$

Therefore we define $X_{m,n} \colon X(m+n) \Rightarrow X(m) \circ X(n)$ by the left (resp. right) string diagram below in the case (iii-1) (resp. (iii-2)).

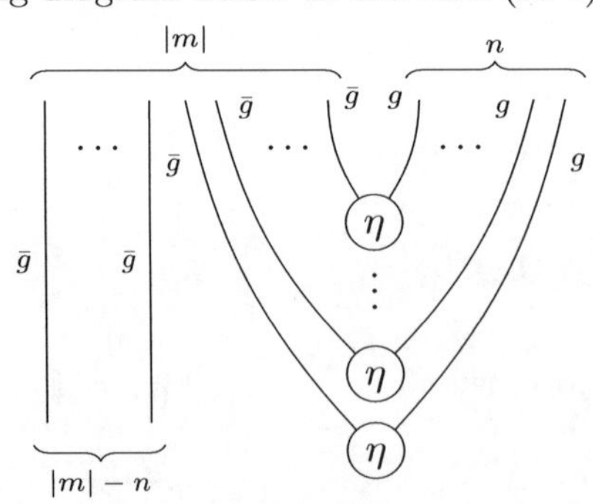

,
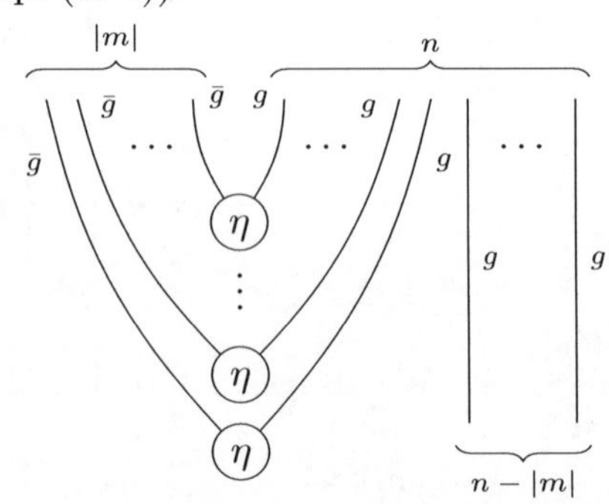

Under the definition of X above, it suffices to show the following formulas:

(a₁)

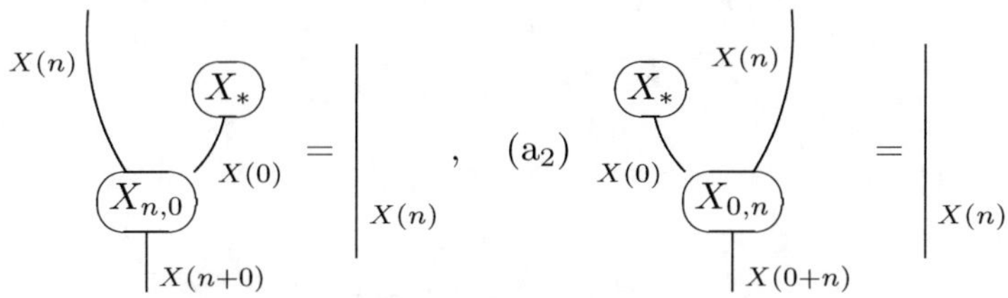

$(n \in \mathbb{Z})$,

(b)

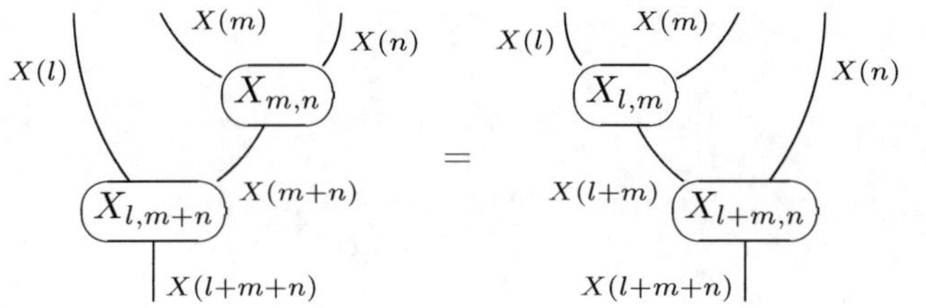

$(l, m, n \in \mathbb{Z})$.

Both (a₁) and (a₂) are clear from definitions. The equality (b) follows from the zigzag equations of adjoints. Here, we show it in the case $l > 0, m < 0, n > 0$ by using the example that $(l, m, n) = (4, -3, 5)$. The same proof works in general.

By definition, (b) is expressed by the following string diagram:

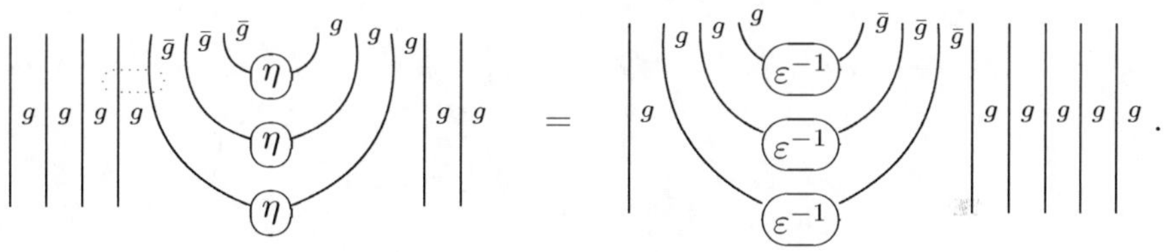

Replace the identity natural transformation drawn by the dotted line on the left-hand side by its equivalent $\varepsilon \circ \varepsilon^{-1}$, and use the zigzag equations. Then the left-hand side becomes:

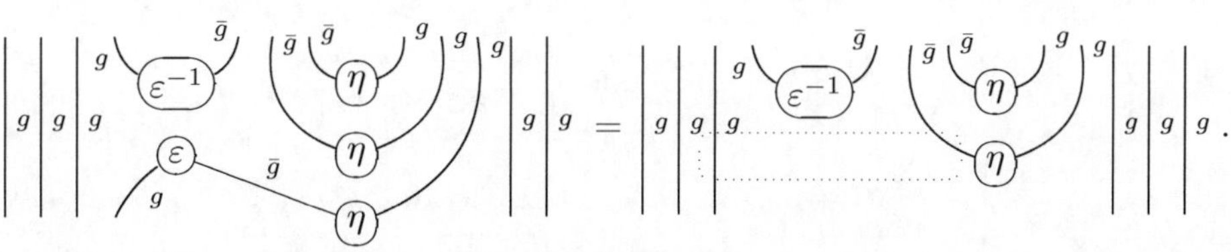

Next, we do the same for the identity natural transformation depicted by the dotted line in the last figure, and then do the same operation again, and the result is finally equal to the right-hand side of (b).

(2) We define a family $\Sigma = (\Sigma_n \colon X'(n) \circ E \Rightarrow E \circ X(n))_{n \in \mathbb{Z}}$ of natural isomorphisms as follows:

(i) When $n = 0$, we set $\Sigma_0 := \mathbb{1}_E$.

(ii) When $n > 0$, we define $\Sigma_n \colon g'^n \circ E \Rightarrow E \circ g^n$ by the following string diagram:

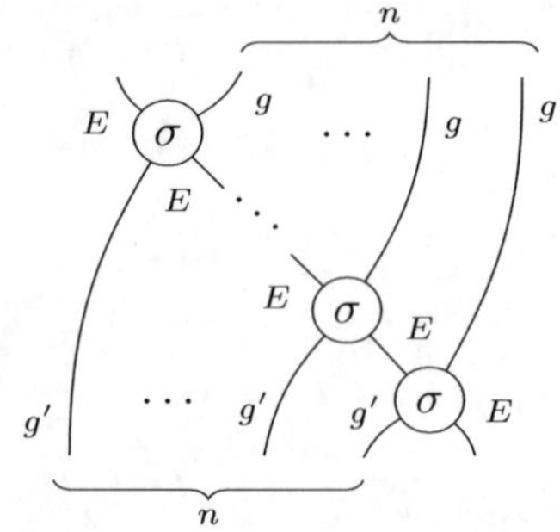

.

(iii) When $n < 0$, we first define $\Sigma_{-1} \colon \bar{g'} \circ E \Rightarrow E \circ \bar{g}$ for $n = -1$ by the left string diagram below. Using this, for each $n < 0$, we define $\Sigma_n \colon X'(n) \circ E \Rightarrow E \circ X(n)$ by the right string diagram below, paying attention to the fact that $X'(n) = (\bar{g'})^{|n|}, X(n) = (\bar{g})^{|n|}$:

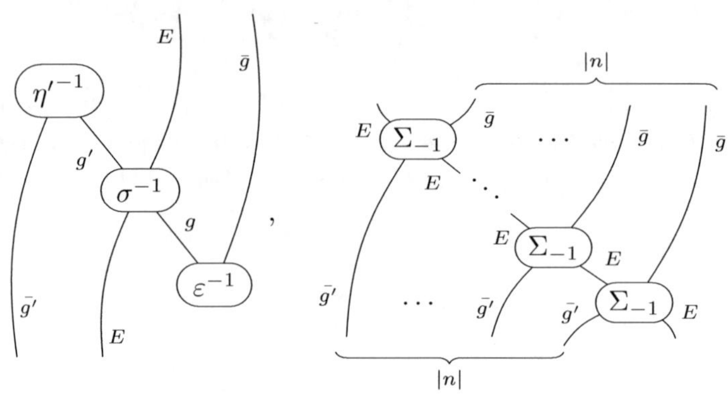

Under the definition above, it is enough to verify the equalities (5.1), (5.2). It is clear that the equality (5.1) holds. The equality (5.2) has the following form in this case:

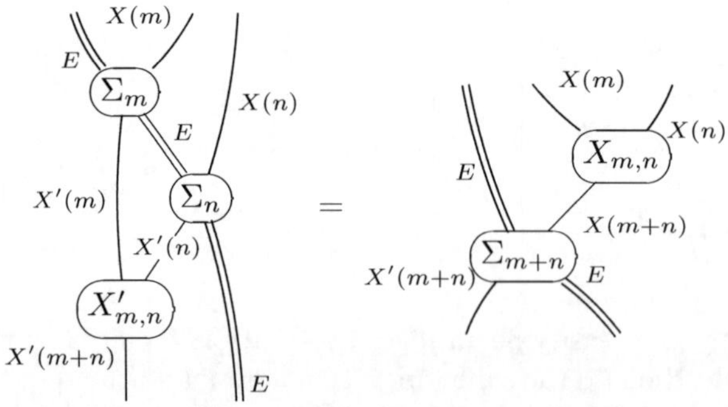

$(m, n \in \mathbb{Z})$.

When $mn \geq 0$, both $X_{m,n}$ and $X'_{m,n}$ are identities, and hence (5.2) holds. When $mn < 0$, we give a proof for the case that $m = 3, n = -2$ below. The other general cases are proved in exactly the same way. In this case the equality in question has the following form:

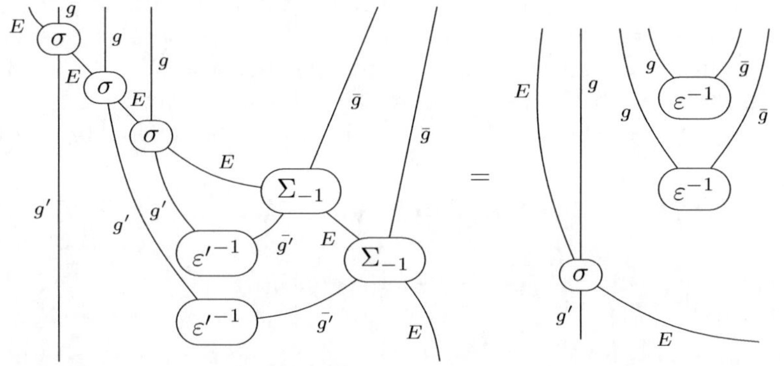

To show this, it is enough to show the following equality:

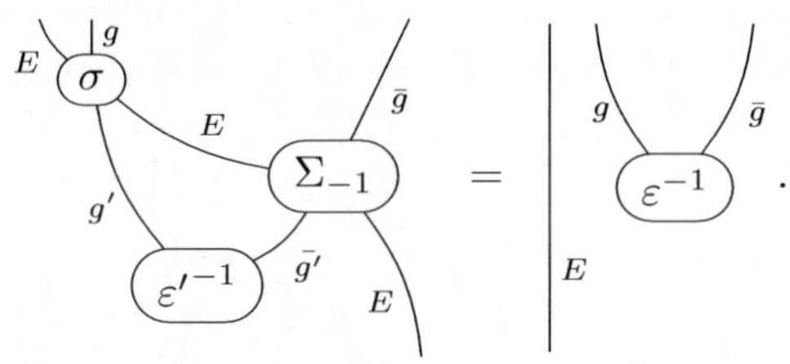

This is verified by the following computation:

(LHS) =

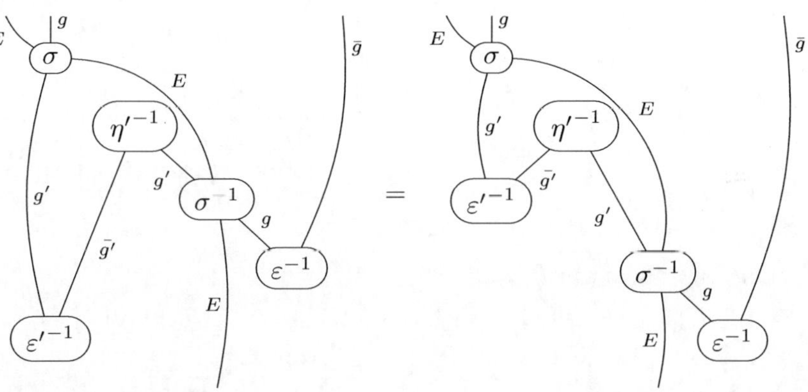

=

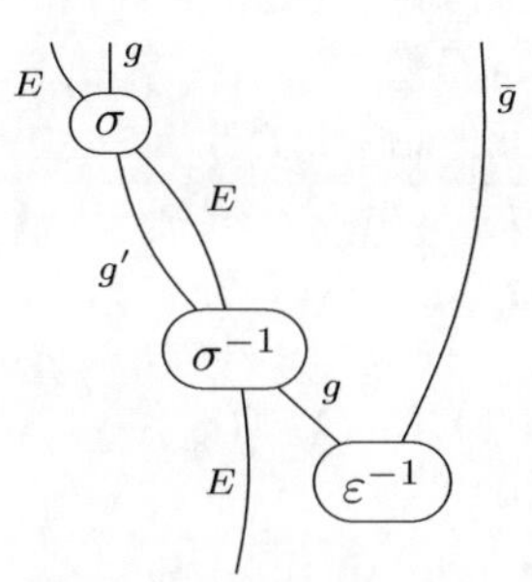

= (RHS). □

5.5. Orbit 2-functors, diagonal 2-functors and their 2-adjoints

By forgetting G-gradings of the orbit category $\mathscr{C}/G$ we regard $\mathscr{C}/G \in \Bbbk\text{-}\mathbf{Cat}_0$ (Remark 5.4.5). This yields a 2-functor $?/G\colon G\text{-}\mathbf{Cat} \to \Bbbk\text{-}\mathbf{Cat}$. In this subsection we show that this gives a strict left 2-adjoint to the diagonal 2-functor $\Delta\colon \Bbbk\text{-}\mathbf{Cat} \to G\text{-}\mathbf{Cat}$. Through the proof we see that the canonical covering functor gives a component of 2-unit of this adjoint, which gives us a characterization of covering functors by universality.

DEFINITION 5.5.1. For each $\mathscr{D} \in \Bbbk\text{-}\mathbf{Cat}$ we define a functor $Q_{\mathscr{D}}\colon \Delta(\mathscr{D})/G \to \mathscr{D}$ as follows:

- For each $x \in (\Delta(\mathscr{D})/G)_0 = \mathscr{D}_0$, we set $Q_{\mathscr{D}}(x) := x$,
- For each $(f_a)_{a\in G} \in (\Delta(\mathscr{D})/G)(x,y) = \coprod_{a\in G} \mathscr{D}(x,y)$ we set $Q_{\mathscr{D}}((f_a)_{a\in G}) := \sum_{a\in G} f_a \quad (x, y \in (\Delta(\mathscr{D})/G)_0)$.

As is easily seen, $Q_{\mathscr{D}}$ is a linear functor.

THEOREM 5.5.2. *Orbit 2-functor $?/G\colon G\text{-}\mathbf{Cat} \to \Bbbk\text{-}\mathbf{Cat}$ is a strict left 2-adjoint to the diagonal 2-functor $\Delta\colon \Bbbk\text{-}\mathbf{Cat} \to G\text{-}\mathbf{Cat}$. The unit is given by the family of canonical covering functor $((P_{\mathscr{C}}, \phi_{\mathscr{C}})\colon \mathscr{C} \to \Delta(\mathscr{C}/G))_{\mathscr{C}\in G\text{-}\mathbf{Cat}_0}$, and counit is given by the family $(Q_{\mathscr{D}}\colon \Delta(\mathscr{D})/G \to \mathscr{D})_{\mathscr{D}\in\Bbbk\text{-}\mathbf{Cat}_0}$ of functors defined as above.*

Therefore by a general theory (Theorem 4.6.3, Remark 4.6.5), *for each $\mathscr{C} \in G\text{-}\mathbf{Cat}, \mathscr{D} \in \Bbbk\text{-}\mathbf{Cat}$, we have a natural isomorphism*

$$\Bbbk\text{-}\mathbf{Cat}(\mathscr{C}/G, \mathscr{D}) \cong G\text{-}\mathbf{Cat}(\mathscr{C}, \Delta(\mathscr{D})), \quad H \mapsto \Delta(H) \circ (P_{\mathscr{C}}, \phi_{\mathscr{C}}) \tag{5.19}$$

of categories. In particular, $(P_{\mathscr{C}}, \phi_{\mathscr{C}})$ has a universality in the following sense. Namely, for each 1-morphism $(F,\psi)\colon \mathscr{C} \to \Delta(\mathscr{D})$, $(\mathscr{D} \in \Bbbk\text{-}\mathbf{Cat})$ in $G\text{-}\mathbf{Cat}$, there exists a unique 1-morphism $H\colon \mathscr{C}/G \to \mathscr{D}$ in $\Bbbk\text{-}\mathbf{Cat}$ that makes the following diagram in $G\text{-}\mathbf{Cat}$ strictly commutative:

$$\begin{array}{ccc} \mathscr{C} & \xrightarrow{(F,\psi)} & \Delta(\mathscr{D}). \\ {\scriptstyle (P_{\mathscr{C}},\phi_{\mathscr{C}})}\downarrow & \nearrow{\scriptstyle \Delta(H)} & \\ \Delta(\mathscr{C}/G) & & \end{array}$$

PROOF. We set $\eta := ((P_{\mathscr{C}}, \phi_{\mathscr{C}}))_{\mathscr{C}\in G\text{-}\mathbf{Cat}_0}$, $\varepsilon := (Q_{\mathscr{D}})_{\mathscr{D}\in\Bbbk\text{-}\mathbf{Cat}_0}$.

CLAIM 1. $(\Delta \circ \varepsilon) \bullet (\eta \circ \Delta) = \mathbb{1}_{\Delta}$.

Indeed, let $\mathscr{D} \in (\Bbbk\text{-}\mathbf{Cat})_0$. Then it is enough to show $\Delta(Q_{\mathscr{D}}) \cdot (P_{\Delta(\mathscr{D})}, \phi_{\Delta(\mathscr{D})}) = \mathbb{1}_{\Delta(\mathscr{D})}$, where

$$\begin{aligned} (\text{LHS}) &= \left(Q_{\mathscr{D}} \circ P_{\Delta(\mathscr{D})}, (Q_{\mathscr{D}}\phi_{\Delta(\mathscr{D})}(a))_{a\in G}\right), \text{ and} \\ (\text{RHS}) &= \left(\mathbb{1}_{\mathscr{D}}, (\mathbb{1}_{\mathbb{1}_{\mathscr{D}}})_{a\in G}\right). \end{aligned}$$

Equality in the first entry: We show that $Q_{\mathscr{D}} \circ P_{\Delta(\mathscr{D})} = \mathbb{1}_{\mathscr{D}}$. For each $x, y \in \mathscr{D}_0$ and $f \in \mathscr{D}(x,y)$, we have $(Q_{\mathscr{D}} \circ P_{\Delta(\mathscr{D})})(x) = Q_{\mathscr{D}}(x) = x$ and $(Q_{\mathscr{D}} \circ P_{\Delta(\mathscr{D})})(f) = (\delta_{a,1} f \cdot (\eta_{\Delta(\mathscr{D})} x))_{a\in G} = \sum_{a\in G} \delta_{a,1} f = f$.

Equality in the second entry: For each $a \in G$, we show $Q_{\mathscr{D}}\phi_{\Delta(\mathscr{D})}(a) = \mathbb{1}_{\mathbb{1}_{\mathscr{D}}}$. For each $x \in \mathscr{D}_0$, we have $Q_{\mathscr{D}}\left(\phi_{\Delta(\mathscr{D})}(a)x\right) = Q_{\mathscr{D}}\left((\delta_{b,a}\mathbb{1}_{\Delta(\mathscr{D})(a)x})_{b\in G}\right) = \sum_{b\in G} \delta_{b,a}\mathbb{1}_x = \mathbb{1}_x = \mathbb{1}_{\mathbb{1}_{\mathscr{D}}x}$. Hence (LHS) = (RHS).

CLAIM 2. $(\varepsilon \circ (?/G)) \bullet ((?/G) \circ \eta) = \mathbb{1}_{?/G}$.

Indeed, let $\mathscr{C} \in G\text{-}\mathbf{Cat}_0$. Then it is enough to show $Q_{\mathscr{C}/G} \cdot (P_{\mathscr{C}}, \phi_{\mathscr{C}})/G = \mathbb{1}_{\mathscr{C}/G}$.

On objects: For each $x \in (\mathscr{C}/G)_0$, we have $Q_{\mathscr{C}/G}((P_{\mathscr{C}}, \phi_{\mathscr{C}})/G)(x)) = Q_{\mathscr{C}/G}(P_{\mathscr{C}}(x)) = x$.

On morphisms: For each morphism $f = (f_a)_{a \in G} \colon X(a)x \to y$ in $\mathscr{C}/G$ we have $Q_{\mathscr{C}/G}((P_{\mathscr{C}}, \phi_{\mathscr{C}})/G)(f) = Q_{\mathscr{C}/G}((P_{\mathscr{C}}(f_a) \circ (\phi_{\mathscr{C}})_a\, x)_{a \in G}) = \sum_{a \in G} P_{\mathscr{C}}(f_a) \circ (\phi_{\mathscr{C}})_a\, x = (P_{\mathscr{C}}, \phi_{\mathscr{C}})^{(1)}_{x,y}(f) = f$. Hence the claim holds.

The two claims above prove the theorem. □

COROLLARY 5.5.3. *For each 1-morphism $(F, \psi) \colon \mathscr{C} \to \Delta(\mathscr{D})$ in G-**Cat**, the following are equivalent:*

(1) *(F, ψ) is a G-covering;*
(2) *There exists an equivalence $H \colon \mathscr{C}/G \to \mathscr{D}$ such that the diagram*

$$\begin{array}{ccc} \mathscr{C} & \xrightarrow{(F,\psi)} & \Delta(\mathscr{D}) \\ {\scriptstyle (P_{\mathscr{C}}, \phi_{\mathscr{C}})} \downarrow & \nearrow {\scriptstyle \Delta(H)} & \\ \Delta(\mathscr{C}/G) & & \end{array}$$

is strictly commutative, i.e., we have an equality $(F, \psi) = \Delta(H) \circ (P_{\mathscr{C}}, \phi_{\mathscr{C}})$.

PROOF. This immediately follows by Theorem 5.5.2 and Lemma 5.4.14. □

REMARK 5.5.4. In the above, we discussed only small categories to fit the theory within the range of light 2-categories. However, we also can define a 2-category G-**CAT** (resp. G-$\mathbf{CAT}^{\mathrm{s}}$) of light linear categories with pseudo-G-actions in the same way for G-**Cat** (resp. G-$\mathbf{Cat}^{\mathrm{s}}$), where we have to note that these 2-categories are 2-moderate (cf. Proposition A.4.2). For each $\mathscr{C} \in G\text{-}\mathbf{CAT}_0$, we may define $\mathscr{C}/G$ in a similar way, and also in this case $?/G$ turns out to be a left adjoint to Δ. Therefore a natural isomorphism (5.19) is extended between these to have a natural isomorphism

$$\mathbb{k}\text{-}\mathbf{CAT}(\mathscr{C}/G, \mathscr{D}) \cong G\text{-}\mathbf{CAT}(\mathscr{C}, \Delta(\mathscr{D})),\ H \mapsto \Delta(H) \circ (P_{\mathscr{C}}, \phi_{\mathscr{C}}). \tag{5.20}$$

EXAMPLE 5.5.5. If $\mathscr{C} = \mathbb{k}$ is a field and the G-action is trivial, then $\mathscr{C}/G = \mathbb{k}G$ is just the usual group algebra. If $\mathscr{D} = \operatorname{Mod} \mathbb{k}$, the category of $\mathbb{k}$-vector spaces, then $\mathbb{k}\text{-}\mathbf{CAT}(\mathbb{k}G, \operatorname{Mod} \mathbb{k}) = \mathbb{k}G\text{-}\operatorname{Mod}$ is the category of left $\mathbb{k}G$-modules, and

$$G\text{-}\mathbf{CAT}(\mathbb{k}, \Delta(\operatorname{Mod} \mathbb{k})) = \operatorname{Rep}_{\mathbb{k}} G$$

is the category of $\mathbb{k}$-representation of G. In this case, the isomorphism of categories (5.20) gives the well-known isomorphism $\mathbb{k}G\text{-}\operatorname{Mod} \cong \operatorname{Rep}_k G$.

EXERCISE 5.5.6. In the case above, show that $G\text{-}\mathbf{CAT}(\mathbb{k}, \Delta(\operatorname{Mod} \mathbb{k}))$ coincides with the category $\operatorname{Rep}_{\mathbb{k}} G$ of $\mathbb{k}$-representations of G.

To give a characterization of G-covering functors we define the following:

DEFINITION 5.5.7. Let $\mathscr{C} = (\mathscr{C}, X)$ be a pseudo-G-category and $\mathscr{D}$ a linear category. Then a G-equivariant functor (F, ψ) from $\mathscr{C}$ to $\Delta(\mathscr{D})$ is called a *G-invariant* functor from $\mathscr{C}$ to $\mathscr{D}$ and denote it by $(F, \psi) \colon \mathscr{C} \to \mathscr{D}$. Hence in particular, ψ is a family $(\psi_a)_{a \in G}$ of natural isomorphisms $\psi_a \colon F \Rightarrow FX(a)$.

The following theorem follows by Theorem 5.5.2 and Corollary 5.5.3, which gives a characterization of G-covering functors.

THEOREM 5.5.8. *For a G-invariant functor $(F,\psi)\colon \mathscr{C} \to \mathscr{D}$ the following are equivalent:*

(1) (F,ψ) *is a G-covering functor;*
(2) (F,ψ) *is a G-precovering and is universal among G-precoverings from $\mathscr{C}$;*
(3) (F,ψ) *is G-invariant, and is universal among G-invariant functors from $\mathscr{C}$;*
(3′) (F,ψ) *is G-invariant, and is 2-universal among G-invariant functors from $\mathscr{C}$, i.e., (F,ψ) induces an isomorphism*

$$(F,\psi)^*\colon \Bbbk\text{-}\mathbf{Cat}(\mathscr{C}/G,\mathscr{D}) \to G\text{-}\mathbf{Cat}(\mathscr{C},\Delta(\mathscr{D})),\quad H \mapsto \Delta(H)\circ(F,\psi)$$

of categories for all linear categories $\mathscr{D}$;
(4) *There exists an equivalence $H\colon \mathscr{C}/G \to \mathscr{D}$ of categories such that $(F,\psi) \cong H\circ(P_{\mathscr{C}},\phi_{\mathscr{C}})$ as G-invariant functors;*
(5) *There exists an equivalence $H\colon \mathscr{C}/G \to \mathscr{D}$ of categories such that $(F,\psi) = H\circ(P_{\mathscr{C}},\phi_{\mathscr{C}})$.*

As a special case of Lemma 5.1.8, we have statement (1) below.

LEMMA 5.5.9.

(1) *Let $\mathscr{C},\mathscr{C}'$ be G-categories and $\mathscr{D}$ a linear category. Consider morphisms $\mathscr{C}' \xrightarrow{(E,\rho)} \mathscr{C} \xrightarrow{(F,\psi)} \mathscr{D}$ between them, where (E,ρ) is a G-equivariant functor and (F,ψ) a G-invariant functor. Then*

$$(F,\psi)\circ(E,\rho) := (F\circ E, ((F\rho_a)(\psi_a E))_{a\in G})\colon \mathscr{C}' \to \mathscr{D}$$

turns out to be a G-invariant functor.
(2) *In the above, if (E,ρ) is a G-equivariant equivalence and (F,ψ) is a G-covering functor, then the composite $(F,\psi)\circ(E,\rho)$ turns out to be a G-covering functor. Therefore $\mathscr{C}'/G$ and $\mathscr{D}$ are equivalent.*

5.6. Smash products

In this section, we define a 2-functor $?\#G\colon G\text{-}\mathbf{GrCat} \to G\text{-}\mathbf{Cat}$, which will be a quasi-inverse of the orbit 2-functor $?/G\colon G\text{-}\mathbf{Cat} \to G\text{-}\mathbf{GrCat}$.

DEFINITION 5.6.1. Let $\mathscr{B}$ be a G-graded category. Then we define a G-category $\mathscr{B}\#G$ called the *smash product* of $\mathscr{B}$ and G.

- $(\mathscr{B}\#G)_0 := \mathscr{B}_0 \times G = \{x^{(a)} := (x,a) \mid x \in \mathscr{B}, a \in G\}$.
- For each $x^{(a)}, y^{(b)} \in (\mathscr{B}\#G)_0$, we set $(\mathscr{B}\#G)(x^{(a)},y^{(b)}) := \mathscr{B}^{b^{-1}a}(x,y)$. To be more precise (namely to make the union of the set of morphisms disjoint) we set

$$\begin{aligned}(\mathscr{B}\#G)(x^{(a)},y^{(b)}) &:= \{b\}\times\mathscr{B}^{b^{-1}a}(x,y)\times\{a\}\\ &= \{{}^{(b)}f^{(a)} := (b,f,a) \mid f \in \mathscr{B}^{b^{-1}a}(x,y)\}.\end{aligned}$$

But if there is no confusion we identify it with $\mathscr{B}^{b^{-1}a}(x,y)$ by the second projection

$$\mathrm{pr}_2\colon (\mathscr{B}\#G)(x^{(a)},y^{(b)}) \to \mathscr{B}^{b^{-1}a}(x,y).\quad {}^{(b)}f^{(a)} \mapsto f.$$

- For each $x^{(a)}, y^{(b)}, z^{(c)} \in (\mathscr{B}\#G)_0$, we define the composition by the following commutative diagram:

$$\begin{array}{ccc}
(\mathscr{B}\#G)(y^{(b)}, z^{(c)}) \times (\mathscr{B}\#G)(x^{(a)}, y^{(b)}) & \longrightarrow & (\mathscr{B}\#G)(x^{(a)}, z^{(c)}) \\
\downarrow{\scriptstyle \mathrm{pr}_2 \times \mathrm{pr}_2} & & \downarrow{\scriptstyle \mathrm{pr}_2} \\
\mathscr{B}^{c^{-1}b}(y, z) \times \mathscr{B}^{b^{-1}a}(x, y) & \longrightarrow & \mathscr{B}^{c^{-1}a}(x, z) \\
\downarrow & & \downarrow \\
\mathscr{B}(y, z) \times \mathscr{B}(x, y) & \longrightarrow & \mathscr{B}(x, z),
\end{array}$$

where the bottom homomorphism is defined by the composition of $\mathscr{B}$. Thus for each $(g, f) \in \mathscr{B}^{c^{-1}b}(y, z) \times \mathscr{B}^{b^{-1}a}(x, y)$, we set

$$ {}^{(c)}g^{(b)} \circ {}^{(b)}f^{(a)} := {}^{(c)}(g \circ f)^{(a)}. \tag{5.21}$$

- We define a G-action $X := X^{\mathscr{B}\#G}$ on $\mathscr{B}\#G$ as follows:

$$\begin{cases} X(c)(x^{(a)}) := x^{(ca)} \\ X(c)({}^{(b)}f^{(a)}) := {}^{(cb)}f^{(ca)} \end{cases} \tag{5.22}$$

$(x^{(a)}, y^{(b)} \in (\mathscr{B}\#G)_0, {}^{(b)}f^{(a)} \in (\mathscr{B}\#G)(x^{(a)}, y^{(b)}), x, y \in \mathscr{B}_0, a, b, c \in G, f \in \mathscr{B}^{b^{-1}a}(x, y)))$. It is obvious by definition that this action is free.

REMARK 5.6.2. On definition of the morphism set of smash products, we remark the following:

(1) The source (resp. target) of ${}^{(b)}f^{(a)}$ is determined as $\mathrm{dom}(f)^{(a)}$ (resp. $\mathrm{cod}(f)^{(b)}$). Therefore using this form the union of morphism sets turns out to be disjoint (cf. Remark 1.1.7).

(2) For each $x^{(a)}, y^{(b)} \in \mathscr{B}_0 \times G$, the morphism set of $(\mathscr{B}\#G)^{\mathrm{op}}$ is given by

$$\begin{aligned}(\mathscr{B}\#G)^{\mathrm{op}}(x^{(a)}, y^{(b)}) &:= (\mathscr{B}\#G)(y^{(b)}, x^{(a)}) = \mathscr{B}^{a^{-1}b}(y, x) \\ &= \{f\colon x \to y \text{ (in } \mathscr{B}^{\mathrm{op}}) \mid f \in \mathscr{B}^{a^{-1}b}(y, x)\}.\end{aligned}$$

Therefore to make the union of the morphism sets disjoint we set

$$(\mathscr{B}\#G)^{\mathrm{op}}(x^{(a)}, y^{(b)}) := \{b\} \times \mathscr{B}^{a^{-1}b}(y, x) \times \{a\}.$$

Note that we do not set $\{a\} \times \mathscr{B}^{a^{-1}b}(y, x) \times \{b\}$ (cf. Proposition 5.10.6).

(3) If $\mathscr{B}$ is a light G-graded category, then by the definition above $\mathscr{B}\#G$ is also a light category.

DEFINITION 5.6.3. For a G-graded category $\mathscr{B}$, we define a functor $Q_{\mathscr{B},G} := Q\colon \mathscr{B}\#G \to \mathscr{B}$ as follows:

- $Q(x^{(a)}) = x \quad (x^{(a)} \in \mathscr{B}\#G)$,
- $Q({}^{(b)}f^{(a)}) := f \quad ({}^{(b)}f^{(a)} \in (\mathscr{B}\#G)(x^{(a)}, y^{(b)}))$.

PROPOSITION 5.6.4. *Let Q be as above. Then we have $Q = QX(a)$ for all $a \in G$. Thus, Q is a strict G-invariant functor, and $Q = (Q, \mathbb{1})\colon \mathscr{B}\#G \to \mathscr{B}$ turns out to be a G-covering functor. Hence in particular, if $(P, \phi)\colon \mathscr{B}\#G \to (\mathscr{B}\#G)/G$ is a canonical covering functor, then there exists a unique equivalence $H\colon (\mathscr{B}\#G)/G \to \mathscr{B}$ such that $Q = H \circ (P, \phi)$.*

DEFINITION 5.6.5. We extend the smash product construction to a 2-functor as follows:

On 1-morphisms: Let $(H, r)\colon \mathscr{B} \to \mathscr{B}'$ be a 1-morphism of G-**GrCat**. Then we define a functor $(H, r)\#G\colon \mathscr{B}\#G \to \mathscr{B}'\#G$, which is a 1-morphism of G-**Cat** as follows:

On objects: For each $x^{(a)} \in \mathscr{B}\#G$, we set

$$((H, r)\#G)(x^{(a)}) := (Hx)^{(ar_x)}.$$

On morphisms: For each $f \in (\mathscr{B}\#G)(x^{(a)}, y^{(b)}) = \mathscr{B}^{b^{-1}a}(x, y)$ by noting that $H(f) \in \mathscr{B}'^{r_y^{-1}b^{-1}ar_x}(Hx, Hy) \cong (\mathscr{B}'\#G)((Hx)^{(ar_x)}, (Hy)^{(br_y)})$, we set

$$((H, r)\#G)({}^{(b)}f^{(a)}) := {}^{(br_y)}H(f)^{(ar_x)}.$$

Then as is easily seen, $(H, r)\#G$ is a strict G-equivariant functor. Hnece $(H, r)\#G = ((H, r)\#G, 1)\colon \mathscr{B}\#G \to \mathscr{B}'\#G$ is a 1-morphism of G-**Cat**.

On 2-morphisms: Next let $(H', r')\colon \mathscr{B} \to \mathscr{B}'$ be a 1-morphism of G-**GrCat** and $\theta\colon (H, r) \Rightarrow (H', r')$ a 2-morphism. Then we define $\theta\#G\colon (H, r)\#G \Rightarrow (H', r')\#G$ by

$$\begin{aligned}(\theta\#G)x^{(a)} &:= \theta x \in \mathscr{B}'^{r_x'^{-1}r_x}(Hx, H'x) \cong (\mathscr{B}'\#G)((Hx)^{(ar_x)}, (H'x)^{(ar'_x)}) \\ &= (\mathscr{B}'\#G)(((H, r)\#G)(x^{(a)}), ((H', r')\#G)(x^{(a)}))\end{aligned}$$

for all $x^{(a)} \in \mathscr{B}\#G$. As is easily verified, $\theta\#G$ is a 2-morphism of G-**Cat**.

LEMMA 5.6.6. *By the definition above, the smash product construction is extended to a 2-functor*

$$?\#G\colon G\text{-}\mathbf{GrCat} \to G\text{-}\mathbf{Cat}.$$

PROOF. Here, we only check that $?\#G$ preserves the horizontal compositions; the remaining conditions for $?\#G$ to be a 2-functor follow immediately from definition. Now consider the following diagram in G-**GrCat**

$$\mathscr{B}'' \underset{(F,\zeta)}{\overset{(F',\zeta')}{\longleftarrow}} \!\!\!\!\! {}^{\theta'\Uparrow}\ \mathscr{B}' \underset{(H,\xi)}{\overset{(H',\xi')}{\longleftarrow}} \!\!\!\!\! {}^{\theta\Uparrow}\ \mathscr{B}.$$

For each $x^{(a)} \in \mathscr{B}\#G$, we have

$$\begin{aligned}((\theta' \circ \theta)\#G)(x^{(a)}) &= (\theta' \circ \theta)x = ((F' \circ \theta) \bullet (\theta' \circ H))x \\ &= (F' \circ \theta)x \circ (\theta' \circ H)x = F'(\theta x) \circ \theta'(Hx)\end{aligned}$$

and

$$\begin{aligned}((\theta'\#G) \circ (\theta\#G))(x^{(a)}) &= (((F', \zeta')\#G) \circ (\theta\#G)) \bullet ((\theta'\#G) \circ ((H, \xi)\#G)))(x^{(a)}) \\ &= ((F', \zeta')\#G) \circ (\theta\#G))(x^{(a)}) \circ ((\theta'\#G) \circ ((H, \xi)\#G))(x^{(a)}) \\ &= ((F', \zeta')\#G)(\theta x) \circ (\theta'\#G)((Hx)^{(a\xi_x)}) \\ &= F'(\theta x) \circ \theta'(Hx).\end{aligned}$$

Therefore we have $(\theta' \circ \theta)\#G = (\theta'\#G) \circ (\theta\#G)$. □

5.7. 2-categorical Cohen–Montgomery duality

Under the preparation above, we state the main result of this chapter. This gives a large extension of the result by Cohen–Montgomery **[20]** that gives a duality between rings with G-action and G-graded rings.

THEOREM 5.7.1. *Both 2-functors $?/G$ and $?\#G$ are 2-equivalences. These are 2-quasi-inverses of each other. Hence the 2-categories G-**Cat** and G-**GrCat** are 2-equivalent* (cf. Definition 4.6.1(3)). *More precisely, there exist four 2-natural transformations*

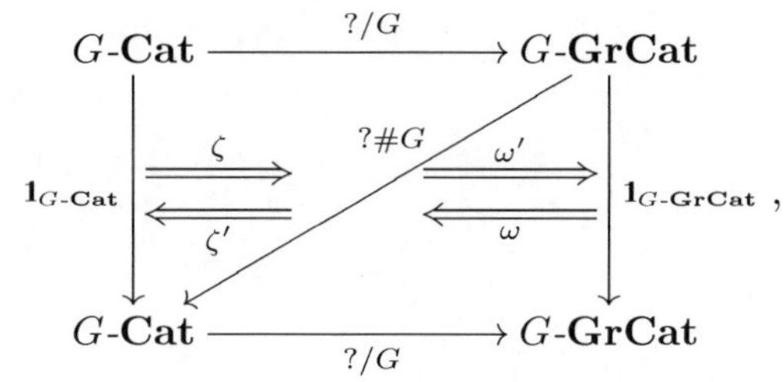

which satisfy the four relations below, and are 2-natural equivalences:

$$\zeta' \bullet \zeta \cong \mathbb{1}_{\mathbf{1}_{G\text{-}\mathbf{Cat}}}, \tag{5.23}$$
$$\zeta \bullet \zeta' \cong \mathbb{1}_{(?\#G)\circ(?/G)}, \tag{5.24}$$
$$\omega' \bullet \omega = \mathbb{1}_{\mathbf{1}_{G\text{-}\mathbf{GrCat}}}, \tag{5.25}$$
$$\omega \bullet \omega' \cong \mathbb{1}_{(?/G)\circ(?\#G)}. \tag{5.26}$$

Moreover, the following hold.

(1) *If $\mathscr{C} \in G\text{-}\mathbf{Cat}_0^{\mathrm{s}}$, then $\zeta'_{\mathscr{C}}$ is a strict G-equivariant functor, and $\omega_{\mathscr{B}}$ is a strict degree-preserving functor for all $\mathscr{B} \in G\text{-}\mathbf{GrCat}_0$.*
(2) *ζ and ω' are strict 2-natural transformations.*
(3) *$(?/G, ?\#G, \zeta, \omega')$ is a strict 2-adjoint system* (cf. Definition 4.6.1(4)).

We prove this theorem in the next section.

5.8. Proof of Theorem 5.7.1

5.8.a. $\zeta\colon \mathbf{1}_{G\text{-}\mathbf{Cat}} \Rightarrow (?\#G)\circ(?/G)$.

DEFINITION 5.8.1. Let $\mathscr{C} = (\mathscr{C}, X)$ be an object of G-**Cat**, $(P, \phi)\colon \mathscr{C} \to \mathscr{C}/G$ the canonical covering functor, and $X' := X^{(\mathscr{C}/G)\#G}$ the free G-action on $(\mathscr{C}/G)\#G$. Then we define a G-equivariant functor $\zeta_{\mathscr{C}}\colon \mathscr{C} \to (\mathscr{C}/G)\#G$ as follows:

On objects. For each $x \in \mathscr{C}_0$, we set

$$\zeta_{\mathscr{C}}(x) := (Px)^{(1)}.$$

On morphisms. For each morphism $f\colon x \to y$ of $\mathscr{C}$, we set

$$\zeta_{\mathscr{C}}(f) := {}^{(1)}(Pf)^{(1)} \in ((\mathscr{C}/G)\#G)((Px)^{(1)}, (Py)^{(1)}). \tag{5.27}$$

Natural isomorphisms. For each $a \in G$ and $x \in \mathscr{C}_0$, we have

$$((\mathscr{C}/G)\#G)((Px)^{(a)}, (PX(a)x)^{(1)}) \overset{\mathrm{pr}_2}{\cong} (\mathscr{C}/G)^a(Px, PX(a)x) \ni \phi_a x,$$

from which we can define a morphism $\overline{\phi}_a x\colon X'(a)\zeta_{\mathscr{C}}x \to \zeta_{\mathscr{C}}X(a)x$ of $(\mathscr{C}/G)\#G$ by the commutative diagram

$$\begin{array}{ccc} X'(a)\zeta_{\mathscr{C}}x & \xrightarrow{\overline{\phi}_a x} & \zeta_{\mathscr{C}}X(a)x \\ \| & & \| \\ (Px)^{(a)} & \xrightarrow[{}^{(1)}(\phi_a x)^{(a)}]{} & (PX(a)x)^{(1)}. \end{array} \tag{5.28}$$

Furthermore by (5.12), we have

$$(\phi_a x)^{-1} \in (\mathscr{C}/G)^{a^{-1}}(PX(a)x, Px).$$

Hence we can take

$$\overline{\phi}'_a x := {}^{(a)}((\phi_a x)^{-1})^{(1)} \in ((\mathscr{C}/G)\#G)((PX(a)x)^{(1)}, (Px)^{(a)}).$$

Here by the definition (5.21) of compositions of smash products, we obtain the right commutative diagram from the left commutative diagram:

$$\begin{array}{ccc} Px & \xrightarrow{\phi_a x} & PX(a)x \\ \| & \swarrow_{(\phi_a x)^{-1}} & \| \\ Px & \xrightarrow[\phi_a x]{} & PX(a)x \end{array}, \qquad \begin{array}{ccc} (Px)^{(a)} & \xrightarrow{\overline{\phi}_a x} & (PX(a)x)^{(1)} \\ \| & \swarrow_{\overline{\phi}'_a x} & \| \\ (Px)^{(a)} & \xrightarrow[\overline{\phi}_a x]{} & (PX(a)x)^{(1)} \end{array}.$$

Thus, $\overline{\phi}_a x$ is also an isomorphism. Then by setting $\overline{\phi}_a := (\overline{\phi}_a x)_{x \in \mathscr{C}_0}$, we obtain a natural isomorphism $\overline{\phi}_a\colon X'(a) \circ \zeta_{\mathscr{C}} \Rightarrow \zeta_{\mathscr{C}} \circ X(a)$. Indeed, again by definition (5.21) of compositions of smash products, for each morphism $f\colon x \to y$ of $\mathscr{C}$ and $a \in G$, the commutative diagram (5.11) yields the commutative diagram

$$\begin{array}{ccc} (Px)^{(a)} & \xrightarrow{\overline{\phi}_a x} & (PX(a)x)^{(1)} \\ {\scriptstyle f \circ X_* x}\downarrow & & \downarrow{\scriptstyle X(a)f \circ X_*(X(a)x)} \\ (Py)^{(a)} & \xrightarrow[\overline{\phi}_a y]{} & (PX(a)y)^{(1)}. \end{array}$$

Then we set $\overline{\phi}_{\mathscr{C}} := (\overline{\phi}_a)_{a \in G}$. In what follows, we identify $\overline{\phi}_a = \phi_a, \overline{\phi}_{\mathscr{C}} = \phi$ if there seems to be no confusion.

LEMMA 5.8.2. *$\zeta_{\mathscr{C}} = (\zeta_{\mathscr{C}}, \overline{\phi}_{\mathscr{C}})$ is a G-equivariant functor for all $\mathscr{C} \in G$-$\mathbf{Cat}_0$.*

PROOF. Noting the equality (5.27), in the same way as in the previous proof, we obtain a commutative diagram that expresses the fact that $(\zeta_{\mathscr{C}}, \overline{\phi}_{\mathscr{C}})$ is a G-equivariant functor from the commutative diagram (5.14) and (5.15) by using the definition (5.21) of compositions of smash products. □

EXERCISE 5.8.3. Verify the proof above.

LEMMA 5.8.4. *ζ is a strict 2-natural transformation.*

PROOF. Let $\mathscr{C} = (\mathscr{C}, X), \mathscr{C}' = (\mathscr{C}', X')$ be objects of G-**Cat** and $(P, \phi)\colon \mathscr{C} \to \mathscr{C}/G, (P', \phi')\colon \mathscr{C}' \to \mathscr{C}'/G$ the canonical covering functor.

(1) Let $(E, \rho) \in (G\text{-}\mathbf{Cat})(\mathscr{C}, \mathscr{C}')$ and set $(H, 1) := (E, \rho)/G$. Then we show that the diagram

$$\begin{array}{ccc} \mathscr{C} & \xrightarrow{(E,\rho)} & \mathscr{C}' \\ {\scriptstyle \zeta_{\mathscr{C}}}\downarrow & & \downarrow{\scriptstyle \zeta_{\mathscr{C}'}} \\ (\mathscr{C}/G)\#G & \xrightarrow[(H,1)\#G]{} & (\mathscr{C}'/G)\#G. \end{array} \tag{5.29}$$

is strictly commutative.

On objects. Let $x \in \mathscr{C}_0$. Then we have

$$\begin{aligned} (((H,1)\#G) \circ \zeta_{\mathscr{C}})(x) &= ((H,1)\#G)((Px)^{(1)}) = (P'Ex)^{(1)} \\ &= (\zeta_{\mathscr{C}'} \circ (E,\rho))(x). \end{aligned}$$

On morphisms. Let $f\colon x \to y$ be a morphism of $\mathscr{C}$. Then we have

$$\begin{aligned} (((H,1)\#G) \circ \zeta_{\mathscr{C}})(f) &= ((H,1)\#G)(f \circ X_*x) = E(f \circ X_*x) \circ \rho_1 x \\ &= Ef \circ EX_*x \circ \rho_1 x = Ef \circ (EX_* \bullet \rho_1)x \\ &= Ef \circ X'_*Ex = (\zeta_{\mathscr{C}'} \circ (E,\rho))(f). \end{aligned}$$

Therefore the diagram (5.29) is strictly commutative.

(2) Let (E, ρ) and (E', ρ') be 1-morphisms of the 2-category G-**Cat** and $\eta\colon (E, \rho) \Rightarrow (E', \rho')$ a 2-morphism. Then we show

$$\zeta_{\mathscr{C}'} \circ \eta = ((\eta/G)\#G) \circ \zeta_{\mathscr{C}}.$$

Set $\theta := \eta/G$ and let $x \in \mathscr{C}$. Then it is enough to show that

$$\zeta_{\mathscr{C}'}(\eta_x) = (\theta\#G)(\zeta_{\mathscr{C}}x)$$

holds. This is verified by definition as follows:

$$(\text{RHS}) = (\theta\#G)((Px)^{(1)}) = \theta(Px) = (\eta/G)(Px) = \eta_x \circ X_*Ex = (\text{LHS}).$$

By (1) and (2) above ζ is a strict 2-natural transformation. □

5.8.b. $\zeta'\colon (?\#G) \circ (?/G) \Rightarrow \mathbb{1}_{G\text{-}\mathbf{Cat}}$.

DEFINITION 5.8.5. Let $\mathscr{C} = (\mathscr{C}, X)$ be an object of G-**Cat**, and $(P, \phi)\colon \mathscr{C} \to \mathscr{C}/G$ the canonical covering functor.

(1) We define a G-equivariant functor $\zeta'_{\mathscr{C}}\colon (\mathscr{C}/G)\#G \to \mathscr{C}$ as follows:

On objects. For each $x \in \mathscr{C}_0$ and $a \in G$, we set

$$\zeta'_{\mathscr{C}}((Px)^{(a)}) := X(a)x.$$

On morphisms. Let $f\colon (Px)^{(a)} \to (Py)^{(b)}$ be a morphism of $(\mathscr{C}/G)\#G$. Then we define $\zeta'_{\mathscr{C}}$ in such a way that the diagram

$$
\begin{array}{ccc}
((\mathscr{C}/G)\#G)((Px)^{(a)},(Py)^{(b)}) & \overset{\zeta'_{\mathscr{C}}}{\dashrightarrow} & \mathscr{C}(X(a)x, X(b)y) \\
\| & & \uparrow{\wr}\ \mathscr{C}(X_{b,b^{-1}a}, X(b)y) \\
\| & & \mathscr{C}(X(b)X(b^{-1}a)x, X(b)y) \\
\| & & \uparrow{\wr}\ X(b) \\
(\mathscr{C}/G)^{b^{-1}a}(Px,Py) & = & \mathscr{C}(X(b^{-1}a)x, y).
\end{array}
$$

is commutative. Namely, we set

$$\zeta'_{\mathscr{C}}(f) := X(b)(f) \circ X_{b,b^{-1}a}.$$

Natural isomorphisms. Set $Y := X^{(\mathscr{C}/G)\#G}$ (cf. (5.22)). For each $c \in G$ we define

$$\theta_c := (\theta_{\mathscr{C}})_c\colon X(c) \circ \zeta'_{\mathscr{C}} \Rightarrow \zeta'_{\mathscr{C}} \circ Y(c)$$

by

$$\theta_c((Px)^{(a)}) := X_{c,a}^{-1}x\colon X(c)X(a)x \to X(ca)x, \quad ((Px)^{(a)} \in (\mathscr{C}/G)\#G).$$

Since $X_{c,b}$ is a natural isomorphism for all $b, c \in G$, and $(\mathscr{C}, X)$ satisfies the axiom (b) of pseudo-G-categories, we see that each θ_c is a natural isomorphism. In particular, if X is a group action, then $\theta_{\mathscr{C}}$ is the identity 2-morphism and $\zeta'_{\mathscr{C}}$ is a strict G-equivariant functor.

(2) Let $(E,\rho)\colon \mathscr{C} \to \mathscr{C}'$ be a 1-morphism of G-**Cat**. We next define a natural transformation $\zeta'_{(E,\rho)}$ in the diagram

$$
\begin{array}{ccc}
(\mathscr{C}/G)\#G & \xrightarrow{((E,\rho)/G)\#G} & (\mathscr{C}'/G)\#G \\
\zeta'_{\mathscr{C}}\downarrow & \overset{\zeta'_{(E,\rho)}}{\swarrow\!\!\swarrow}\ \cong & \downarrow\zeta'_{\mathscr{C}'} \\
\mathscr{C} & \xrightarrow[(E,\rho)]{} & \mathscr{C}'.
\end{array}
$$

To this end, first take any $(Px)^{(a)} \in ((\mathscr{C}/G)\#G)_0$, and let $(P,\phi)\colon \mathscr{C} \to \mathscr{C}/G, (P',\phi')\colon \mathscr{C}' \to \mathscr{C}'/G$ be the canonical covering functors. Then the clockwise composite sends $(Px)^{(a)}$ to

$$[\zeta'_{\mathscr{C}'} \circ (((E,\rho)/G)\#G)]((Px)^{(a)}) = \zeta'_{\mathscr{C}'}((P'(Ex))^{(a)}) = X'(a)Ex$$

because $[((E,\rho)/G)\#G]((Px)^{(a)}) = [(E,\rho)/G]((Px)^{(a)}) = (P'(Ex))^{(a)}$. Whereas the counter-clockwise composite sends it to

$$[(E,\rho) \circ \zeta'_{\mathscr{C}}]((Px)^{(a)}) = EX(a)x.$$

Then we define

$$(\zeta'_{(E,\rho)})((Px)^{(a)}) := \rho_a x \in \mathscr{C}'(X'(a)Ex, EX(a)x).$$

LEMMA 5.8.6. *$\zeta'_{\mathscr{C}} := (\zeta'_{\mathscr{C}}, \theta_{\mathscr{C}})\colon (\mathscr{C}/G)\#G \to \mathscr{C}$ is a 1-morphism of G-**Cat** for all $\mathscr{C} \in G$-$\mathbf{Cat}_0$.*

PROOF. This immediately follows from the fact that $(\mathscr{C}, X)$ satisfies the axioms of pseudo-G-categories. □

LEMMA 5.8.7. *$\zeta' := ((\zeta'_{\mathscr{C}})_{\mathscr{C}\in G\text{-}\mathbf{Cat}_0}, (\zeta'_{(E,\rho)})_{(E,\rho)\in G\text{-}\mathbf{Cat}_1})$ is a 2-natural transformation.*

PROOF. For each 1-morphism (E, ρ) of G-**Cat**, it follows from the fact that (E, ρ) is a 1-morphism (resp. by the definition of $\theta_{\mathscr{C}}$ and from the fact that (E, ρ) is a 1-morphism) that $\zeta'_{(E,\rho)}$ is a natural transformation (resp. a 2-isomorphism) in G-**Cat** (cf. Definition 5.1.7)). Thus ζ' has the 1-naturality. Next let $(E', \rho') \colon \mathscr{C} \to \mathscr{C}'$ be another 1-morphism and $\eta\colon (E, \rho) \Rightarrow (E', \rho')$ a 2-morphism of G-**Cat**. Then it is easily verified that the diagram

$$\begin{array}{ccc} \zeta'_{\mathscr{C}'} \circ ((E,\rho)/G)\#G & \overset{\zeta'_{(E,\rho)}}{\underset{\cong}{\Longrightarrow}} & (E,\rho)\circ\zeta'_{\mathscr{C}} \\ \Big\Downarrow{\scriptstyle \zeta'_{\mathscr{C}'}\circ((\eta/G)\#G)} & & \Big\Downarrow{\scriptstyle \eta\circ\zeta'_{\mathscr{C}}} \\ \zeta'_{\mathscr{C}'} \circ ((E',\rho')/G)\#G & \overset{\cong}{\underset{\zeta'_{(E',\rho')}}{\Longrightarrow}} & (E',\rho')\circ\zeta'_{\mathscr{C}} \end{array}$$

is commutative. Thus ζ' has the 2-naturality. □

5.8.c. $\omega\colon \mathbb{1}_{G\text{-}\mathbf{GrCat}} \Rightarrow (?/G)\circ(?\#G)$.

DEFINITION 5.8.8. Let $\mathscr{B} \in G$-$\mathbf{GrCat}_0$, and $(P, \phi)\colon \mathscr{B}\#G \to (\mathscr{B}\#G)/G$ be the canonical covering functor. We then define a 1-morphism $\omega_{\mathscr{B}}\colon \mathscr{B} \to (\mathscr{B}\#G)/G$ of G-**GrCat** as follows:

On objects. For each $x \in \mathscr{B}_0$, we set

$$\omega_{\mathscr{B}}(x) := P(x^{(1)}).$$

On morphisms. We set $X := X^{\mathscr{B}\#G}$. By noting that

$$\begin{aligned} ((\mathscr{B}\#G)/G)(P(x^{(1)}), P(y^{(1)})) &= \bigoplus_{a\in G}(\mathscr{B}\#G)(X(a)x^{(1)}, y^{(1)}) \\ &= \bigoplus_{a\in G}(\mathscr{B}\#G)(x^{(a)}, y^{(1)}) = \bigoplus_{a\in G}\mathscr{B}^a(x,y) \\ &= \mathscr{B}(x,y), \end{aligned}$$

we set

$$\omega_{\mathscr{B}}(f) := (P,\phi)^{(1)}_{x^{(1)},y^{(1)}}(f) = f\colon P(x^{(1)}) \to P(y^{(1)})$$

for each morphism $f\colon x \to y$ of $\mathscr{B}$.

LEMMA 5.8.9. *$\omega_{\mathscr{B}}$ defined above is a strict degree-preserving equivalence of G-graded categories for all $\mathscr{B} \in (G\text{-}\mathbf{GrCat})_0$.*

PROOF. First of all, it is clear that $\omega_{\mathscr{B}}$ is fully faithful by construction. Let $x, y \in \mathscr{B}_0$, $a \in G$ and set $X := X^{\mathscr{B}\#G}$. Then since $X(a)(x^{(1)}) = x^{(a)}$, we have $P(x^{(a)}) \cong P(x^{(1)})$ in $(\mathscr{B}\#G)/G$, from which we see that $\omega_{\mathscr{B}}$ is dense. It remains to show that

$$\omega_{\mathscr{B}}(\mathscr{B}^a(x,y)) \subseteq ((\mathscr{B}\#G)/G)^a(\omega_{\mathscr{B}}(x), \omega_{\mathscr{B}}(y)).$$

Indeed more strongly, we can show the equality by the following computation (the first equality follows by (5.16)):

$$\begin{aligned}(\text{RHS}) &= (\mathscr{B}\#G)(X(a)(x^{(1)}), y^{(1)}) = (\mathscr{B}\#G)(x^{(a)}, y^{(1)}) \\ &= \mathscr{B}^a(x, y) = (\text{LHS}).\end{aligned}$$

□

LEMMA 5.8.10. *ω is a 2-natural transformation.*

PROOF. Let $(H, r)\colon \mathscr{B} \to \mathscr{B}'$ be a 1-morphism of G-**GrCat** and $(P', \phi')\colon \mathscr{B}'\#G \to (\mathscr{B}'\#G)/G$ the canonical covering functor. We define a natural transformation $\omega_{(H,r)}$ in the diagram

$$\begin{array}{ccc} \mathscr{B} & \xrightarrow{\quad (H,r) \quad} & \mathscr{B}' \\ {\scriptstyle \omega_{\mathscr{B}}}\Big\downarrow & \overset{\omega_{(H,r)}}{\underset{\cong}{\Swarrow}} & \Big\downarrow{\scriptstyle \omega_{\mathscr{B}'}} \\ (\mathscr{B}\#G)/G & \xrightarrow[\quad ((H,r)\#G)/G \quad]{} & (\mathscr{B}'\#G)/G \end{array}$$

by

$$\omega_{(H,r)}x := \phi'_{r_x}(Hx)^{(1)} \quad (x \in \mathscr{B}).$$

Then it is not hard to verify that $\omega_{(H,r)}$ is a 2-isomorphism of G-**GrCat**, by which we see that ω has 1-naturality. Now let $(H', r')\colon \mathscr{B} \to \mathscr{B}'$ be another 1-morphism of G-**GrCat** and $\theta\colon (H, r) \Rightarrow (H', r')$ a 2-morphism. Then it is easy to verify the commutativity of the diagram

$$\begin{array}{ccc} \omega_{\mathscr{B}'} \cdot (H, r) & \overset{\omega_{(H,r)}}{\underset{\cong}{\Longrightarrow}} & ((H, r)\#G)/G \cdot \omega_{\mathscr{B}} \\ \Big\Downarrow{\scriptstyle \omega_{\mathscr{B}'}\cdot\theta} & & \Big\Downarrow{\scriptstyle (\theta\#G)/G\cdot\omega_{\mathscr{B}}} \\ \omega_{\mathscr{B}'} \cdot (H', r') & \overset{\cong}{\underset{\omega_{(H',r')}}{\Longrightarrow}} & ((H', r')\#G)/G \cdot \omega_{\mathscr{B}}. \end{array}$$

Hence ω has 2-naturality. □

5.8.d. $\omega'\colon (?/G) \circ (?\#G) \Rightarrow \mathbb{1}_{G\text{-}GrCat}$.

DEFINITION 5.8.11 (cf. Proposition 5.6.4). Let $\mathscr{B} \in G$-**GrCat**$_0$ and $(P, \phi)\colon \mathscr{B}\#G \to (\mathscr{B}\#G)/G$ be the canonical covering functor. We define a functor $\omega'_{\mathscr{B}}\colon (\mathscr{B}\#G)/G \to \mathscr{B}$ as the unique functor that makes the following diagram strictly commutative:

$$\begin{array}{ccc} \mathscr{B}\#G & \xrightarrow{\qquad Q \qquad} & \mathscr{B}, \\ & {\scriptstyle (P,\phi)}\searrow \quad \nearrow{\scriptstyle \omega'_{\mathscr{B}}} & \\ & (\mathscr{B}\#G)/G & \end{array}$$

where Q is the canonical G-covering functor determined by the smash product. Thus the explicit form of $\omega'_{\mathscr{B}}$ is as follows:

On objects. For each object $P(x^{(a)})$ of $(\mathscr{B}\#G)/G$ (where $x \in \mathscr{B}_0, a \in G$) we set

$$\omega'_{\mathscr{B}}(P(x^{(a)})) := x. \tag{5.30}$$

On morphisms. For each $P(x^{(a)}), P(y^{(b)}) \in ((\mathscr{B}\#G)/G)_0$, note that

$$\begin{aligned}((\mathscr{B}\#G)/G)(P(x^{(a)}), P(y^{(b)})) &= \bigoplus_{c\in G}(\mathscr{B}\#G)(x^{(ca)}, y^{(b)})\\ &= \bigoplus_{c\in G}\mathscr{B}^{b^{-1}ca}(x,y) = \mathscr{B}(x,y).\end{aligned}$$

Then for each

$$u = (u_c)_{c\in G} \in \bigoplus_{c\in G}(\mathscr{B}\#G)(x^{(ca)}, y^{(b)}) = ((\mathscr{B}\#G)/G)(P(x^{(a)}), P(y^{(b)}))$$

we set

$$\omega'_{\mathscr{B}}(u) := u = \sum_{c\in G} u_c \in \bigoplus_{c\in G}\mathscr{B}^{b^{-1}ca}(x,y) = \mathscr{B}(x,y). \tag{5.31}$$

Degree adjuster. Finally, we set $\omega'_{\mathscr{B}} := (\omega'_{\mathscr{B}}, r_{\mathscr{B}})$, where we define $r_{\mathscr{B}}$ as follows:

$$r_{\mathscr{B}}(P(x^{(a)})) := a \quad (P(x^{(a)}) \in ((\mathscr{B}\#G)/G)_0).$$

Then note that ω' is dense by (5.30) and is fully faithful by (5.31), and hence it is an equivalence.

LEMMA 5.8.12. *$\omega'_{\mathscr{B}} = (\omega'_{\mathscr{B}}, r_{\mathscr{B}})$ is a degree-preserving functor, i.e., a 1-morphism of G-**GrCat** for all $\mathscr{B} \in G$-$\mathbf{GrCat}_0$.*

PROOF. It is not hard to verify that $\omega'_{\mathscr{B}}$ is a functor, and is left to the reader. We show that $\omega'_{\mathscr{B}} = (\omega'_{\mathscr{B}}, r_{\mathscr{B}})$ is a degree-preserving functor. (cf. Definition 5.2.1). Set $X := X^{\mathscr{B}\#G}$ and let $P(x^{(a)}), P(y^{(b)})$ be an object of $(\mathscr{B}\#G)/G$ and take $c \in G$. Then we have

$$\begin{aligned}&\omega'_{\mathscr{B}}((\mathscr{B}\#G/G)^{r_{\mathscr{B}}(y^{(b)})\cdot c}(P(x^{(a)}), P(y^{(b)})))\\ &= \omega'_{\mathscr{B}}((\mathscr{B}\#G)(X(bc)x^{(a)}, y^{(b)}))\\ &= (\mathscr{B}\#G)(x^{(bca)}, y^{(b)})\\ &= \mathscr{B}^{b^{-1}bca}(x,y) = \mathscr{B}^{ca}(x,y)\\ &= \mathscr{B}^{c\cdot r_{\mathscr{B}}(x^{(a)})}(\omega'_{\mathscr{B}}(P(x^{(a)})), \omega'_{\mathscr{B}}(P(y^{(b)}))).\end{aligned}$$

□

REMARK 5.8.13. As we have seen above, $\omega'_{\mathscr{B}}$ is not a strictly degree-preserving functor in general. For this reason, we needed to take this weaker form as the definition of degree-preserving functor.

LEMMA 5.8.14. *ω' is a strict 2-natural transformation.*

PROOF. Let $(H, r)\colon \mathscr{B} \to \mathscr{B}'$ be a 1-morphism of G-**GrCat** and $(P, \phi)\colon \mathscr{B}\#G \to (\mathscr{B}\#G)/G$, $(P', \phi')\colon \mathscr{B}'\#G \to (\mathscr{B}'\#G)/G$ the canonical covering functors. We first show that ω' has a strict 1-naturality, namely that the following diagram is commutative:

$$\begin{array}{ccc}(\mathscr{B}\#G)/G & \xrightarrow{((H,r)\#G)/G} & (\mathscr{B}'\#G)/G\\ {\scriptstyle \omega'_{\mathscr{B}}}\downarrow & & \downarrow{\scriptstyle \omega'_{\mathscr{B}'}}\\ \mathscr{B} & \xrightarrow[(H,r)]{} & \mathscr{B}'.\end{array}$$

To this end let $u\colon P(x^{(a)}) \to P(y^{(b)})$ be a morphism of $(\mathscr{B}\#G)/G$ and set $f := \omega'_{\mathscr{B}}(u)$. Then we have

$$[\omega'_{\mathscr{B}'} \circ ((H,r)\#G)/G](P(x^{(a)})) = \omega'_{\mathscr{B}'}(P'((Hx)^{(ar_x)})) = Hx = [(H,r)\circ\omega'_{\mathscr{B}}](P(x^{(a)})),$$

and

$$\begin{aligned}[\omega'_{\mathscr{B}'} \circ ((H,r)\#G)/G](u) &\overset{\text{(a)}}{=} [((P',\phi')((H,r)\#G))^{(1)}_{x^{(a)},y^{(b)}}](f)\\ &\overset{\text{(b)}}{=} (P',\phi')^{(1)}_{(Hx)^{(ar_x)},(Hy)^{(br_y)}}(Hf)\\ &= Hf\\ &= [(H,r)\circ\omega'_{\mathscr{B}}](u).\end{aligned}$$

In the above, the equality (a) follows by the definition of $((H,r)\#G)/G$, and the equality (b) follows by the fact that $(H,r)\#G$ is a strictly G-equivariant functor.

To show that ω' has the strict 2-naturality, take another 1-morphism $(H',r')\colon \mathscr{B} \to \mathscr{B}'$ and a 2-morphism $\theta\colon (H,r) \Rightarrow (H',r')$ of G-**GrCat**. Then it is enough to verify that

$$\omega'_{\mathscr{B}'}((\theta\#G)/G) = \theta\omega'_{\mathscr{B}}.$$

This is shown by the following computation: For each $P(x^{(a)}) \in ((\mathscr{B}\#G)/G)_0$ we have

$$\begin{aligned}[\omega'_{\mathscr{B}'}((\theta\#G)/G)]P(x^{(a)}) &= \omega'_{\mathscr{B}'}((\theta\#G)/G)P(x^{(a)}))\\ &= \omega'_{\mathscr{B}'}(P'((\theta\#G)(x^{(a)})))\\ &= \omega'_{\mathscr{B}'}(P'(\theta x))\\ &= \omega'_{\mathscr{B}'}(P'^{(1)}_{(Hx)^{(ar_x)},(H'x)^{(ar'_x)}}(\theta x))\\ &= \theta x\\ &= \theta\omega'_{\mathscr{B}}(P(x^{(a)})).\end{aligned}$$

□

5.8.e. Verifications of relations.

Verification of (5.23). For each $\mathscr{C} = (\mathscr{C}, X^{\mathscr{C}}) \in G\text{-}\mathbf{Cat}_0$, we have

$$((\zeta' \bullet \zeta)_{\mathscr{C}})(x) = (\zeta'_{\mathscr{C}} \circ \zeta_{\mathscr{C}})(x) = X^{\mathscr{C}}(1)x$$

by definitions of ζ and ζ'. Therefore by setting $X^{\mathscr{C}}_* := (X^{\mathscr{C}}_* x)_{x\in\mathscr{C}_0}$, we have a 2-isomorphism $X^{\mathscr{C}}_*\colon (\zeta' \bullet \zeta)\mathscr{C} \Rightarrow \mathbb{1}_{\mathscr{C}}$ in G-**Cat**. Then $X_* := (X^{\mathscr{C}}_*)_{\mathscr{C}\in G\text{-}\mathbf{Cat}_0}\colon \zeta' \bullet \zeta \rightsquigarrow \mathbb{1}_{\mathbf{1}_{G\text{-}\mathbf{Cat}}}$ turns out to be an isomorphic modification, and the formula (5.23) is verified.

Verification of (5.24). Let $\mathscr{C} \in G\text{-}\mathbf{Cat}_0$ and $(P,\phi)\colon \mathscr{C} \to \mathscr{C}/G$ be the canonical covering functor. For each $(Px)^{(a)} \in ((\mathscr{C}/G)\#G)_0$, note that $(\zeta_{\mathscr{C}} \circ \zeta'_{\mathscr{C}})((Px)^{(a)}) = (P(X(a)x))^{(1)}$ by definition. Then we define a modification $\Theta\colon \mathbb{1}_{(?\#G)\circ(?/G)} \rightsquigarrow \zeta\bullet\zeta'$ by

$$\begin{aligned}\Theta &:= (\Theta_{\mathscr{C}}\colon \mathbb{1}_{(\mathscr{C}/G)\#G} \Rightarrow \zeta_{\mathscr{C}} \circ \zeta'_{\mathscr{C}})_{\mathscr{C}\in G\text{-}\mathbf{Cat}_0}\\ \Theta_{\mathscr{C}}((Px)^{(a)}) &:= \overline{\phi}_a x\colon (Px)^{(a)} \to (P(X(a)x))^{(1)} \quad ((Px)^{(a)} \in ((\mathscr{C}/G)\#G)_0)\end{aligned}$$

(cf. (5.28)). It is enough to verify that this is an isomorphic modification. First, $\Theta_{\mathscr{C}}((Px)^{(a)})$ is already shown to be an isomorphism for each $(Px)^{(a)} \in ((\mathscr{C}/G)\#G)_0$ some lines after (5.28). It remains to show that this is a modification, i.e., that for each diagram $\mathscr{C} \underset{(F,\psi)}{\overset{(E,\rho)}{\Downarrow\alpha}} \mathscr{D}$ in G-**Cat**, the following diagram is commutative (cf. Definition 4.1.20):

$$
\begin{array}{ccc}
((E,\rho)/G)\#G & = & ((E,\rho)/G)\#G \\
{\scriptstyle \Theta_{\mathscr{D}}\circ((\alpha/G)\#G)}\Downarrow & & \Downarrow{\scriptstyle ((\alpha/G)\#G)\circ\Theta_{\mathscr{C}}} \\
(\zeta\bullet\zeta')_{\mathscr{D}}\circ((F,\psi)/G)\#G & \underset{(\zeta\bullet\zeta')_{(F,\psi)}}{\Longrightarrow} & ((F,\psi)/G)\#G\circ(\zeta\bullet\zeta')_{\mathscr{C}}.
\end{array}
$$

Here by Definition 4.1.17, we have

$$
\begin{gathered}
(\zeta\bullet\zeta')_{\mathscr{D}} = \zeta_{\mathscr{D}}\circ\zeta'_{\mathscr{D}}, (\zeta\bullet\zeta')_{\mathscr{C}} = \zeta_{\mathscr{C}}\circ\zeta'_{\mathscr{C}}, \\
(\zeta\bullet\zeta')_{(F,\psi)} = (\zeta_{(F,\psi)}\circ\zeta'_{\mathscr{C}})\bullet(\zeta_{\mathscr{D}}\circ\zeta'_{(F,\psi)}) \overset{*}{=} \zeta_{\mathscr{D}}\circ\zeta'_{(F,\psi)}
\end{gathered}
$$

(note: the equality $\overset{*}{=}$ holds because $\zeta_{(F,\psi)}$ is the identity 2-morphism by Lemma 5.8.4). Also since the left vertical morphism $\Theta_{\mathscr{D}}\circ((\alpha/G)\#G)$ is the horizontal composite of the two 2-morphisms

$$
(\mathscr{D}/G)\#G \;\underset{\mathbf{1}}{\overset{\zeta_{\mathscr{D}}\circ\zeta'_{\mathscr{D}}}{\longleftarrow\!\Theta_{\mathscr{D}}\Uparrow\!\longleftarrow}}\; (\mathscr{D}/G)\#G \;\underset{((E,\rho)/G)\#G}{\overset{((F,\psi)/G)\#G}{\longleftarrow\!\Uparrow(\alpha/G)\#G\!\longleftarrow}}\; (\mathscr{C}/G)\#G,
$$

we have

$$
\Theta_{\mathscr{D}}\circ((\alpha/G)\#G) = (\Theta_{\mathscr{D}}\circ((F,\psi)/G)\#G)\bullet((\alpha/G)\#G).
$$

Whereas since the right vertical morphism $((\alpha/G)\#G)\circ\Theta_{\mathscr{C}}$ is the horizontal composite of the two 2-morphisms

$$
(\mathscr{D}/G)\#G \;\underset{((E,\rho)/G)\#G}{\overset{((F,\psi)/G)\#G}{\longleftarrow\!(\alpha/G)\#G\Uparrow\!\longleftarrow}}\; (\mathscr{C}/G)\#G \;\underset{\mathbf{1}}{\overset{\zeta_{\mathscr{C}}\circ\zeta'_{\mathscr{C}}}{\longleftarrow\!\Uparrow\Theta_{\mathscr{C}}\!\longleftarrow}}\; (\mathscr{C}/G)\#G,
$$

we have

$$
((\alpha/G)\#G)\circ\Theta_{\mathscr{C}} = (((F,\psi)/G)\#G\circ\Theta_{\mathscr{C}})\bullet((\alpha/G)\#G).
$$

Therefore it is enough to show the commutativity of the diagram

$$
\begin{array}{ccc}
((E,\rho)/G)\#G & = & ((E,\rho)/G)\#G \\
{\scriptstyle (\alpha/G)\#G}\Downarrow & & \Downarrow{\scriptstyle (\alpha/G)\#G} \\
((F,\psi)/G)\#G & = & ((F,\psi)/G)\#G \\
{\scriptstyle \Theta_{\mathscr{D}}\circ((F,\psi)/G)\#G}\Downarrow & & \Downarrow{\scriptstyle ((F,\psi)/G)\#G\circ\Theta_{\mathscr{C}}} \\
\zeta_{\mathscr{D}}\circ\zeta'_{\mathscr{D}}\circ((F,\psi)/G)\#G & \underset{\zeta_{\mathscr{D}}\circ\zeta'_{(F,\psi)}}{\Longrightarrow} & ((F,\psi)/G)\#G\circ\zeta_{\mathscr{C}}\circ\zeta'_{\mathscr{C}}.
\end{array}
$$

Since the commutativity of the upper square is obvious, it suffices to show the commutativity of the lower square. To this end, we have only to show the commutativity of the diagram

$$
\begin{array}{ccc}
[((F,\psi)/G)\#G]((Px)^{(a)}) & \xlongequal{\qquad} & [((F,\psi)/G)\#G]((Px)^{(a)}) \\
\Big\downarrow {\scriptstyle [\Theta_{\mathscr{D}}\circ((F,\psi)/G)\#G]((Px)^{(a)})} & & \Big\downarrow {\scriptstyle [((F,\psi)/G)\#G\circ\Theta_{\mathscr{C}}]((Px)^{(a)})} \\
[\zeta_{\mathscr{D}}\circ\zeta'_{\mathscr{D}}\circ((F,\psi)/G)\#G]((Px)^{(a)}) & \xrightarrow[{[\zeta_{\mathscr{D}}\circ\zeta'_{(F,\psi)}]((Px)^{(a)})}]{} & [((F,\psi)/G)\#G\circ\zeta_{\mathscr{C}}\circ\zeta'_{\mathscr{C}}]((Px)^{(a)})
\end{array}
\tag{5.32}
$$

in $(\mathscr{D}/G)\#G$ for all $(Px)^{(a)} \in ((\mathscr{C}/G)\#G)_0$. Here let $(Q,\chi)\colon \mathscr{D} \to \mathscr{D}/G$ be the canonical covering functor, and set $\mathscr{D} = (\mathscr{D}, Y)$, $(F,\psi)/G = (H,1)$. Then the diagram

$$
\begin{array}{ccc}
(\mathscr{C}, X) & \xrightarrow{(F,\psi)} & (\mathscr{D}, Y) \\
{\scriptstyle (P,\phi)}\Big\downarrow & & \Big\downarrow{\scriptstyle (Q,\chi)} \\
\Delta(\mathscr{C}/G) & \xrightarrow[\Delta(H)]{} & \Delta(\mathscr{D}/G)
\end{array}
\tag{5.33}
$$

is strictly commutative in G-**Cat**, which we use to compute each term of the diagram (5.32).

On objects: we have

$$
\begin{aligned}
&[((F,\psi)/G)\#G]((Px)^{(a)}) = (HPx)^{(a)} = (QFx)^{(a)}, \\
&[\zeta_{\mathscr{D}}\circ\zeta'_{\mathscr{D}}\circ((F,\psi)/G)\#G]((Px)^{(a)}) \\
&= \zeta_{\mathscr{D}}(\zeta'_{\mathscr{D}}((QFx)^{(a)})) = \zeta_{\mathscr{D}}(Y(a)Fx) = (QY(a)Fx)^{(1)}, \\
&[((F,\psi)/G)\#G\circ\zeta_{\mathscr{C}}\circ\zeta'_{\mathscr{C}}]((Px)^{(a)}) \\
&= ((F,\psi)/G)\#G(\zeta_{\mathscr{C}}(X(a)x)) = ((F,\psi)/G)\#G((PX(a)x)^{(1)}) \\
&= ((H,1)\#G)((PX(a)x)^{(1)}) = (HPX(a)x)^{(1)} = (QFX(a)x)^{(1)}.
\end{aligned}
$$

On morphisms: we have

$$
\begin{aligned}
&[\Theta_{\mathscr{D}}\circ((F,\psi)/G)\#G]((Px)^{(a)}) = \Theta_{\mathscr{D}}((QFx)^{(a)}) \\
&= \chi_a(Fx)\colon QFx \to QY(a)Fx, \\
&[((F,\psi)/G)\#G\circ\Theta_{\mathscr{C}}]((Px)^{(a)}) = ((F,\psi)/G)\#G(\phi_a x) = H(\phi_a x) \\
&[\zeta_{\mathscr{D}}\circ\zeta'_{(F,\psi)}]((Px)^{(a)}) = \zeta_{\mathscr{D}}(\psi_a x) = Q(\psi_a x).
\end{aligned}
$$

Consequently, diagram (5.32) is equal to the following:

$$
\begin{array}{ccc}
(QFx)^{(a)} & \xlongequal{\qquad} & (QFx)^{(a)} \\
{\scriptstyle \chi_a(Fx)}\Big\downarrow & & \Big\downarrow{\scriptstyle H(\phi_a x)} \\
(QY(a)Fx)^{(1)} & \xrightarrow[Q(\psi_a x)]{} & (QFX(a)x)^{(1)}
\end{array}.
$$

The commutativity of this diagram follows from that of diagram (5.33).

Verification of (5.25). It is obvious from the definitions of ω and ω' that equality (5.25) holds.

Verification of (5.26). Let $\mathscr{B} \in G\text{-}\mathbf{GrCat}_0$ and $(P, \phi)\colon \mathscr{B}\#G \to (\mathscr{B}\#G)/G$ be the canonical covering functor. Then it is easily verified that we can define a natural transformation $\Xi_{\mathscr{B}}\colon \omega_{\mathscr{B}} \circ \omega'_{\mathscr{B}} \Rightarrow \mathbb{1}_{(\mathscr{B}\#G)/G}$ by

$$\Xi_{\mathscr{B}}(P(x^{(a)})) := \phi_a(x^{(1)}) \quad (P(x^{(a)}) \in ((\mathscr{B}\#G)/G)_0),$$

using which we can define a modification $\Xi\colon \omega \bullet \omega' \rightsquigarrow \mathbb{1}_{(?/G)\circ(?\#G)}$ by $\Xi := (\Xi_{\mathscr{B}})_{\mathscr{B}\in G\text{-}\mathbf{GrCat}_0}$. It is not hard to verify that this is an isomorphism, and is left to the reader.

5.8.f. Verifications of claims.

Verification of Claim (1). The first statement is shown by the construction (Definition 5.8.5) of natural isomorphism $\theta_{\mathscr{C}}$, and the second one is shown by Lemma 5.8.9.

Verification of Claim (2). The first statement is shown by Lemma 5.8.4, and the second one by Lemma 5.8.14.

Verification of Claim (3). It is enough to show the following two zigzag equations:

$$(\omega' \circ \mathbb{1}_{?/G}) \bullet (\mathbb{1}_{?/G} \circ \zeta) = \mathbb{1}_{?/G}, \text{ and} \tag{5.34}$$

$$(\mathbb{1}_{?\#G} \circ \omega') \bullet (\zeta \circ \mathbb{1}_{?\#G}) = \mathbb{1}_{?\#G}. \tag{5.35}$$

Proof of (5.34). It suffices to show the following equality in G-**GrCat** for each $\mathscr{C} = (\mathscr{C}, X) \in G\text{-}\mathbf{Cat}_0$:

$$\omega'_{\mathscr{C}/G} \circ ((\zeta_{\mathscr{C}}, \phi_{\mathscr{C}})/G) = \mathbb{1}_{\mathscr{C}/G}. \tag{5.36}$$

To show this let $(P, \phi)\colon \mathscr{C} \to \mathscr{C}/G$ and $(P', \phi')\colon (\mathscr{C}/G)\#G \to ((\mathscr{C}/G)\#G)/G$ be the canonical covering functors, and $f = (f_a)_{a\in G}\colon Px \to Py$ be any morphism of $\mathscr{C}/G$. Then we have $[\omega'_{\mathscr{C}/G} \circ ((\zeta_{\mathscr{C}}, \phi_{\mathscr{C}})/G)](Px) = \omega'_{\mathscr{C}/G}(P'((Px)^{(1)})) = Px$, and by the axiom (a$_2$) of pseudo-$G$-categories, we have $((\zeta_{\mathscr{C}}, \phi_{\mathscr{C}})/G)(f) = (\zeta_{\mathscr{C}}(f_a) \circ \phi_a x)_{a\in G} = f$. Therefore we have also $[\omega'_{\mathscr{C}/G} \circ ((\zeta_{\mathscr{C}}, \phi_{\mathscr{C}})/G)](f) = f$. Hence as functors, the equation (5.36) holds. Moreover, we set $(\omega'_{\mathscr{C}/G} \circ (\zeta_{\mathscr{C}}, \phi_{\mathscr{C}})/G, s) := (\omega'_{\mathscr{C}/G}, r_{\mathscr{C}/G}) \circ ((\zeta_{\mathscr{C}}, \phi_{\mathscr{C}})/G, 1)$. Then for each $x \in \mathscr{C}_0$, we have

$$s_{Px} = 1 \cdot (r_{\mathscr{C}/G})_{((\zeta_{\mathscr{C}},\phi_{\mathscr{C}})/G)(Px)} = (r_{\mathscr{C}/G})_{P'((Px)^{(1)})} = 1.$$

Thus also as degree-preserving functors, the equation (5.36) holds.

Proof of (5.35). It is enough to show that for each $\mathscr{B} \in G\text{-}\mathbf{GrCat}_0$ the equation

$$(\omega'_{\mathscr{B}}\#G) \circ \zeta_{\mathscr{B}\#G} = \mathbb{1}_{\mathscr{B}\#G} \tag{5.37}$$

holds in G-**Cat**. To show this let $(P, \phi)\colon \mathscr{B}\#G \to (\mathscr{B}\#G)/G$ be the canonical covering functor, and $f\colon x^{(a)} \to y^{(b)}$ any morphism of $\mathscr{B}\#G$. Then we have

$$[(\omega'_{\mathscr{B}}\#G) \circ \zeta_{\mathscr{B}\#G}](x^{(a)}) = [(\omega'_{\mathscr{B}}\#G)]((Px^{(a)})^{(1)}) = (\omega'_{\mathscr{B}}(Px^{(a)}))^{(1\cdot a)} = x^{(a)},$$

and

$$[(\omega'_{\mathscr{B}}\#G) \circ \zeta_{\mathscr{B}\#G}](f) = \omega'_{\mathscr{B}}(P(f)) = \omega'_{\mathscr{B}}(\sigma_1^{\mathscr{B}}(f)) = f.$$

Therefore as functors, the equation (5.37) holds. Moreover, we have

$$(\omega'_{\mathscr{B}}\#G, \mathbb{1}) \circ (\zeta_{\mathscr{B}\#G}, \phi) = (\omega'_{\mathscr{B}}\#G \circ \zeta_{\mathscr{B}\#G}, ((\omega'_{\mathscr{B}}\#G) \circ \phi_c)_{c\in G}),$$

and for each $c \in G, x^{(a)} \in (\mathscr{B}\#G)_0$, we have $[(\omega'_{\mathscr{B}}\#G) \circ \phi_c](x^{(a)}) = \mathbb{1}_{x^{(ca)}}$. Thus we see that the left-hand side of (5.35) is also a strictly G-equivariant functor. Hence also as G-equivariant functors, the equation (5.35) holds. This completes the proof of Theorem 5.7.1. □

REMARK 5.8.15. In the above we discussed only small categories that fit the theory within the range of light 2-categories. However, we also can define a 2-category G-**GrCAT** of light linear G-graded categories in the same way for G-**GrCat**. By Remark 5.6.2(3) $?\#G$ is extended to a 2-functor $?\#G\colon G\text{-}\mathbf{GrCAT} \to G\text{-}\mathbf{CAT}$. Also by these proofs, Theorem 5.7.1 is extended to a theorem between G-**CAT** and G-**GrCAT** by noting that both G-**CAT** and G-**GrCAT** are 2-moderate 2-categories (cf. Proposition A.4.2).

5.9. Equivalences in the 2-categories G-Cat and G-GrCat

To distinguish several kinds of equivalences (resp. isomorphisms) we call equivalences (resp. isomorphisms) between categories *category equivalences* (resp. *category isomorphisms*) here. In this section, we give characterizations of equivalences in the 2-categories G-**Cat** and G-**GrCat**, and investigate relationships

(a) between G-equivariant functors that are category equivalences and equivalences in the 2-category G-**Cat** (see Theorem 5.9.1); and

(b) between degree-preserving functors that are category equivalences and equivalences in the 2-category G-**GrCat** (see Theorems 5.9.5 and 5.9.10).

Note that a category equivalence was characterized by one half of a pair of functors in mutually reverse directions, namely a functor is a category equivalence if and only if it is fully faithful and dense. We give similar characterizations of equivalences in both 2-categories G-**Cat** and G-**GrCat**.

5.9.a. Equivalences in the 2-category G-Cat.

We first characterize equivalences in G-**Cat** in the following theorem.

THEOREM 5.9.1. *For a G-equivariant functor $(E, \rho)\colon (\mathscr{C}, X) \to (\mathscr{C}', X')$ in G-**Cat** the following are equivalent:*

*(1) (E, ρ) is an equivalence in G-**Cat**;*

(2) E is fully faithful and dense (i.e., E is a categorical equivalence).

Therefore what we called G-equivariant equivalences *are just equivalences in* G-**Cat**.

PROOF. Since the implication (1) $\Rightarrow$ (2) is trivial, we show (2) $\Rightarrow$ (1). Let E be a categorical equivalence. Then by Corollary 4.4.5, there exists an adjoint equivalence system $(E, F, \eta, \varepsilon)$. Here $F\colon \mathscr{C}' \to \mathscr{C}$ is a quasi-inverse of E, and both the unit $\eta\colon \mathbb{1}_{\mathscr{C}} \Rightarrow F \circ E$ and the counit $\varepsilon\colon E \circ F \Rightarrow \mathbb{1}_{\mathscr{C}'}$ are natural isomorpisms. Since (E, ρ) is G-equivariant, ρ_a is also a natural isomorphism for each $a \in G$. Therefore we can define a family $\lambda = (\lambda_a)_{a \in G}$ of natural isomorphisms as follows:

$\lambda_a :=$ 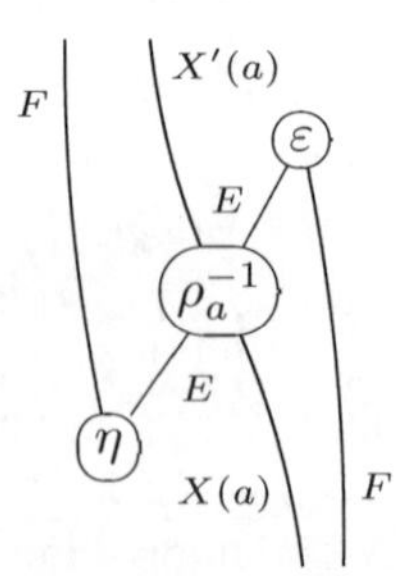 .

CLAIM 1. *$(F, \lambda)\colon \mathscr{C}' \to \mathscr{C}$ is a 1-morphism of G-**Cat**.*

By Definition 5.1.5 it is enough to show the following for each $a, b \in G$:

(i)

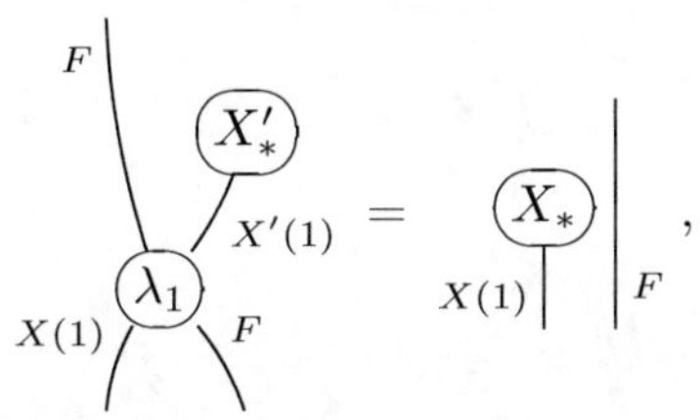

(ii)

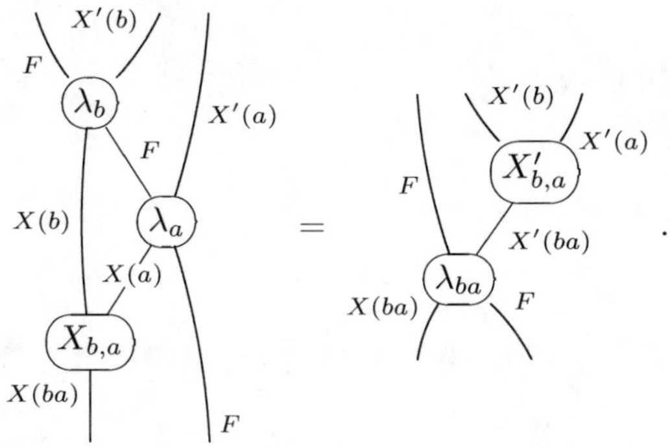

Proof of (i). Compute (5.1) $\bullet\rho_1^{-1}$ and exchange LHS and RHS to obtain

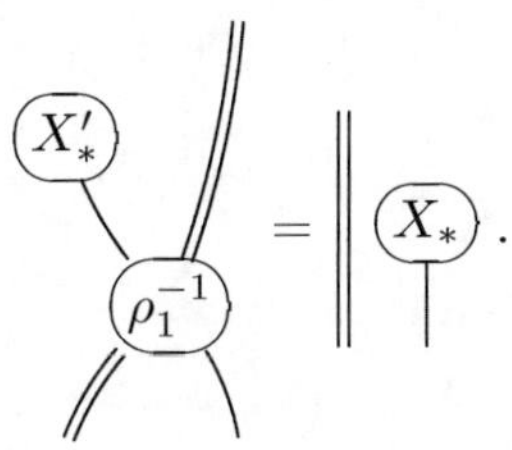

By this equality and the zigzag equations, we have

(LHS) = 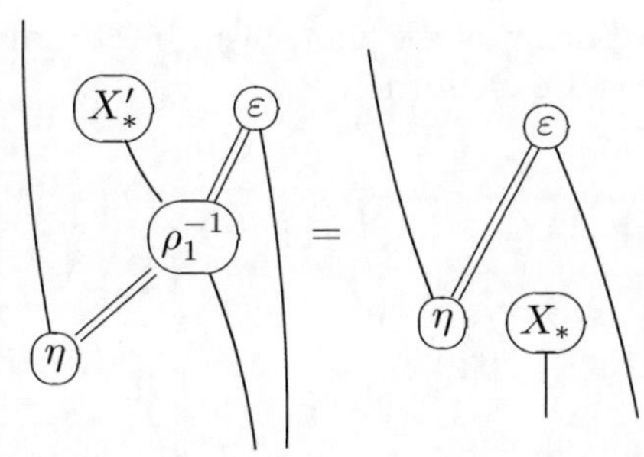 = (RHS).

Proof of (ii). Make vertical compositions to the formula (5.2) from the bottom with ρ_{ba}^{-1}, and from the top with ρ_b^{-1}, ρ_a^{-1} in this order, then exchange LHS and

RHS to obtain

(5.38)

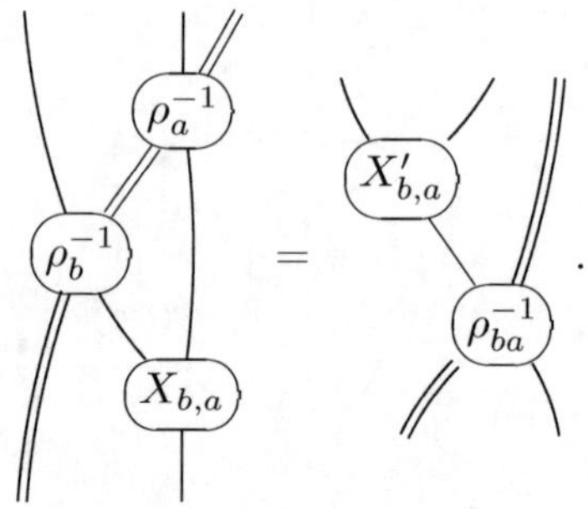

Using the zigzag equations and this, we obtain

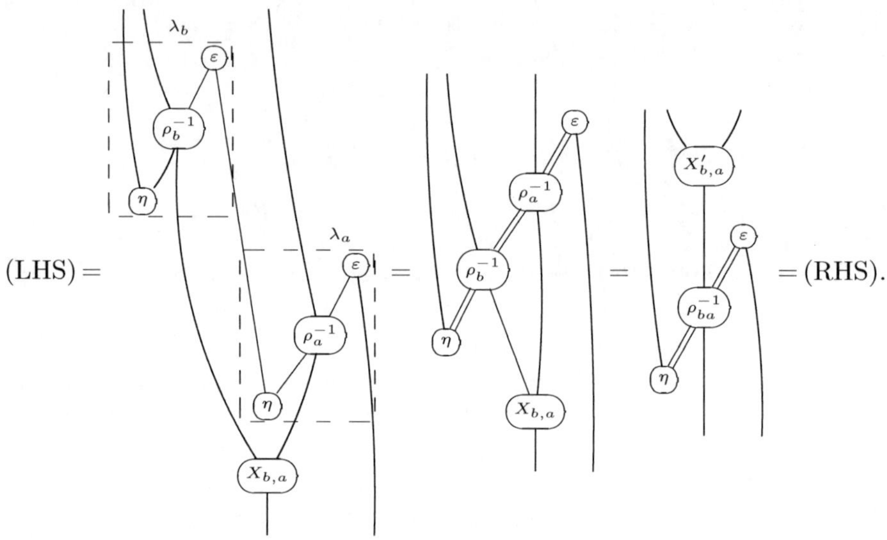

CLAIM 2. $\varepsilon\colon (E,\rho)\circ(F,\lambda) \Rightarrow (\mathbb{1}_{\mathscr{C}'}, (\mathbb{1}_{X'(a)})_{a\in G})$ *is a* 2*-isomorphism in* G-**Cat**.

It is enough to show that ε is a 2-morphism of G-**Cat**, namely, it suffices to show the following equality by Definition 5.1.7:

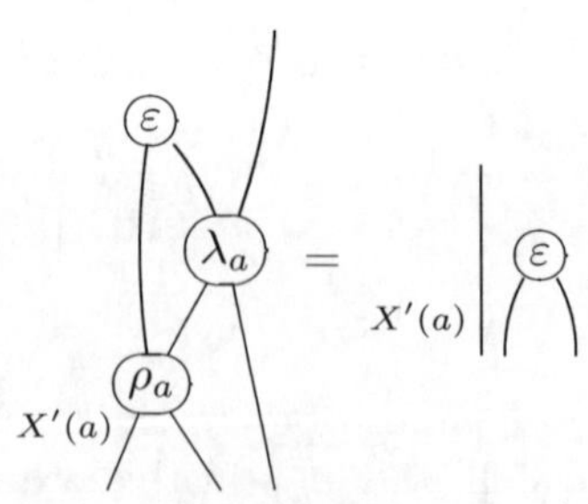

This is verified as follows:

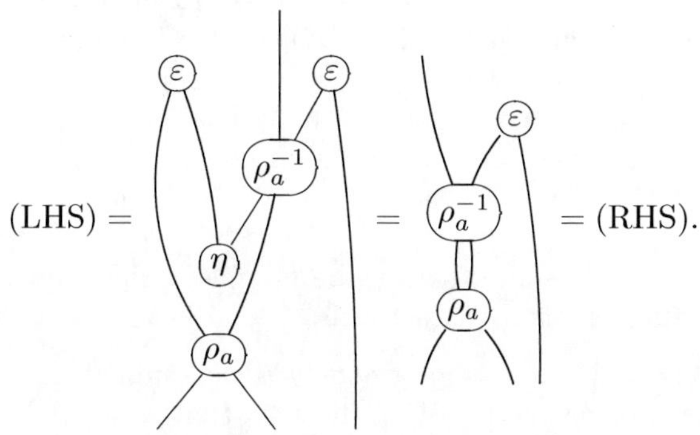

CLAIM 3. *η: $(\mathbb{1}_{\mathscr{C}}, (\mathbb{1}_{X(a)})_{a\in G}) \Rightarrow (F,\lambda)\circ(E,\rho)$ is a 2-isomorphism in G-**Cat**.*

It is enough to show that η is a 2-morphism in G-**Cat**, namely, it suffices to show the following equality by Definition 5.1.7:

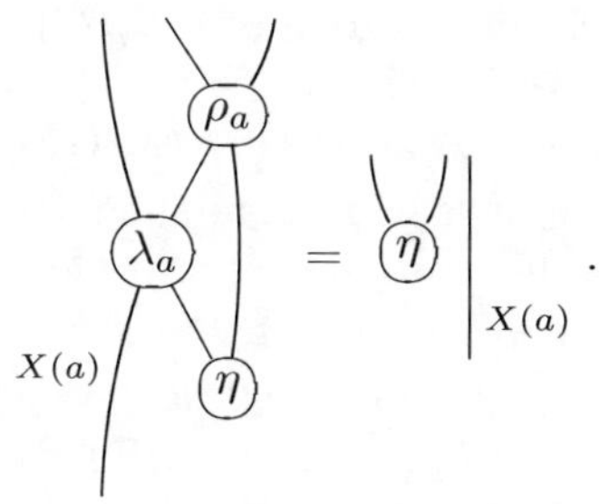

This is verified as follows:

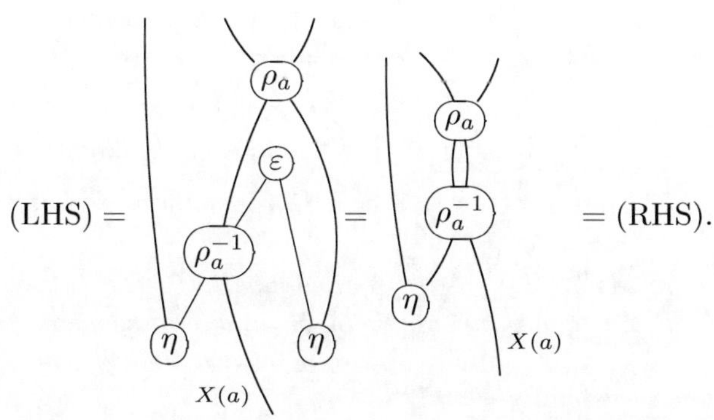

By the three claims above, (E,ρ) is an equivalence in G-**Cat**. □

REMARK 5.9.2. By Theorem 5.9.1 above the functor $\varepsilon_{\mathscr{C}}\colon \mathscr{C} \to (\mathscr{C}/G)\#G$ is an equivalence in G-**Cat** as a G-equivariant categorical equivalence for all $\mathscr{C} \in G\text{-}\mathbf{Cat}_0$.

5.9.b. Equivalences in G-GrCat. Next we characterize equivalences in the 2-category G-**GrCat**. We first define necessary terminologies.

DEFINITION 5.9.3. Let $\mathscr{A}$ be a category and $\mathscr{B}$ a G-graded category.

(1) Let $E, F\colon \mathscr{A} \to \mathscr{B}$ be functors. Then a natural transformation $\varepsilon\colon E \Rightarrow F$ is said to be *homogeneous* if $\varepsilon_x\colon Ex \to Fx$ is homogeneous in $\mathscr{B}$ for all $x \in \mathscr{A}_0$.
(2) Let $\mathscr{S}$ be a subset of $\mathscr{B}_0$, and $\mathscr{B}'$ the full subcategory of $\mathscr{B}$ with $\mathscr{B}'_0 = \mathscr{S}$. Then $\mathscr{S}$ (or $\mathscr{B}'$) is said to be *homogeneously dense* in $\mathscr{B}$ if for each $x \in \mathscr{B}_0$ there exists an $x' \in \mathscr{S}$ such that there exists a homogeneous isomorphism $x \to x'$.
(3) A functor $F\colon \mathscr{A} \to \mathscr{B}$ is said to be *homogeneously dense* if the object set $F(\mathscr{A}_0)$ is homogeneously dense in $\mathscr{B}$.

We give two examples of homogeneously dense subcategories, the latter will be used to give an alternative proof of the fact that $\omega_{\mathscr{B}}\colon \mathscr{B} \to (\mathscr{B}\#G)/G$ is an equivalence in G-**GrCat** in Remark 5.9.7(1).

EXAMPLE 5.9.4. Let $\mathscr{B}$ be a G-graded category and $(P, \psi)\colon \mathscr{B}\#G \to (\mathscr{B}\#G)/G$ the canonical functor.

(1) If $\mathscr{B}(x,x)$ are local $\Bbbk$-algebras (cf. Definition 2.3.20) for all $x \in \mathscr{B}_0$, then any dense full subcategory $\mathscr{B}'$ of $\mathscr{B}$ is homogeneously dense.
(2) Let $\mathscr{B}'$ be the full subcategory of $(\mathscr{B}\#G)/G$ with $\mathscr{B}'_0 := \omega_{\mathscr{B}}(\mathscr{B}_0) = \{P(x^{(1)}) \mid x \in \mathscr{B}\}$ (see Definition 5.8.8). Then $\mathscr{B}'$ is homogeneously dense in $(\mathscr{B}\#G)/G$. Hence $\omega_{\mathscr{B}}\colon \mathscr{B} \to (\mathscr{B}\#G)/G$ is homogeneously dense.

Indeed, to show statement (1) it is enough to show that if $x \cong y$ in $\mathscr{B}$, then there exists a homogeneous isomorphism in $\mathscr{B}(x,y)$. Now let $f\colon x \to y$ be an isomorphism in $\mathscr{B}$. We may assume that $x \neq 0$. Write f and f^{-1} as finite sums: $f = \sum_{a\in G} f_a$ and $f^{-1} = \sum_{b\in G} g_b$ with $f_a \in \mathscr{B}^a(x,y)$ and $g_b \in \mathscr{B}^b(y,x)$ for all $a, b \in G$. Then $\sum_{a,b\in G} g_b f_a = \mathbb{1}_x$ shows that $h := g_b f_a$ is an automorphism of x for some $a, b \in G$ because $\mathscr{B}(x,x)$ is a local algebra. Thus $(h^{-1}g_b)f_a = \mathbb{1}_x$ and $e := f_a(h^{-1}g_b)$ is an idempotent in $\mathscr{B}(x,x)$, and hence $e = \mathbb{1}_x$ or $e = 0$. But $(h^{-1}g_b)ef_a = \mathbb{1}_x \neq 0$ shows that $e \neq 0$. Hence $f_a\colon x \to y$ is a homogeneous isomorphism.

By Lemma 5.4.9, $\psi_{a,\,x}\colon P(x^{(1)}) \to P(x^{(a)})$ are homogeneous isomorphisms of degree a in $(\mathscr{B}\#G)/G$ for all $x \in \mathscr{B}_0$ and all $a \in G$, from which statement (2) follows.

We now give a characterization of equivalences in the 2-category G-**GrCat**.

THEOREM 5.9.5. *Let $(H, r)\colon \mathscr{B} \to \mathscr{A}$ be a degree-preserving functor in G-**GrCat**. Then the following are equivalent.*

(1) *The 1-morphism (H, r) is an equivalence in G-**GrCat**.*
(2) *There exists a quasi-inverse I of H and homogeneous natural isomorphisms $\eta\colon \mathbb{1}_{\mathscr{A}} \Rightarrow HI$ and $\varepsilon\colon IH \Rightarrow \mathbb{1}_{\mathscr{B}}$ such that $(I, H, \eta, \varepsilon)$ is an adjoint equivalence system.*
(3) *H is fully faithful and homogeneously dense.*

In (2) above, there exists a unique degree adjuster s such that $((I, s), (H, r), \eta, \varepsilon)$ is an adjoint equivalence system. This s is given by:

$$s = (s_x)_{x\in\mathscr{A}_0} \text{ with } s_x := (\deg \eta_x)^{-1} r_{Ix}^{-1} \in G \text{ for all } x \in \mathscr{A}_0. \tag{5.39}$$

PROOF. (2) $\Rightarrow$ (1). Assume statement (2). Set $t_x := \deg \varepsilon_x$ for all $x \in \mathscr{B}_0$, and $t'_x := \deg \eta_x$ for all $x \in \mathscr{A}_0$ for simplicity. Define s as in (5.39), i.e., $s_x := {t'_x}^{-1} r_{Ix}^{-1} \in G$ for all $x \in \mathscr{A}_0$.

CLAIM 1. *$(I, s)\colon \mathscr{A} \to \mathscr{B}$ is a 1-morphism in G-**GrCat**.*

Indeed, let $x, y \in \mathscr{A}_0$, $a \in G$ and $f \in \mathscr{A}^a(x, y)$. It is enough to show that $If \in \mathscr{B}^{s_y^{-1} a s_x}(Ix, Iy)$. Since η is a natural transformation, we have $HIf = \eta_y f \eta_x^{-1} \in \mathscr{A}^{t'_y}(y, HIy)\mathscr{A}^a(x, y)\mathscr{A}^{t'^{-1}_x}(HIx, x) \subseteq \mathscr{A}^{t'_y a t'^{-1}_x}(HIx, HIy)$. Since H is fully faithful, H induces a bijection $\mathscr{B}(Ix, Iy) \to \mathscr{A}(HIx, HIy)$, which also induces bijections

$$\mathscr{B}^b(Ix, Iy) \to \mathscr{A}^{r_{Iy}{}^{-1} b r_{Ix}}(HIx, HIy)$$

for all $b \in G$. Applying this to the element b of G satisfying $r_{Iy}{}^{-1} b\, r_{Ix} = t'_y a\, t'^{-1}_x$, we have

$$If \in \mathscr{B}^{r_{Iy} t'_y a t'^{-1}_x r_{Ix}^{-1}}(Ix, Iy) = \mathscr{B}^{s_y^{-1} a s_x}(Ix, Iy).$$

CLAIM 2. *$\varepsilon\colon (I, s)(H, r) \Rightarrow (\mathbb{1}_{\mathscr{B}}, 1)$ is a 2-isomorphism in G-**GrCat**.*

Indeed, it is enough to show that ε is a 2-morphism in G-**GrCat**. This is equivalent to saying that $t_x = r_x s_{Hx}$ for all $x \in \mathscr{B}_0$ because $(I, s)(H, r) = (IH, (r_x s_{Hx})_{x\in\mathscr{B}_0})$ (cf. Definition 5.2.1(4)). Let $x \in \mathscr{B}_0$. Then since $(H\varepsilon x)(\eta Hx) = \mathbb{1}_{Hx}$, we have $1 = \deg(H\varepsilon x)\deg(\eta Hx) = r_x^{-1} t_x r_{IHx} t'_{Hx}$. Hence

$$r_x s_{Hx} = r_x t'_{Hx}{}^{-1} r_{IHx}^{-1} = r_x r_x^{-1} t_x = t_x,$$

as desired.

CLAIM 3. *$\eta\colon (\mathbb{1}_{\mathscr{A}}, 1) \Rightarrow (H, r)(I, s)$ is a 2-isomorphism in G-**GrCat**.*

Indeed, it is enough to show that η is a 2-morphism in G-**GrCat**. This is equivalent to saying that $t'_x = r_{Ix}^{-1} s_x^{-1}$ for all $x \in \mathscr{A}_0$ because $(H, r)(I, s) = (HI, (s_x r_{Ix})_{x\in\mathscr{A}_0})$. By definition, we have $r_{Ix}^{-1} s_x^{-1} = r_{Ix}^{-1} r_{Ix} t'_x = t'_x$, as desired. These three claims show that (H, r) is an equivalence in G-**GrCat**. By looking at the proof of Claim 3, we see that the degree adjuster s of I is uniquely determined as in (5.39) by η and r.

(1) $\Rightarrow$ (3). Assume statement (1), and let (I, s) be a quasi-inverse of (H, r) with 2-isomorphisms $\varepsilon\colon (I, s)(H, r) \Rightarrow \mathbb{1}_{\mathscr{B}}$ and $\eta\colon \mathbb{1}_{\mathscr{A}} \Rightarrow (H, r)(I, s)$. Then it is clear that H is fully faithful. Moreover, since $\eta\colon \mathbb{1}_{\mathscr{A}} \Rightarrow HI$ is a homogeneous natural isomorphism, $HI(\mathscr{A}_0)$ is homogeneously dense in $\mathscr{A}$. Hence H is also homogeneously dense in $\mathscr{A}$ because $H(\mathscr{B}_0) \supseteq HI(\mathscr{A}_0)$.

(3) $\Rightarrow$ (2). Assume statement (3). We can imitate the proof of (iii) $\Rightarrow$ (ii) in [**34**, p. 93, Theorem 1] ((3) $\Rightarrow$ (2) in Remark 4.4.6) to construct a quasi-inverse $I\colon \mathscr{A} \to \mathscr{B}$ as a left adjoint to H and a pair of a counit $\varepsilon\colon IH \Rightarrow \mathbb{1}_{\mathscr{B}}$ and a unit $\eta\colon \mathbb{1}_{\mathscr{A}} \Rightarrow HI$. Here we just give definitions of them. Then it is enough to show that both ε and η are homogeneous natural isomorphisms.

Definition of I and η. Let $x \in \mathscr{A}_0$. Since H is homogeneously dense, there exists a $y_x \in \mathscr{B}_0$ such that there is a homogeneous isomorphism $\eta_x\colon x \to Hy_x$. Choose a pair (y_x, η_x) once for all x, and define $Ix := y_x$. Then $\eta_x\colon x \to HIx$ is a homogeneous isomorphism, and define $\eta := (\eta_x)_{x\in\mathscr{A}_0}$. Let $f \in \mathscr{A}(x, x')$. We define

If as follows: Since H is fully faithful, H induces a bijection $H_{Ix,Ix'}\colon \mathscr{B}(Ix,Ix') \to \mathscr{A}(HIx,HIx')$. Then define $If := H_{Ix,Ix'}^{-1}(\eta_{x'}f\eta_x^{-1})$ as in the following diagram:

$$\begin{array}{ccc}
x & \xrightarrow{\quad f \quad} & x' \\
{\scriptstyle \eta_x}\downarrow & \circlearrowleft & \downarrow{\scriptstyle \eta_{x'}} \\
HIx & \xrightarrow{\ \eta_{x'}f\eta_x^{-1}\ } & HIx' \\
{\scriptstyle H}\uparrow & \uparrow{\scriptstyle H} & \uparrow{\scriptstyle H} \\
Ix & \xrightarrow[\quad If \quad]{} & Ix'.
\end{array}$$

Definition of ε. Let $y \in \mathscr{B}_0$. Then $Hy \in \mathscr{A}_0$, and $\eta_{Hy} \in \mathscr{A}(Hy, HIHy)$. H induces a bijection $H_{IHy,\,y}\colon \mathscr{B}(IHy,y) \to \mathscr{A}(HIHy,Hy)$. Then define $\varepsilon_y := H_{IHy,\,y}^{-1}(\eta_{Hy}^{-1})$. Then the proof in [**34**, p. 93] works (or it is straightforward) to show that (I,H,η,ε) is an adjoint equivalence system. Now by definition, η is a homogeneous natural isomorphism. It remains to show that ε_y are homogeneous isomorphisms for all $y \in \mathscr{B}_0$. Since η_{Hy} is a homogeneous isomorphism, so is $\eta_{Hy}^{-1} \in \mathscr{A}(HIHy,Hy)$. Set $a := \deg \eta_{Hy}$. Since (H,r) is a degree-preserving functor, the bijection $H_{IHy,\,y}$ induces a bijection

$$\mathscr{B}^{r_y a\, r_{IHy}^{-1}}(IHy,y) \to \mathscr{A}^{a}(HIHy,Hy) \ni \eta_{Hy}^{-1}.$$

Hence $\varepsilon_y = H_{IHy,\,y}^{-1}(\eta_{Hy}^{-1}) \in \mathscr{B}^{r_y b\, r_{IHy}^{-1}}(IHy,y)$ and is a homogeneous isomorphism. □

The following is immediate by Theorem 5.9.5.

COROLLARY 5.9.6. *Let $(H,r)\colon \mathscr{B} \to \mathscr{A}$ be a 1-morphism in G-**GrCat**. Then the following are equivalent.*

(1) *(H,r) is an isomorphism in G-**GrCat**.*
(2) *$H\colon \mathscr{B} \to \mathscr{A}$ is a category isomorphism.*

If this is the case, then the inverse of (H,r) is given by

$$(H,r)^{-1} = (H^{-1}, (r_{H^{-1}x}^{-1})_{x\in\mathscr{A}_0}).$$

REMARK 5.9.7. Let $\mathscr{B} \in G$-**GrCat**$_0$.

(1) Theorem 5.9.5 and Example 5.9.4(2) give an immediate alternative proof of the fact that $\omega_{\mathscr{B}}\colon \mathscr{B} \to (\mathscr{B}\#G)/G$ is an equivalence in G-**GrCat**.
(2) Also by Theorem 5.9.5, a degree-preserving functor $(H,r)\colon \mathscr{A} \to \mathscr{B}$ in G-**GrCat** with H a category equivalence is an equivalence in G-**GrCat** if and only if H is homogeneously dense. In particular, if $\mathscr{B}(x,x)$ are local algebras for all $x \in \mathscr{B}_0$, then all degree-preserving functors that are category equivalences are equivalences in G-**GrCat** by Example 5.9.4(1).

Next we will give one more characterization of an equivalence in G-**GrCat** using the composite of three degree preserving functors which are surjective, bijective, and injective on objects, respectively. First we add necessary terminologies.

DEFINITION 5.9.8. Let $\mathscr{B}$ be a G-graded category.

(1) For each $x, y \in \mathscr{B}$, we say that x and y are *homogeneously isomorphic* (and write $x \cong_H y$) if there exists a homogeneous isomorphism $x \to y$. Since the set of homogeneous isomorphisms in $\mathscr{B}$ is closed under composition and taking inverses, the relation $\cong_H$ on $\mathscr{B}_0$ is an equivalence relation, whose equivalence classes are called *homogeneous isoclasses*.

(2) Let $\mathscr{B}'$ be a full subcategory of $\mathscr{B}$. Then $\mathscr{B}'$ is called a *homogeneous skeleton* of $\mathscr{B}$ if $\mathscr{B}'_0$ forms a complete set of representatives of homogeneous isoclasses in $\mathscr{B}_0$. Note that $\mathscr{B}'$ is homogeneously dense in $\mathscr{B}$ if and only if $\mathscr{B}'$ contains a homogeneous skeleton of $\mathscr{B}$.

LEMMA 5.9.9. *Let $\mathscr{B} \in G$-$\mathbf{GrCat}_0$. If $\mathscr{B}'$ is a homogeneously dense full subcategory of $\mathscr{B}$, then the inclusion functor $S\colon \mathscr{B}' \hookrightarrow \mathscr{B}$ induces an equivalence $(S,1)\colon \mathscr{B}' \to \mathscr{B}$ in G-**GrCat**.*

PROOF. Note that $\mathscr{B}'$ is again a G-graded category by setting $\mathscr{B}'^a(x,y) := \mathscr{B}^a(x,y)$ for all $x, y \in \mathscr{B}'_0$ and all $a \in G$, and hence $(S,1)\colon \mathscr{B}' \to \mathscr{B}$ is a degree-preserving functor. Then the assertion follows by Theorem 5.9.5. $\square$

PROPOSITION 5.9.10. *Let $(H,r)\colon \mathscr{B} \to \mathscr{A}$ be a degree-preserving functor in G-**GrCat**. Then the following are equivalent.*

(1) *(H,r) is an equivalence in G-**GrCat**.*

(2) *There exist homogeneously dense full subcategories $\mathscr{B}'$ and $\mathscr{A}'$ of $\mathscr{B}$ and $\mathscr{A}$, respectively with an isomorphism $(H',r')\colon \mathscr{B}' \to \mathscr{A}'$ in G-**GrCat**, and a homogeneous natural isomorphism*

$$\zeta\colon (H,r) \Rightarrow (S',1)(H',r')(N,s),$$

*where $S\colon \mathscr{B}' \hookrightarrow \mathscr{B}$ and $S'\colon \mathscr{A}' \hookrightarrow \mathscr{A}$ are inclusion functors, and (N,s) is a quasi-inverse of the equivalence $(S,1)$ in G-**GrCat**.*

PROOF. (2) $\Rightarrow$ (1). This immediately follows by Theorem 5.9.5.

(1) $\Rightarrow$ (2). Assume statement (1). Then there exist degree-preserving functors $(H,r)\colon \mathscr{B} \to \mathscr{A}$ and $(I,s)\colon \mathscr{A} \to \mathscr{B}$, and 2-isomorphisms $\varepsilon\colon (I,s)(H,r) \Rightarrow (\mathbb{1}_{\mathscr{B}},1)$ and $\eta\colon (\mathbb{1}_{\mathscr{A}},1) \Rightarrow (H,r)(I,s)$ in G-**GrCat**. Let $\mathscr{B}'$ be a homogeneous skeleton of $\mathscr{B}$. Then $\mathscr{B}'$ is homogeneously dense in $\mathscr{B}$. Let $\mathscr{A}'$ be the full subcategory of $\mathscr{A}$ with $\mathscr{A}'_0 := H(\mathscr{B}'_0)$. Then we first show the following.

CLAIM. *$\mathscr{A}'$ is a homogeneous skeleton of $\mathscr{A}$.*

Indeed, let $x \in \mathscr{A}_0$. Then by construction, there exist an $x' \in \mathscr{B}'$ and a homogeneous isomorphism $f\colon Ix \to x'$ in $\mathscr{B}$. Hence we have homogeneous isomorphisms $x \xrightarrow{\eta_x} HIx \xrightarrow{Hf} Hx'$ in $\mathscr{A}$. Thus $x \cong_H Hx' \in \mathscr{A}'_0$, which shows that $\mathscr{A}'$ is homogeneously dense in $\mathscr{A}$. Next assume that there exists a homogeneous isomorphism $g\colon Hx \to Hy$ for some $x, y \in \mathscr{B}'_0$. Then we have homogeneous isomorphisms $x \xleftarrow{\varepsilon_x} IHx \xrightarrow{Ig} IHy \xrightarrow{\varepsilon_y} y$. Thus $x \cong_H y$, and hence $x = y$. As a consequence

$$x \neq y \text{ implies } Hx \not\cong_H Hy. \tag{5.40}$$

This proves the claim.

Now let $S\colon \mathscr{B}' \hookrightarrow \mathscr{B}$ and $S'\colon \mathscr{A}' \hookrightarrow \mathscr{A}$ be inclusion functors, and as in the proof of Theorem 5.9.5, construct a quasi-inverse (N,s) of the equivalence $(S,1)$ in G-**GrCat** such that

$$((N,s),(S,1),\nu\colon \mathbb{1}_{\mathscr{B}} \overset{\sim}{\Rightarrow} SN, \mathbb{1}_{\mathbb{1}_{\mathscr{B}'}}\colon NS = \mathbb{1}_{\mathscr{B}'})$$

is an adjoint equivalence system. Then (5.39) shows that $s_x = (\deg \nu_x)^{-1}$ for all $x \in \mathscr{B}_0$. The implication (5.40) also shows that H induces a bijection $\mathscr{B}'_0 \to \mathscr{A}'_0$. As H is fully faithful, H induces a category isomorphism $H' \colon \mathscr{B}' \to \mathscr{A}'$ that satisfies $S'H' = HS$. Let r' be the restriction of r to $\mathscr{B}'_0$. Then $(H', r') \colon \mathscr{B}' \to \mathscr{A}'$ is a degree-preserving functor, which turns out to be an isomorphism in G-**GrCat** by Corollary 5.9.6. Now $\xi := H\nu$ is a homogeneous natural isomorphism $H \Rightarrow HSN = S'H'N$. It remains to show that ξ is a 2-morphism $(H, r) \Rightarrow (S', 1)(H', r')(N, s) = (S'H'N, (s_x r'_{Nx})_{x \in \mathscr{B}_0})$ in G-**GrCat**. For this it is enough to show that $\deg H\nu_x = (s_x r'_{Nx})^{-1} r_x$ for all $x \in \mathscr{B}_0$. Now since $\deg \nu_x = s_x^{-1}$ and $\nu_x \colon x \to SNx = Nx$, we have $\deg H\nu_x = r'^{-1}_{Nx} s_x^{-1} r_x = (s_x r'_{Nx})^{-1} r_x$, as desired. □

The following is immediate by Proposition 5.9.10 and Lemma 5.9.9.

COROLLARY 5.9.11. *Let $\mathscr{B}, \mathscr{A} \in G$-$\mathbf{GrCat}_0$. Then the following are equivalent.*

(1) *$\mathscr{B}$ and $\mathscr{A}$ are equivalent in G-**GrCat**.*
(2) *There exist homogeneously dense full subcategories $\mathscr{B}'$ and $\mathscr{A}'$ of $\mathscr{B}$ and $\mathscr{A}$, respectively such that $\mathscr{B}' \cong \mathscr{A}'$ in G-**GrCat**.*

5.10. On the second orbit categories and the right smash products

Let $\mathscr{C}$ be a pseudo-G-category. In Definition 5.4.1, as a definition of the local morphism set of the orbit category $\mathscr{C}/G$, we took the direct sum

$$(\mathscr{C}/G)(x, y) := \coprod_{a \in G} \mathscr{C}(X(a)x, y)$$

by applying the group action on the first variable of the functor $\mathscr{C}(\text{-}, ?)$ with two variables for all $x, y \in (\mathscr{C}/G)_0 = \mathscr{C}_0$. Beside this, it is possible to consider a definition that takes the direct sum by applying the group action on the second variable. To distinguish these, we denote the latter by $\mathscr{C}/_2 G$, where $\mathscr{C}$ needs to have a pseudo-G^{op}-action in order that $\mathscr{C}/_2 G$ is a G-graded category.

In the following, we explain this definition and its relation to the original definition, as well as the corresponding smash products and their relations.

DEFINITION 5.10.1. G^{op}-categories (resp. pseudo-G^{op}-categories, lax G^{op}-categories, colax G^{op}-categories, lax pseudo-G^{op}-categories) are called *right G-categories* (resp. *right pseudo-G-categories*, *right lax G-categories*, *right colax G-categories*, *right lax pseudo-G-categories*).

REMARK 5.10.2. Let $(\mathscr{C}, X)$ be a colax G-category.

(1) If we define $(\mathscr{C}^{\text{op}}, X_{\text{op}})$ as follows, then it turns out to be a lax G-category:
   (a) A linear category $\mathscr{C}^{\text{op}}$.
   (b) A family $(X_{\text{op}}(a) \colon \mathscr{C}^{\text{op}} \to \mathscr{C}^{\text{op}})_{a \in G}$ of functors defined by $X_{\text{op}}(a) := X(a)$ for all $a \in G$.
   (c) A natural transformation $(X_{\text{op}})_* \colon \mathbb{1}_{\mathscr{C}^{\text{op}}} \Rightarrow X_{\text{op}}(1)$ defined by
   $$(X_{\text{op}})_* x := X_* x \colon x \to X(1)x, \ (x \in \mathscr{C}_0^{\text{op}}).$$
   (d) A family $((X_{\text{op}})_{b,a} \colon X_{\text{op}}(b) X_{\text{op}}(a) \Rightarrow X_{\text{op}}(b \circ a))_{a,b \in G}$ of natural transformations defined by
   $$(X_{\text{op}})_{b,a} x := X_{b,a} x \colon X(b)X(a)x \to X(ba)x, \ (x \in \mathscr{C}_0^{\text{op}}).$$

Indeed, if we regard X_{op} as $G \to \Bbbk\text{-}\mathbf{Cat}(\{\mathscr{C}^{\mathrm{op}}\})^{\mathrm{co}}$, then by construction, it clearly satisfies the axioms of colax functors (see Definition 5.1.1). In particular, if $(\mathscr{C}, X)$ is a G-category, then so is $(\mathscr{C}^{\mathrm{op}}, X_{\mathrm{op}})$.

(2) If we define $(\mathscr{C}, X^{\mathrm{op}})$ using the group anti-isomorphism $?^{-1}\colon G \to G, x \mapsto x^{-1}$ as follows, then it turns out to be a right colax G-category:
   (a) A linear category $\mathscr{C}$.
   (b) A family $(X^{\mathrm{op}}(a)\colon \mathscr{C} \to \mathscr{C})_{a\in G}$ of functors defined by $X^{\mathrm{op}}(a) := X(a^{-1})$ for all $a \in G$.
   (c) A natural transformation $X^{\mathrm{op}}_{*}\colon X^{\mathrm{op}}(1) \Rightarrow \mathbb{1}_{\mathscr{C}}$ defined by $X^{\mathrm{op}}_{*} := X_{*}$.
   (d) A family $(X^{\mathrm{op}}_{b,a}\colon X^{\mathrm{op}}(ba) \Rightarrow X^{\mathrm{op}}(a) \circ X^{\mathrm{op}}(b))_{a,b\in G}$ of natural transformations defined by

$$
\begin{aligned}
X^{\mathrm{op}}_{b,a} := X_{a^{-1},b^{-1}}\colon X^{\mathrm{op}}(ba) = X((ba)^{-1}) &= X(a^{-1}b^{-1}) \\
&\Rightarrow X(a^{-1}) \circ X(b^{-1}) = X^{\mathrm{op}}(a) \circ X^{\mathrm{op}}(b).
\end{aligned}
$$

(3) By the above, $(\mathscr{C}^{\mathrm{op}}, X^{\mathrm{op}}_{\mathrm{op}})$ turns out to be a right lax G-category, where we set $X^{\mathrm{op}}_{\mathrm{op}} := (X^{\mathrm{op}})_{\mathrm{op}}$. Namely, first regard $(\mathscr{C}, X^{\mathrm{op}})$ in (2) as a colax G^{op}-category, and apply (1) to it. Then $(\mathscr{C}^{\mathrm{op}}, X^{\mathrm{op}}_{\mathrm{op}})$ turns out to be a lax G^{op}-category, i.e., a right lax G-category.

DEFINITION 5.10.3. For a right lax G-category $\mathscr{C}$, we define a linear category $\mathscr{C}/_2 G$ as follows, and call it the *second orbit category* of $\mathscr{C}$ by G:

- $(\mathscr{C}/_2 G)_0 := \mathscr{C}_0$.
- For each $x, y \in (\mathscr{C}/_2 G)_0$, we set $(\mathscr{C}/_2 G)(x,y) := \coprod_{a\in G} \mathscr{C}(x, X(a)y)$ Here if we let

$$\sigma_a^{\mathscr{C}/_2 G}\colon \mathscr{C}(x, X(a)y) \to (\mathscr{C}/_2 G)(x,y), \quad f \mapsto (\delta_{b,a} f)_{b\in G}$$

be the canonical injection, and set $(\mathscr{C}/_2 G)^a(x,y) := \sigma_a^{\mathscr{C}/_2 G}(\mathscr{C}(x, X(a)y))$, $(x, y \in \mathscr{C}_0 = (\mathscr{C}/_2 G)_0, a \in G)$, then we have the following inner direct sum:

$$(\mathscr{C}/_2 G)(x,y) = \bigoplus_{a\in G} (\mathscr{C}/_2 G)^a(x,y). \tag{5.41}$$

- For each $f = (f_a)_{a\in G} \in (\mathscr{C}/_2 G)(x,y), g = (g_b)_{b\in G} \in (\mathscr{C}/_2 G)(y,z)$, we set

$$g \circ f := \left(\sum_{\substack{a,b\in G \\ ba=c}} X_{b,a}z \circ X(a)g_b \circ f_a \right)_{c\in G},$$

where each entry of right-hand side is given by the following composite:

$$x \xrightarrow{f_a} X(a)y \xrightarrow{X(a)g_b} (X(a) \circ X(b))z \xrightarrow{X_{b,a}z} X(ba)z.$$

- For each $x \in (\mathscr{C}/_2 G)_0$, the identity of x in $\mathscr{C}/_2 G$ is given by

$$\mathbb{1}_x^{\mathscr{C}/_2 G} = \sigma_1^{\mathscr{C}/_2 G}(X_* x) = (\delta_{a,1} X_* x)_{a\in G} \in \coprod_{a\in G} \mathscr{C}(x, X(a)x).$$

- By the formula (5.41) and the composition given above, $\mathscr{C}/_2 G$ becomes a G-graded category.

Conversely, from a G-graded category a right G-category $(= \text{a } G^{\mathrm{op}}\text{-category})$ is defined as follows:

DEFINITION 5.10.4. Let $\mathscr{B}$ be a G-graded category. Then we define as follows a right G-category $\mathscr{B}\#^{\mathrm{op}}G$ which is called the *right smash product* of $\mathscr{B}$ and G:

- $(\mathscr{B}\#^{\mathrm{op}}G)_0 := \{x^{(a)} := (x, a) \mid x \in \mathscr{B}_0, a \in G\}$.
- $(\mathscr{B}\#^{\mathrm{op}}G)(x^{(a)}, y^{(b)}) := \mathscr{B}^{ba^{-1}}(x, y)$ $(x^{(a)}, y^{(b)} \in (\mathscr{B}\#^{\mathrm{op}}G)_0)$, where the precise definition (to let the union of all local morphism sets be a disjoint union) is given by $(\mathscr{B}\#^{\mathrm{op}}G)(x^{(a)}, y^{(b)}) := \{b\} \times \mathscr{B}^{ba^{-1}}(x, y) \times \{a\} = \{{}^{(b)}f^{(a)} := (b, f, a) \mid f \in \mathscr{B}^{ba^{-1}}(x, y)\}$, but if there seems to be no risk of confusion, then we identify the latter with the former by the second projection.
- The composition of $\mathscr{B}\#^{\mathrm{op}}G$ is defined by the following commutative diagram:

$$\begin{array}{ccc} (\mathscr{B}\#^{\mathrm{op}}G)(y^{(b)}, z^{(c)}) \times (\mathscr{B}\#^{\mathrm{op}}G)(x^{(a)}, y^{(b)}) & \longrightarrow & (\mathscr{B}\#^{\mathrm{op}}G)(x^{(a)}, z^{(c)}) \\ \| & & \| \\ \mathscr{B}^{cb^{-1}}(y, z) \times \mathscr{B}^{ba^{-1}}(x, y) & \longrightarrow & \mathscr{B}^{ca^{-1}}(x, z) \end{array}$$

  for all $x^{(a)}, y^{(b)}, z^{(c)} \in (\mathscr{B}\#^{\mathrm{op}}G)_0$, where the composition in the lower row is given by that of $\mathscr{B}$. Thus the composition can be written in terms of entries as follows:

$${}^{(c)}g^{(b)} \circ {}^{(b)}f^{(a)} := {}^{(c)}(g \circ f)^{(a)}, \quad ((g, f) \in \mathscr{B}^{cb^{-1}}(y, z) \times \mathscr{B}^{ba^{-1}}(x, y)).$$

- The right action of G is defined as follows:

$$\begin{cases} x^{(a)} \cdot c := x^{(ac)}, \\ {}^{(b)}f^{(a)} \cdot c := {}^{(bc)}f^{(ac)} \end{cases} \quad \begin{pmatrix} x^{(a)}, y^{(b)} \in (\mathscr{B}\#^{\mathrm{op}}G)_0, \\ c \in G, f \in \mathscr{B}^{ba^{-1}}(x, y) \end{pmatrix}.$$

REMARK 5.10.5. Let $(\mathscr{B}, D)$ be a G-graded category, and denote the composition of $\mathscr{B}^{\mathrm{op}}$ by $\circ^{\mathrm{op}}$ and the operation of G^{op} by $\circ$.

(1) If we define D_{op} as follows, then $(\mathscr{B}^{\mathrm{op}}, D_{\mathrm{op}})$ turns out to be a G^{op}-graded category:

$$D^a_{\mathrm{op}}(x, y) := D^a(y, x) \quad (x, y \in \mathscr{B}_0, a \in G).$$

Indeed, at first, the direct sum decomposition $\mathscr{B}(y, x) = \bigoplus_{a \in G} D^a(y, x)$

gives the direct sum decomposition $\mathscr{B}^{\mathrm{op}}(x,y) = \bigoplus_{a\in G} D^a_{\mathrm{op}}(x,y)$, and also, we have

$$\begin{aligned} D^b_{\mathrm{op}}(y,z) \circ^{\mathrm{op}} D^a_{\mathrm{op}}(x,y) &= D^a(y,x) \circ D^b(z,y) \\ &\subseteq D^{ab}(z,x) = D^{b\circ a}_{\mathrm{op}}(x,z) \ (x,y,z \in \mathscr{B}_0, a,b \in G). \end{aligned}$$

(2) If we define D^{op} as follows, then $(\mathscr{B}, D^{\mathrm{op}})$ turns out to be a G^{op}-graded category:

$$(D^{\mathrm{op}})^a(x,y) := D^{a^{-1}}(x,y) \ (x,y \in \mathscr{B}_0, a \in G).$$

Indeed, at first, the direct sum decomposition $\mathscr{B}(x,y) = \bigoplus_{a\in G} D^a(x,y)$ gives the direct sum decomposition $\mathscr{B}(x,y) = \bigoplus_{a\in G} D^{a^{-1}}(x,y) = \bigoplus_{a\in G}(D^{\mathrm{op}})^a(x,y)$, and also, we have

$$\begin{aligned} (D^{\mathrm{op}})^b(y,z) \circ (D^{\mathrm{op}})^a(x,y) &= D^{b^{-1}}(y,z) \circ D^{a^{-1}}(x,y) \\ &\subseteq D^{b^{-1}a^{-1}}(x,z) = D^{(ab)^{-1}}(x,z) \\ &= (D^{\mathrm{op}})^{b\circ a}(x,z) \ (x,y,z \in \mathscr{B}_0, a,b \in G). \end{aligned}$$

(3) By the above, $(\mathscr{B}^{\mathrm{op}}, D^{\mathrm{op}}_{\mathrm{op}})$ turns out to be a G-graded category, where $D^{\mathrm{op}}_{\mathrm{op}} := (D^{\mathrm{op}})_{\mathrm{op}} = (D_{\mathrm{op}})^{\mathrm{op}}$ is defined as follows:

$$(D^{\mathrm{op}}_{\mathrm{op}})^a(x,y) := D^{a^{-1}}(y,x) \quad (x,y \in \mathscr{B}_0, a \in G).$$

The relation between two smash products $\#$ and $\#^{\mathrm{op}}$ is as follows: If $(\mathscr{B}, D)$ is a G-graded category, then by Remark 5.10.5(1), $(\mathscr{B}^{\mathrm{op}}, D_{\mathrm{op}})$ becomes a G^{op}-graded category, and by Definition 5.6.1, $((\mathscr{B}^{\mathrm{op}}, D_{\mathrm{op}})\#G^{\mathrm{op}}, X^{(\mathscr{B}^{\mathrm{op}}, D_{\mathrm{op}})\#G^{\mathrm{op}}})$ becomes a G^{op}-category. Therefore $(((\mathscr{B}^{\mathrm{op}}, D_{\mathrm{op}})\#G^{\mathrm{op}})^{\mathrm{op}}, (X^{(\mathscr{B}^{\mathrm{op}}, D_{\mathrm{op}})\#G^{\mathrm{op}}})_{\mathrm{op}})$ becomes a G^{op}-category, i.e., a right G-category by Remark 5.10.2(1). By this remark, statement (1) below makes sense.

PROPOSITION 5.10.6. *Let $(\mathscr{B}, D)$ be a G-graded category, and $(\mathscr{A}, C)$ a G^{op}-graded category. The the following hold.*

(1) $(\mathscr{B}, D)\#^{\mathrm{op}}G = (((\mathscr{B}^{\mathrm{op}}, D_{\mathrm{op}})\#G^{\mathrm{op}})^{\mathrm{op}}, (X^{(\mathscr{B}^{\mathrm{op}}, D_{\mathrm{op}})\#G^{\mathrm{op}}})_{\mathrm{op}})$ *(as right G-categories),*

(2) $(\mathscr{A}, C^{\mathrm{op}})\#G \cong ((\mathscr{A}, C)\#G^{\mathrm{op}}, (X^{(\mathscr{A},C)\#G})^{\mathrm{op}})$ *(as colax G-categories, natural in $(\mathscr{A}, C)$),*

(3) $(\mathscr{B}^{\mathrm{op}}, D^{\mathrm{op}}_{\mathrm{op}})\#G \cong (((\mathscr{B}, D)\#^{\mathrm{op}}G)^{\mathrm{op}}, (X^{(\mathscr{B},D)\#^{\mathrm{op}}G})^{\mathrm{op}}_{\mathrm{op}})$ *(as colax G-categories, natural in $(\mathscr{B}, D)$).*

PROOF. Statement (3) follows from (1) and (2). We will only show (1) because (2) is proved similarly.

The object sets of both sides coincide with $\mathscr{B}_0 \times G$. Let $x^{(a)}, y^{(b)} \in \mathscr{B}_0 \times G$. Then the local morphism set of the left-hand side is given by

$$((\mathscr{B}, D)\#^{\mathrm{op}}G)(x^{(a)}, y^{(b)}) = \{b\} \times D^{ba^{-1}}(x,y) \times \{a\},$$

and by Remark 5.6.2, that of the right-hand side is given by

$$\begin{aligned} ((\mathscr{B}^{\mathrm{op}}, D_{\mathrm{op}})\#G^{\mathrm{op}})^{\mathrm{op}}(x^{(a)}, y^{(b)}) &= (\mathscr{B}^{\mathrm{op}}, D_{\mathrm{op}})\#G^{\mathrm{op}})(y^{(b)}, x^{(a)}) \\ &= \{b\} \times D^{a^{-1}\circ b}_{\mathrm{op}}(y,x) \times \{b\} \\ &= \{b\} \times D^{ba^{-1}}(x,y) \times \{a\}, \end{aligned}$$

and hence they coincide. For a pair of morphisms $x^{(a)} \xrightarrow{{}^{(b)}f^{(a)}} y^{(b)} \xrightarrow{{}^{(c)}g^{(b)}} z^{(c)}$ on both sides of (1), both composites are given by ${}^{(c)}g^{(b)} \circ {}^{(b)}f^{(a)} = {}^{(c)}(g \circ f)^{(a)}$. Finally, the right action of G on both sides are given by $x^{(a)} \cdot c = x^{(ac)}$, ${}^{(b)}f^{(a)} \cdot c = {}^{(bc)}f^{(ac)}$ for all $a, b, c \in G, x, y \in \mathscr{B}_0, f \in \mathscr{B}^{ba^{-1}}(x, y)$. □

EXERCISE 5.10.7. Prove Proposition 5.10.6 (2) and (3). (Hint: For both (2) and (3), the natural isomorphism is given by $x^{(a)} \mapsto x^{(a^{-1})}$ and ${}^{(b)}f^{(a)} \mapsto {}^{(b^{-1})}f^{(a^{-1})}$.)

The relationship between $?/G$ and $?/_2G$ is given below. Let $(\mathscr{C}, X)$ be a colax G-category. Then $(\mathscr{C}^{\mathrm{op}}, X^{\mathrm{op}}_{\mathrm{op}})$ becomes a right lax G-category by Remark 5.10.2 (3), and $(\mathscr{C}^{\mathrm{op}}, X^{\mathrm{op}}_{\mathrm{op}})/_2G$ becomes a G-graded category by Definition 5.10.3. On the other hand, we obtain a G-graded category $(\mathscr{C}/G, D_{\mathscr{C}/G})$ by Remark 5.4.10, and therefore a G-graded category $((\mathscr{C}/G)^{\mathrm{op}}, (D_{\mathscr{C}/G})^{\mathrm{op}}_{\mathrm{op}})$ by Remark 5.10.5(3). Hence statement (3) below makes sense.

PROPOSITION 5.10.8. *Let $(\mathscr{C}, X)$ be a colax G-category, $(\mathscr{D}, Y)$ a colax G^{op}-category. Then the following equalities hold:*

(1) $(\mathscr{C}, X^{\mathrm{op}})/G^{\mathrm{op}} = (\mathscr{C}/G, D^{\mathrm{op}}_{\mathscr{C}/G})$ *(as G^{op}-graded categories);*
(2) $(\mathscr{D}^{\mathrm{op}}, Y_{\mathrm{op}})/_2G = ((\mathscr{D}/G^{\mathrm{op}})^{\mathrm{op}}, (D_{\mathscr{C}/G^{\mathrm{op}}})_{\mathrm{op}})$ *(as G-graded categories);*
(3) $(\mathscr{C}^{\mathrm{op}}, X^{\mathrm{op}}_{\mathrm{op}})/_2G = ((\mathscr{C}/G)^{\mathrm{op}}, (D_{\mathscr{C}/G})^{\mathrm{op}}_{\mathrm{op}})$ *(as G-graded categories).*

PROOF. Statement (3) follows from statements (1) and (2), the proofs of which are left to the reader. We will only give a proof of statement (3). The object sets of both sides are equal to $\mathscr{C}^{\mathrm{op}}_0$. Let $x, y \in \mathscr{C}^{\mathrm{op}}_0$. Then the local morphism set at (x, y) of the left-hand side is given by

$$((\mathscr{C}^{\mathrm{op}}, X^{\mathrm{op}}_{\mathrm{op}})/_2G)(x, y) = \coprod_{a \in G} \mathscr{C}^{\mathrm{op}}(x, X^{\mathrm{op}}_{\mathrm{op}}(a)y) = \coprod_{a \in G} \mathscr{C}^{\mathrm{op}}(x, X(a^{-1})y)$$
$$= \coprod_{a \in G} \mathscr{C}(X(a^{-1})y, x),$$

and that of the right-hand side is given by

$$((\mathscr{C}/G)^{\mathrm{op}}, (D_{\mathscr{C}/G})^{\mathrm{op}}_{\mathrm{op}})(x, y) = \coprod_{a \in G} ((D_{\mathscr{C}/G})^{\mathrm{op}}_{\mathrm{op}})^a(x, y)$$
$$= \coprod_{a \in G} (D_{\mathscr{C}/G})^{a^{-1}}(y, x) = \coprod_{a \in G} \mathscr{C}(X(a^{-1})y, x),$$

and hence they coincide. For a pair of morphisms $x \xrightarrow{f} y \xrightarrow{g} z$, $f = (f_a)_{a \in G}, g = (g_b)_{b \in G}$ $(f_a \in \mathscr{C}(X(a^{-1})y, x), g_b \in \mathscr{C}(X(b^{-1})z, y))$ of both sides, its composite in both sides is given by

$$g \circ f = \left(\sum_{\substack{a,b \in G \\ c = a^{-1}b^{-1}}} f_a \circ X(a^{-1})g_b \circ X_{a^{-1}, b^{-1}} \right)_{c \in G},$$

and hence the compositions of both sides coincide. □

EXERCISE 5.10.9. Prove statements (1) and (2) in Proposition 5.10.8.

REMARK 5.10.10. If we define a 2-category **Cat**-G of all small, right lax pseudo-G-categories in the same way as the 2-category G-**Cat** of all small, pseudo-G-categories, then the construction of the second orbit categories is extended to a

2-functor $?/_2G\colon \mathbf{Cat}\text{-}G \to G\text{-}\mathbf{GrCat}$, and the construction of right smash products is extended to a 2-functor $?\#^{\mathrm{op}}G\colon G\text{-}\mathbf{GrCat} \to \mathbf{Cat}\text{-}G$. The constructions in Remark 5.10.2 (1), (2) and Remark 5.10.5 (1), (2) are extended to isomorphisms of 2-categories

$$\begin{aligned}
((\text{-})^{\mathrm{op}}, (\text{-})_{\mathrm{op}})\colon \mathbf{Cat}\text{-}G \to G^{\mathrm{op}}\text{-}\mathbf{Cat},\ (\mathscr{C}, X) \mapsto (\mathscr{C}^{\mathrm{op}}, X^{\mathrm{op}}_{\mathrm{op}}),\\
((\text{-}), (\text{-})^{\mathrm{op}})\colon G^{\mathrm{op}}\text{-}\mathbf{Cat} \to G\text{-}\mathbf{Cat},\ (\mathscr{C}, X) \mapsto (\mathscr{C}, X^{\mathrm{op}}),\ \text{and}\\
((\text{-})^{\mathrm{op}}, (\text{-})_{\mathrm{op}})\colon G\text{-}\mathbf{GrCat} \to G^{\mathrm{op}}\text{-}\mathbf{GrCat},,\ (\mathscr{B}, D) \mapsto (\mathscr{B}^{\mathrm{op}}, D^{\mathrm{op}}_{\mathrm{op}}),\\
((\text{-}), (\text{-})^{\mathrm{op}})\colon G^{\mathrm{op}}\text{-}\mathbf{GrCat} \to G\text{-}\mathbf{GrCat},\ (\mathscr{B}, D) \mapsto (\mathscr{B}, D^{\mathrm{op}}),
\end{aligned}$$

respectively $(((\text{-})^{\mathrm{op}}, (\text{-})_{\mathrm{op}}) \circ ((\text{-})^{\mathrm{op}}, (\text{-})_{\mathrm{op}}) = \mathbb{1}, ((\text{-}), (\text{-})^{\mathrm{op}}) \circ ((\text{-}), (\text{-})^{\mathrm{op}}) = \mathbb{1})$. Moreover, these form the following commutative diagrams (they are strictly commutative except for one):

$$\begin{array}{ccccc}
G\text{-}\mathbf{Cat} & \xrightarrow[\cong]{((\text{-}),(\text{-})^{\mathrm{op}})} & G^{\mathrm{op}}\text{-}\mathbf{Cat} & \xrightarrow[\cong]{((\text{-})^{\mathrm{op}},(\text{-})^{\mathrm{op}}_{\mathrm{op}})} & \mathbf{Cat}\text{-}G\\
{\scriptstyle ?/G}\big\downarrow & & \big\downarrow{\scriptstyle ?/G^{\mathrm{op}}} & & \big\downarrow{\scriptstyle ?/_2G}\\
G\text{-}\mathbf{GrCat} & \xrightarrow[((\text{-}),(\text{-})^{\mathrm{op}})]{\cong} & G^{\mathrm{op}}\text{-}\mathbf{GrCat} & \xrightarrow[((\text{-})^{\mathrm{op}},(\text{-})^{\mathrm{op}}_{\mathrm{op}})]{\cong} & G\text{-}\mathbf{GrCat},
\end{array}$$

$$\begin{array}{ccccc}
G\text{-}\mathbf{Cat} & \xrightarrow[\cong]{((\text{-}),(\text{-})^{\mathrm{op}})} & G^{\mathrm{op}}\text{-}\mathbf{Cat} & \xrightarrow[\cong]{((\text{-})^{\mathrm{op}},(\text{-})^{\mathrm{op}}_{\mathrm{op}})} & \mathbf{Cat}\text{-}G\\
{\scriptstyle ?\#G}\big\uparrow & \text{up to natural isomorphism} & \big\uparrow{\scriptstyle ?\#G^{\mathrm{op}}} & & \big\uparrow{\scriptstyle ?\#^{\mathrm{op}}G}\\
G\text{-}\mathbf{GrCat} & \xrightarrow[((\text{-}),(\text{-})^{\mathrm{op}})]{\cong} & G^{\mathrm{op}}\text{-}\mathbf{GrCat} & \xrightarrow[((\text{-})^{\mathrm{op}},(\text{-})^{\mathrm{op}}_{\mathrm{op}})]{\cong} & G\text{-}\mathbf{GrCat}.
\end{array}$$

Therefore for 2-functors $?/_2G\colon \mathbf{Cat}\text{-}G \to G\text{-}\mathbf{GrCat}$ and $?\#^{\mathrm{op}}G\colon G\text{-}\mathbf{GrCat} \to \mathbf{Cat}\text{-}G$, the theorem corresponding to Theorem 5.7.1 holds by Theorem 5.7.1 for G^{op}.

CHAPTER 6

Computations of Orbit Categories and Smash Products

In this chapter, we describe how to compute the orbit category and the smash product, where for the orbit category, we restrict ourselves to orbit categories for categories with G-action rather than general pseudo-G-categories, and for the smash product, we restrict ourselves to categories with G-grading whose gradings are defined by G-weight maps for quivers of the category rather than general G-graded categories. This is because these forms appear most frequently in actual calculations and are easy to compute. They can also be used to compute modules in Chapter 7. The computation of the orbit category is given more generally in the paper **[5]** for the case where G is a monoid (which may contain 0). The following computation is a specialization of this theorem to the case where G is a group. The computation of the smash product given here is a rewriting of the method and proof of the right smash product given in **[11]**.

6.1. Computation of orbit categories

In this section, when a $\Bbbk$-category $\mathscr{C}$ with a strict G-action is presented by a quiver with relations, and G is presented as a monoid, we use these two presentations to compute the quiver presentation of the orbit category $\mathscr{C}/G$. For the monoid presentation, see Howie's book **[29]**. Note that the quiver presentation given here is not necessarily admissible, even if that of $\mathscr{C}$ is admissible. If an admissible one is required, the presentation given here should be modified (cf. for example **[40]**). Throughout this section, we assume the following setting.

(1) $\Bbbk$ is a field;
(2) $Q := (Q_0, Q_1, \mathbf{s}, \mathbf{t})$ is a locally finite quiver;
(3) (Q, I) is a quiver with relations, where I is generated by a set ρ of relations.
(4) G is a group having a monoid presentation $G = \langle S \mid R \rangle$;
(5) $\mathscr{C} := \Bbbk[Q, I] := \Bbbk[Q]/I$, $\Phi\colon \Bbbk[Q] \to \mathscr{C}$ is the canonical functor, and we set $\tilde{\mu} := \Phi(\mu) := \mu + I \in (\Bbbk[Q]/I)_1$ $(\mu \in \Bbbk[Q]_1)$;
(6) The action of G on $\mathscr{C}$ is given by a monomorphism $X\colon G \rightarrowtail \mathrm{Aut}(\mathscr{C})$. (The general case can be reduced to this case by taking $G/\operatorname{Ker} X$ instead of G.) In the following, we use the abbreviation $ax := X(a)x$ $(a \in G, x \in \mathscr{C}_0 \cup \mathscr{C}_1)$.
(7) $(P, \phi)\colon \mathscr{C} \to \mathscr{C}/G$ is the canonical covering functor.

In the above, the monoid presentation $\langle S \mid R \rangle$ in (4) is defined as follows. The free monoid generated by a set S is denoted by S^*. For a subset R of $S^* \times S^*$, the smallest equivalence relation on S^* containing $R^c := \{(bga, bha) \mid (g, h) \in R, a, b \in S^*\}$ is denoted by $R^\#$. Then the factor monoid $S^*/R^\#$ is denoted by $\langle S \mid R \rangle$. For each $g \in S^*$, we set $\bar{g}$ to be the element of $\langle S \mid R \rangle$ containing g. If $g \neq h$ $(g, h \in S)$, then we may assume that $\bar{g} \neq \bar{h}$ (because if S' is a complete set of representatives of $S/(S \cap R^{\#})$, then it is possible to present G by using S', and we may replace S by S'). In the above, we regard G as a monoid, and give it a monoid presentation. For example, an infinite cyclic group generated by g is presented by $\langle g \rangle$ as a group, but it is presented by $\langle g, h \mid (gh, 1), (hg, 1) \rangle$ as a monoid.

Under this setting, we define a new quiver with relations (Q', I') as follows. We construct Q' by adding the new arrows

$$S \times Q_0 := \{(g, x) \colon x \to gx \mid g \in S, x \in Q_0\}$$

to Q. Namely, we define a quiver $Q' = (Q'_0, Q'_1, \mathbf{s}', \mathbf{t}')$ as follows:

$$\begin{aligned} Q'_0 &:= Q_0, \\ Q'_1 &:= Q_1 \sqcup (S \times Q_0), \\ (\mathbf{s}'(\alpha), \mathbf{t}'(\alpha)) &:= (\mathbf{s}(\alpha), \mathbf{t}(\alpha)), \quad (\alpha \in Q_1), \\ (\mathbf{s}'(g,x), \mathbf{t}'(g,x)) &:= (x, gx), \quad ((g,x) \in S \times Q_0). \end{aligned}$$

We also define an ideal I' of $\Bbbk[Q']$ as follows:

$$\begin{aligned} I' := \langle \rho \rangle + \langle (g,y)\alpha - (g\alpha)(g,x) \mid (x \xrightarrow{\alpha} y) \in Q_1, g \in S \rangle \\ + \langle \pi(g,x) - \pi(h,x) \mid (g,h) \in R, x \in Q_0 \rangle. \end{aligned}$$

Here in the second term, $g\alpha$ is taken so that the condition $g\alpha \in \bar{g}\tilde{\alpha}$ (i.e., $\widetilde{g\alpha} = \bar{g}\tilde{\alpha}$) is satisfied. Note that I' does not depend on the choice of it because I' contains $\langle \rho \rangle$ as its first term. In the third term, for each $x \in Q_0$ and for each $g \in \langle S \rangle \setminus \{1\}$, if g is expressed as $g = g_t \cdots g_1$ ($g_1, \dots, g_t \in S$, $t \geq 1$), then $\pi(g,x)$ is defined to be the path $\pi(g,x) := (g_t, g_{t-1} \cdots g_1 x) \cdots (g_2, g_1 x)(g_1, x)$ in Q. Namely, it is displayed as follows:

$$gx \xleftarrow{(g_t, g_{t-1} \cdots g_1 x)} \cdots \xleftarrow{(g_3, g_2 g_1 x)} g_2 g_1 x \xleftarrow{(g_2, g_1 x)} g_1 x \xleftarrow{(g_1, x)} x.$$

We also set $\pi(1, x) := e_x$.

REMARK 6.1.1. The generators of I' correspond the following three types of relations:

(1) The relations in $\mathscr{C}$: $\mu = 0$ ($\mu \in \rho$);
(2) The following commutativity relations corresponding to the shifting of morphisms in the multiplication of $\mathscr{C}/G$:

$$\begin{array}{ccc} x & \xrightarrow{\alpha} & y \\ {\scriptstyle (g,x)}\downarrow & & \downarrow{\scriptstyle (g,y)} \\ gx & \xrightarrow[g\alpha]{} & gy \end{array} ;$$

(3) The relations in G: $\pi(g,x) = \pi(h,x)$ ($(g,h) \in R$).

When Q_0 is a finite set, $\mathscr{C}$ is regarded as an algebra, and $\mathscr{C}/G = \Bbbk(Q,\rho) * G$ turns out to be a skew group algebra by Example 5.4.6. Therefore the theorem below gives also a computation of skew group algebras.

THEOREM 6.1.2. *Under the setting above, the orbit category $\mathscr{C}/G$ is isomorphic to the factor category $\Bbbk[Q']/I'$.*

PROOF. Note first that for each $g \in S^*$ and $x \in Q_0$, the vertex $gx \in Q_0$ is determined by $gx := \bar{g}x$ because S^* acts on $\mathscr{C}$ by $S^* \xrightarrow{\text{canonical epi.}} G \rightarrowtail \operatorname{Aut}(\mathscr{C})$

and $\mathscr{C}_0 = \mathbb{k}[Q]_0 = Q_0$. The following correspondence defines a quiver morphism $Q' \to \mathscr{C}/G$:

$$\begin{aligned} x &\mapsto x, \quad (x \in Q_0), \\ \alpha &\mapsto P\tilde{\alpha} := \sigma_1(\tilde{\alpha}) = (\delta_{a,1}\tilde{\alpha})_{a\in G} \in (\mathscr{C}/G)^1(Px, Py). \quad (\alpha \in Q_1), \\ (g, x) &\mapsto \sigma_{\bar{g}}(\mathbb{1}_{\bar{g}x}) \in (\mathscr{C}/G)^{\bar{g}}(Px, P\bar{g}x), \quad ((g, x) \in S \times Q_0). \end{aligned}$$

Therefore this is uniquely extended to a $\mathbb{k}$-functor $\Psi\colon \mathbb{k}[Q'] \to \mathscr{C}/G$.

CLAIM 1. $\Psi(g\alpha) = P(\bar{g}\tilde{\alpha})$ $(g \in S, \alpha \in Q_1)$.

Indeed, by definition of Ψ, in the diagram

$$\begin{array}{ccccc} Q & \hookrightarrow & \mathbb{k}[Q] & \xrightarrow{\Phi} & \mathscr{C} \\ & & {\scriptstyle\tau}\downarrow & & \downarrow{\scriptstyle P} \\ & & \mathbb{k}[Q'] & \xrightarrow[\Psi]{} & \mathscr{C}/G \end{array},$$

we have $(P \circ \Phi)|_Q = (\Psi \circ \tau)|_Q$, and hence it holds that $P \circ \Phi = \Psi \circ \tau$ because $\mathbb{k}[Q]$ is the path category of Q. Therefore we have

$$\Psi(g\alpha) = (\Psi \circ \tau)(g\alpha) = (P \circ \Phi)(g\alpha) = P(\widetilde{g\alpha}) = P(\bar{g}\tilde{\alpha}).$$

CLAIM 2. $\Psi(I') = 0$.

Indeed, it first follows from $\Psi(\rho) = 0$ that $\Psi(\langle\rho\rangle) = 0$. Next, for each arrow $\alpha\colon x \to y$ of Q and each $g \in S$, we have

$$\begin{aligned} \Psi((g, y)\alpha - g(\alpha)(g, x)) &= \sigma_{\bar{g}}(\mathbb{1}_{\bar{g}y})P(\tilde{\alpha}) - P(\bar{g}(\tilde{\alpha}))(\sigma_{\bar{g}}(\mathbb{1}_{\bar{g}x})) \\ &= (\delta_{\bar{g},b}\mathbb{1}_{\bar{g}y})_{b\in G} \cdot (\delta_{1,a}\tilde{\alpha})_{a\in G} \\ &\qquad - (\delta_{1,b}\bar{g}(\tilde{\alpha}))_{b\in G} \cdot (\delta_{\bar{g},a}\mathbb{1}_{\bar{g}x})_{a\in G} \\ &= (\delta_{\bar{g},c}\mathbb{1}_{\bar{g}y}\bar{g}\tilde{\alpha})_{c\in G} - (\delta_{\bar{g},c}\bar{g}\tilde{\alpha}\mathbb{1}_{cx})_{c\in G} = 0. \end{aligned}$$

Finally for each $(g, h) \in R$ and $x \in Q_0$, we show that $\Psi(\pi(g, x) - \pi(h, x)) = 0$. As a preparation, we show that for each $(g, x) \in S^* \times Q_0$ the following holds:

$$\Psi(\pi(g, x)) = \sigma_{\bar{g}}(\mathbb{1}_{\bar{g}x}). \tag{6.1}$$

First, when $g = 1$, (LHS) $= \Psi(e_x) = \mathbb{1}_x^{\mathscr{C}/G} = \sigma_1(\mathbb{1}_x) =$ (RHS). In the other case, we can write $g = g_t \cdots g_1$ for some $g_1, \dots, g_t \in S$ and $t \geq 1$. We show the equality by induction on t.

For $t = 1$, since $g = g_1 \in S$ and $\pi(g, x) = (g, x)$, (6.1) is clear by definition of Ψ.

For $t \geq 2$, set $h := g_{t-1} \cdots g_2 g_1$, $y := hx$. Then we have

$$\pi(g, x) = (g_t, g_{t-1} \dots g_1 x) \cdots (g_2, g_1 x)(g_1, x) = (g_t, y)\pi(h, x).$$

Use the definition of Ψ to (g_t, y), and apply the induction hypothesis to (h, x) to have

$$\begin{aligned} \Psi(\pi(g, x)) &= \Psi(g_t, y)\Psi(\pi(h, x)) = \sigma_{\bar{g}_t}(\mathbb{1}_{\bar{g}_t y})\sigma_{\bar{h}}(\mathbb{1}_{\bar{h}x}) \\ &= (\delta_{\bar{g}_t,b}\mathbb{1}_{\bar{g}_t y})_{b\in G} \cdot (\delta_{\bar{h},a}\mathbb{1}_{\bar{h}x})_{a\in G} = (\delta_{\bar{g},c}\mathbb{1}_{\bar{g}x} \cdot \mathbb{1}_{\bar{g}x})_{c\in G} = \sigma_{\bar{g}}(\mathbb{1}_{\bar{g}x}). \end{aligned}$$

This finishes the proof of the formula (6.1). Now, if $(g, h) \in R$, $x \in Q_0$, then since $\bar{g} = \bar{h}$, we have by (6.1) that

$$\Psi(\pi(g, x) - \pi(h, x)) = \Psi(\pi(g, x)) - \Psi(\pi(h, x)) = \sigma_{\bar{g}}(\mathbb{1}_{\bar{g}x}) - \sigma_{\bar{h}}(\mathbb{1}_{\bar{h}x}) = 0.$$

Accordingly, we have $\Psi(I') = 0$.

By Claim 2, the $\Bbbk$-functor Ψ induces the $\Bbbk$-functor $\bar{\Psi}\colon \Bbbk[Q']/I' \to \mathscr{C}/G$. For each morphism $\mu \in \Bbbk[Q'](x,y)$ $(x, y \in Q'_0 = Q_0)$ of $\Bbbk[Q']$, we set $[\mu] := \mu + I'(x,y) \in (\Bbbk[Q']/I')(x,y)$. It remains to show that $\bar{\Psi}$ is an isomorphism. Since $\bar{\Psi}$ is the identity on the objects, it is obvious that $\bar{\Psi}_0$ is a bijection, thus, $\bar{\Psi}$ is dense. Therefore it is enough to show that $\bar{\Psi}$ is fully faithful. To this end take any $x, y \in Q_0 = (\Bbbk[Q']/I')_0$. Below we show that $\bar{\Psi}\colon (\Bbbk[Q']/I')(x,y) \to (\mathscr{C}/G)(x,y)$ is an isomorphism by dividing the arguments into several steps.

For each $u, v \in Q_0$, we have $\mathscr{C}(u,v) = \sum_{\mu\in\mathbb{P}Q(u,v)} \Bbbk\tilde{\mu}$ because $\Bbbk[Q](u,v) = \bigoplus_{\mu\in\mathbb{P}Q(u,v)} \Bbbk\mu$. Therefore there exists some $\mathscr{M}_{u,v} \subseteq \mathbb{P}Q(u,v)$ such that $\tilde{\mathscr{M}}_{u,v} := \{\tilde{\mu} \mid \mu \in \mathscr{M}_{u,v}\}$ is a basis of $\mathscr{C}(u,v)$.

CLAIM 3. *$\tilde{\mathscr{M}} := \{\sigma_a(\tilde{\mu}) \mid a \in G, \mu \in \mathscr{M}_{ax,y}\}$ forms a basis of $(\mathscr{C}/G)(x,y)$.*

Indeed, $(\mathscr{C}/G)(x,y) = \bigoplus_{a\in G} \sigma_a(\mathscr{C}(ax,y)) = \bigoplus_{a\in G}\bigoplus_{\mu\in\mathscr{M}_{ax,y}} \Bbbk\,\sigma_a(\tilde{\mu})$.

CLAIM 4. *For each $g, h \in S^*$ and $u \in Q_0$, if $\bar{g} = \bar{h}$ in G, then $[\pi(g,u)] = [\pi(h,u)]$ in $\Bbbk[Q']/I'$.*

Indeed, $\bar{g} = \bar{h}$ in G if and only if $(g,h) \in R^{\#}$. If $g = h$, the the claim is obvious. Otherwise, by definition of $R^{\#}$, it is enough to show the statement in the case that there exist some $(a,b) \in R$ and $c, d \in S^*$ such that $g = cad, h = cbd$. Then noting that $adx := \bar{a}\bar{d}x = \bar{b}\bar{d}x =: bdx$ because $\bar{a} = \bar{b}$ in this case, we have

$$\begin{aligned}\pi(g,x) - \pi(h,x) &= \pi(cad,x) - \pi(cbd,x)\\ &= \pi(c,adx)\pi(a,dx)\pi(d,x) - \pi(c,bdx)\pi(b,dx)\pi(d,x)\\ &= \pi(c,adx)(\pi(a,dx) - \pi(b,dx))\pi(d,x) \in I',\end{aligned}$$

which proves the claim.

By the claim above, we may define

$$[\pi(\bar{g},x)] := [\pi(g,x)]$$

for all $\bar{g} \in G$ $(g \in S^*)$.

CLAIM 5. *For each $\eta \in \mathbb{P}Q'(x,y)$, $[\eta]$ is expressed as a linear combination of elements of the form $[\lambda][\pi(a,x)] \in (\Bbbk Q'/I')(x,y)$ with $a \in G, \lambda \in \mathscr{M}_{ax,y}$. Hence in particular, the set $\mathscr{S} := \{[\mu][\pi(a,x)] \mid a \in G, \mu \in \mathscr{M}_{ax,y}\}$ spans $(\Bbbk[Q']/I')(x,y)$ over $\Bbbk$.*

Indeed, by definition of I', the following holds in $\Bbbk[Q']/I'$:

$$[(g,v)][\alpha] = [g\alpha][(g,u)] \quad (g \in S, u, v \in Q_0, \alpha \in Q(u,v)). \tag{6.2}$$

Using this formula (6.2), we can transport factors of the form $[(g,v)]$ with $(g,v) \in S \times Q_0$ in $[\eta]$ to the right. If we transport all of them, then it has the following form for some $g_0, g_1, \dots, g_t \in S$ and $\mu \in (\mathbb{P}Q)(g_t \cdots g_1 g_0 x, y)$

$$[\eta] = [\mu(g_t, g_{t-1}\cdots g_1 g_0 x)\cdots(g_1, g_1 x)(g_0, x)].$$

See the commutative diagram in $\Bbbk[Q']/I'$ below to understand this argument, where the broken arrows stand for new arrows $\in Q'_1 \setminus Q_1$.

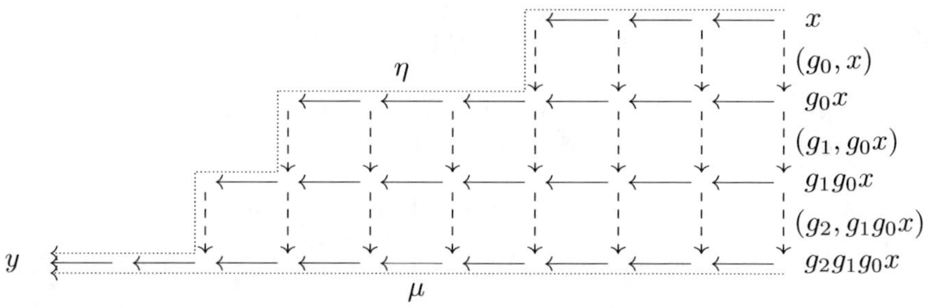

Here if we set $g := g_t \cdots g_1 \in S^*$, $a := \bar{g}$, then we have

$$[\eta] = [\mu\pi(g, x)] = [\mu][\pi(a, x)] \ (\mu \in \mathbb{P}Q(ax, y), a \in G). \tag{6.3}$$

On the other hand, since $\tilde{\mathscr{M}}_{ax,y}$ is a basis of $\mathscr{C}(ax, y)$, we can write $\tilde{\mu}$ as a linear combination of elements of the form $\tilde{\lambda}$ with $\lambda \in \mathscr{M}_{ax,y}$. Therefore $[\mu]$ is expressed as a linear combination of $[\lambda]$ with $\lambda \in \mathscr{M}_{ax,y}$ because $\langle \rho \rangle \leq I'$. Substituting this into the expression (6.3) shows the claim.

CLAIM 6. *For each $a \in G$ and $\mu \in \mathscr{M}_{ax,y}$, the following holds:*

$$\bar{\Psi}([\mu][\pi(a, x)]) = \sigma_a(\tilde{\mu}).$$

Indeed, (LHS) $= \Psi(\mu)\Psi(\pi(a, x)) = \sigma_1(\tilde{\mu}) \cdot \sigma_a(\mathbb{1}_{ax}) =$ (RHS).

Therefore, since $\bar{\Psi}$ maps $\mathscr{S}$ into the linearly independent set $\tilde{\mathscr{M}}$, we see that $\mathscr{S}$ is linearly independent, and hence it forms a basis of $(\Bbbk[Q']/I')(x, y)$. As a consequence, $\bar{\Psi}\colon (\Bbbk[Q']/I')(x, y) \to (\mathscr{C}/G)(x, y)$ maps the basis $\mathscr{S}$ into the basis $\tilde{\mathscr{M}}$, and it is an isomorphism. □

PROPOSITION 6.1.3. *For a G-category $(\mathscr{C}, X)$, assume that $\mathscr{C}$ is a spectroid. Then for each $x, y \in \mathscr{C}_0$, the following are equivalent:*

(1) *$x \cong y$ in $\mathscr{C}/G$; and*
(2) *$Gx = Gy$.*

Therefore a complete set of representatives of G-orbits in $\mathscr{C}$ is nothing but a complete set of representatives of isoclasses of $\mathscr{C}/G$.

PROOF. (2) $\Rightarrow$ (1). Assume statement (2). Then we have $y = ax$ for some $a \in G$. Lemma 5.4.9 shows that $\phi_a x\colon x \to ax = y$ is an isomorphism in $\mathscr{C}/G$. Therefore statement (1) holds.

(1) $\Rightarrow$ (2). Assume statement (1). Then there exist some $f = (f_a)_a \in (\mathscr{C}/G)(x, y)$ and some $g = (g_b)_b \in (\mathscr{C}/G)(y, x)$ such that $g \circ f = \mathbb{1}_x^{\mathscr{C}/G}$. Namely, we have

$$\left(\sum_{\substack{a,b \in G \\ ba=c}} g_b \circ X(b) f_a \right)_{c \in G} = (\delta_{c,1} \mathbb{1}_x)_{c \in G}.$$

Comparing the terms for $c = 1$, we have $\sum_{b \in G} g_b \circ X(b) f_{b^{-1}} = \mathbb{1}_x \in \mathscr{C}(x, x)$. Here since $\mathscr{C}(x, x)$ is local by assumption, there exists some $b \in G$ such that $g_b \circ X(b) f_{b^{-1}}$ is an isomorphism in $\mathscr{C}$. Thus $g_b \circ h = \mathbb{1}_x$ for some $h \in \mathscr{C}(x, by)$. Then $h \circ g_b$ turns out to be an idempotent in $\mathscr{C}(by, by)$ because $(h \circ g_b)^2 = h \circ g_b \circ h \circ g_b = h \circ g_b$. Here again since $\mathscr{C}(by, by)$ is local, it holds that $h \circ g_b$ is equal to $\mathbb{1}_{by}$ or 0. If $h \circ g_b = 0$, then $\mathbb{1}_x = g_b \circ h \circ g_b \circ h = 0$, a contradiction. Hence $h \circ g_b = \mathbb{1}_{by}$, and g_b turns out to be an isomorphism in $\mathscr{C}$. Therefore $by \cong x$ in $\mathscr{C}$, and hence $by = x$ because $\mathscr{C}$ is basic. As a consequence, statement (2) holds. $\square$

EXAMPLE 6.1.4. Let G be an infinite cyclic group with a generator a. Then as a monoid presentation of G we can take the following:

$$G = \langle a, a^{-1} \mid (aa^{-1}, 1), (a^{-1}a, 1) \rangle.$$

Here consider the G-category $(\mathscr{C}, X)$ in Example 3.2.4, which was given as follows:

$$\begin{aligned} \tilde{Q} &:= (\cdots -1 \xrightarrow{\alpha_{-1}} 0 \xrightarrow{\alpha_0} 1 \xrightarrow{\alpha_1} \cdots), \\ \tilde{I} &:= \langle \alpha_{i+2}\alpha_{i+1}\alpha_i \mid i \in \mathbb{Z} \rangle, \\ \mathscr{C} &:= \mathbb{k}[\tilde{Q}, \tilde{I}], \\ X(a)(i) &:= i+2, \quad X(a)(\alpha_i) := \alpha_{i+2} \ (i \in \mathbb{Z}). \end{aligned}$$

Theorem 6.1.2 shows that $\mathscr{C}/G$ is given by the following Q' and the ideal I' generated by the relations below:

Q' :

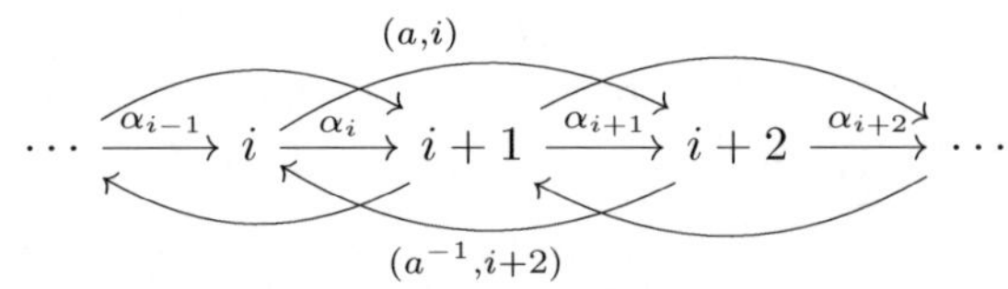

A generating set for I':

- $\alpha_{i+2}\alpha_{i+1}\alpha_i \quad (i \in \mathbb{Z})$,
- $(a, i+1)\alpha_i - \alpha_{i+2}(a, i), \quad (a^{-1}, i+1)\alpha_i - \alpha_{i-2}(a^{-1}, i) \quad (i \in \mathbb{Z})$,
- $(a^{-1}, i+2)(a, i) - \mathbb{1}_i, \quad (a, i-2)(a^{-1}, i) - \mathbb{1}_i \quad (i \in \mathbb{Z})$.

We next compute a skeleton $\mathscr{C}/_{\!c}G$ of $\mathscr{C}/G$, which is uniquely determined up to isomorphisms and is equivalent to $\mathscr{C}/G$ by Proposition 1.4.9. Proposition 6.1.3 insists that $\mathscr{C}/_{\!c}G$ is given as a full subcategory of $\mathscr{C}/G$ whose object set is a complete set of representatives of G-orbits in $\mathscr{C}$. It is obvious that the set $\{1, 2\}$ forms a complete set of representatives of the G-orbits in $\mathscr{C}$. The local morphism sets are given as follows:

$$\begin{aligned} (\mathscr{C}/G)(1,1) &= \sigma_1\mathscr{C}(1,1) \oplus \sigma_{a^{-1}}\mathscr{C}(-1,1) = \mathbb{k}\mathbb{1}_1 \oplus \mathbb{k}\alpha_0\alpha_{-1}(a^{-1},1), \\ (\mathscr{C}/G)(1,2) &= \sigma_1\mathscr{C}(1,2) = \mathbb{k}\alpha_1, \\ (\mathscr{C}/G)(2,1) &= \sigma_{a^{-1}}\mathscr{C}(0,1) = \mathbb{k}\alpha_0(a^{-1},2), \\ (\mathscr{C}/G)(2,2) &= \sigma_1\mathscr{C}(2,2) \oplus \sigma_{a^{-1}}\mathscr{C}(0,2) = \mathbb{k}\mathbb{1}_2 \oplus \mathbb{k}\alpha_1\alpha_0(a^{-1},2), \end{aligned}$$

where the G-degrees are given by the indices of σ. Namely, the degrees of morphisms except for the identity are given as follows:

$$\begin{aligned} \deg \alpha_0\alpha_{-1}(a^{-1},1) = a^{-1},\ \deg \alpha_1 = 1, \\ \deg \alpha_0(a^{-1},2) = a^{-1},\ \deg \alpha_1\alpha_0(a^{-1},2) = a^{-1}. \end{aligned} \tag{6.4}$$

Therefore using a way to present categories explained in Example 2.3.29, the category $\mathscr{C}/_{\mathrm{c}}G$ is presented as follows:

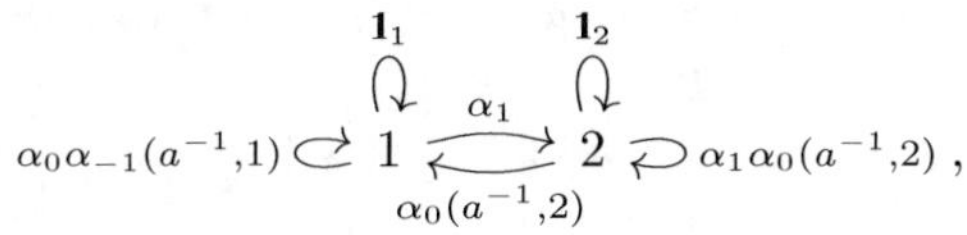

and the composition is given as follows:

$$\begin{aligned} \alpha_0(a^{-1},2)\circ\alpha_1 &= \alpha_0\alpha_{-1}(a^{-1},1), \\ \alpha_1\circ\alpha_0(a^{-1},2) &= \alpha_1\alpha_0(a^{-1},2), \\ \alpha_1\circ(\alpha_0\alpha_{-1}(a^{-1},1)) &= (a^{-1},3)\alpha_3\alpha_2\alpha_1 = 0, \text{ similarly} \\ (\alpha_0\alpha_{-1}(a^{-1},1))\circ\alpha_0(a^{-1},2) &= 0, \\ (\alpha_1\alpha_0(a^{-1},2))\circ\alpha_1 &= 0, \\ \alpha_0(a^{-1},2)\circ(\alpha_1\alpha_0(a^{-1},2)) &= 0. \end{aligned}$$

As a consequence, $\mathscr{C}/_{\mathrm{c}}G$ is given by the following bound quiver (Q, I):

$$Q := \quad 1 \underset{\beta}{\overset{\alpha}{\rightleftarrows}} 2 \quad , \qquad I := \langle \alpha\beta\alpha, \beta\alpha\beta \rangle.$$

We note here that the G-grading is given by the following *weight* W (see Definition 6.2.8) by the formula (6.4):

$$W(\alpha) = 1, W(\beta) = a^{-1}.$$

REMARK 6.1.5. In Theorem 6.1.2, the definition of Ψ in the proof shows the following: Define a map $W\colon Q'_1 \to G$ by

$$\begin{cases} W(\alpha) := 1 & (\alpha \in Q_1), \\ W((g,x)) := g & ((g,x) \in S \times Q_0). \end{cases}$$

Then W turns out to be a *homogeneous G-weight* on (Q', I'), and gives an isomorphism $\mathscr{C}/G \cong \Bbbk(Q', I', W)$ as G-graded categories (see Definition 6.2.9).

6.2. Computation of smash products

DEFINITION 6.2.1. Let $\mathscr{B}$ be a G-graded category and I its ideal. If we have

$$I(x,y) = \bigoplus_{a \in G} (I(x,y) \cap \mathscr{B}^a(x,y))$$

for all $x, y \in \mathscr{B}_0$, then I is said to be *homogeneous*.

LEMMA 6.2.2. *Let $\mathscr{B}$ be a G-graded category and I its ideal. Then I is homogeneous if and only if I is generated by homogeneous morphisms.*

EXERCISE 6.2.3. Prove the lemma above.

LEMMA 6.2.4. *Let $\mathscr{B}$ be a G-graded category and I its homogeneous ideal. Then $\mathscr{B}/I$ turns out to be a G-graded category if we set*

$$(\mathscr{B}/I)^a(x,y) := (\mathscr{B}^a(x,y) + I(x,y))/I(x,y) \ (\cong \mathscr{B}^a(x,y)/I^a(x,y))$$

for all $a \in G$. Henceforth, we regard $\mathscr{B}/I$ as a G-graded category in this way.

LEMMA 6.2.5. *Let $(\mathscr{C}, X)$ be a G-category, I an ideal of $\mathscr{C}$, and $\pi\colon \mathscr{C} \to \mathscr{C}/I$ the canonical functor. Then I is called a* G-invariant ideal *of $(\mathscr{C}, X)$ if $X(a)(I(x,y)) \subseteq I(X(a)x, X(a)y)$ for all $x, y \in \mathscr{C}_0$ and $a \in G$. When this is the case, a G-action X/I on $\mathscr{C}/I$ is defined as follows:*

$$(6.5)\qquad \begin{aligned}(X/I)(a)(x) &:= X(a)(x) \quad (x \in (\mathscr{C}/I)_0 = \mathscr{C}_0),\\ (X/I)(a)(\pi(f)) &:= \pi(X(a)(f)) \quad (f \in \mathscr{C}_1).\end{aligned}$$

LEMMA 6.2.6. *Let $\mathscr{B}$ be a G-graded category and I its homogeneous ideal. For each $x^{(a)}, y^{(b)} \in (\mathscr{B}\#G)_0$ $(x, y \in \mathscr{B}_0, a, b \in G)$, we define $(I\#G)(x^{(a)}, y^{(b)}) := \{{}^{(b)}f^{(a)} \mid f \in I^{b^{-1}a}(x,y)\}$ as a $\Bbbk$-subspace of $(\mathscr{B}\#G)(x^{(a)}, y^{(b)})$. Then $I\#G$ turns out to be a G-invariant ideal of $\mathscr{B}\#G$, and there exists a natural isomorphism*

$$(\mathscr{B}\#G)/(I\#G) \cong (\mathscr{B}/I)\#G$$

between G-categories, by which we identify these in the sequel.

PROOF. Let $\pi^{\#}\colon \mathscr{B}\#G \to (\mathscr{B}\#G)/(I\#G)$ and $\pi\colon \mathscr{B} \to \mathscr{B}/I$ be canonical functors. We first note that the objects sets of both sides coincide. Indeed,

$$((\mathscr{B}\#G)/(I\#G))_0 = (\mathscr{B}\#G)_0 = \mathscr{B}_0 \times G = (\mathscr{B}/I)_0 \times G = ((\mathscr{B}/I)\#G)_0.$$

To prove the statement using Theorem 2.6.3, we define a functor $F\colon \mathscr{B}\#G \to (\mathscr{B}/I)\#G$. First of all, by using the fact above, we set

$$F_0\colon (\mathscr{B}\#G)_0 \to ((\mathscr{B}/I)\#G)_0$$

to be the identity map. Next, for each $(x,a), (y,b) \in \mathscr{B}_0 \times G$, we define an surjective linear map

$$F\colon (\mathscr{B}\#G)(x^{(a)}, y^{(b)}) \to ((\mathscr{B}/I)\#G)(x^{(a)}, y^{(b)})$$

as the composite of the following natural homomorphisms:

$$\begin{aligned}(\mathscr{B}\#G)(x^{(a)}, y^{(b)}) &\xrightarrow{\sim} \mathscr{B}^{b^{-1}a}(x,y) \twoheadrightarrow (\mathscr{B}^{b^{-1}a}(x,y) + I(x,y))/I(x,y)\\ &\xrightarrow{\sim} (\mathscr{B}/I)^{b^{-1}a}(x,y) \xrightarrow{\sim} ((\mathscr{B}/I)\#G)(x^{(a)}, y^{(b)}).\end{aligned}$$

This has the following explicit form:

$${}^{(b)}f^{(a)} \mapsto {}^{(b)}\pi(f)^{(a)} \quad (f \in \mathscr{B}^{b^{-1}a}(x,y)).$$

Then $\operatorname{Ker} F$ is given by

$$\begin{aligned}(\operatorname{Ker} F)(x^{(a)}, y^{(b)}) &= \{{}^{(b)}f^{(a)} \mid F(f) = 0, f \in \mathscr{B}^{b^{-1}a}(x,y)\}\\ &= \{{}^{(b)}f^{(a)} \mid f \in \mathscr{B}^{b^{-1}a}(x,y) \cap I(x,y)\}\\ &= (I\#G)(x^{(a)}, y^{(b)}).\end{aligned}$$

Therefore, by Theorem 2.6.3, $I\#G$ $(= \operatorname{Ker} F)$ is an ideal of $\mathscr{B}\#G$, and F induces a natural isomorphism

$$(6.6)\qquad \begin{aligned}\bar{F}\colon (\mathscr{B}\#G)/(I\#G) &\xrightarrow{\sim} (\mathscr{B}/I)\#G,\\ \pi^{\#}({}^{(b)}f^{(a)}) &\mapsto {}^{(b)}\pi(f)^{(a)} \quad (f \in \mathscr{B}^{b^{-1}a}(x,y))\end{aligned}$$

between linear categories. By the form of $I\#G$ and the formula (5.22), it follows that $I\#G$ is a G-invariant ideal of $\mathscr{B}\#G$. Therefore, by Lemma 6.2.5, a G-action is defined on $(\mathscr{B}\#G)/(I\#G)$.

We finally show that this $\bar{F}$ commutes with the G-action. Let $c \in G$. The action of c on objects is defined both for $(\mathscr{B}\#G)/(I\#G)$ and $(\mathscr{B}/I)\#G$ as follows:

$$x^{(a)} \mapsto x^{(ca)} \quad ((x, a) \in \mathscr{B}_0 \times G).$$

Thus $\bar{F}$ commutes with the G-action on objects. Next, by formulas (6.5) and (5.22), the action of c on morphisms is defined for $(\mathscr{B}\#G)/(I\#G)$ and $(\mathscr{B}/I)\#G$ as follows:

$$\begin{aligned} &\pi^{\#}({}^{(b)}f^{(a)}) \mapsto \pi^{\#}({}^{(cb)}f^{(ca)}), \\ &{}^{(b)}\pi(f)^{(a)} \mapsto {}^{(cb)}\pi(f)^{(ca)} \\ &\qquad ((x,a),(y,b) \in \mathscr{B}_0 \times G, f \in \mathscr{B}^{b^{-1}a}(x,y)). \end{aligned} \tag{6.7}$$

By formulas (6.6) and (6.7), we see that $\bar{F}$ commutes with the G-action on morphisms as well. □

Definition 6.2.7. Let Q be a quiver. The a map $W\colon Q_1 \to G$ is called a *G-weight* on Q, and the pair (Q, W) a *quiver with G-weight*.

Definition 6.2.8. Let (Q, W) be a quiver with G-weight, and μ a path in Q. If $|\mu| := n \geq 1$, then since it is uniquely written as $\mu = \alpha_n \cdots \alpha_1$ $(\alpha_1, \ldots, \alpha_n \in Q_1)$, we set

$$W(\mu) := W(\alpha_n) \cdots W(\alpha_1).$$

If $|\mu| = 0$, we set $W(\mu) := 1$. Then it is obvious that the path category $\Bbbk[Q]$ turns out to be a G-graded category $\Bbbk(Q, W)$ if we set

$$\Bbbk(Q,W)^a(x,y) := \bigoplus_{\substack{\mu \in \mathbb{P}Q(x,y) \\ W(\mu)=a}} \Bbbk\mu \qquad (a \in G, x, y \in Q_0).$$

Definition 6.2.9. Let (Q, I) be a quiver with relations, and W a G-weight on Q.

(1) An element $\rho = \sum_{i=1}^n t_i\mu_i$ $(t_i \in \Bbbk, \mu_i \in \mathbb{P}Q(x,y))$ in $I(x,y)$ $(x, y \in Q_0)$ is called a *minimal relation* in I if we have $\sum_{i\in J} t_i\mu_i \notin I$ for all proper subsets $J \neq \emptyset$.

(2) This W is called a *homogeneous G-weight* on (Q, I), and the triple (Q, I, W) a *G-weighted quiver with relations* if for each minimal relation $\sum_{i=1}^n t_i\mu_i$ $(x, y \in Q_0, t_i \in \Bbbk, \mu_i \in \mathbb{P}Q(x,y))$ in I, we have

$$W(\mu_i) = W(\mu_1)$$

for all $i = 1, \ldots, n$.

(3) When the condition above holds, I becomes a homogeneous ideal of the G-graded category $\Bbbk(Q, W)$ by Lemma 6.2.2. Then we define a G-graded category $\Bbbk(Q, I, W)$ by

$$\Bbbk(Q, I, W) := \Bbbk(Q, W)/I. \tag{6.8}$$

Exercise 6.2.10. In Definition 6.2.9 (1), each element $\rho = \sum_{i=1}^n t_i\mu_i$ in $I(x,y)$ can be written in the sum of n or less of minimal relations in I. Prove this fact (by induction on n).

EXAMPLE 6.2.11. Let G be the additive group $\mathbb{Z}$, Q a quiver

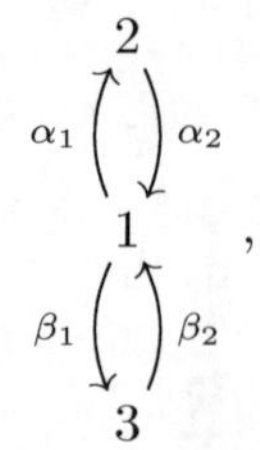

and I the ideal $\langle \alpha_2\alpha_1 - \beta_2\beta_1, \beta_1\alpha_2, \alpha_1\beta_2, \alpha_2\alpha_1\alpha_2, \beta_2\beta_1\beta_2 \rangle$ of $\Bbbk[Q]$. If we define a G-weight W on Q by $W(\alpha_1) = 0 = W(\beta_1), W(\alpha_2) = 1 = W(\beta_2)$, then W becomes a homogeneous G-weight on (Q, I).

DEFINITION 6.2.12. Let (Q, I) be a quiver with relations and W a homogeneous G-weight on (Q, I).

(1) We define a quiver $Q_{G,W} = ((Q_{G,W})_0, (Q_{G,W})_1, s_{G,W}, t_{G,W})$ as follows:

$$(Q_{G,W})_0 := \{x^{(a)} := (x, a) \mid x \in Q_0, a \in G\} = Q_0 \times G,$$
$$(Q_{G,W})_1 := \{\alpha^{(a)} \colon x^{(aW(\alpha))} \to y^{(a)} \mid (x \xrightarrow{\alpha} y) \in Q_1, a \in G\}, \text{and}$$
$$s_{G,W}(\alpha^{(a)}) := s(\alpha)^{(aW(\alpha))}, \quad t_{G,W}(\alpha^{(a)}) := t(\alpha)^{(a)} \quad (\alpha \in Q_1, a \in G).$$

(2) For each path $\mu = \alpha_n \cdots \alpha_1$ of length n (≥ 2) and $a \in G$, we define a path $\mu^{(a)}$ from $x^{(aW(\mu))}$ to $y^{(a)}$ by setting

$$\mu^{(a)} := \alpha_n^{(a)} \alpha_{n-1}^{(aa_n)} \cdots \alpha_2^{(aa_n \cdots a_3)} \alpha_1^{(aa_n \cdots a_2)},$$

where we set $a_i := W(\alpha_i)$ for all $i = 1, \ldots, n$, and thus note that $W(\mu) := a_n \cdots a_1$. By giving symbols to vertices, this is visually expressed as follows: If μ is a path of the form

$$y = x_n \xleftarrow{\alpha_n} x_{n-1} \xleftarrow{\alpha_{n-1}} \cdots \xleftarrow{\alpha_2} x_1 \xleftarrow{\alpha_1} x_0 = x,$$

then $\mu^{(a)}$ is a path of the form

$$x_n^{(a)} \xleftarrow{\alpha_n^{(a)}} x_{n-1}^{(aa_n)} \xleftarrow{\alpha_{n-1}^{(aa_n)}} \cdots \xleftarrow{\alpha_2^{(aa_n \cdots a_3)}} x_1^{(aa_n \cdots a_2)} \xleftarrow{\alpha_1^{(aa_n \cdots a_2)}} x_0^{(aa_n \cdots a_1)}.$$

(3) For each minimal relation $\rho = \sum_i k_i \mu_i$ ($k_i \in \Bbbk, \mu_i \in \mathbb{P}Q(x, y), x, y \in Q_0$) in I and $a \in G$, we set

$$\rho^{(a)} := \sum_i k_i \mu_i^{(a)}.$$

Note here that $\mu_i^{(a)} \in \mathbb{P}Q_{G,W}(x^{(aW(\mu_1))}, y^{(a)})$ because $W(\mu_i) = W(\mu_1)$ for all i. Then a G-action X on $\Bbbk[Q_{G,W}]$ is defined by the following quiver morphism:

$$X_c \colon (x^{(aW(\alpha))} \xrightarrow{\alpha^{(a)}} y^{(a)}) \mapsto (x^{(caW(\alpha))} \xrightarrow{\alpha^{(ca)}} y^{(ca)})$$
$$((x^{(aW(\alpha))} \xrightarrow{\alpha^{(a)}} y^{(a)}) \in (Q_{G,W})_1, a, c \in G, x, y \in Q_0, \alpha \in Q_1).$$

(4) We define an ideal $I_{G,W}$ of $\Bbbk[Q_{G,W}]$ by

$$I_{G,W} := \langle \rho^{(a)} \mid a \in G, \rho \text{ is a minimal relation in } I \rangle.$$

(5) We call $(Q_{G,W}, I_{G,W})$ the *smash product* of (Q, I, W) and G.

REMARK 6.2.13. In the setting above, it is obvious that $I_{G,W}$ is a G-invariant ideal. Therefore the G-action X induces a G-action $\overline{X} := X/I_{G,W}$ on $\Bbbk(Q_{G,W}, I_{G,W}) := \Bbbk[Q_{G,W}]/I_{G,W}$, by which we regard $\Bbbk(Q_{G,W}, I_{G,W})$ as a G-category.

We here introduce a covering of a quiver, which will be used later in the proof (5) of Theorem 6.2.18.

DEFINITION 6.2.14. In general, for a quiver $Q = (Q_0, Q_1, s, t)$ and each $x \in Q_0$, we set

$$x^+ := \{\alpha \in Q_1 \mid s(\alpha) = x\},\ x^- := \{\alpha \in Q_1 \mid t(\alpha) = x\}.$$

A *covering* of the quiver Q is a quiver morphism $F\colon \tilde{Q} \to Q$ satisfying the following two conditions:

(1) $F_0\colon \tilde{Q}_0 \to Q_0$ is a surjection, and
(2) F induces a bijection

$$x^+ \to (Fx)^+ \quad \text{and} \quad x^- \to (Fx)^-$$

for all $x \in \tilde{Q}_0$.

By definition of a covering of a quiver, the following is easily shown.

LEMMA 6.2.15. *Let $F\colon \tilde{Q} \to Q$ be a covering of a quiver, and $\mu = \beta_n \cdots \beta_2\beta_1 \in \mathbb{P}Q(x,y)$ $(x, y \in Q_0)$. Then for each $\tilde{y} \in \tilde{Q}_0$ such that $F(\tilde{y}) = y$, there exists a unique path $\lambda = \alpha_n \ldots \alpha_2\alpha_1$ in $\tilde{Q}$ such that $F(\lambda) = \mu, t(\lambda) = \tilde{y}$, where we set $F(\lambda) := F(\alpha_n)\cdots F(\alpha_2)F(\alpha_1)$. This λ is called the* lift *of μ to $\tilde{y}$.*

EXERCISE 6.2.16. Prove the lemma above. We note that the statement holds more generally also for the case that μ above is a "walk".

PROPOSITION 6.2.17. *Define a quiver morphism $F_{G,W}\colon Q_{G,W} \to Q$ by $F_{G,W}(x^{(a)}) := x, F_{G,W}(\alpha^{(a)}) := \alpha$ $(x \in Q_0, \alpha \in Q_1, a \in G)$. Then this turns out to be a covering of the quiver Q.*

PROOF. For simplicity, we set $F := F_{G,W}$. It is obvious that F is a quiver morphism by definition of $Q_{G,W}$. F_0 is surjective because $F(x^{(1)}) = x$ for all $x \in Q_0$. Let $x \in Q_0$ and $b \in G$. Then we have

$$\begin{aligned}
(x^{(b)})^+ &= \{\alpha^{(a)} \mid a \in G, \alpha \in Q_1, s_{G,W}(\alpha^{(a)}) = x^{(b)}\}\\
&= \{\alpha^{(a)} \mid a \in G, \alpha \in Q_1, s(\alpha)^{(aW(\alpha))} = x^{(b)}\}\\
&= \{\alpha^{(a)} \mid a \in G, \alpha \in Q_1, s(\alpha) = x, aW(\alpha) = b\}\\
&= \{\alpha^{(bW(\alpha)^{-1})} \mid \alpha \in Q_1, s(\alpha) = x\}\\
&= \{\alpha^{(bW(\alpha)^{-1})} \mid \alpha \in x^+\}.
\end{aligned}$$

Therefore, F induces a bijection $(x^{(b)})^+ \to x^+$. Similarly we see that $(x^{(b)})^- = \{\alpha^{(b)} \mid \alpha \in x^-\}$, and F induces a bijection $(x^{(b)})^- \to x^-$ as well. □

The following theorem gives a way to compute the smash product.

THEOREM 6.2.18. *Let (Q, I) be a quiver with relations, and W a homogeneous G-weight on $\Bbbk(Q, I)$. Then the smash product $\Bbbk(Q, I, W)\#G$ is presented by the quiver with relations $(Q_{G,W}, I_{G,W})$, i.e., there exists an isomorphism*

$$\Bbbk(Q, I, W)\#G \cong \Bbbk(Q_{G,W}, I_{G,W})$$

of G-categories.

PROOF. Note that we have an isomorphism

$$\Bbbk(Q, I, W)\#G = (\Bbbk(Q, W)\#G)/(I\#G)$$

of G-categories by Lemma 6.2.6 and the defining formula (6.8). We first construct a functor $\phi\colon \Bbbk(Q, W)\#G \to \Bbbk(Q_{G,W})$ as follows:

On objects: For each $x^{(a)} \in (\Bbbk(Q, W)\#G)_0$ $(x \in Q_0, a \in G)$, we set

$$\phi(x^{(a)}) := x^{(a)} \in (Q_{G,W})_0.$$

On morphisms: Let $x^{(a)}, y^{(b)} \in (\Bbbk(Q, W)\#G)_0$, and ${}^{(b)}f^{(a)} \in (\Bbbk(Q, W)\#G)(x^{(a)}, y^{(b)})$, with $f \in \Bbbk(Q, W)^{b^{-1}a}(x, y)$. Then there exists a unique $(k_i, \mu_i)_{i=1}^n$ with $k_i \in \Bbbk, \mu_i \in \mathbb{P}Q(x, y)$ such that $f = \sum_{i=1}^n k_i\mu_i$, where $W(\mu_i) = b^{-1}a$ for all i. Therefore each $\mu_i^{(b)}$ is a path from $x^{(a)}$ to $y^{(b)}$ by Definition 6.2.12(2), and we can set

$$\phi(f) := \sum_{i=1}^{n} k_i\mu_i^{(b)} \in \Bbbk Q_{G,W}(x^{(a)}, y^{(b)}).$$

(1) ϕ is a $\Bbbk$-functor. Indeed, let $x^{(a)} \in (\Bbbk(Q, W)\#G)_0$. Then

$$\mathbb{1}_{x^{(a)}} \in (\Bbbk(Q, W)\#G)(x^{(a)}, x^{(a)}) = \Bbbk(Q, W)^1(x, x),$$

which is a linear combination of paths from x to x, and hence we have $\mathbb{1}_{x^{(a)}} = \mathbb{1}_{x^{(a)}}e_x = e_x$. Therefore, we have $\phi(\mathbb{1}_{x^{(a)}}) = \phi(e_x) = e_x^{(a)} = \mathbb{1}_{x^{(a)}}$ in $\Bbbk(Q_{G,W}, I_{G,W})$.

Next, take morphisms $x^{(a)} \xrightarrow{f} y^{(b)} \xrightarrow{g} z^{(c)}$ in $\Bbbk(Q, W)\#G$. Then there exist $k_i, l_j \in \Bbbk$, $\lambda_i \in \mathbb{P}Q(x, y)$, and $\mu_j \in \mathbb{P}Q(y, z)$ such that $f = \sum_{i=1}^m k_i\lambda_i$, $g = \sum_{j=1}^n l_j\mu_j$, where $W(\mu_j) = c^{-1}b$ and $W(\lambda_i) = b^{-1}a$ for all i, j. Here since $W(\mu_j\lambda_i) = c^{-1}a$, $cW(\mu_j) = b$, and $bW(\lambda_i) = a$, we have $(\mu_j\lambda_i)^{(c)} = \mu_j^{(c)}\lambda_i^{(b)}$. Therefore it holds that

$$\begin{aligned}\phi(g \cdot f) &= \phi(\sum_{i,j}(l_jk_i)\mu_j\lambda_i) = \sum_{i,j}(l_jk_i)(\mu_j\lambda_i)^{(c)}\\ &= \sum_{j=1}^{n} l_j\mu_j^{(c)} \sum_{i=1}^{m} k_i\lambda_i^{(b)} = \phi(g) \cdot \phi(f).\end{aligned}$$

It is obvious that ϕ is $\Bbbk$-linear.

(2) $\phi(I\#G) \subseteq I_{G,W}$. Thus ϕ induces a functor $\overline{\phi}\colon \Bbbk(Q, I, W)\#G \to \Bbbk(Q_{G,W}, I_{G,W})$. Indeed, let $x^{(a)}, y^{(b)} \in (\Bbbk(Q, I, W)\#G)_0$. Then it is enough to show that $\phi((I\#G)(x^{(a)}, y^{(b)})) \subseteq I_{G,W}(x^{(a)}, y^{(b)})$. Take ${}^{(b)}f^{(a)} \in (I\#G)(x^{(a)}, y^{(b)})$ with $f \in I^{b^{-1}a}(x, y)$. Then we have $f = \sum_{i=1}^m k_i\mu_i$ for some $k_i \in \Bbbk$ and $\mu_i \in \mathbb{P}Q(x, y)$, where $W(\mu_i) = b^{-1}a$ for all i. There exists a partition $S_1 \sqcup S_2 \sqcup \cdots \sqcup S_t = \{1, 2, \ldots, n\}$ of the set $\{1, 2, \ldots, n\}$ such that f is expressed as the sum $f = \rho_1 + \rho_2 + \cdots + \rho_t$ of minimal relations $\rho_j = \sum_{i\in S_j} k_i\mu_i$ $(j = 1, 2, \ldots, t)$. Then we have $\phi(f) = \sum_{i=1}^m k_i\mu_i^{(b)} = \sum_{j=1}^t \sum_{i\in S_j} k_i\mu_i^{(b)} = \rho_1^{(b)} + \rho_2^{(b)} + \cdots + \rho_t^{(b)} \in I_{G,W}(x^{(a)}, y^{(b)})$.

(3) $\overline{\phi}$ is bijective on objects. Indeed, this is obvious because ϕ is the identity on objects.

(4) $\overline{\phi}$ commutes with the G-action. Indeed, let $x, y \in Q_0$ and $a, b, c \in G$. Then it is enough to show the commutativity of the following diagram:

$$\begin{array}{ccc} (\Bbbk(Q,I,W)\#G)(x^{(a)},y^{(b)}) & \xrightarrow{\overline{\phi}} & \Bbbk(Q_{G,W},I_{G,W})(x^{(a)},y^{(b)}) \\ {\scriptstyle X'_c}\downarrow & & \downarrow{\scriptstyle \overline{X}_c} \\ (\Bbbk(Q,I,W)\#G)(x^{(ca)},y^{(cb)}) & \xrightarrow[\overline{\phi}]{} & \Bbbk(Q_{G,W},I_{G,W})(x^{(ca)},y^{(cb)}), \end{array} \tag{6.9}$$

where X' denotes the G-action on $\Bbbk(Q,I,W)\#G$. We have only to show the commutativity for the elements of the form $\bar{\mu} := \mu + (I\#G)(x^{(a)},y^{(b)})$ for some path $\mu \in \mathbb{P}Q(x,y)$ satisfying $W(\mu) = b^{-1}a$. This is verified by the following equalities:

$$\overline{\phi}(X'_c({}^{(b)}\bar{\mu}^{(a)})) = \overline{\phi}({}^{(cb)}\bar{\mu}^{(ca)}) = \widetilde{\mu^{(cb)}} = \overline{X}_c(\widetilde{\mu^{(b)}}) = \overline{X}_c(\overline{\phi}({}^{(b)}\bar{\mu}^{(a)})),$$

where $\widetilde{(\text{-})}$ stands for the coset in $\Bbbk(Q_{G,W},I_{G,W})$.

(5) $\overline{\phi}$ is fully faithful. Indeed, let $x^{(a)}, y^{(b)} \in (\Bbbk(Q,I,W)\#G)_0$. Then there exists a commutative diagram

$$\begin{array}{ccccccccc} 0 \to & (I\#G)(x^{(a)},y^{(b)}) & \hookrightarrow & (\Bbbk(Q,W)\#G)(x^{(a)},y^{(b)}) & \to & (\Bbbk(Q,I,W)\#G)(x^{(a)},y^{(b)}) & \to 0 \\ & {\scriptstyle \phi|I\#G}\downarrow & & \downarrow{\scriptstyle \phi} & & \downarrow{\scriptstyle \overline{\phi}} & \\ 0 \to & I_{G,W}(x^{(a)},y^{(b)}) & \hookrightarrow & \Bbbk Q_{G,W}(x^{(a)},y^{(b)}) & \to & \Bbbk(Q_{G,W},I_{G,W})(x^{(a)},y^{(b)}) & \to 0 \end{array}$$

with exact rows. Therefore, it is enough to show that both ϕ and $\phi|I\#G$ in the diagram above are isomorphisms by 5-lemma (add one more vertical isomorphism $0 \to 0$ on the right to have $\overline{\phi}$ at the center of five vertical maps).

We first show that ϕ above is an isomorphism. For each $c \in G$ we set $\mathbb{P}Q^c(x,y) := \{\mu \in \mathbb{P}Q(x,y) \mid W(\mu) = c\}$. By the second projection $(\Bbbk(Q,W)\#G)(x^{(a)},y^{(b)}) \xrightarrow{\sim} \Bbbk(Q,W)^{b^{-1}a}(x,y)$, we identify both sides. Then $(\Bbbk(Q,W)\#G)(x^{(a)},y^{(b)})$ and $\Bbbk Q_{G,W}(x^{(a)},y^{(b)})$ have bases $\mathbb{P}Q^{b^{-1}a}(x,y)$ and $\mathbb{P}Q_{G,W}(x^{(a)},y^{(b)})$, respectively, and ϕ induces a map

$$\phi' \colon \mathbb{P}Q^{b^{-1}a}(x,y) \to \mathbb{P}Q_{G,W}(x^{(a)},y^{(b)}), \mu \mapsto \mu^{(b)}$$

between these bases. Therefore it is enough to show that ϕ' is bijective. We denote by F the covering $F_{G,W} \colon Q_{G,W} \to Q$ of the quiver Q defined in Proposition 6.2.17. By restricting F, define a map $F' \colon \mathbb{P}Q_{G,W}(x^{(a)},y^{(b)}) \to \mathbb{P}Q^{b^{-1}a}(x,y)$. We here verify that ϕ' and F' are inverses of each other, which shows the bijectivity of ϕ'. Since $F'(\phi'(\mu)) = F(\mu^{(b)}) = \mu$ for all $\mu \in \mathbb{P}Q^{b^{-1}a}(x,y)$, we see that $F'\phi' = \mathbb{1}_{\mathbb{P}Q^{b^{-1}a}(x,y)}$. Next let $\xi \in \mathbb{P}Q_{G,W}(x^{(a)},y^{(b)})$ and set $\mu := F(\xi)\ (= F'(\xi))$. Then we have $F(\mu^{(b)}) = \mu = F(\xi)$ and $t_{G,W}(\mu^{(b)}) = y^{(b)} = t_{G,W}(\xi)$. Therefore by Lemma 6.2.15 (uniqueness of the lift), we have $\mu^{(b)} = \xi$. Hence $\phi'(F'(\xi)) = \mu^{(b)} = \xi$, and we have $\phi'F' = \mathbb{1}_{\mathbb{P}Q_{G,W}(x^{(a)},y^{(b)})}$. As a consequence, ϕ' is bijective, and we see that ϕ above is an isomorphism.

Finally we show that $\phi|I\#G$ is an isomorphism. The commutativity of the left square in the diagram above and the injectivity of ϕ show that $\phi|I\#G$ is injective. To show the surjectivity of $\phi|I\#G$, let $b \in G$, $x, y \in Q_0$, and ρ a minimal relation in $I(x,y)$, and take $\rho^{(b)} \in I_{G,W}$. Then we have $t_{G,W}(\rho^{(b)}) = y^{(b)}$, and there exists some $a \in G$ such that $s_{G,W}(\rho^{(b)}) = x^{(a)}$. Here ρ is expressed as $\rho = \sum_{i=1}^{m} k_i\mu_i$ (with

$0 \neq k_i \in \Bbbk$, $\mu_i \in \mathbb{P}Q(x,y)$), and W is a homogeneous weight, which shows that $W(\mu_i) = W(\mu_1) = b^{-1}a$ for all i. Therefore, $\rho \in I^{b^{-1}a}(x,y) = (I\#G)(x^{(a)}, y^{(b)})$ and we have $\rho^{(b)} = \phi(\rho) \in \phi((I\#G)(x^{(a)}, y^{(b)}))$. Thus $\phi|I\#G$ is surjective, and hence an isomorphism. $\square$

The following is easy to show.

COROLLARY 6.2.19. *Let (Q,I) and W be as in Theorem 6.2.18, and $F\colon \Bbbk(Q,I,W)\#G \to \Bbbk(Q,I)$ the canonical G-covering. Then we have the strictly commutative diagram*

$$\begin{array}{ccc} \Bbbk(Q,I,W)\#G & \xrightarrow{\ \overline{\phi}\ } & \Bbbk(Q_{G,W}, I_{G,W}) \\ & {\scriptstyle F}\searrow \quad \swarrow{\scriptstyle \Bbbk F_{G,W}} & \\ & \Bbbk(Q,I) & \end{array}.$$

Therefore we can regard $F_{G,W}$ as a presentation of F.

EXERCISE 6.2.20. Prove the corollary above.

EXAMPLE 6.2.21. Let $G = \mathbb{Z}$ and (Q,I,W) be the G-weighted quiver with relations in Example 6.2.11. Then the smash product $\Bbbk(Q,I,W)\#G$ is given by the quiver with relations $(Q_{G,W}, I_{G,W})$, where $Q_{G,W}$ is the quiver

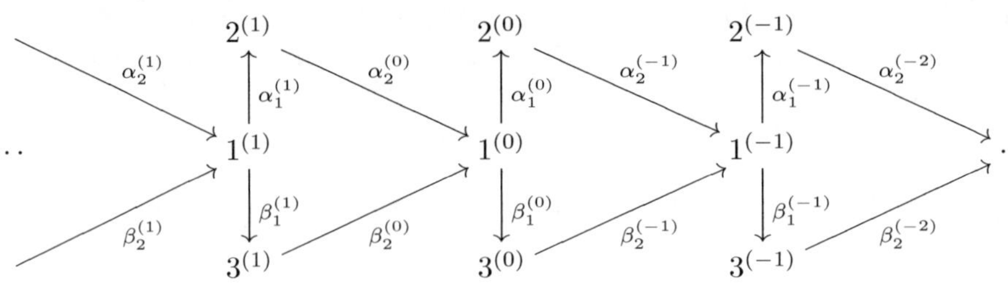

and $I_{G,W}$ is generated by the set

$$\{\alpha_2^{(i-1)}\alpha_1^{(i)} - \beta_2^{(i-1)}\beta_1^{(i)}, \beta_1^{(i)}\alpha_2^{(i)}, \alpha_1^{(i)}\beta_2^{(i)}, \alpha_2^{(i-1)}\alpha_1^{(i)}\alpha_2^{(i)}, \beta_2^{(i-1)}\beta_1^{(i)}\beta_2^{(i)} \mid i \in \mathbb{Z}\}.$$

EXERCISE 6.2.22. Let $G = \langle a \rangle$ be an infinite cyclic group, and consider the G-weighted quiver with relations (Q,I,W) computed in Example 6.1.4. Set $\mathscr{B} := \Bbbk(Q,I,W) = \mathscr{C}/_{\!c}G$. Then verify that $\mathscr{B}\#G$ is isomorphic to $\mathscr{C} = \Bbbk[\tilde{Q}, \tilde{I}]$ in Example 6.1.4. This gives a simple example of the Cohen–Montgomery duality.

6.3. Computation of right smash products

In the following, we use Proposition 5.10.6 (1) to translate the computation of smash products into the computation of right smash products.

DEFINITION 6.3.1. Let (Q,I) be a quiver with relations, and W a homogeneous G-weight on $\Bbbk(Q,I)$.

(1) We define a quiver $Q'_{G,W} = ((Q'_{G,W})_0, (Q'_{G,W})_1, s'_{G,W}, t'_{G,W})$ as follows:

$$(Q'_{G,W})_0 := \{x^{(a)} := (x,a) \mid x \in Q_0, a \in G\} = Q_0 \times G,$$
$$(Q'_{G,W})_1 := \{\alpha^{(a)} : x^{(a)} \to y^{(W(\alpha)a)} \mid (x \xrightarrow{\alpha} y) \in Q_1, a \in G\}, \text{and}$$
$$s'_{G,W}(\alpha^{(a)}) := s(\alpha)^{(a)}, \ \ t'_{G,W}(\alpha^{(a)}) := t(\alpha)^{(W(\alpha)a)} \quad (\alpha \in Q_1, a \in G).$$

(2) For each path $\mu = \alpha_n \cdots \alpha_1$ of length n (≥ 2) and $a \in G$, we set a path $\mu^{(a)}$ from $x^{(a)}$ to $y^{(W(\mu)a)}$ to be

$$\mu^{(a)} := \alpha_n^{(a_{n-1}\cdots a_1 a)} \cdots \alpha_2^{(a_1 a)} \alpha_1^{(a)},$$

where we set $a_i := W(\alpha_i)$ $(i = 1, \dots, n)$. Therefore note that $W(\mu) := a_n \cdots a_1$. This is visually expressed as follows: If μ is a path of the form $y = x_n \xleftarrow{\alpha_n} \cdots \xleftarrow{\alpha_2} x_1 \xleftarrow{\alpha_1} x_0 = x$, then $\mu^{(a)}$ is a path of the form

$$x_n^{(a_n\cdots a_1 a)} \xleftarrow{\alpha_n^{(a_{n-1}\cdots a_1 a)}} \cdots \xleftarrow{\alpha_2^{(a_1 a)}} x_1^{(a_1 a)} \xleftarrow{\alpha_1^{(a)}} x_0^{(a)}.$$

(3) For each minimal relation $\rho = \sum_i k_i \mu_i$ $(k_i \in \Bbbk, \mu_i \in \mathbb{P}Q(x,y), x, y \in Q_0)$ in I and $a \in G$, we set

$$\rho^{(a)} := \sum_i k_i \mu_i^{(a)}.$$

Note here that $\mu_i^{(a)} \in \mathbb{P}Q_{G,W}(x^{(a)}, y^{(W(\mu_1)a)})$ because $W(\mu_i) = W(\mu_1)$ for all i. Then a right G-action X' on $\Bbbk[Q'_{G,W}]$ is defined by the following quiver morphism:

$$X'_c : (x^{(a)} \xrightarrow{\alpha^{(a)}} y^{(W(\alpha)a)}) \mapsto (x^{(ac)} \xrightarrow{\alpha^{(ac)}} y^{(W(\alpha)ac)})$$
$$((x^{(a)} \xrightarrow{\alpha^{(a)}} y^{(W(\alpha)a)}) \in (Q'_{G,W})_1, a, c \in G, x, y \in Q_0, \alpha \in Q_1).$$

(4) We define an ideal $I'_{G,W}$ of $\Bbbk[Q'_{G,W}]$ by

$$I'_{G,W} := \langle \rho^{(a)} \mid a \in G, \rho \text{ is a minimal relation in } I \rangle.$$

(5) We call $(Q'_{G,W}, I'_{G,W})$ the *right smash product* of (Q, I, W) and G.

Using the notations above, Theorem 6.2.18 can be translated as follows:

THEOREM 6.3.2. *Let (Q, I) be a quiver with relations, and W a homogeneous G-weight on $\Bbbk(Q, I)$. Then the right smash product $\Bbbk(Q, I, W)\#^{\mathrm{op}}G$ is presented by the quiver with relations $(Q'_{G,W}, I'_{G,W})$, i.e., there exists an isomorphism*

$$\Bbbk(Q, I, W)\#^{\mathrm{op}}G \cong \Bbbk(Q'_{G,W}, I'_{G,W})$$

of right G-categories.

For comparison, the right smash product corresponding to Example 6.2.21 is computed as follows:

EXAMPLE 6.3.3. Let $G = \mathbb{Z}$ and (Q, I, W) be the G-weighted quiver with relations in Example 6.2.11. Then the right smash product $\Bbbk(Q, I, W)\#G$ is given by the quiver with relations $(Q'_{G,W}, I'_{G,W})$, where $Q'_{G,W}$ is the quiver

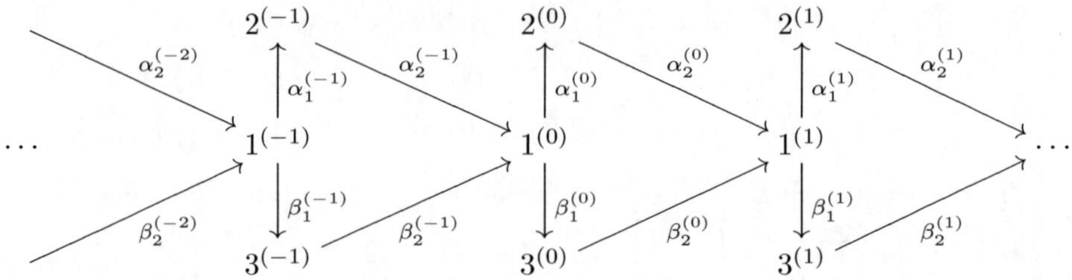

and $I'_{G,W}$ is generated by the set

$$\{\alpha_2^{(i)}\alpha_1^{(i)} - \beta_2^{(i)}\beta_1^{(i)}, \beta_1^{(i+1)}\alpha_2^{(i)}, \alpha_1^{(i+1)}\beta_2^{(i)}, \alpha_2^{(i+1)}\alpha_1^{(i+1)}\alpha_2^{(i)}, \beta_2^{(i+1)}\beta_1^{(i+1)}\beta_2^{(i)} \mid i \in \mathbb{Z}\}.$$

CHAPTER 7

Relationships between module categories

In this chapter, we investigate the relationships between the module categories $\operatorname{Mod}\mathscr{C}$ and $\operatorname{Mod}\mathscr{D}$ for a G-covering functor $(F,\psi)\colon \mathscr{C} \to \mathscr{D}$. Without loss of generality, we may take the canonical covering functor $(P,\phi)\colon \mathscr{C} \to \mathscr{C}/G$ as (F,ψ) by Theorem 5.5.8. In the following, the category $\mathscr{C}$ we start with is assumed to be small. Even in this case, however, $\operatorname{Mod}\mathscr{C}$ is not a small category anymore but a light category. Therefore, the following arguments cannot be directly applied to $\operatorname{Mod}\mathscr{C}$. However, using the tensor product over light categories described in the appendix, the arguments can be applied to $\operatorname{Mod}\mathscr{C}$ as well. When $(\mathscr{C},X)$ is a pseudo-G-category, a natural structure of a pseudo-G-category can be induced in $\operatorname{Mod}\mathscr{C}$, but to prove it we need a general theory on 2-categories (see Remark 8.2.10 and [**9**, Theorem 6.5, §9]). Therefore, in this section we will restrict ourselves to the case where X is a strict G-action. Even with this restriction, the following theory can be applied when $\mathscr{C}$ is skeletal because in this case, all auto-equivalences of $\mathscr{C}$ are automorphisms, and usually it is enough to consider only strict G-actions on $\mathscr{C}$.

7.1. Tensor products over a small category and left Kan extensions

For a small linear category $\mathscr{C}$ we first define a tensor product functor

$$\text{-}\otimes_{\mathscr{C}}?\colon \operatorname{Mod}\mathscr{C}\times\mathscr{C}\text{-}\operatorname{Mod}\to\operatorname{Mod}\Bbbk$$

over $\mathscr{C}$, and give it an explicit form. For small linear categories $\mathscr{C}$ and $\mathscr{D}$ we next define $\mathscr{C}$-$\mathscr{D}$-bimodules as bilinear functors $\mathscr{D}^{\mathrm{op}}\times\mathscr{C}\to\operatorname{Mod}\Bbbk$. For instance, let $F\colon\mathscr{C}\to\mathscr{D}$ be a functor. Then ${}_F\mathscr{D} := ((?,\text{-})\mapsto\mathscr{D}(\text{-},F(?)))$ turns out to be a $\mathscr{C}$-$\mathscr{D}$-bimodule. We then define a category ${}_{\mathscr{C}}\operatorname{Mod}_{\mathscr{D}}$ formed by the $\mathscr{C}$-$\mathscr{D}$-bimodules, and induce a functor $\text{-}\otimes_{\mathscr{C}}?\colon\operatorname{Mod}\mathscr{C}\times{}_{\mathscr{C}}\operatorname{Mod}_{\mathscr{D}}\to\operatorname{Mod}\mathscr{D}$ by using the tensor product functor above. This constructs a *left Kan extension* $M\otimes_{\mathscr{C}}{}_F\mathscr{D}$ along F of M for each $M\in(\operatorname{Mod}\mathscr{C})_0$. Then the left adjoint $F_{\bullet}\colon\operatorname{Mod}\mathscr{C}\to\operatorname{Mod}\mathscr{D}$ to the functor $F^{\bullet}:=\operatorname{Mod}F\cong\operatorname{Mod}\mathscr{D}({}_F\mathscr{D},\text{-})\colon\operatorname{Mod}\mathscr{D}\to\operatorname{Mod}\mathscr{C}$ defined by the 2-functor Mod in Example 4.1.13 is constructed as $F_{\bullet}:=\text{-}\otimes_{\mathscr{C}}({}_F\mathscr{D})$. In the next section, by applying this construction to the canonical covering functor $(P,\phi)\colon\mathscr{C}\to\mathscr{C}/G$ for a G-category $(\mathscr{C},X)$, we define the pushdown functor $P_{\bullet}\colon\operatorname{Mod}\mathscr{C}\to\operatorname{Mod}\mathscr{C}/G$, and compute its explicit form.

DEFINITION 7.1.1. Let $\mathscr{C},\mathscr{D}$ be small linear categories, and $F\colon\mathscr{C}\times\mathscr{D}\to\operatorname{Mod}\Bbbk$ a functor with two variables. Then F is said to be *bilinear* if the map

$$\begin{aligned}(\mathscr{C}\times\mathscr{D})((x_1,y_1),(x_2,y_2)) &= \mathscr{C}(x_1,x_2)\times\mathscr{D}(y_1,y_2)\\ &\to\operatorname{Mod}\Bbbk(F(x_1,y_1),F(x_2,y_2))\end{aligned}$$

induced from F is bilinear for all $(x_1, y_1), (x_2, y_2) \in (\mathscr{C} \times \mathscr{D})_0$. For two bilinear functors $F, F'\colon \mathscr{C} \times \mathscr{D} \to \operatorname{Mod} \Bbbk$, a *morphism* from F to F' is defined to be a morphism from F to F' as functors of two variables.

We here give a way to compute the colimit $\varinjlim F$ for a functor $F\colon I \to \operatorname{Mod}\mathscr{C}$ from a small category I to the module category of a small linear category $\mathscr{C}$.

PROPOSITION 7.1.2. *Let $\mathscr{C}$ be a small linear category, I a small category, and $F\colon I \to \operatorname{Mod}\mathscr{C}$ a functor. Then there exists the following exact sequence in* $\operatorname{Mod}\mathscr{C}$*:*

$$\bigoplus_{f\in I_1} F(\operatorname{dom}(f)) \xrightarrow{(\sigma_{\operatorname{cod}(f)}\circ F(f) - \sigma_{\operatorname{dom}(f)})_{f\in I_1}} \bigoplus_{i\in I_0} F(i) \to \varinjlim F \to 0,$$

where $\sigma_j\colon F(j) \to \bigoplus_{i\in I_0} F(i)$ are the canonical injections for all $j \in I_0$. Therefore in particular, $\varinjlim F$ exists in $\operatorname{Mod}\mathscr{C}$, and a linear functor $R\colon \operatorname{Mod}\mathscr{C} \to \operatorname{Mod}\Bbbk$ is small-cocontinuous if and only if R is right exact and preserves coproducts for small index sets.

PROOF. By Remark A.1.5(6), since I_0 is a small set, both sides of

$$\bigoplus_{i\in I_0} (F(i))(x) \subseteq \prod_{i\in I_0} (F(i))(x)$$

are small for all $x \in \mathscr{C}_0$. Therefore, we have $\bigoplus_{i\in I_0} F(i) \in (\operatorname{Mod}\mathscr{C})_0$. Similarly, since I_1 is also a small set, we have $\bigoplus_{f\in I_1} F(\operatorname{dom}(f)) \in (\operatorname{Mod}\mathscr{C})_0$. By definition of colimits, it is easy to verify that $\varinjlim F = \operatorname{Coker}(\sigma_{\operatorname{cod}(f)} \circ F(f) - \sigma_{\operatorname{dom}(f)})_{f\in I_1}$. Therefore by Remark A.1.5 (10), we have $\varinjlim F \in (\operatorname{Mod}\mathscr{C})_0$. Hence the canonical morphism to the cokernel gives us the desired exact sequence in $\operatorname{Mod}\mathscr{C}$. □

THEOREM 7.1.3. *Let $\mathscr{C}$ be a small linear category. Then there exists a bilinear functor $\text{-} \otimes_{\mathscr{C}} ?\colon \operatorname{Mod}\mathscr{C} \times \mathscr{C}\text{-}\operatorname{Mod} \to \operatorname{Mod}\Bbbk$ satisfying the following conditions, which is uniquely determined up to isomorphisms, and is called the* tensor product functor *over $\mathscr{C}$:*

(1) *For each $N \in \mathscr{C}\text{-}\operatorname{Mod}$,*
   (a) *$\text{-} \otimes_{\mathscr{C}} N\colon \operatorname{Mod}\mathscr{C} \to \operatorname{Mod}\Bbbk$ is small-cocontinuous (equivalently, right exact and preserves small coproducts); and*
   (b) *For each $x \in \mathscr{C}_0$, there exists an isomorphism $s_{x,N}\colon \mathscr{C}(\text{-}, x) \otimes_{\mathscr{C}} N \xrightarrow{\sim} N(x)$ in $\operatorname{Mod}\Bbbk$ that is natural in x and in N.*

(2) *For each $M \in \operatorname{Mod}\mathscr{C}$,*
   (a) *$M \otimes_{\mathscr{C}} \text{-}\colon \mathscr{C}\text{-}\operatorname{Mod} \to \operatorname{Mod}\Bbbk$ is small-cocontinuous; and*
   (b) *For each $x \in \mathscr{C}_0$, there exists an isomorphism $t_{x,M}\colon M \otimes_{\mathscr{C}} \mathscr{C}(x, \text{-}) \xrightarrow{\sim} M(x)$ in $\operatorname{Mod}\Bbbk$ that is natural in x and in M.*

PROOF. First of all, we explicitly construct the functor

$$\text{-} \otimes_{\mathscr{C}} ?\colon \operatorname{Mod}\mathscr{C} \times \mathscr{C}\text{-}\operatorname{Mod} \to \operatorname{Mod}\Bbbk$$

as follows:

On objects: Let $M \in (\operatorname{Mod}\mathscr{C})_0$ and $N \in (\mathscr{C}\text{-}\operatorname{Mod})_0$. Then we set

$$\begin{aligned} M \otimes_{\mathscr{C}} N &:= \left(\bigoplus_{x\in\mathscr{C}_0} M(x) \otimes_{\Bbbk} N(x)\right) \Big/ I(M, N), \quad \text{where} \\ I(M, N) &:= \langle mM(f) \otimes n - m \otimes N(f)n \mid x, y \in \mathscr{C}_0, \\ &\qquad m \in M(y), f \in \mathscr{C}(x, y), n \in N(x)\rangle. \end{aligned} \tag{7.1}$$

In the above, $\otimes_{\Bbbk}$ denotes the tensor product as $\Bbbk$-modules, and for each set S, $\langle S\rangle$ denotes the $\Bbbk$-module generated by S, and also we set $mM(f) := [M(f)](m)$. For each $x, y \in \mathscr{C}_0, m \in M(x), n \in N(x)$, we set

$$m \otimes_{\mathscr{C}} n := m \otimes n + I(M,N) \in M \otimes_{\mathscr{C}} N.$$

Then by the definition above, we have

$$m \otimes_{\mathscr{C}} N(f)n = mM(f) \otimes_{\mathscr{C}} n$$

for all $x, y \in \mathscr{C}_0, m \in M(y), f \in \mathscr{C}(x,y), n \in N(x)$. We next verify that $M \otimes_{\mathscr{C}} N \in (\operatorname{Mod}\Bbbk)_0$. Since $\mathscr{C}_0$ is a small set, and for each $x \in \mathscr{C}_0$ the set $M(x) \otimes_{\Bbbk} N(x)$ is also small, it holds that both sides of $\bigoplus_{x\in\mathscr{C}_0} M(x) \otimes_{\Bbbk} N(x) \subseteq \prod_{x\in\mathscr{C}_0} M(x) \otimes_{\Bbbk} N(x)$ are small (Remark A.1.5 (6)). Therefore, by Remark A.1.5 (10), the set $M \otimes_{\mathscr{C}} N$ is small. Thus $M \otimes_{\mathscr{C}} N \in (\operatorname{Mod}\Bbbk)_0$.

On morphisms: Let $f\colon M \to M'$ be a morphism in $\operatorname{Mod}\mathscr{C}$ and $g\colon N \to N'$ a morphism in $\mathscr{C}$-Mod. Then we define $f \otimes_{\mathscr{C}} g\colon M \otimes_{\mathscr{C}} N \to M' \otimes_{\mathscr{C}} N'$ as follows. It is easy to verify that the map

$$F := \bigoplus_{x\in\mathscr{C}_0} f_x \otimes_{\Bbbk} g_x \colon \bigoplus_{x\in\mathscr{C}_0} M(x) \otimes_{\Bbbk} N(x) \to \bigoplus_{x\in\mathscr{C}_0} M'(x) \otimes_{\Bbbk} N'(x)$$

satisfies $F(I(M,N)) \subseteq I(M',N')$. Therefore we can define $f \otimes_{\mathscr{C}} g$ as the map between $\Bbbk$-modules induced from F:

$$\begin{array}{ccc} \bigoplus_{x\in\mathscr{C}_0} M(x) \otimes_{\Bbbk} N(x) & \xrightarrow{F} & \bigoplus_{x\in\mathscr{C}_0} M'(x) \otimes_{\Bbbk} N'(x) \\ {\scriptstyle\text{can.}}\downarrow & & \downarrow{\scriptstyle\text{can.}} \\ M \otimes_{\mathscr{C}} N & \xrightarrow{f\otimes_{\mathscr{C}} g} & M' \otimes_{\mathscr{C}} N'. \end{array}$$

It is easy to check that the quiver morphism $\text{-} \otimes_{\mathscr{C}} ?\colon \operatorname{Mod}\mathscr{C} \times \mathscr{C}\text{-}\operatorname{Mod} \to \operatorname{Mod}\Bbbk$ defined above is a bilinear functor.

We now verify that this satisfies the required conditions (1) and (2). Since both are shown similarly, we only show the condition (1). To this end let $N \in (\mathscr{C}\text{-Mod})_0$.

CLAIM 1. *There exists the following adjoint:*

$$\operatorname{Mod}\mathscr{C} \underset{\operatorname{Hom}_{\Bbbk}(N,\text{-})}{\overset{\text{-}\otimes_{\mathscr{C}} N}{\rightleftarrows}} \operatorname{Mod}\Bbbk, \quad (\text{-}\otimes_{\mathscr{C}} N) \dashv \operatorname{Hom}_{\Bbbk}(N,\text{-}).$$

To show this we define a unit $\eta\colon \mathbb{1}_{\operatorname{Mod}\mathscr{C}} \Rightarrow \operatorname{Hom}_{\Bbbk}(N,\text{-}) \circ (\text{-}\otimes_{\mathscr{C}} N)$ and a counit $\varepsilon\colon (\text{-}\otimes_{\mathscr{C}} N) \circ \operatorname{Hom}_{\Bbbk}(N,\text{-}) \Rightarrow \mathbb{1}_{\operatorname{Mod}\Bbbk}$ as follows:

Construction of η: For each $M \in (\operatorname{Mod}\mathscr{C})_0$ and $x \in \mathscr{C}_0$ we set

$$\eta_{M,x}\colon M(x) \to \operatorname{Hom}_{\Bbbk}(N(x), M \otimes_{\mathscr{C}} N), m \mapsto m \otimes_{\mathscr{C}} \text{-}$$

and $\eta_M := (\eta_{M,x})_{x\in\mathscr{C}_0}$. Then it is easy to verify that for each morphism $F\colon M \to M'$ in $\operatorname{Mod}\mathscr{C}$ the following is commutative:

$$\begin{array}{ccc} M(x) & \xrightarrow{\eta_{M,x}} & \operatorname{Hom}_{\Bbbk}(N(x), M \otimes_{\mathscr{C}} N) \\ {\scriptstyle F_x}\downarrow & & \downarrow{\scriptstyle \operatorname{Hom}_{\Bbbk}(N(x), F\otimes_{\mathscr{C}} N)} \\ M'(x) & \xrightarrow[\eta_{M',x}]{} & \operatorname{Hom}_{\Bbbk}(N(x), M' \otimes_{\mathscr{C}} N). \end{array}$$

Therefore, $\eta := (\eta_M)_{M\in(\mathrm{Mod}\,\mathscr{C})_0}$ turns out to be a natural transformation

$$\mathbb{1}_{\mathrm{Mod}\,\mathscr{C}} \Rightarrow \mathrm{Hom}_{\Bbbk}(N,\text{-}) \circ (\text{-}\otimes_{\mathscr{C}} N).$$

Construction of ε: For each $V \in (\mathrm{Mod}\,\Bbbk)_0$ we define a morphism

$$\varepsilon'_V \colon \bigoplus_{x\in\mathscr{C}_0} \mathrm{Hom}_{\Bbbk}(N(x),V)\otimes_{\Bbbk} N(x) \to V$$

in $\mathrm{Mod}\,\Bbbk$ by $(u_x\otimes n_x)_{x\in\mathscr{C}_0} \mapsto \sum_{x\in\mathscr{C}_0} u_x(n_x)$ $(u_x \in \mathrm{Hom}_{\Bbbk}(N(x),V), n_x \in N(x))$. Then it is easy to check that $\varepsilon'_V(I(\mathrm{Hom}_{\Bbbk}(N,V),N)) = 0$. Therefore, ε'_V induces a morphism

$$\varepsilon_V \colon \mathrm{Hom}_{\Bbbk}(N,V)\otimes_{\mathscr{C}} N \to V$$

in $\mathrm{Mod}\,\Bbbk$. Then it is immediate that $\varepsilon := (\varepsilon_V)_{V\in(\mathrm{Mod}\,\Bbbk)_0}$ turns out to be a natural transformation

$$(\text{-}\otimes_{\mathscr{C}} N)\circ \mathrm{Hom}_{\Bbbk}(N,\text{-}) \Rightarrow \mathbb{1}_{\mathrm{Mod}\,\Bbbk}.$$

It is also immediate from definition that η, ε satisfy the zigzag equations, i.e., that both of the two diagrams

$$\begin{array}{ccc} \text{-}\otimes_{\mathscr{C}} N & & \\ \Big\Downarrow{\scriptstyle \eta(\text{-})\otimes_{\mathscr{C}} N} & \searrow\!\!= & \\ \mathrm{Hom}_{\Bbbk}(N,\text{-}\otimes_{\mathscr{C}} N)\otimes_{\mathscr{C}} N & \xrightarrow[\varepsilon_{\text{-}\otimes_{\mathscr{C}} N}]{} & \text{-}\otimes_{\mathscr{C}} N, \end{array}$$

(7.2)

$$\begin{array}{ccc} \mathrm{Hom}_{\Bbbk}(N,\text{-}) & & \\ \Big\Downarrow{\scriptstyle \eta_{\mathrm{Hom}_{\Bbbk}(N,\text{-})}} & \searrow\!\!= & \\ \mathrm{Hom}_{\Bbbk}(N,\mathrm{Hom}_{\Bbbk}(N,\text{-})\otimes_{\mathscr{C}} N) & \xrightarrow[\mathrm{Hom}_{\Bbbk}(N,\varepsilon)]{} & \mathrm{Hom}_{\Bbbk}(N,\text{-}) \end{array}$$

are commutative. Hence we have an adjoint $\text{-}\otimes_{\mathscr{C}} N \dashv \mathrm{Hom}_{\Bbbk}(N,\text{-})$. By Theorem 4.4.7, it also holds that there exists an isomorphism

$$\omega_{M,V} \colon \mathrm{Hom}_{\Bbbk}(M\otimes_{\mathscr{C}} N, V) \xrightarrow{\sim} (\mathrm{Mod}\,\mathscr{C})(M, \mathrm{Hom}_{\Bbbk}(N,V)) \tag{7.3}$$

natural in $M \in (\mathrm{Mod}\,\mathscr{C})_0$ and in $V \in (\mathrm{Mod}\,\Bbbk)_0$.

Using the claim above we prove statement (1).

(a) Let I be a small category. For each $F \in ((\mathrm{Mod}\,\mathscr{C})^I)_0$, there exist $\varinjlim F$ and the colimit of $F(\text{-})\otimes_{\mathscr{C}} N \in ((\mathrm{Mod}\,\Bbbk)^I)_0$ by Proposition 7.1.2. Therefore, Proposition 4.5.8 (2) shows that $\text{-}\otimes_{\mathscr{C}} N$ is cocontinuous with respect to I. Hence $\text{-}\otimes_{\mathscr{C}} N$ is small-cocontinuous.

(b) We show that for each $x \in \mathscr{C}_0$, there exists an isomorphism $\mathscr{C}(\text{-},x)\otimes_{\mathscr{C}} N \xrightarrow{\sim} N(x)$ in $\mathrm{Mod}\,\Bbbk$ that is natural in x. Apply the formula (7.3) to $M = \mathscr{C}(\text{-},x)$. Then by the Yoneda lemma, we have

$$\begin{aligned} \mathrm{Hom}_{\Bbbk}(\mathscr{C}(\text{-},x)\otimes_{\mathscr{C}} N, ?) &\xrightarrow{\sim} \mathrm{Mod}\,\mathscr{C}(\mathscr{C}(\text{-},x), \mathrm{Hom}_{k}(N,?)) \\ &\xrightarrow{\sim} \mathrm{Hom}_{\Bbbk}(N,?)(x) = \mathrm{Hom}_{\Bbbk}(N(x),?). \end{aligned} \tag{7.4}$$

Therefore, by Corollary 2.5.11, there exists an isomorphism $\mathscr{C}(\text{-},x)\otimes_{\mathscr{C}} N \xrightarrow{\sim} N(x)$ in $\mathrm{Mod}\,\Bbbk$. It is easy to check that this is natural both in $x \in \mathscr{C}_0$ and in N by using the Yoneda lemma. This finishes the proof of the existence of $\text{-}\otimes_{\mathscr{C}} ?$.

Uniqueness: Consider a bilinear functor

$$-\bar{\otimes}_{\mathscr{C}}?\colon \operatorname{Mod}\mathscr{C} \times \mathscr{C}\text{-Mod} \to \operatorname{Mod}\Bbbk$$

that has the same property as $-\otimes_{\mathscr{C}}?$, and take $N \in (\mathscr{C}\text{-Mod})_0$. Then by the property (1)(b) of $-\otimes_{\mathscr{C}}?$ and of $-\bar{\otimes}_{\mathscr{C}}?$, there exist isomorphisms

$$\mathscr{C}(\text{-},x)\bar{\otimes}_{\mathscr{C}}N \xrightarrow[\sim]{\bar{s}_x} N(x) \xleftarrow[\mathscr{C}]{s_x} (\text{-},x)\otimes_{\mathscr{C}} N$$

that are natural in $x \in \mathscr{C}_0$.

(i) The case where M is of the form $M = \mathscr{C}(\text{-},x)$ for some $x \in \mathscr{C}_0$: We set $\phi_{M,N} := s_x^{-1}\circ \bar{s}_x\colon M\bar{\otimes}_{\mathscr{C}}N \xrightarrow{\sim} M\otimes_{\mathscr{C}} N$.

(ii) The case where M is of the form $M = \bigoplus_{x\in\mathscr{C}_0}\mathscr{C}(\text{-},x)^{(I_x)}$ for some small sets I_x $(x \in \mathscr{C}_0)$: By the property (1)(a) of $-\otimes_{\mathscr{C}}?$ and of $-\bar{\otimes}_{\mathscr{C}}?$, the vertical canonical morphisms are isomorphisms in the diagram

$$\begin{array}{ccccc}
\bigoplus_{x\in\mathscr{C}_0}\mathscr{C}(\text{-},x)^{(I_x)}\bar{\otimes}_{\mathscr{C}}N & \xrightarrow[\sim]{\bigoplus_{x\in\mathscr{C}_0}\bar{s}_x^{(I_x)}} & \bigoplus_{x\in\mathscr{C}_0}N(x)^{(I_x)} & \xleftarrow[\sim]{\bigoplus_{x\in\mathscr{C}_0}s_x^{(I_x)}} & \bigoplus_{x\in\mathscr{C}_0}\mathscr{C}(\text{-},x)^{(I_x)}\otimes_{\mathscr{C}}N \\
\wr\downarrow & & & & \downarrow\wr \\
\left(\bigoplus_{x\in\mathscr{C}_0}\mathscr{C}(\text{-},x)^{(I_x)}\right)\bar{\otimes}_{\mathscr{C}}N & & \xrightarrow[\sim]{\quad\quad} & & \left(\bigoplus_{x\in\mathscr{C}_0}\mathscr{C}(\text{-},x)^{(I_x)}\right)\otimes_{\mathscr{C}}N.
\end{array}$$

We then define $\phi_{M,N}\colon M\bar{\otimes}_{\mathscr{C}}N \xrightarrow{\sim} M\otimes_{\mathscr{C}}N$ as the unique isomorphism making this diagram commutative. Note that $\phi_{M,N}$ is natural in M for those M having this form.

(iii) The general case: First of all we show the following for all $M \in (\operatorname{Mod}\mathscr{C})_0$.

CLAIM 2. *There exists a family $(I_x)_{x\in\mathscr{C}_0}$ of small sets such that we have an epimorphism $\phi_M\colon \bigoplus_{x\in\mathscr{C}_0}\mathscr{C}(\text{-},x)^{(I_x)} \to M$ in* $\operatorname{Mod}\mathscr{C}$.

Indeed by the Yoneda lemma, for each $x \in \mathscr{C}_0$, there exists an isomorphism

$$\psi_x\colon M(x) \to (\operatorname{Mod}\mathscr{C})(\mathscr{C}(\text{-},x), M)$$

such that $\psi_x(a)(\mathbb{1}_x) = a$ for all $a \in M(x)$. Therefore,

$$(\psi_x(a))_{a\in M(x)}\colon \bigoplus_{a\in M(x)}\mathscr{C}(\text{-},x) \to M$$

becomes an epimorphism $\bigoplus_{a\in M(x)}\mathscr{C}(x,x) \to M(x)$ in $\operatorname{Mod}\Bbbk$ at each $x \in \mathscr{C}_0$. Thus

$$((\psi_x(a))_{a\in M(x)})_{x\in\mathscr{C}_0}\colon \bigoplus_{x\in\mathscr{C}_0}\bigoplus_{a\in M(x)}\mathscr{C}(\text{-},x) \to M$$

is an epimorphism in $\operatorname{Mod}\mathscr{C}$ (note that $\bigoplus_{x\in\mathscr{C}_0}\bigoplus_{a\in M(x)}\mathscr{C}(y,x) \in \operatorname{Mod}\Bbbk$ for all $y \in \mathscr{C}_0$). Hence we can take $I_x := M(x)$.

CLAIM 3. *There exist families $(I_x)_{x\in\mathscr{C}_0}$ and $(J_y)_{y\in\mathscr{C}_0}$ of small sets such that we have an exact sequence*

$$\bigoplus_{y\in\mathscr{C}_0}\mathscr{C}(\text{-},y)^{(J_y)} \xrightarrow{F} \bigoplus_{x\in\mathscr{C}_0}\mathscr{C}(\text{-},x)^{(I_x)} \xrightarrow{E} M \to 0 \tag{7.5}$$

in $\operatorname{Mod}\mathscr{C}$.

Indeed, take ϕ_M in Claim 2 as E, and set $K := \operatorname{Ker}\phi_M$. Then noting that $J_y := K(y) \in \operatorname{Mod}\Bbbk$ for all $y \in \mathscr{C}_0$ and $\bigoplus_{y\in\mathscr{C}_0}\mathscr{C}(z,y)^{(J_y)} \in \operatorname{Mod}\Bbbk$ for all $z \in \mathscr{C}_0$, the argument above constructs an epimorphism $\phi_K\colon \bigoplus_{y\in\mathscr{C}_0}\mathscr{C}(\text{-},y)^{(J_y)} \to K$ in $\operatorname{Mod}\mathscr{C}$. Therefore we can take ϕ_K as F.

In the claim above, note that we have isomorphisms

$$P_0 := \bigoplus_{x\in\mathscr{C}_0}\mathscr{C}(\text{-},x)^{(I_x)} \cong \bigoplus_{(x,a)\in\bigsqcup_{x\in\mathscr{C}_0} I_x}\mathscr{C}(\text{-},x),$$

$$P_1 := \bigoplus_{y\in\mathscr{C}_0}\mathscr{C}(\text{-},y)^{(J_y)} \cong \bigoplus_{(y,b)\in\bigsqcup_{y\in\mathscr{C}_0} J_y}\mathscr{C}(\text{-},y)$$

and that both $\bigsqcup_{x\in\mathscr{C}_0} I_x$ and $\bigsqcup_{y\in\mathscr{C}_0} J_y$ are small sets. Then by (ii) above, both $\phi_{P_0,N}\colon P_0\bar{\otimes}_{\mathscr{C}}N \to P_0 \otimes_{\mathscr{C}} N$ and $\phi_{P_1,N}\colon P_1\bar{\otimes}_{\mathscr{C}}N \to P_1 \otimes_{\mathscr{C}} N$ are isomorphisms, and the diagram

$$(7.6)\qquad \begin{array}{ccccccc}
P_1\bar{\otimes}_{\mathscr{C}}N & \xrightarrow{F\bar{\otimes}_{\mathscr{C}}N} & P_0\bar{\otimes}_{\mathscr{C}}N & \xrightarrow{E\bar{\otimes}_{\mathscr{C}}N} & M\bar{\otimes}_{\mathscr{C}}N & \longrightarrow & 0\\
{\scriptstyle\phi_{P_1,N}}\downarrow\wr & & {\scriptstyle\phi_{P_0,N}}\downarrow\wr & & \downarrow{\scriptstyle\phi_{M,N}} & & \\
P_1\otimes_{\mathscr{C}}N & \xrightarrow[F\otimes_{\mathscr{C}}N]{} & P_0\otimes_{\mathscr{C}}N & \xrightarrow[E\otimes_{\mathscr{C}}N]{} & M\otimes_{\mathscr{C}}N & \longrightarrow & 0
\end{array}$$

is commutative, which has exact rows by the property (1)(a). Therefore, by the universality of cokernels and by the five lemma, we obtain an isomorphism $\phi_{M,N}\colon M\bar{\otimes}_{\mathscr{C}}N \to M \otimes_{\mathscr{C}} N$ that is natural in M. To show that it is also natural in N, let $f\colon N \to N'$ be a morphism in $\mathscr{C}$-Mod and consider the diagram

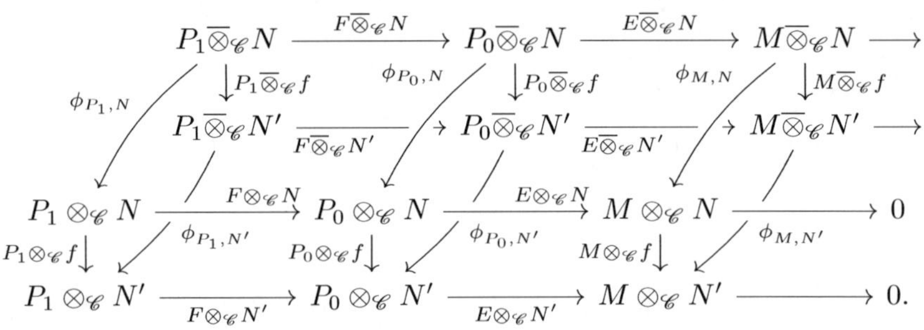

The front and back faces are commutative because $\text{-}\otimes_{\mathscr{C}}?$ and $\text{-}\bar{\otimes}_{\mathscr{C}}?$ are functors with two variables. The upper and lower surfaces are commutative from the beginning by the construction of $\phi_{M,N}$. The left face and the middle partition are commutative because of the naturality of property (1)(b) with respect to N. Therefore, the right face is also commutative. Thus $\phi_{M,N}$ is natural in N. As a result, $(\phi_{M,N})_{M,N}\colon \text{-}\bar{\otimes}_{\mathscr{C}}? \to \text{-}\otimes_{\mathscr{C}}?$ is an isomorphism of functors with two variables. $\square$

DEFINITION 7.1.4. A light category $\mathscr{C}$ is said to be *skeletally small* if the set of isoclasses of objects is a small set, i.e., if it has a small skeleton (see Definition 1.4.7).

EXAMPLE 7.1.5. If $\mathscr{C}$ is a small linear category, then the full subcategory $\operatorname{mod}\mathscr{C}$ of $\operatorname{Mod}\mathscr{C}$ consisting of finitely generated $\mathscr{C}$-modules is skeletally small.

REMARK 7.1.6. Let $\mathscr{C}$ be a skeletally small linear category, and $\mathscr{C}'$ a skeleton of $\mathscr{C}$. If we define a tensor product functor over $\mathscr{C}$ by replacing $\mathscr{C}_0$ with $\mathscr{C}'_0$ in the formula (7.1), then Theorem 7.1.3 above holds also in the case that $\mathscr{C}$ is a skeletally small category.

7.2. Tensor product functor by a bimodule

DEFINITION 7.2.1. Let $\mathscr{C}$ and $\mathscr{D}$ be small linear categories. Then a bilinear functor $M: \mathscr{D}^{\mathrm{op}} \times \mathscr{C} \to \operatorname{Mod} \Bbbk$ is called a *$\mathscr{C}$-$\mathscr{D}$-bimodule*[1]. The category formed by the $\mathscr{C}$-$\mathscr{D}$-bimodules and the bilinear functors between them is denoted by ${}_{\mathscr{C}}\mathrm{Mod}_{\mathscr{D}}$.

REMARK 7.2.2. Note that the positions of $\mathscr{C}$ and $\mathscr{D}$ in the category $\mathscr{D}^{\mathrm{op}} \times \mathscr{C}$ from which the bilinear functor starts and those in the expression "$\mathscr{C}$-$\mathscr{D}$-bimodule" are reversed. This was determined by taking the following points into account: left modules are covariant functors, right modules are contravariant functors, and the functor $\mathscr{C}(\text{-}, ?)$ is contravariant in the first variable, and is covariant in the second variable. To resolve this reversed position of the variables, we use the following notation.

NOTATION 7.2.3. Let $\mathscr{C}$ and $\mathscr{D}$ be small linear categories, and M a $\mathscr{C}$-$\mathscr{D}$-bimodule. For each $x \in \mathscr{C}_0 \cup \mathscr{C}_1, y \in \mathscr{D}_0 \cup \mathscr{D}_1$ we set ${}_xM_y := M(y,x), {}_xM := M(\text{-},x), M_y := M(y,\text{-})$.

LEMMA 7.2.4. *Let $\mathscr{C}$ and $\mathscr{D}$ be small linear categories, M a $\mathscr{C}$-$\mathscr{D}$-bimodule, $f: x \to x'$ a morphism in $\mathscr{C}$, and $g: y' \to y$ a morphism in $\mathscr{D}$. Then ${}_fM: {}_xM \to {}_{x'}M$ and $M_g: M_y \to M_{y'}$ are morphisms in* $\operatorname{Mod} \mathscr{D}$ *and* $\mathscr{C}$-Mod, *respectively.*

EXERCISE 7.2.5. Prove the lemma above.

EXAMPLE 7.2.6. Let $\mathscr{C}$ be a small linear category. Then $\mathscr{C}(\text{-}, ?): \mathscr{C}^{\mathrm{op}} \times \mathscr{C} \to \operatorname{Mod} \Bbbk$, $(y,x) \mapsto \mathscr{C}(y,x)$ $((y,x) \in (\mathscr{C}_0^{\mathrm{op}} \times \mathscr{C}_0) \cup (\mathscr{C}_1^{\mathrm{op}} \times \mathscr{C}_1))$ turns out to be a $\mathscr{C}$-$\mathscr{C}$-bimodule, which is denoted by ${}_{\mathscr{C}}\mathscr{C}_{\mathscr{C}}$ or simply by $\mathscr{C}$ if there seems to be no confusion. By applying the general notation above, we write ${}_x\mathscr{C}_y := \mathscr{C}(y,x)$, ${}_x\mathscr{C} := \mathscr{C}(\text{-},x)$, $\mathscr{C}_y := \mathscr{C}(y,\text{-})$ for all $x, y \in \mathscr{C}_0 \cup \mathscr{C}_1$.

Let $\mathscr{D}$ be another small linear category, and M a $\mathscr{C}$-$\mathscr{D}$-bimodule. Then a multilinear map

$$ {}_{x'}\mathscr{C}_x \times {}_xM_y \times {}_y\mathscr{D}_{y'} \to {}_{x'}M_{y'} $$

is defined by $(c, m, d) \mapsto cmd := M(d,c)(m)$ for all $x, x' \in \mathscr{C}_0, y, y' \in \mathscr{D}_0$.

In accordance with the notation for bimodules above, it is convenient to use the following notations for the left modules and the right modules.

NOTATION 7.2.7. Let $\mathscr{C}$ be a small linear category. Then we set

$$ {}_{\mathscr{C}}\mathrm{Mod} := \mathscr{C}\text{-Mod}, \mathrm{Mod}_{\mathscr{C}} := \mathrm{Mod}\,\mathscr{C}, \text{and} $$
$$ {}_{\mathscr{C}}\mathrm{mod} := \mathscr{C}\text{-mod}, \mathrm{mod}_{\mathscr{C}} := \mathrm{mod}\,\mathscr{C}. $$

Let $M \in (\mathrm{Mod}_{\mathscr{C}})_0, N \in ({}_{\mathscr{C}}\mathrm{Mod})_0$. Then we set

$$ M_x := M(x), M_f := M(f): M_y \to M_x, \text{and} $$
$$ {}_xN := N(x), {}_fN := N(f): {}_xN \to {}_yN. $$

for all $x, y \in \mathscr{C}_0, f \in \mathscr{C}(x,y)$.

---

[1] As in the case of the direct product of categories, if we define the linear category $\mathscr{D}^{\mathrm{op}} \otimes_{\Bbbk} \mathscr{C}$ by taking the tensor product over $\Bbbk$ instead of the direct product as each of its local morphism sets, then this can be viewed as a left $\mathscr{D}^{\mathrm{op}} \otimes_{\Bbbk} \mathscr{C}$-module.

DEFINITION 7.2.8. Let $\mathscr{C}$ and $\mathscr{D}$ be small linear categories, and N a $\mathscr{C}$-$\mathscr{D}$-bimodule. Then we define a linear functor

$$\text{-} \otimes_{\mathscr{C}} N \colon \operatorname{Mod}\mathscr{C} \to \operatorname{Mod}\mathscr{D}$$

as follows:

On objects: For each $M \in (\operatorname{Mod}_{\mathscr{C}})_0$, we set $(\text{-} \otimes_{\mathscr{C}} N)(M) := M \otimes_{\mathscr{C}} N \in (\operatorname{Mod}_{\mathscr{D}})_0$, where

$$\begin{aligned}&(M \otimes_{\mathscr{C}} N)(x) := M \otimes_{\mathscr{C}} N_x \ (x \in \mathscr{D}_0), \text{and}\\ &(M \otimes_{\mathscr{C}} N)(f) := M \otimes_{\mathscr{C}} N_f \colon M \otimes_{\mathscr{C}} N_x \to M \otimes_{\mathscr{C}} N_y, \ (f \in \mathscr{D}(y,x)).\end{aligned}$$

In the above note that $N_f \colon N_x \to N_y$ is a morphism in ${}_{\mathscr{C}}\operatorname{Mod}$ by Lemma 7.2.4.

On morphisms: For each morphism $F \colon M \to M'$ in $\operatorname{Mod}_{\mathscr{C}}$, we set

$$\begin{aligned}&(\text{-} \otimes_{\mathscr{C}} N)(F) := F \otimes_{\mathscr{C}} N \colon M \otimes_{\mathscr{C}} N \to M' \otimes_{\mathscr{C}} N, \text{where}\\ &(F \otimes_{\mathscr{C}} N)_x := F \otimes_{\mathscr{C}} N_x \colon M \otimes_{\mathscr{C}} N_x \to M' \otimes_{\mathscr{C}} N_x \ (x \in \mathscr{D}_0).\end{aligned}$$

REMARK 7.2.9. Let $\mathscr{C}$ and $\mathscr{D}$ be small linear categories. Then as a natural extension of the definition above, the bilinear functor

$$\text{-}\otimes_{\mathscr{C}}?\colon \operatorname{Mod}_{\mathscr{C}} \times {}_{\mathscr{C}}\operatorname{Mod}_{\mathscr{D}} \to \operatorname{Mod}_{\mathscr{D}}$$

is defined.

7.3. Hom **functor by a bimodule and its left adjoint**

DEFINITION 7.3.1. Let $\mathscr{C}$ and $\mathscr{D}$ be small linear categories, and N a $\mathscr{C}$-$\mathscr{D}$-bimodule. Then we define a linear functor

$$(\operatorname{Mod}\mathscr{D})(N,\text{-})\colon \operatorname{Mod}\mathscr{D} \to \operatorname{Mod}\mathscr{C}$$

as follows:

On objects: For each $L \in (\operatorname{Mod}\mathscr{D})_0$, we set

$$(\operatorname{Mod}\mathscr{D})(N,\text{-})(L) := (\operatorname{Mod}\mathscr{D})(N,L) \in (\operatorname{Mod}\mathscr{C})_0,$$

where

$$\begin{aligned}&(\operatorname{Mod}\mathscr{D})(N,L)(x) := (\operatorname{Mod}\mathscr{D})({}_xN, L) \ (x \in \mathscr{C}_0), \text{and}\\ &(\operatorname{Mod}\mathscr{D})(N,L)(f) := (\operatorname{Mod}\mathscr{D})({}_fN, L)\colon\\ &\qquad\qquad (\operatorname{Mod}\mathscr{D})({}_xN, L) \to (\operatorname{Mod}\mathscr{D})({}_yN, L), \ (f \in \mathscr{C}(y,x)).\end{aligned}$$

In the above, note that ${}_fN\colon {}_yN \to {}_xN$ is a morphism in $\operatorname{Mod}\mathscr{D}$ by Lemma 7.2.4.

On morphisms: For each morphism $F\colon M \to M'$ in $\operatorname{Mod}\mathscr{D}$, we set

$$\begin{aligned}&(\operatorname{Mod}\mathscr{D})(N,\text{-})(F) := (\operatorname{Mod}\mathscr{D})(N,F)\colon \operatorname{Mod}\mathscr{D}(N,M) \to \operatorname{Mod}\mathscr{D}(N,M'), \text{and}\\ &(\operatorname{Mod}\mathscr{D})(N,F)_x := (\operatorname{Mod}\mathscr{D})({}_xN,F)\colon\\ &\qquad\qquad \operatorname{Mod}\mathscr{D}({}_xN,M) \to \operatorname{Mod}\mathscr{D}({}_xN,M') \ (x \in \mathscr{C}_0).\end{aligned}$$

PROPOSITION 7.3.2. *Let $\mathscr{C}$ and $\mathscr{D}$ be small linear categories, and N a $\mathscr{C}$-$\mathscr{D}$-bimodule. Then we have the following adjoint:*

$$\operatorname{Mod}\mathscr{C} \underset{(\operatorname{Mod}\mathscr{D})(N,\text{-})}{\overset{\text{-}\otimes_{\mathscr{C}}N}{\rightleftarrows}} \operatorname{Mod}\mathscr{D}, \qquad \text{-}\otimes_{\mathscr{C}}N \dashv (\operatorname{Mod}\mathscr{D})(N,\text{-}).$$

PROOF. We define a unit $\eta\colon \mathbb{1}_{\operatorname{Mod}\mathscr{C}} \Rightarrow \operatorname{Mod}\mathscr{D}(N,\text{-}) \circ (\text{-}\otimes_{\mathscr{C}}N)$ and a counit $\varepsilon\colon (\text{-}\otimes_{\mathscr{C}}N) \circ \operatorname{Mod}\mathscr{D}(N,\text{-}) \Rightarrow \mathbb{1}_{\operatorname{Mod}\mathscr{D}}$ as follows:

Construction of η: For each $M \in (\operatorname{Mod}\mathscr{C})_0$ and $x \in \mathscr{C}_0$, we define

$$\eta_{M,x}\colon M_x \to \operatorname{Mod}\mathscr{D}({}_xN, M \otimes_{\mathscr{C}} N), m \mapsto m \otimes_{\mathscr{C}} \text{-}$$

and set $\eta_M := (\eta_{M,x})_{x\in\mathscr{C}_0}$. Then for each morphism $F\colon M \to M'$ in $\operatorname{Mod}\mathscr{C}$, it is easy to verify the commutativity of the diagram

$$\begin{array}{ccc} M_x & \xrightarrow{\ \eta_{M,x}\ } & \operatorname{Hom}_{\Bbbk}({}_xN, M \otimes_{\mathscr{C}} N) \\ {\scriptstyle F_x}\downarrow & & \downarrow{\scriptstyle \operatorname{Hom}_{\Bbbk}({}_xN, F\otimes_{\mathscr{C}} N)} \\ M'_x & \xrightarrow[\ \eta_{M',x}\ ]{} & \operatorname{Hom}_{\Bbbk}({}_xN, M' \otimes_{\mathscr{C}} N). \end{array}$$

Therefore, $\eta := (\eta_M)_{M\in(\operatorname{Mod}\mathscr{C})_0}$ turns out to be a natural transformation

$$\mathbb{1}_{\operatorname{Mod}\mathscr{C}} \Rightarrow \operatorname{Mod}\mathscr{D}(N, \text{-}) \circ (\text{-} \otimes_{\mathscr{C}} N).$$

Construction of ε: For each $V \in (\operatorname{Mod}\mathscr{D})_0$, we define a morphism

$$\varepsilon'_V\colon \bigoplus_{x\in\mathscr{C}_0} \operatorname{Mod}\mathscr{D}({}_xN, V) \otimes_{\Bbbk} {}_xN \to V$$

in $\operatorname{Mod}\mathscr{D}$ by $(u_x \otimes n_x)_{x\in\mathscr{C}_0} \mapsto \sum_{x\in\mathscr{C}_0} u_x(n_x)$ $(u_x \in \operatorname{Mod}\mathscr{D}({}_xN, V), n_x \in {}_xN)$. Then it is easy to see that $\varepsilon'_V(I(\operatorname{Mod}\mathscr{D}(N,V), N)) = 0$. Therefore, ε'_V induces a morphism

$$\varepsilon_V\colon \operatorname{Mod}\mathscr{D}(N, V) \otimes_{\mathscr{C}} N \to V$$

in $\operatorname{Mod}\mathscr{D}$. Then, it is immediate from definition that $\varepsilon := (\varepsilon_V)_{V\in(\operatorname{Mod}\mathscr{D})_0}$ turns out to be a natural transformation

$$(\text{-} \otimes_{\mathscr{C}} N) \circ \operatorname{Mod}\mathscr{D}(N, \text{-}) \Rightarrow \mathbb{1}_{\operatorname{Mod}\mathscr{D}}.$$

Here the natural transformations η, ε above satisfy the zigzag equations, i.e., both of the diagrams

$$\begin{array}{ccc} \text{-} \otimes_{\mathscr{C}} N & & \\ {\scriptstyle \eta(\text{-})\otimes_{\mathscr{C}} N}\Downarrow & \searrow\!\!\!\!= & \\ \operatorname{Mod}\mathscr{D}(N, \text{-} \otimes_{\mathscr{C}} N) \otimes_{\mathscr{C}} N & \xRightarrow[\ \varepsilon_{\text{-}\otimes_{\mathscr{C}} N}\ ]{} & \text{-} \otimes_{\mathscr{C}} N \end{array} \quad \text{, and}$$

$$\begin{array}{ccc} \operatorname{Mod}\mathscr{D}(N, \text{-}) & & \\ {\scriptstyle \eta_{\operatorname{Mod}\mathscr{D}(N,\text{-})}}\Downarrow & \searrow\!\!\!\!= & \\ \operatorname{Mod}\mathscr{D}(N, \operatorname{Hom}_{\Bbbk}(N, \text{-}) \otimes_{\mathscr{C}} N) & \xRightarrow[\ \operatorname{Mod}\mathscr{D}(N,\varepsilon)\ ]{} & \operatorname{Mod}\mathscr{D}(N, \text{-}) \end{array}$$

are commutative. Indeed, this follows from the fact that for each $u \in \mathscr{D}_0, M \in (\operatorname{Mod}\mathscr{C})_0$ and $x \in \mathscr{C}_0, V \in \operatorname{Mod}\mathscr{D}$, both of the following diagrams turn out to be commutative in $\operatorname{Mod}\Bbbk$ by the commutativity of diagram (7.2):

$$\begin{array}{ccc} M \otimes_{\mathscr{C}} N_u & & \\ {\scriptstyle \eta_{M\otimes_{\mathscr{C}} N_u}}\downarrow & & \\ \operatorname{Mod}\mathscr{D}(N, M \otimes_{\mathscr{C}} N) \otimes_{\mathscr{C}} N_u & \xrightarrow[\varepsilon_{M\otimes_{\mathscr{C}} N_u}]{} & M \otimes_{\mathscr{C}} N_u \end{array}, \text{ and}$$

$$\begin{array}{ccc} \operatorname{Mod}\mathscr{D}({}_xN, V) & & \\ {\scriptstyle \eta_{\operatorname{Mod}\mathscr{D}({}_xN,V)}}\downarrow & & \\ \operatorname{Mod}\mathscr{D}({}_xN, \operatorname{Mod}\mathscr{D}(N,V) \otimes_{\mathscr{C}} N) & \xrightarrow[\operatorname{Mod}\mathscr{D}({}_xN,\varepsilon)]{} & \operatorname{Mod}\mathscr{D}({}_xN, V). \end{array}$$

□

COROLLARY 7.3.3. *Let $\mathscr{C}$ and $\mathscr{D}$ be small linear categories, and N a $\mathscr{C}$-$\mathscr{D}$-bimodule. Then there exists an isomorphism*

$$\mathscr{C}(\text{-}, x) \otimes_{\mathscr{C}} N \cong {}_xN$$

in $\operatorname{Mod}\mathscr{D}$ *for all* $x \in \mathscr{C}$.

PROOF. By Proposition 7.3.2 above, there exist isomorphisms obtained from (7.4) in the proof of Theorem 7.1.3 by replacing $\operatorname{Hom}_\Bbbk$ with $\operatorname{Mod}\mathscr{D}$, which clearly shows the assertion. It is also possible to prove it by using the natural isomorphism in Theorem 7.1.3 (1)(b). □

LEMMA 7.3.4. *Let $\mathscr{C}$ and $\mathscr{D}$ be small linear categories, and $F\colon \mathscr{C} \to \mathscr{D}$ a linear functor. Then ${}_F\mathscr{D}_{\mathscr{D}} := \mathscr{D}(\text{-}, F(?))\colon \mathscr{D}^{\mathrm{op}} \times \mathscr{C} \to \operatorname{Mod}\Bbbk$ is a $\mathscr{C}$-$\mathscr{D}$-bimodule, and the functor $F^{\boldsymbol{\cdot}} := \operatorname{Mod}F\colon \operatorname{Mod}\mathscr{D} \to \operatorname{Mod}\mathscr{C}$ is isomorphic to the functor $(\operatorname{Mod}\mathscr{D})({}_F\mathscr{D}_{\mathscr{D}}, \text{-})$. Therefore, by Proposition 7.3.2, the functor $\text{-}\otimes_{\mathscr{C}} {}_F\mathscr{D}_{\mathscr{D}}\colon \operatorname{Mod}\mathscr{C} \to \operatorname{Mod}\mathscr{D}$ turns out to be a left adjoint to $F^{\boldsymbol{\cdot}}$.*

PROOF. It is obvious that ${}_F\mathscr{D}_{\mathscr{D}}$ turns out to be a $\mathscr{C}$-$\mathscr{D}$-bimodule. Then by the Yoneda lemma, we have an isomorphism

$$(\operatorname{Mod}\mathscr{D})({}_F\mathscr{D}_{\mathscr{D}}, M) = (\operatorname{Mod}\mathscr{D})(\mathscr{D}(\text{-}, F(?)), M) \cong M(F(?)) = F^{\boldsymbol{\cdot}}(M)$$

that is natural in $M \in (\operatorname{Mod}\mathscr{D})_0$. □

7.4. The pullup functor and the pushdown functor

Throughout this section, let $\mathscr{C}$ be a small linear category, $(\mathscr{C}, X)$ a G-category, and $(P, \phi)\colon \mathscr{C} \to \mathscr{C}/G$ the canonical covering functor. We refer the reader to Definition 3.2.1 for the definitions of modules over $\mathscr{C}$ and the category $\operatorname{Mod}\mathscr{C}$, and to Section 2.5 for finitely generated projective $\mathscr{C}$-modules. After reviewing the definitions of pullup and pushdown functors, we give a concrete form for the pushdown functor.

DEFINITION 7.4.1.

(1) By applying the 2-functor $\operatorname{Mod}(\text{-})\colon \Bbbk\text{-}\mathbf{Cat} \to \Bbbk\text{-}\mathbf{CAT}^{\mathrm{coop}}$ defined in Example 4.1.13 to $P\colon \mathscr{C} \to \mathscr{C}/G$, we define a linear functor

$$P^{\boldsymbol{\cdot}} := \operatorname{Mod}P\colon \operatorname{Mod}\mathscr{C}/G \to \operatorname{Mod}\mathscr{C},$$

which is called the *pullup* of P. By Lemma 7.3.4, $P^{\bullet}$ is isomorphic to the functor $(\operatorname{Mod}\mathscr{C}/G)({}_P(\mathscr{C}/G)_{\mathscr{C}/G}, \text{-})$, and therefore, the functor

$$P_{\bullet} := \text{-} \otimes_{\mathscr{C}} {}_P(\mathscr{C}/G)_{\mathscr{C}/G} \colon \operatorname{Mod}\mathscr{C} \to \operatorname{Mod}\mathscr{C}/G$$

turns out to be a left adjoint to $P^{\bullet}$, which is called the *pushdown* of P.

(2) We define a G-action $\operatorname{Mod}X \colon G \to \operatorname{Aut}(\operatorname{Mod}\mathscr{C})$ on $\operatorname{Mod}\mathscr{C}$ as follows: For each $a \in G$, consider the functor $X(a^{-1}) \colon \mathscr{C} \to \mathscr{C}$. By applying the 2-functor $\operatorname{Mod}(\text{-})$ to it, we set

$${}^{a}(\text{-}) := (\operatorname{Mod}X)(a) := \operatorname{Mod}(X(a^{-1})) \colon \operatorname{Mod}\mathscr{C} \to \operatorname{Mod}\mathscr{C}.$$

Thus for each $M \in (\operatorname{Mod}\mathscr{C})_0 \cup (\operatorname{Mod}\mathscr{C})_1$, we set

$$(\operatorname{Mod}X)(a)M := {}^{a}M := M \circ X(a^{-1}).$$

Using this, we define the G-category $\operatorname{Mod}(\mathscr{C}, X) := (\operatorname{Mod}\mathscr{C}, \operatorname{Mod}X)$.

REMARK 7.4.2. In the definition above, we have the following remarks.

(1) By the adjoint $P_{\bullet} \dashv P^{\bullet}$, the functor $P^{\bullet}$ is small continuous, in particular, left exact, and the functor $P_{\bullet}$ is small cocontinuous, in particular right exact (in fact, these are exact functors because ${}_P(\mathscr{C}/G)$ is projective both as a left $\mathscr{C}$-module and as a right $\mathscr{C}/G$-module). By applying Corollary 7.3.3 to $N = {}_P(\mathscr{C}/G)$, we see that for each $x \in \mathscr{C}_0$ there exists an isomorphism $P_{\bullet}\mathscr{C}(\text{-}, x) \cong \mathscr{C}/G(\text{-}, Px)$. This, together with the right exactness of $P_{\bullet}$, shows that the functor $P_{\bullet}$ induces the functor $P_{\bullet} \colon \operatorname{mod}\mathscr{C} \to \operatorname{mod}\mathscr{C}/G$.

(2) Let $a \in G$. Then we have ${}^{a}\mathscr{C}(\text{-}, x) = \mathscr{C}(X(a^{-1})(\text{-}), x) \cong \mathscr{C}(\text{-}, X(a)x)$ for all $x \in \mathscr{C}_0$. Therefore by noting that the functor ${}^{a}(\text{-}) \colon \operatorname{Mod}\mathscr{C} \to \operatorname{Mod}\mathscr{C}$ is right exact as a left adjoint to ${}^{a^{-1}}(\text{-})$, we see that ${}^{a}(\text{-})$ sends finitely generated $\mathscr{C}$-modules to themselves. Thus it induces an isomorphism $(\operatorname{mod}X)(a) \colon \operatorname{mod}\mathscr{C} \to \operatorname{mod}\mathscr{C}$, by which a G-category $\operatorname{mod}(\mathscr{C}, X) := (\operatorname{mod}\mathscr{C}, \operatorname{mod}X)$ is defined.

DEFINITION 7.4.3. Let $\mathscr{C}$ and $\mathscr{D}$ be small linear categories, and $F \colon \operatorname{Mod}\mathscr{C} \to \operatorname{Mod}\mathscr{D}$ a linear functor. Then F is said to *preserve locally column-finite families* if for each small set I and locally column-finite family $(L \xrightarrow{p_i} L_i)_{i \in I}$ in $\operatorname{Mod}\mathscr{C}$, the family $(FL \xrightarrow{Fp_i} FL_i)_{i \in I}$ is locally column-finite in $\operatorname{Mod}\mathscr{D}$.

LEMMA 7.4.4. *Let $\mathscr{C}$ and $\mathscr{D}$ be small linear categories. If a linear functor $F \colon \operatorname{Mod}\mathscr{C} \to \operatorname{Mod}\mathscr{D}$ is small cocontinuous, then F preserves locally column-finite families.*

PROOF. Let I be a small set, $(p_i \colon M \to M_i)_{i \in I}$ a locally column-finite family in $\operatorname{Mod}\mathscr{C}$, $(\pi_i \colon \prod_{j \in I} M_j \to M_i)_{i \in I}$ the canonical projection family of the product, and $(\sigma_i \colon M_i \to \coprod_{j \in I} M_j)_{i \in I}$ the canonical injection family of the coproduct. Furthermore, let $\sigma \colon \coprod_{i \in I} M_i \hookrightarrow \prod_{i \in I} M_i$ be the inclusion, and $p \colon M \to \prod_{i \in I} M_i$ be the unique morphism such that $p_i = \pi_i \circ p$ $(i \in I)$ (i.e., $p := [p_i)_{i \in I}$). Then since the family $(p_i)_{i \in I}$ is locally column-finite, there exists a morphism $p' \colon M \to \coprod_{i \in I} M_i$ such that $p = \sigma \circ p'$. Therefore, we obtain the following commutative diagram:

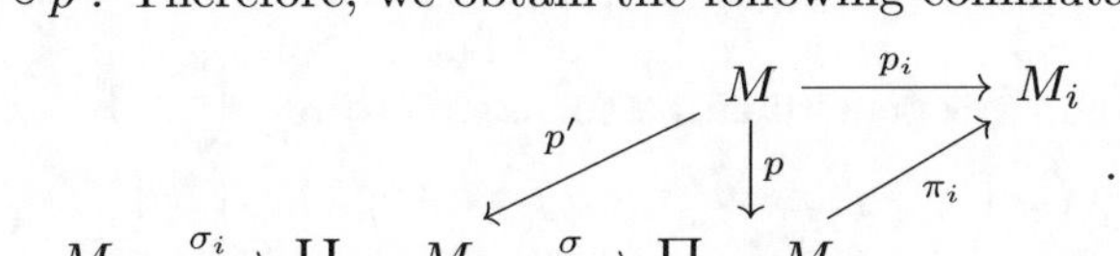

Send it by F, and consider the following diagram:

$$(7.7)\qquad \begin{array}{ccccccc} & & & & FM & \xrightarrow{Fp_i} & FM_i \\ & & & \swarrow^{Fp'} & \downarrow{\scriptstyle Fp} & \nearrow_{F\pi_i} & \\ FM_i & \xrightarrow{F\sigma_i} & F(\coprod_{i\in I} M_i) & \xrightarrow{F\sigma} & F(\prod_{i\in I} M_i) & & \\ & \searrow_{\overline{\sigma_i}} & \wr\uparrow{\scriptstyle u} & & \downarrow{\scriptstyle v} & \nearrow_{\overline{\pi_i}} & \\ & & \coprod_{i\in I} FM_i & \xrightarrow[\overline{\sigma}]{} & \prod_{i\in I} FM_i & & \end{array},$$

where $(\overline{\sigma_i}\colon FM_i \to \coprod_{j\in I} FM_j)_{i\in I}$ is the canonical injection family of direct sum, $(\overline{\pi_i}\colon \prod_{j\in I} FM_j \to FM_i)_{i\in I}$ is the canonical projection family of product, and $\overline{\sigma}\colon \coprod_{i\in I} FM_i \to \prod_{i\in I} FM_i$ is the inclusion; and v is the unique morphism satisfying $F\pi_i = \overline{\pi_i}\circ v$ $(i \in I)$ (i.e., $v := [F\pi_i]_{i\in I}$), and u is the unique morphism satisfying $F\sigma_i = u\circ\overline{\sigma_i}$ $(i\in I)$ (i.e., $u := {}^t[F\pi_i]_{i\in I}$). Here, since F is small cocontinuous, u is an isomorphism. Note that all of the four triangular regions in the diagram (7.7) are commutative by construction. Then, since we have $Fp_i = \overline{\pi_i}\circ(v\circ Fp)$ $(i\in I)$ (i.e., $v\circ Fp = [Fp_i]_{i\in I}$), it suffices to show that $v\circ Fp$ factors through $\overline{\sigma}$ to complete the proof. To this end it is enough to show the commutativity of the rectangular region, i.e., the equality

$$v\circ F\sigma\circ u = \overline{\sigma}. \tag{7.8}$$

Indeed, if this is shown, then the diagram (7.7) becomes commutative, and we have $v\circ Fp = \overline{\sigma}\circ(u^{-1}\circ Fp')$ as desired. Now by Proposition 1.1.27, to show the equality (7.8), it is enough to show that both sides have the same matrix expression with respect to $(\overline{\sigma_i})_{i\in I}, (\overline{\pi_i})_{i\in I}$. Note that the matrix expression of $\overline{\sigma}$ (resp. of σ with respect to $(\sigma_i)_{i\in I}, (\pi_i)_{i\in I}$) is given by $[\delta_{j,i}\mathbb{1}_{FM_i}]_{j,i\in I}$ (resp. $[\delta_{j.i}\mathbb{1}_{M_i}]_{j,i\in I}$). Then the commutativity of the triangular regions in the diagram (7.7) shows that for each $i, j\in I$, the (j,i)-entry of the matrix expression of the left-hand side is given by

$$\overline{\pi_j}\circ v\circ F\sigma\circ u\circ\overline{\sigma_i} = F(\pi_j\circ\sigma\circ\sigma_i) = F(\delta_{i,j}\mathbb{1}_{M_i}) = \delta_{i,j}\mathbb{1}_{FM_i},$$

which coincides with the (j,i)-entry of $\overline{\sigma}$. Therefore the equality (7.8) holds. □

PROPOSITION 7.4.5. *Let $\mathscr{D}$ be a small linear category, I, J small sets, and $F\colon \operatorname{Mod}\mathscr{C} \to \operatorname{Mod}\mathscr{D}$ a linear functor preserving locally column-finite families. Assume that both $(L_i \xrightarrow{q_i} L \xrightarrow{p_i} L_i)_{i\in I}$ and $(M_j \xrightarrow{s_j} M \xrightarrow{r_j} M_j)_{j\in J}$ are direct sum systems in $\operatorname{Mod}\mathscr{C}$, and that a morphism $f\colon L\to M$ in $\operatorname{Mod}\mathscr{C}$ has a matrix expression $f = [f_{j,i}]_{(j,i)\in J\times I}$ with respect to these. Then both $(FL_i \xrightarrow{Fq_i} FL \xrightarrow{Fp_i} FL_i)_{i\in I}$ and $(FM_j \xrightarrow{Fs_j} FM \xrightarrow{Fr_j} FM_j)_{j\in J}$ are direct sum systems in $\operatorname{Mod}\mathscr{D}$, and the matrix expression of $Ff\colon FL\to FM$ with respect to these is given by $Ff = [Ff_{j,i}]_{(j,i)\in J\times I}$.*

PROOF. It is clear that both $(FL_i \xrightarrow{Fq_i} FL \xrightarrow{Fp_i} FL_i)_{i\in I}$ and $(FM_j \xrightarrow{Fs_j} FM \xrightarrow{Fr_j} FM_j)_{j\in J}$ are direct sum systems in $\operatorname{Mod}\mathscr{D}$ by definition and by the fact that both families $(Fp_i)_{i\in I}$ and $(Fr_j)_{j\in J}$ are locally column finite. Then the matrix expression of Ff with respect to these turns out to be

$$[Fr_j\circ Ff\circ Fq_i]_{(j,i)\in J\times I} = [F(r_j\circ f\circ q_i)]_{(j,i)\in J\times I} = [Ff_{j,i}]_{(j,i)\in J\times I}.$$

□

LEMMA 7.4.6 (Matrix expression of right composition in orbit category). *If $f \in \mathscr{C}/G(x,y)$, then the following diagram is commutative:*

$$\begin{array}{ccc}
\mathscr{C}/G(y,P(\text{-})) & \xrightarrow{\mathscr{C}/G(f,P(\text{-}))} & \mathscr{C}/G(x,P(\text{-})) \\
\| & & \| \\
\coprod_{b\in G}\mathscr{C}(X(b)y,\text{-}) & \xrightarrow{[\mathscr{C}(X(b)f_{b^{-1}a},\text{-})]_{a,b\in G}} & \coprod_{a\in G}\mathscr{C}(X(a)x,\text{-}).
\end{array}$$

PROOF. For each $z \in (\mathscr{C}/G)_0 = \mathscr{C}_0$ and $g = (g_b)_{b\in G} \in \mathscr{C}/G(y,z)$, we have

$$(\mathscr{C}/G)(f,P(z))(g) = g\circ f = \left(\sum_{\substack{b,c\in G\\ bc=a}} g_b \circ X(b)f_c\right)_{a\in G}$$

$$= \left(\sum_{b\in G} g_b \circ X(b)f_{b^{-1}a}\right)_{a\in G} = \left(\sum_{b\in G} \mathscr{C}(X(b)f_{b^{-1}a}, z)(g_b)\right)_{a\in G}$$

$$= [\mathscr{C}(X(b)f_{b^{-1}a}, z)]_{a,b\in G}(g).$$

For the last equality see Lemma 2.4.9 (1). □

DEFINITION 7.4.7. Here we give an alternative concrete form for the pushdown.

(1) We define a (covariant) linear functor $P_*\colon \operatorname{Mod}\mathscr{C} \to \operatorname{Mod}\mathscr{C}/G$ as follows:

(a) If $M \in (\operatorname{Mod}\mathscr{C})_0$, then we define a right $\mathscr{C}/G$-module $P_*(M)$, which is a contravariant functor $\mathscr{C}/G \to \operatorname{Mod}\Bbbk$, by the commutativity of the following diagram for all morphisms $f\colon x\to y$ in $\mathscr{C}/G$:

$$\begin{array}{ccc}
(P_*M)(y) & \xrightarrow{(P_*M)(f)} & (P_*M)(x) \\
\| & & \| \\
\coprod_{b\in G} M(X(b)y) & \xrightarrow[{[M(X(b)f_{b^{-1}a})]_{a,b\in G}}]{} & \coprod_{a\in G} M(X(a)x)
\end{array}.$$

(b) If $u\colon M \to M'$ is a morphism in $\operatorname{Mod}\mathscr{C}$, then we define $(P_*u)_x$ by the commutativity of the following diagram for all $x \in \mathscr{C}_0$, and set $P_*u := ((P_*u)_x)_{x\in(\mathscr{C}/G)_0}$:

$$\begin{array}{ccc}
(P_*M)(x) & \xrightarrow{(P_*u)_x} & (P_*M')(x) \\
\| & & \| \\
\coprod_{a\in G} M(X(a)x) & \xrightarrow[\coprod_{a\in G} u_{X(a)x}]{} & \coprod_{a\in G} M'(X(a)x)
\end{array}. \tag{7.9}$$

(2) Next, for each $M \in (\operatorname{Mod}\mathscr{C})_0$ and $x \in (\mathscr{C}/G)_0 = \mathscr{C}_0$, we define a morphism $\xi_{M,x}\colon (P_{\bullet}M)(x) \to (P_*M)(x)$ in $\operatorname{Mod}\Bbbk$ by the composite of the following natural isomorphisms, and set $\xi_M := (\xi_{M,x})_{x\in(\mathscr{C}/G)_0}$ and $\xi := (\xi_M)_{M\in(\operatorname{Mod}\mathscr{C})_0}$:

$$(P_{\bullet}M)(x) = M\otimes_{\mathscr{C}} (\mathscr{C}/G)(x,P(?)) = M\otimes_{\mathscr{C}} \coprod_{a\in G}\mathscr{C}(X(a)x,?)$$

$$\xrightarrow{\sim} \coprod_{a\in G} M\otimes_{\mathscr{C}}\mathscr{C}(X(a)x,?) \xrightarrow{\coprod_{a\in G} s_{X(a)x,M}} \coprod_{a\in G} M(X(a)x) = (P_*)(x).$$

PROPOSITION 7.4.8. *The family ξ defined above turns out to be a natural isomorphism $P_{\bullet} \to P_*$.*

PROOF. (1) For each $M \in (\operatorname{Mod}\mathscr{C})_0$, we show that $\xi_M \colon P_{\bullet}M \to P_*M$ is an isomorphism in $\operatorname{Mod}\mathscr{C}/G$. To this end, it is enough to show that the following diagram is commutative for all morphisms $f \colon x \to y$ in $\mathscr{C}/G$.

$$\begin{array}{ccc}
M \otimes_{\mathscr{C}} (\mathscr{C}/G)(y, P(?)) & \xrightarrow{M\otimes_{\mathscr{C}}(\mathscr{C}/G))(f,P(?))} & M \otimes_{\mathscr{C}} (\mathscr{C}/G)(x, P(?)) \\
\| & & \| \\
M \otimes_{\mathscr{C}} \coprod_{b\in G} \mathscr{C}(X(b)y, ?) & \xrightarrow{M\otimes_{\mathscr{C}}[\mathscr{C}(X(b)f_{b^{-1}a},?)]_{a,b\in G}} & M \otimes_{\mathscr{C}} \coprod_{a\in G} \mathscr{C}(X(a)x, ?) \\
\wr\downarrow & & \downarrow\wr \\
\coprod_{b\in G} M \otimes_{\mathscr{C}} \mathscr{C}(X(b)y, ?) & \xrightarrow{[M\otimes_{\mathscr{C}}\mathscr{C}(X(b)f_{b^{-1}a},?)]_{a,b\in G}} & \coprod_{a\in G} M \otimes_{\mathscr{C}} \mathscr{C}(X(a)x, ?) \\
\wr\downarrow & & \downarrow\wr \\
\coprod_{b\in G} M(X(b)y) & \xrightarrow[{[M(X(b)f_{b^{-1}a})]_{a,b\in G}}]{} & \coprod_{a\in G} M(X(a)x).
\end{array}$$

Now, the commutativity of the upper rectangle follows by Lemma 7.4.6. That of the middle one follows by Lemma 7.4.4 and Proposition 7.4.5. Finally, that of the lower one follows by Theorem 7.1.3 (2)(b).

(2) The naturally of ξ follows from the commutativity of the following diagram for all morphisms $u \colon M \to M'$ in $\operatorname{Mod}\mathscr{C}$ and $x \in \mathscr{C}_0$:

$$\begin{array}{ccc}
M \otimes_{\mathscr{C}} (\mathscr{C}/G)(x, P(?)) & \xrightarrow{u\otimes_{\mathscr{C}}(\mathscr{C}/G)(x,P(?))} & M' \otimes_{\mathscr{C}} (\mathscr{C}/G)(x, P(?)) \\
\| & & \| \\
M \otimes_{\mathscr{C}} \coprod_{a\in G} \mathscr{C}(X(a)x, ?) & \xrightarrow{u\otimes_{\mathscr{C}}\coprod_{a\in G}\mathscr{C}(X(a)x,?)} & M' \otimes_{\mathscr{C}} \coprod_{a\in G} \mathscr{C}(X(a)x, ?) \\
\wr\downarrow & & \downarrow\wr \\
\coprod_{a\in G} M \otimes_{\mathscr{C}} \mathscr{C}(X(a)x, ?) & \xrightarrow{\coprod_{a\in G} u\otimes_{\mathscr{C}}\mathscr{C}(X(a)x,?)} & \coprod_{a\in G} M' \otimes_{\mathscr{C}} \mathscr{C}(X(a)x, ?) \\
\wr\downarrow & & \downarrow\wr \\
\coprod_{a\in G} M(X(a)x) & \xrightarrow{\coprod_{a\in G} u_{X(a)x}} & \coprod_{a\in G} M'(X(a)x).
\end{array}$$

□

By the proposition above, also P_* becomes a left adjoint to $P^{\bullet}$. Through the natural isomorphism ξ, we can compute $P_{\bullet}$ as P_*.

7.4.a. Precovering between categories of finitely generated modules.

LEMMA 7.4.9. *It holds that $P^{\bullet} \circ P_* = \coprod_{a\in G} {}^a(\text{-})$. Therefore,*

$$P^{\bullet} \circ \xi \colon P^{\bullet} \circ P_{\bullet} \overset{\sim}{\Rightarrow} \coprod_{a\in G} {}^a(\text{-})$$

is a natural isomorphism.

PROOF. Let $f \in \mathscr{C}(x, y)$. Then by noting that $Pf = (\delta_{a,1}f)_{a\in G} \colon Px \to Py$, we obtain the following commutative diagram for all $M \in (\operatorname{Mod}\mathscr{C})_0$, which finishes the proof:

$$
\begin{array}{ccc}
(P^{\cdot}P_{*}M)(y) & \xrightarrow{\qquad (P^{\cdot}P_{*}M)(f) \qquad} & (P^{\cdot}P_{*}M)(x) \\
\Vert & & \Vert \\
(P_{*}M)(Py) & \xrightarrow{\qquad (P_{*}M)(Pf) \qquad} & (P_{*}M)(Px) \\
\Vert & & \Vert \\
\coprod_{b\in G} M(X(b)y) & \xrightarrow{[M(X(b)\delta_{b^{-1}a,1}f)]_{a,b\in G}} & \coprod_{a\in G} M(X(a)x) \\
\Vert & & \Vert \\
\coprod_{b\in G} M(X(b)y) & \xrightarrow{[\delta_{a,b}M(X(a)f)]_{a,b\in G}} & \coprod_{a\in G} M(X(a)x) \\
\Vert & & \Vert \\
\coprod_{a\in G} M(X(a)y) & \xrightarrow{\coprod_{a\in G} M(X(a)f)} & \coprod_{a\in G} M(X(a)x) \\
\Vert & & \Vert \\
\coprod_{a\in G} {}^{a^{-1}}M(y) & \xrightarrow{\coprod_{a\in G} {}^{a^{-1}}M(f)} & \coprod_{a\in G} {}^{a^{-1}}M(x) \\
\Vert & & \Vert \\
\coprod_{a\in G} {}^{a}M(y) & \xrightarrow{\qquad \coprod_{a\in G} {}^{a}M(f) \qquad} & \coprod_{a\in G} {}^{a}M(x).
\end{array}
$$

□

From now on, $P_{\cdot}$ and P_{*} are identified by ξ.

LEMMA 7.4.10. *Let $M \in (\operatorname{Mod}\mathscr{C})_0$, and assume that M is finitely generated. Then M is compact.*

PROOF. Let I be a small set, $M_i \in (\operatorname{Mod}\mathscr{C})_0$ for all $i \in I$, and $(\coprod_{i\in I} M_i, (\sigma_i)_{i\in I})$ their direct sum. Then it suffices to show that the canonical morphism

$$\phi\colon \coprod_{i\in I}(\operatorname{Mod}\mathscr{C})(M, M_i) \to (\operatorname{Mod}\mathscr{C})(M, \coprod_{i\in I} M_i)$$

with respect to the canonical injection family

$$(\overline{\sigma_i}\colon (\operatorname{Mod}\mathscr{C})(M, M_i) \to \coprod_{i\in I}(\operatorname{Mod}\mathscr{C})(M, M_i))_{i\in I}$$

is an isomorphism. In general, we already know that ϕ is a monomorphism by Lemma 1.5.17. Assume now that M is finitely generated. Then it is enough to show that ϕ is an epimorphism. By definition, there exist some $1 \leq n \in \mathbb{N}$, $x_1, \ldots, x_n \in \mathscr{C}_0$, and $v_1 \in M(x_1), \ldots, v_n \in M(x_n)$ such that $M = \langle v_1, \ldots, v_n\rangle$. To show that ϕ is an epimorphism, take any $f \in (\operatorname{Mod}\mathscr{C})(M, \coprod_{i\in I} M_i)$. For each $1 \leq j \leq n$, since f_{x_j} is a morphism $M(x_j) \to \coprod_{i\in I} M_i(x_j)$, we have $f_{x_j}(v_j) \in \coprod_{i\in I_j} M_i$ for some finite subset I_j of I, where we regard $\coprod_{i\in I_j} M_i$ as a subset of $\coprod_{i\in I} M_i$ by a natural embedding. Therefore, $f(\langle v_j\rangle) \subseteq \langle f_{x_j}(v_j)\rangle \subseteq \coprod_{i\in I_j} M_i$. Hence if we set $I_f := \bigcup_{j=1}^{n} I_j$, then I_f is a finite subset of I, and we have $\operatorname{Im} f = \sum_{j=1}^{n} f(\langle v_j\rangle) \subseteq \coprod_{i\in I_f} M_i$. Thus f can be written as the composite $f = \sigma_f \circ f'$ for some $f'\colon M \to \coprod_{i\in I_f} M_i$ and the inclusion $\sigma_f\colon \coprod_{i\in I_f} M_i \hookrightarrow \coprod_{i\in I} M_i$. Since I_f is finite, Corollary 1.5.15 gives us an isomorphism

$$\coprod_{i\in I_f}(\operatorname{Mod}\mathscr{C})(M,M_i)\to(\operatorname{Mod}\mathscr{C})(M,\coprod_{i\in I_f}M_i),\ (f_i)_{i\in I_f}^n\mapsto\sum_{i\in I_f}\sigma_i\circ f_i.$$

Therefore there exists some $(f_i)_{i\in I_f}\in\prod_{i\in I_f}(\operatorname{Mod}\mathscr{C})(M,M_i)$ such that $f'=\sum_{i\in I_f}\sigma_i\circ f_i$. Set here $f_i:=0$ for all $i\in I\setminus I_f$. Then we have $\phi((f_i)_{i\in I})=\sum_{i\in I}\sigma_i\circ f_i=\sigma_f\circ\sum_{i\in I_f}\sigma_i\circ f_i=\sigma_f\circ f'=f$. Hence ϕ is an epimorphism. □

THEOREM 7.4.11. *The pushdown functor $P_{\bullet}$ can be extended to a G-precovering $(P_{\bullet},\phi_{\bullet})\colon \operatorname{mod}(\mathscr{C},X)\to\operatorname{mod}\mathscr{C}/G$.*

PROOF. We first define $\phi_{\bullet}$ as follows: For each $c\in G$, we define $\phi_{\bullet c}\colon P_{\bullet}\Rightarrow P_{\bullet}\circ{}^c(\text{-})$ by $\phi_{\bullet c}:=(\phi_{\bullet c,M})_{M\in(\operatorname{Mod}\mathscr{C})_0}$, where for each $M\in\operatorname{Mod}\mathscr{C}$ and $x\in\mathscr{C}_0$, $\phi_{\bullet c,M,x}$ is defined by the commutative diagram

$$\begin{array}{ccc}(P_{\bullet}M)(x) & \xrightarrow{\phi_{\bullet c,M,x}} & (P_{\bullet}({}^cM))(x)\\ \| & & \| \\ \coprod_{a\in G}M(X(a)x) & \xrightarrow[{[\delta_{a,c^{-1}b}\mathbf{1}_{M(X(a)x)}]_{b,a\in G}}]{} & \coprod_{b\in G}M(X(c^{-1}b)x)\end{array},$$

and set $\phi_{\bullet c,M}:=(\phi_{\bullet c,M,x})_{x\in\mathscr{C}}$.

CLAIM 1. *$\phi_{\bullet c}$ ($c\in G$) is a natural isomorphism.*

Indeed, by definition, it is immediate that we have

$$\phi_{\bullet c,M,x}((m_a)_{a\in G})=(m_{c^{-1}a})_{a\in G}\quad((m_a)_{a\in G}\in(P_{\bullet}M)(x))\tag{7.10}$$

for all $M\in(\operatorname{Mod}\mathscr{C})_0$ and $x\in\mathscr{C}_0$. Using this, it is easy to verify that for each morphism $u\colon M\to N$ in $\operatorname{Mod}\mathscr{C}$, the diagram

$$\begin{array}{ccc}(P_{\bullet}M)(x) & \xrightarrow{\phi_{\bullet c,M,x}} & (P_{\bullet}({}^cM))(x)\\ {\scriptstyle\coprod_{a\in G}u_{X(a)x}=(P_{\bullet}u)_x}\downarrow & & \downarrow{\scriptstyle(P_{\bullet}({}^cu))_x=\coprod_{a\in G}u_{X(c^{-1}a)x}}\\ (P_{\bullet}N)(x) & \xrightarrow{\phi_{\bullet c,N,x}} & (P_{\bullet}({}^cN))(x)\end{array}$$

is commutative.

CLAIM 2. *Set $\phi_{\bullet}:=(\phi_{\bullet c})_{c\in G}$. Then $(P_{\bullet},\phi_{\bullet})$ turns out to be a G-invariant functor.*

Indeed, it is enough to verify the equalities (5.1) and (5.2). Since X is a G-action, these equalities have the following forms: For each $M\in(\operatorname{Mod}\mathscr{C})_0, x\in\mathscr{C}_0$,

(5.1): $\phi_{\bullet 1,M,x}=\mathbb{1}_{(P_{\bullet}M)(x)}$, and

(5.2): The following diagram is commutative:

$$\begin{array}{ccc}(P_{\bullet}M)(x) & \xrightarrow{\phi_{\bullet c,M,x}} & (P_{\bullet}({}^cM))(x)\\ {\scriptstyle\phi_{\bullet dc,M,x}}\downarrow & & \downarrow{\scriptstyle\phi_{\bullet d,{}^cM,x}}\\ (P_{\bullet}^{(dc)}M)(x) & = & (P_{\bullet}^{d}({}^cN))(x)\end{array}\quad(c,d\in G).$$

Both can be checked easily by using the formula (7.10).

Now for each $M, N \in (\operatorname{mod}\mathscr{C})_0$, we have the following diagram consisting of natural morphisms:

$$
\begin{array}{ccc}
\coprod_{a\in G}(\operatorname{mod}\mathscr{C})(M, {}^aN) & \xrightarrow[\phi]{\sim} & (\operatorname{Mod}\mathscr{C})(M, \coprod_{a\in G}{}^aN) \\
\Big\downarrow {\scriptstyle (P_\cdot,\phi_\cdot)^{(2)}_{M,N}} & & \Big\downarrow {\scriptstyle (\operatorname{Mod}\mathscr{C})(M,t)}\ \wr \\
 & & (\operatorname{Mod}\mathscr{C})(M, \coprod_{a\in G}{}^aN) \\
 & & \Vert \\
(\operatorname{mod}\mathscr{C}/G)(P_\cdot M, P_\cdot N) & \xrightarrow[\omega_{M,P_\cdot N}]{\sim} & (\operatorname{Mod}\mathscr{C})(M, P^\cdot P_\cdot N),
\end{array}
\tag{7.11}
$$

where $t \in (\operatorname{Mod}\mathscr{C})(\coprod_{a\in G}{}^aN, \coprod_{a\in G}{}^aN)$ is an isomorphism defined by

$$t_x((n_a)_{a\in G}) := (n_{a^{-1}})_{a\in G} \quad ((n_a)_{a\in G} \in \coprod_{a\in G}{}^aN(x))$$

for all $x \in \mathscr{C}_0$. In this diagram, note that ϕ is an isomorphism because M is compact by Lemma 7.4.10, that $\omega_{M,P_\cdot N}$ is the adjoint isomorphism, and that the equality in the right column follows by Lemma 7.4.9.

CLAIM 3. *The diagram* (7.11) *above is commutative.*

Indeed, note first that for each $M \in (\operatorname{Mod}\mathscr{C})_0$ and $L \in (\operatorname{Mod}(\mathscr{C}/G))_0$, the adjoint isomorphism $\omega_{M,L} \colon (\operatorname{mod}\mathscr{C}/G)(P_\cdot M, L) \to (\operatorname{Mod}\mathscr{C})(M, P^\cdot L)$ is given by

$$\omega_{M,L}(u)_x = u_x \circ \sigma_1, \quad (u \in \operatorname{Mod}(\mathscr{C}/G)(P_\cdot M, L), x \in \mathscr{C}_0), \tag{7.12}$$

where $\sigma_1 \colon M(x) \to \coprod_{a\in G} M(X(a)x) = (P_\cdot M)(x)$ is a canonical injection. Here take any $f := (f_a)_{a\in G} \in \coprod_{a\in G} M(X(a)x)$, and let $\sigma_a^N \colon {}^aN \to \coprod_{b\in G}{}^bN$ be a canonical injection. Then the image of f by the clockwise composite in the diagram (7.11) becomes $\sum_{a\in G}\sigma_a^N \circ f_{a^{-1}}$ by Lemma 1.5.17 (1). On the other hand, since by definition we have $(P_\cdot, \phi_\cdot)^{(2)}_{M,N}(f) = \sum_{a\in G}\phi_{\cdot a,N}^{-1} \circ P_\cdot f_a$, formulas (7.12) and (7.10) give us that the image of f by the counterclockwise composite in the diagram (7.11) also becomes $\sum_{a\in G}\sigma_a^N \circ f_{a^{-1}}$. Therefore, the diagram (7.11) is commutative.

By Claim 3, each $(P_\cdot, \phi_\cdot)^{(2)}_{M,N}$ becomes an isomorphism. Therefore, this together with Claim 2 shows that $(P_\cdot, \phi_\cdot)$ turns out to be a G-precovering. □

EXERCISE 7.4.12. In the proof above, verify that the adjoint isomorphism $\omega_{M,L}$ is given by the formula (7.12), or give the inverse of $\omega_{M,L}$ defined by this formula.

REMARK 7.4.13. We note the following.

(1) By the theorem above, we can examine the category $\operatorname{mod}A$ of finitely generated modules over an algebra A of the form $A \cong \mathscr{C}/G$ for some small G-category $\mathscr{C}$ by using the category of finitely generated modules over $\mathscr{C}$. For instance, when $\mathscr{C}$ is a locally bounded spectroid (roughly speaking, given by bound quivers), if $\mathscr{C}$ is locally support-finite[2], then the $(P_\cdot, \phi_\cdot)$ above becomes a G-covering. By this fact, we know that if $\mathscr{C}$ is locally representation-finite[3], then A is representation-finite[4].

[2]i.e., for each $x \in \mathscr{C}_0$, when M runs through all those indecomposable representations such that $M(x) \neq 0$, the union of their support $\operatorname{supp} M := \{y \in \mathscr{C}_0 \mid M(y) \neq 0\}$ becomes finite.

[3]i.e., for each $x \in \mathscr{C}_0$, there exists only a finite number of isoclasses of indecomposable representations M of $\mathscr{C}$ such that $M(x) \neq 0$.

[4]i.e. it has only a finite number of isoclasses of indecomposable modules.

(2) An analogous theorem to the one above holds for the bounded homotopy category $\mathscr{K}^{\mathrm{b}}(\mathrm{prj}\,\mathscr{C})$ of $\mathrm{prj}\,\mathscr{C}$. Namely, the pushdown functor $P_{\bullet}\colon \mathrm{Mod}\,\mathscr{C} \to \mathrm{Mod}(\mathscr{C}/G)$ induces a G-precovering functor

$$\mathscr{K}^{\mathrm{b}}(\mathrm{prj}(\mathscr{C}, X)) \to \mathscr{K}^{\mathrm{b}}(\mathrm{prj}\,\mathscr{C}/G).$$

Using this functor, a way for constructing a tilting complex of $\mathscr{C}/G$ from a tilting complex of $\mathscr{C}$ is given, and a way is given to induce a derived equivalence between $\mathscr{C}/G$ and $\mathscr{C}'/G$ from a derived equivalence between two small G-categories $\mathscr{C}$ and $\mathscr{C}'$ (cf. **[2]**, **[3]**, **[4]**, **[5]**, **[8]**, **[9]**).

7.5. The module category of the orbit category and the category of invariant modules

In this section, we show that the pullup functor $P^{\bullet}\colon \mathrm{Mod}\,\mathscr{C}/G \to \mathrm{Mod}\,\mathscr{C}$ induces an equivalence (more strongly, an isomorphism) between $\mathrm{Mod}\,\mathscr{C}/G$ and the category $\mathrm{Mod}^G\,\mathscr{C}$ formed by all "G-invariant" $\mathscr{C}$-modules.

DEFINITION 7.5.1. Set $\mathrm{Mod}^G\,\mathscr{C} := G\text{-}\mathbf{CAT}^{\mathrm{s}}(\mathscr{C}^{\mathrm{op}}, \Delta(\mathrm{Mod}\,\Bbbk))$. Then its objects (M, ϕ) are called *G-invariant* $\mathscr{C}$-modules, and a forgetful functor $\mathrm{Mod}^G\,\mathscr{C} \to \mathrm{Mod}\,\mathscr{C}$ is defined by $(M, \phi) \mapsto M$. However, this is not necessarily injective on the object set (see Example 5.5.5). Therefore we cannot regard $\mathrm{Mod}^G\,\mathscr{C}$ to be a subcategory of $\mathrm{Mod}\,\mathscr{C}$ in general.

THEOREM 7.5.2. *The pullup functor* $P^{\bullet}\colon \mathrm{Mod}\,\mathscr{C}/G \to \mathrm{Mod}\,\mathscr{C}$ *induces an* isomorphism $\mathrm{Mod}\,\mathscr{C}/G \to \mathrm{Mod}^G\,\mathscr{C}$ *of categories.*

PROOF. The formula (5.20) in Remark 5.5.4 gives us an isomorphism

$$\Bbbk\text{-}\mathbf{CAT}(\mathscr{C}/G, \mathscr{D}) \cong G\text{-}\mathbf{CAT}(\mathscr{C}, \Delta(\mathscr{D})), \quad H \mapsto \Delta(H) \circ (P_{\mathscr{C}}, \phi_{\mathscr{C}})$$

of categories. Then the conclusion follows by replacing $\mathscr{C}$ and $\mathscr{D}$ above with $\mathscr{C}^{\mathrm{op}}$ and $\mathrm{Mod}\,\Bbbk$, respectively. □

In particular, when $\mathscr{C} = A$ is an algebra, the following holds.

COROLLARY 7.5.3. *If A is an algebra with a G-action, then we have an isomorphism* $\mathrm{Mod}(A * G) \cong \mathrm{Mod}^G A$ *of categories.* □

REMARK 7.5.4. Let us consider the case where a G-action is free. Then by Theorem 7.5.2, the classical covering functor $(P, \mathbb{1})\colon \mathscr{C} \to \mathscr{C}/_{\mathrm{c}}G$ (note that we have $(\mathscr{C}/_{\mathrm{c}}G)_0 = \{Gx \mid x \in \mathscr{C}_0\}$ because of the classical case) induces an isomorphism $\mathrm{Mod}(\mathscr{C}/_{\mathrm{c}}G) \to \mathrm{Mod}^G\,\mathscr{C}$ of categories (see **[19**, Theorem 4.3**]**). In this case, it holds that $P^{\bullet}(M) = (M, \mathbb{1})$ $(M \in \mathrm{Mod}\,\mathscr{C}/_{\mathrm{c}}G)$, and we can identify $\mathrm{Mod}^G\,\mathscr{C}$ with (the right version of) the category $(\mathrm{Mod}\,\mathscr{C})^G$ defined in **[19]** (i.e., the full subcategory of $\mathrm{Mod}\,\mathscr{C}$ with the object set $\{M \in (\mathrm{Mod}\,\mathscr{C})_0 \mid {}^aM = M\ (a \in G)\}$). Therefore, in this case in particular, $\mathrm{Mod}^G\,\mathscr{C}$ can be regarded as a full subcategory of $\mathrm{Mod}\,\mathscr{C}$.

7.6. The module category and the category of graded modules of the orbit category

In this section, we show that the pushdown functor $P_{\bullet}\colon \mathrm{Mod}\,\mathscr{C} \to \mathrm{Mod}\,\mathscr{C}/G$ induces an equivalence between the category $\mathrm{Mod}\,\mathscr{C}$ and the category $\mathrm{Mod}_G\,\mathscr{C}/G$ formed by the G-graded modules and the degree-preserving morphisms.

DEFINITION 7.6.1. Let $\mathscr{B}$ be a G-graded category.

(1) A *G-graded $\mathscr{B}$-module* is a pair of the following data that satisfies the axiom below.

Data:
- a $\mathscr{B}$-module M,
- a family of direct sum decompositions $d = (M(x) = \bigoplus_{a \in G} M^a(x))_{x \in \mathscr{B}_0}$.

Axiom: $M(f)(M^a(x)) \subseteq M^{ab}(y)$ $(f \in \mathscr{B}^b(y,x),\ x, y \in \mathscr{B}_0,\ b \in G)$.

(2) Let M and N be G-graded $\mathscr{B}$-module. Then a morphism $u \colon M \to N$ is called a *degree-preserving* morphism if $u_x M^a(x) \subseteq N^a(x)$ for all $x \in \mathscr{B}$ and $a \in G$.

(3) We denote by $\operatorname{Mod}_G \mathscr{B}$ the linear category formed by the G-graded modules and the degree-preserving morphisms between them. Here also we can consider a forgetful functor $\operatorname{Fgt} \colon \operatorname{Mod}_G \mathscr{B} \to \operatorname{Mod} \mathscr{B}$, $(M, d) \mapsto M$. Since this is not necessarily injective on the object set (thus, there can be multiple gradings d for the same M), note that $\operatorname{Mod}_G \mathscr{B}$ cannot be regarded as a subcategory of $\operatorname{Mod} \mathscr{B}$.

(4) We denote by $\operatorname{mod}_G \mathscr{B}$ the full subcategory of $\operatorname{Mod}_G \mathscr{B}$ consisting of those $(M, d) \in (\operatorname{Mod}_G \mathscr{B})_0$ such that $M \in (\operatorname{mod} \mathscr{B})_0$.

THEOREM 7.6.2. *The pushdown functor $P_{.} \colon \operatorname{Mod} \mathscr{C} \to \operatorname{Mod} \mathscr{C}/G$ induces an equivalence $P'_{.} \colon \operatorname{Mod} \mathscr{C} \to \operatorname{Mod}_G \mathscr{C}/G$. Thus, there exists an equivalence $P'_{.}$ that makes the following diagram strictly commutative:*

$$\begin{array}{ccc} \operatorname{Mod} \mathscr{C} & \xrightarrow{\quad P_{.} \quad} & \operatorname{Mod} \mathscr{C}/G \\ & {\scriptstyle P'_{.}} \searrow \qquad \nearrow {\scriptstyle \operatorname{Fgt}} & \\ & \operatorname{Mod}_G \mathscr{C}/G & \end{array}.$$

PROOF. First of all, we check that $P_{.}$ maps each morphism $u \colon M \to N$ in $\operatorname{Mod} \mathscr{C}$ into $\operatorname{Mod}_G \mathscr{C}/G$. By Definition 7.4.7, we have

$$(P_{.}M)(Px) = \coprod_{a \in G} M(X(a)x) = \bigoplus_{a \in G} \sigma_a(M(X(a)x))$$

for all $x \in \mathscr{C}_0$, where $(\sigma_b \colon M(X(a)x) \to \coprod_{a \in G} M(X(a)x))_{b \in G}$ is the canonical injection family. By using this, we set

$$(P_{.}M)^a(Px) := \sigma_a(M(X(a)x)), \tag{7.13}$$

which makes $P_{.}M$ a G-graded $\mathscr{C}/G$-module.

Indeed, let $x, y \in \mathscr{C}_0, c, d \in G$, and take any $m = (m_b)_{b \in G} \in (P_{.}M)^c(Py), f = (f_a)_{a \in G} \in (\mathscr{C}/G)^d(x, y)$. Then since $m_b = 0$ $(b \neq c)$, $f_a = 0$ $(a \neq d)$, we have

$$\begin{aligned} (P_{.}M)(f)(m) &= [M(X(b)f_{b^{-1}a})]_{a,b \in G}((m_b)_{b \in G}) \\ &= \left(\sum_{b \in G} M(X(b)f_{b^{-1}a})(m_b) \right)_{a \in G} \\ &= (\delta_{a,cd} M(X(c)f_d)(m_c))_{a \in G} \in (P_{.}M)^{cd}(Px), \end{aligned}$$

which shows that $P_{.}M \in (\operatorname{Mod}_G \mathscr{C}/G)_0$.

Moreover, the formula (7.9) shows that $P_{.}u$ is degree-preserving, i.e., that $P_{.}u \colon P_{.}M \to P_{.}N$ is a morphism in $\operatorname{Mod}_G \mathscr{C}/G$. Therefore, the pushdown functor $P_{.}$

induces a functor $P'_{\bullet}\colon \operatorname{Mod}\mathscr{C} \to \operatorname{Mod}_G \mathscr{C}/G$. Note here that we have $P'_{\bullet}M = P_{\bullet}M$ and $P'_{\bullet}u = P_{\bullet}u$ for all objects M and morphisms u in $\operatorname{Mod}\mathscr{C}$. Thus the diagram above is strictly commutative.

Faithfulness of $P'_{\bullet}$: Let $u\colon M \to N$ be a morphism in $\operatorname{Mod}\mathscr{C}$, and assume that $P_{\bullet}u = 0$. Then it is enough to show that $u = 0$. From this assumption, we have $0 = (P_{\bullet}u)_x = \coprod_{a\in G} u_{X(a)x}$ for all $x \in \mathscr{C}_0$ by (7.9). Thus in particular, $u_x = 0$, and we have $u = (u_x)_{x\in\mathscr{C}_0} = 0$. Hence $P'_{\bullet}$ is faithful.

Fullness of $P'_{\bullet}$: Let $M, N \in (\operatorname{Mod}\mathscr{C})_0$ and take any $v \in \operatorname{Mod}_G \mathscr{C}/G(P_{\bullet}M, P_{\bullet}N)$. Then it is enough to construct some $u \in (\operatorname{Mod}\mathscr{C})(M,N)$ such that $P_{\bullet}u = v$. For each $x \in (\mathscr{C}/G)_0 = \mathscr{C}_0$, since v is degree-preserving, $v_x\colon \coprod_{a\in G} M(X(a)x) \to \coprod_{a\in G} N(X(a)x)$ satisfies the condition that $v_x(M(X(a)x)) \subseteq N(X(a)x)$ $(a \in G)$. Therefore v_x induces a morphism $v_{a,x}\colon M(X(a)x) \to N(X(a)x)$ in $\operatorname{Mod}\Bbbk$ for all $a \in G$, and hence it has the form $v_x = \coprod_{a\in G} v_{a,x}$. Then we define a morphism $u\colon M \to N$ in $\operatorname{Mod}\mathscr{C}$ as follows:

$$u_x := v_{1,x}\colon M(x) \to N(x) \quad (x \in \mathscr{C}_0), \quad u := (u_x)_{x\in\mathscr{C}_0}. \tag{7.14}$$

CLAIM 1. *u is a morphism in* $\operatorname{Mod}\mathscr{C}$.

Indeed, for each morphism $f\colon x \to y$ in $\mathscr{C}$ it is enough to show the commutativity of the diagram

$$\begin{array}{ccc} M(y) & \xrightarrow{u_y} & N(y) \\ {\scriptstyle M(f)}\downarrow & & \downarrow{\scriptstyle N(f)} \\ M(x) & \xrightarrow[u_x]{} & N(x) \end{array} \tag{7.15}$$

We first show the following:

$$(P_{\bullet}M)(Pf) = \coprod_{a\in G} M(X(a)f).$$

This follows from the following computation:

$$\begin{aligned}(P_{\bullet}M)(Pf) &= [M(X(b)(Pf)_{b^{-1}a})]_{a,b\in G} = [M(X(b)\delta_{b^{-1}a,1}f)]_{a,b\in G} \\ &= [M(X(b)\delta_{a,b}f)]_{a,b\in G} = \coprod_{a\in G} M(X(a)f).\end{aligned}$$

Using this, the fact that v is a morphism in $\operatorname{Mod}_G(\mathscr{C}/G)$ yields the commutative diagram

$$\begin{array}{ccc} \coprod_{a\in G} M(X(a)y) & \xrightarrow{\coprod_{a\in G} v_{a,y}} & \coprod_{a\in G} N(X(a)y) \\ {\scriptstyle \coprod_{a\in G} M(X(a)f)}\downarrow & & \downarrow{\scriptstyle \coprod_{a\in G} N(X(a)f)} \\ \coprod_{a\in G} M(X(a)x) & \xrightarrow[\coprod_{a\in G} v_{a,x}]{} & \coprod_{a\in G} N(X(a)x). \end{array}$$

By looking at the component for $a = 1$, we see the commutativity of (7.15).

CLAIM 2. $P_{\bullet}u = v$.

Indeed, this equality is equivalent to saying that $(P_{\bullet}u)_x = v_x$ for all $x \in \mathscr{C}_0$, which means that $\coprod_{a\in G} u_{X(a)x} = \coprod_{a\in G} v_{a,x}$. Therefore, by definition (7.14) of u it is enough to show the following for all $x \in \mathscr{C}$ and $c \in G$:

$$v_{1,X(c)x} = v_{c,x}. \tag{7.16}$$

Now, the isomorphism $\phi_{c,x}\colon Px \to PX(c)x$ in $\mathscr{C}/G$ is given as $\phi_c x = (\delta_{a,c}\mathbb{1}_{X(c)x})_{a\in G}$ by Definition 5.4.7 (3). Therefore we have

$$\begin{aligned}(P_{\bullet}M)(\phi_c x) &= [M(X(b)(\phi_c x)_{b^{-1}a})]_{a,b\in G} = [M(X(b)\delta_{b^{-1}a,c}\mathbb{1}_{X(c)x})]_{a,b\in G} \\ &= [\delta_{a,bc}\mathbb{1}_{M(X(b)X(c)x)}]_{a,b\in G} = [\delta_{a,bc}\mathbb{1}_{M(X(a)x)}]_{a,b\in G}.\end{aligned}$$

Thus the following diagram is commutative:

$$\begin{array}{ccc} P_{\bullet}M(X(c)x) & \xrightarrow{P_{\bullet}M(\phi_{c,x})} & P_{\bullet}M(x) \\ \| & & \| \\ \coprod_{b\in G} M(X(b)X(c)x) & \xrightarrow[{[\delta_{a,bc}\mathbf{1}_{M(X(a)x)})]_{a,b\in G}}]{} & \coprod_{a\in G} M(X(a)x). \end{array}$$

Hence $v \in \operatorname{Mod}(\mathscr{C}/G)(P_{\bullet}M, P_{\bullet}N)$ gives a commutative diagram

$$\begin{array}{ccc} \coprod_{b\in G} M(X(b)X(c)x) & \xrightarrow{\coprod_{b\in G} v_{b,X(c)x}} & \coprod_{b\in G} N(X(b)X(c)x) \\ {\scriptstyle [\delta_{a,bc}\mathbf{1}_{M(X(a)x)})]_{a,b\in G}}\downarrow & & \downarrow{\scriptstyle [\delta_{a,bc}\mathbf{1}_{N(X(a)x)})]_{a,b\in G}} \\ \coprod_{a\in G} M(X(a)x) & \xrightarrow[\coprod_{a\in G} v_{a,x}]{} & \coprod_{a\in G} N(X(a)x). \end{array}$$

This means that

$$\delta_{a,bc}\, v_{a,x} = \delta_{a,bc}\, v_{b,X(c)x} \quad (a, b, c \in G).$$

In particular when $a = c, b = 1$, this gives the equality (7.16).

Denseness of $P'_{\bullet}$: Let (N, d) be an object of $\operatorname{Mod}_G \mathscr{C}/G$, and set $d = (N(x) = \bigoplus_{a\in G} N^a(x))_{x\in(\mathscr{C}/G)_0}$. Then we define $M \in \operatorname{Mod}\mathscr{C}$ as follows:

On objects: For each $x \in \mathscr{C}_0$, we set $M(x) := N^1(x)$.

On morphisms: For each morphisms $f\colon x \to y$ in $\mathscr{C}$, we have $\deg Pf = 1$ by Definition 5.4.7 (2). Therefore $N(Pf)(N^1(y)) \subseteq N^1(x)$, and then we set $M(f) := N(Pf)|_{N^1(y)}\colon N^1(x) \to N^1(x)$.

The following finishes the proof.

CLAIM 3. *It holds that $P_{\bullet}M \cong N$ in $\operatorname{Mod}_G \mathscr{C}/G$.*

We now give a proof of it. Take any $a \in G$ and $x \in \mathscr{C}_0$. Note first that since X is a G-action, the formula (5.12) has the form

$$(\phi_a x)^{-1} = (\delta_{b,a^{-1}}\mathbb{1}_x)_{b\in B} = \phi_{a^{-1}}(X(a)x).$$

By Definition 5.4.7 (3), we also have

$$\deg \phi_a x = a,$$

which shows $\deg \phi_{a^{-1}}(X(a)x) = a^{-1}$ as well. Therefore, the isomorphism $\phi_a x\colon Px \to P(X(a)x)$ in $\mathscr{C}/G$ induces an isomorphism

$$F_{a,Px} := N(\phi_a x)|_{N^1(PX(a)x)}\colon M(X(a)x) = N^1(PX(a)x) \to N^a(Px)$$

in $\mathrm{Mod}\,\Bbbk$. ($N(\phi_{\alpha^{-1}}(X(a)x))|_{N^a(Px)}$ gives its inverse.) Using this for each $x \in \mathscr{C}$, we define an isomorphism F_{Px} in $\mathrm{Mod}\,\Bbbk$ by the commutative diagram

$$\begin{array}{ccc} (P_{.}M)(Px) & \xrightarrow{F_{Px}} & N(Px) \\ \| & & \| \\ \coprod_{a\in G} M(X(a)x) & \xrightarrow[\coprod_{a\in G} F_{a,Px}]{} & \coprod_{a\in G} N^a(Px). \end{array}$$

To show this claim, it suffices to show that $F := (F_{Px})_{Px\in(\mathscr{C}/G)_0}$ belongs to $\mathrm{Mod}_G\,\mathscr{C}/G(P_{.}M, N)$, which means that the following diagram is commutative:

$$\begin{array}{ccc} (P_{.}M)(Py) & \xrightarrow{F_{Py}} & N(Py) \\ {\scriptstyle (P_{.}M)(f)}\downarrow & & \downarrow{\scriptstyle N(f)} \\ (P_{.}M)(Px) & \xrightarrow[F_{Px}]{} & N(Px) \end{array}$$

for all $x, y \in \mathscr{C}_0$ and $f = (f_a)_{a\in G} \in (\mathscr{C}/G)(Px, Py)$. Now since $f = \sum_{a\in G} \sigma_a^{\mathscr{C}/G}(f_a)$, it is enough to show the commutativity of the diagram above for each term $\sigma_a^{\mathscr{C}/G}(f_a)$ with $a \in G$ instead of f itself. Thus we may assume that $\deg f = b$ for some $b \in G$. Therefore, it is enough to show the commutativity of the diagram

$$\begin{array}{ccc} (P_{.}M)^a(Py) & \xrightarrow{F_{a,Py}} & N^a(Py) \\ {\scriptstyle (P_{.}M)(f)|_{(P_{.}M)^a(Py)}}\downarrow & & \downarrow{\scriptstyle N(f)|_{N^a(Py)}} \\ (P_{.}M)^{ab}(Px) & \xrightarrow[F_{ab,Px}]{} & N^{ab}(Px) \end{array} \tag{7.17}$$

for all $a \in G$. Here since we have

$$(P_{.}M)(f) = [M(b)f_{b^{-1}a}]_{a,b\in G}\colon \coprod_{b\in G} M(X(b)y) \to \coprod_{a\in G} M(X(a)x),$$

the left vertical morphism is equal to $M(X(a)f_b)\colon M(X(a)y) \to M(X(ab)x)$, and hence to

$$N(P(X(a)f_b))|_{N^1(X(a)y)}\colon N^1(X(a)y) \to N^1(X(ab)x).$$

Therefore the commutativity of the diagram (7.17) is implied by

$$N(\phi_{ab,x}) \circ N(P(X(a)f_b)) = N(f) \circ N(\phi_{a,y}),$$

which is implied by

$$P(X(a)f_b) \circ \phi_{ab,x} = \phi_{a,y} \circ f.$$

Now, since $\deg f = b$, we have $f = (\delta_{c,b}f_b)_{c\in G}$. By using this, we see that the both sides are equal to $(\delta_{c,ab}X(a)f_b)_{c\in G}$. □

COROLLARY 7.6.3. *If A is an algebra with a G-action, then there exists an equivalence* $\mathrm{Mod}\,R \simeq \mathrm{Mod}_G\,R * G$ *of categories.* □

COROLLARY 7.6.4. *The pushdown functor $P_{.}$ induces a G-covering functor*

$$(P_{.}, \phi_{.})\colon \mathrm{mod}\,\mathscr{C} \to \mathrm{mod}_{(G)}(\mathscr{C}/G),$$

where $\mathrm{mod}_{(G)}(\mathscr{C}/G)$ *is the full subcategory of* $\mathrm{mod}(\mathscr{C}/G)$ *consisting of the objects* $\mathrm{Fgt}(M, d)$ *with* $(M, d) \in \mathrm{mod}_G(\mathscr{C}/G)_0$ *(i.e., G-gradable $\mathscr{C}/G$-modules). Therefore in particular, we have*

$$(\mathrm{mod}\,\mathscr{C})/G \simeq \mathrm{mod}_{(G)}(\mathscr{C}/G), \quad \mathrm{mod}\,\mathscr{C} \simeq (\mathrm{mod}_{(G)}(\mathscr{C}/G))\#G.$$

PROOF. By Theorem 7.4.11, $(P_{\bullet}, \phi_{\bullet})\colon \operatorname{mod}\mathscr{C} \to \operatorname{mod}(\mathscr{C}/G)$ is a precovering. It becomes dense if we restrict its target $\operatorname{mod}(\mathscr{C}/G)$ to $\operatorname{mod}_{(G)}(\mathscr{C}/G)$ by Theorem 7.6.2, and the statement follows. From this fact by (the G-**CAT** version of) Corollary 5.5.3, we have $(\operatorname{mod}\mathscr{C})/G \simeq \operatorname{mod}_{(G)}(\mathscr{C}/G)$. The remaining equivalence follows from Theorem 5.7.1 and Remark 5.8.15. □

CHAPTER 8

2-categorical covering theory under colax actions of a category

So far we have considered actions and pseudo-actions of a group G on a small linear category, which is a notion equivalent to that of functors and pseudofunctors $X: G \to \Bbbk\text{-}\mathbf{Cat}$ when the group itself is viewed as a singleton (say with the object $*$). Thus, X was regarded as the action and pseudo-action of G on $X(*)$. Then, if we extend the range of G from groups to small categories, we can consider actions and pseudo-actions of a small category I on (a set of) linear categories. This can be viewed in another way as a "2-representation" of I because a representation of I is nothing but a functor $I \to \operatorname{Mod}\Bbbk$, and in the setting above the category $\operatorname{Mod}\Bbbk$ is replaced by the 2-category $\Bbbk\text{-}\mathbf{Cat}$.

On the other hand, one merit of considering the action of a group G on a small linear category $\mathscr{C}$ is the fact that one can obtain information about the other from one of the representations of $\mathscr{C}$ and $\mathscr{C}/G$ by relating them using the pushdown functor $\operatorname{Mod}\mathscr{C} \to \operatorname{Mod}(\mathscr{C}/G)$ or the pullup functor induced by the covering functor $G \to \mathscr{C}/G$. Although not explained in this book, as mentioned in Remark 7.4.13 (2), the pushdown functor induces a G-precovering functor $\mathscr{K}^{\mathrm{b}}(\operatorname{prj}\mathscr{C}) \to \mathscr{K}^{\mathrm{b}}(\operatorname{prj}\mathscr{C}/G)$ between bounded homotopy categories, which is an important tool for inducing derived equivalences.

By replacing the group G with the small category I, applications in this area will be further expanded. Therefore, in this Chapter 8, we outline a generalization of the previous covering theory to colax functors $X: I \to \Bbbk\text{-}\mathbf{Cat}$. The already well-known Grothendieck construction gives a generalization of the orbit category construction in this setting, but the concept corresponding to the smash product is new. The generalization of the Cohen-Montgomery duality in this setting has just been completed, and the paper is still in preparation. As for applications to derived equivalences, we have completed a part using the Grothendieck construction, but another part using the smash product is still in progress. Therefore, at this stage, we limit ourselves to explaining the part using the Grothendieck construction. Throughout this chapter, I denotes a small category, and $\Bbbk$ is a commutative ring unless otherwise stated.

8.1. Grothendieck constructions

In this section, we give the definition of the Grothendieck construction ([**27**]) that is a generalization of the orbit category construction (see [**10**] for its quiver presentation).

DEFINITION 8.1.1. Let $X: I \to \Bbbk\text{-}\mathbf{Cat}$ be a colax functor. Then a linear category $\operatorname{Gr}(X)$ is defined as follows:

- $\operatorname{Gr}(X)_0 := \bigcup_{i\in I_0}\{i\} \times X(i)_0 = \{{}_ix := (i,x) \mid i \in I_0, x \in X(i)_0\}$.

- For each ${}_ix, {}_jy \in \mathrm{Gr}(X)_0$, we set
$$\mathrm{Gr}(X)({}_ix, {}_jy) := \coprod_{a\in I(i,j)} X(j)(X(a)x, y).$$
- For each ${}_ix, {}_jy, {}_kz \in \mathrm{Gr}(X)_0$ and $f = (f_a)_{a\in I(i,j)} \in \mathrm{Gr}(X)({}_ix, {}_jy)$, $g = (g_b)_{b\in I(j,k)} \in \mathrm{Gr}(X)({}_jy, {}_kz)$, we set
$$g \circ f := \left(\sum_{\substack{a\in I(i,j)\\ b\in I(j,k)\\ c=ba}} g_b \circ X(b)f_a \circ X_{b,a}x \right)_{c\in I(i,k)}.$$
Thus, each term in the summation is the composite
$$X(ba)x \xrightarrow{X_{b,a}x} X(b)X(a)x \xrightarrow{X(b)f_a} X(b)y \xrightarrow{g_b} z.$$
- For each ${}_ix \in \mathrm{Gr}(X)_0$, the identity $\mathbb{1}_{{}_ix}$ is given by
$$\mathbb{1}_{{}_ix} = (\delta_{a,\mathbf{1}_i} X_i\, x)_{a\in I(i,i)} \in \bigoplus_{a\in I(i,i)} X(i)(X(a)x, x).$$

EXAMPLE 8.1.2. Let A be a $\mathbb{k}$-algebra (regard it as a linear singleton). Define a functor $X := \Delta(A)\colon I \to \mathbb{k}$-**Cat** by $X(i) := A$ for all $i \in I_0$ and $X(a) := \mathbb{1}_A$ for all $a \in I_1$. Then the following hold:

(1) If I is a free category of the quiver $(1 \to 2)$, then $\mathrm{Gr}(X)$ turns out to be the upper triangular matrix algebra over A: $\mathrm{Gr}(X) \cong \begin{bmatrix} A & 0 \\ A & A \end{bmatrix}$. More generally, statements (2) and (3) hold.
(2) If I is a free category of a quiver Q, then $\mathrm{Gr}(X) \cong AQ$ (the path-category of Q over A).
(3) If $I = S$ is a poset, then $\mathrm{Gr}(X) \cong AS$ (the incidence algebra of S over A).
(4) If $I = G$ is a monoid, then $\mathrm{Gr}(X) \cong AG$ (the monoid algebra of G over A).

8.2. Pseudo-actions induced on module categories

Here we need colax functors from a 2-category to another 2-category, and a 2-category consisting of all colax functors between a fixed pair of 2-categories. Note that we can regard a small category I as a 2-category as follows: Regard the set $I(x, y)$ as a discrete category for all objects x, y of I, and for each $x, y, z \in I_0$, define a composition $I(y, z) \times I(x, y) \to I(x, z)$ in a trivial way, i.e., if $f \in I(x, y)_0, g \in I(y, z)_0$, then $\mathbb{1}_g \circ \mathbb{1}_f := \mathbb{1}_{g\circ f}$. When I is regard as a 2-category in this way, the colax functor from I (as a 2-category) to the 2-category $\mathbb{k}$-**Cat** defined below coincides with that from I (as a category) to $\mathbb{k}$-**Cat** defined in Definition 4.3.1.

DEFINITION 8.2.1. Let **B** and **C** be 2-categories.

(1) A *colax functor* X from **B** to **C** (denoted by $X\colon \mathbf{B} \to \mathbf{C}$) consists of the following data (parts different from Definition 4.3.1 are underlined) that satisfies the axioms below.

Data:

- a map $X\colon \mathbf{B}_0 \to \mathbf{C}_0$,
- a functor $X\colon \mathbf{B}(i,j) \to \mathbf{C}(X(i), X(j))$ $(i, j \in \mathbf{B}_0)$,
- a 2-morphism $X_i\colon X(\mathbb{1}_i) \Rightarrow \mathbb{1}_{X(i)}$ $(i \in \mathbf{B}_0)$ in $\mathbf{C}$,
- a 2-morphism $X_{b,a}\colon X(ba) \Rightarrow X(b)X(a)$ in $\mathbf{C}$
  $((b,a) \in \mathbf{B}_1 \times \mathbf{B}_1$ with $\mathrm{dom}(b) = \mathrm{cod}(a)$, natural in $a, b)$,

where "natural in a, b" means that the following diagram is commutative:

$$\begin{array}{ccc} X(ba) & \xRightarrow{X_{b,a}} & X(b)X(a) \\ {\scriptstyle X(\beta*\alpha)}\Big\Downarrow & & \Big\Downarrow{\scriptstyle X(\beta)*X(\alpha)} \\ X(b'a') & \xRightarrow[]{X_{b',a'}} & X(b')X(a') \end{array}$$

$((i \xrightarrow{a,a'} j), (j \xrightarrow{b,b'} k) \in \mathbf{B}_1,\ (a \overset{\alpha}{\Rightarrow} a'), (b \overset{\beta}{\Rightarrow} b') \in \mathbf{B}_2)$.

Axioms. Conditions obtained from Definition 4.3.1 by replacing I_1 with $\mathbf{B}_1$.

(2) A *lax functor* $\mathbf{B} \to \mathbf{C}$ is a colax functor $\mathbf{B} \to \mathbf{C}^{\mathrm{co}}$.
(3) A *pseudofunctor* $\mathbf{B} \to \mathbf{C}$ is a colax functor $X\colon \mathbf{B} \to \mathbf{C}$, where all $X_i, X_{b,a}$ are 2-isomorphisms.
(4) A *2-functor* $\mathbf{B} \to \mathbf{C}$ is a colax functor $X\colon \mathbf{B} \to \mathbf{C}$, where all $X_i, X_{b,a}$ are identity 2-morphisms.

DEFINITION 8.2.2. Let $\mathbf{B}$ and $\mathbf{C}$ be 2-categories. Then a 2-category $\mathrm{Colax}(\mathbf{B}, \mathbf{C})$ having all the colax functors $\mathbf{B} \to \mathbf{C}$ as its objects is defined as follows:

1-morphisms. Let X and X' be colax functors $\mathbf{B} \to \mathbf{C}$. Then a *1-morphism* (weak transformation) from X to X' is a pair (F, ψ) of the following data that satisfies the axioms below:

Data.

- a family $F := (F(i))_{i \in \mathbf{B}_0}$ of 1-morphisms $F(i)\colon X(i) \to X'(i)$ in $\mathbf{C}$,
- a family $\psi := (\psi(a))_{a \in \mathbf{B}_1}$ of 2-morphisms $\psi(a)\colon X'(a)F(i) \Rightarrow F(j)X(a)$

$$\begin{array}{ccc} X(i) & \xrightarrow{F(i)} & X'(i) \\ {\scriptstyle X(a)}\Big\downarrow & \overset{\psi(a)}{\swarrow\!\!\!\swarrow} & \Big\downarrow{\scriptstyle X'(a)} \\ X(j) & \xrightarrow[F(j)]{} & X'(j) \end{array}$$

in $\mathbf{C}$, where for each $\alpha\colon a \Rightarrow b$ in $\mathbf{B}(i,j)$, the following diagram is commutative (no conditions when $\mathbf{B} = I$):

$$\begin{array}{ccc} X'(a)F(i) & \xRightarrow{X'(\alpha)F(i)} & X'(b)F(i) \\ {\scriptstyle \psi(a)}\Big\Downarrow & & \Big\Downarrow{\scriptstyle \psi(b)} \\ F(j)X(a) & \xRightarrow[F(j)X(\alpha)]{} & F(j)X(b). \end{array}$$

Thus, ψ is given by a family of natural transformations

$$\begin{array}{ccc} \mathbf{B}(i,j) & \xrightarrow{X'} & \mathbf{C}(X'(i),X'(j)) \\ {\scriptstyle X}\downarrow & \overset{\psi_{ij}}{\swarrow\!\!\!\!\swarrow} & \downarrow{\scriptstyle \mathbf{C}(F(i),X'(j))} \\ \mathbf{C}(X(i),X(j)) & \xrightarrow[\mathbf{C}(X(i),F(j))]{} & \mathbf{C}(X(i),X'(j)) \end{array} \qquad (i,j \in \mathbf{B}_0).$$

Axioms:

(a) For each $i \in \mathbf{B}_0$, the following is commutative (cf. the string diagram (5.1)):

$$\begin{array}{ccc} X'(\mathbb{1}_i)F(i) & \xRightarrow{\psi(\mathbb{1}_i)} & F(i)X(\mathbb{1}_i) \\ {\scriptstyle X'_iF(i)}\Downarrow & & \Downarrow{\scriptstyle F(i)X_i} \\ \mathbb{1}_{X'(i)}F(i) & = & F(i)\mathbb{1}_{X(i)}. \end{array}$$

(b) For each path $i \xrightarrow{a} j \xrightarrow{b} k$ in $\mathbf{B}_1$, the following is commutative (cf. the string diagram (5.2)):

$$\begin{array}{ccccc} X'(ba)F(i) & \xRightarrow{X'_{b,a}F(i)} & X'(b)X'(a)F(i) & \xRightarrow{X'(b)\psi(a)} & X'(b)F(j)X(a) \\ {\scriptstyle \psi(ba)}\Downarrow & & & & \Downarrow{\scriptstyle \psi(b)X(a)} \\ F(k)X(ba) & & \xRightarrow[F(k)\,X_{b,a}]{} & & F(k)X(b)X(a). \end{array}$$

A 1-morphism (F,ψ) is called I-*equivariant* if $\psi(a)$ is a 2-isomorphism in $\mathbf{C}$ for all $a \in \mathbf{B}_1$.

2-morphisms: Let $X, X' \colon \mathbf{B} \to \mathbf{C}$ be colax functors, and (F,ψ), (F',ψ') 1-morphisms $X \to X'$. Then a 2-*morphism* (modification) from (F,ψ) to (F',ψ') is a family $\zeta = (\zeta(i))_{i\in\mathbf{B}_0}$ of 2-morphisms $\zeta(i)\colon F(i) \Rightarrow F'(i)$ in $\mathbf{C}$ that makes the following diagram commutative for all $(i \xrightarrow{a} j) \in \mathbf{B}_1$:

$$\begin{array}{ccc} X'(a)F(i) & \xRightarrow{X'(a)\zeta(i)} & X'(a)F'(i) \\ {\scriptstyle \psi(a)}\Downarrow & & \Downarrow{\scriptstyle \psi'(a)} \\ F(j)X(a) & \xRightarrow{\zeta(j)X(a)} & F'(j)X(a). \end{array}$$

REMARK 8.2.3. Regard I as a 2-category as explained above, and consider $\mathrm{Colax}(I,\mathbf{C})$. Then its 2-category structure allows us to define colax functors X, $X'\colon I \to \mathbf{C}$ to be equivalent according to the general theory as follows: Colax functors X and X' are said to be *equivalent* if there exists a pair of 1-morphisms $(F,\psi)\colon X \to X'$ and $(F',\psi')\colon X' \to X$, and a pair of 2-isomorphisms $(F',\psi')(F,\psi) \cong \mathbb{1}_X$ and $(F,\psi)(F',\psi') \cong \mathbb{1}_{X'}$.

DEFINITION 8.2.4. Let $\mathscr{C} \in \Bbbk\text{-}\mathbf{Cat}_0$ and $(F, \psi)\colon X \to \Delta(\mathscr{C})$ be a 1-morphism in $\mathrm{Colax}(I, \Bbbk\text{-}\mathbf{Cat})$.

(1) (F, ψ) is called an *I-precovering* (of $\mathscr{C}$) if the homomorphism

$$(F,\psi)^{(1)}_{x,y}\colon \coprod_{a\in I(i,j)} X(j)(X(a)x, y) \to \mathscr{C}(F(i)x, F(j)y)$$

$$(f_a\colon X(a)x \to y)_{a\in I(i,j)} \mapsto \sum_{a\in I(i,j)} F(j)(f_a) \circ \psi(a)(x)$$

of $\Bbbk$-modules is an isomorphism for all $i, j \in I_0$ and $x \in X(i)_0$, $y \in X(j)_0$.

(2) An I-precovering (F, ψ) is called an *I-covering* if it is *dense*, i.e., for each $c \in \mathscr{C}_0$, there exists an $i \in I_0$ and $x \in X(i)_0$ such that $F(i)(x)$ is isomorphic to c in $\mathscr{C}$.

REMARK 8.2.5. The Grothendieck construction can be extended to a 2-functor $\mathrm{Gr}\colon \mathrm{Colax}(I, \Bbbk\text{-}\mathbf{Cat}) \to \Bbbk\text{-}\mathbf{Cat}$, which becomes a left adjoint to $\Delta\colon \Bbbk\text{-}\mathbf{Cat} \to \mathrm{Colax}(I, \Bbbk\text{-}\mathbf{Cat})$. As in Theorem 5.5.2, the component at X of the 2-unit

$$\mathbb{1}_{\mathrm{Colax}(I,\Bbbk\text{-}\mathbf{Cat})} \Rightarrow \Delta \cdot \mathrm{Gr}$$

of this adjoint gives an I-covering $(P_X, \phi_X)\colon X \to \Delta(\mathrm{Gr}\, X)$ for all $X \in \mathrm{Colax}(I, \Bbbk\text{-}\mathbf{Cat})_0$. Thus the Grothendieck construction gives an essential construction of an I-covering.

We will use the following pseudofunctor to define the "module category" of a colax functor.

EXAMPLE 8.2.6. The following correspondence Mod' becomes a 2-functor, and Mod a pseudofunctor:

- $\mathrm{Mod}'\colon \Bbbk\text{-}\mathbf{Cat} \to \Bbbk\text{-}\mathbf{Ab}^{\mathrm{coop}}$, $\mathrm{Mod}' := \Bbbk\text{-}\mathbf{Cat}((\text{-})^{\mathrm{op}}, \mathrm{Mod}\,\Bbbk)$,
  $(\mathscr{C} \xrightarrow{F} \mathscr{C}') \mapsto (\mathrm{Mod}\,\mathscr{C}' \xrightarrow{(\text{-})\circ F^{\mathrm{op}}} \mathrm{Mod}\,\mathscr{C})$.
- $\mathrm{Mod}\colon \Bbbk\text{-}\mathbf{Cat} \to \Bbbk\text{-}\mathbf{Ab}$, $(\mathscr{C} \xrightarrow{F} \mathscr{C}') \mapsto (\mathrm{Mod}\,\mathscr{C} \xrightarrow{\text{-}\otimes_{\mathscr{C}}\overline{F}} \mathrm{Mod}\,\mathscr{C}')$,
  where $\overline{F} := \mathscr{C}'(\text{-}, F(?))$.

We cite the following theorem from [**9**, Theorem 6.5].

THEOREM 8.2.7. *Let $\mathbf{B}, \mathbf{C}, \mathbf{D}$ be 2-categories, $V\colon \mathbf{C} \to \mathbf{D}$ a pseudofunctor. Then the composition with V induces a pseudofunctor:*

$$\mathrm{Colax}(\mathbf{B}, V)\colon \mathrm{Colax}(\mathbf{B}, \mathbf{C}) \to \mathrm{Colax}(\mathbf{B}, \mathbf{D}).$$

REMARK 8.2.8.

(1) In the above, if V is just a colax functor, then $\mathrm{Colax}(\mathbf{B}, V)$ may not be well defined.

(2) If we replace the 2-category $\mathrm{Colax}(\mathbf{B}, \mathbf{C})$ by a bicategory consisting of the pseudofunctors, the strong transformations (transformations (F, ψ) with ψ a 2-isomorphism), and the modifications, then a statement corresponding to the theorem above follows from a result of Gordon–Power–Street [**25**].

By applying the theorem above to the sequence

$$I \xrightarrow{X} \Bbbk\text{-}\mathbf{Cat} \xrightarrow{\mathrm{Mod}} \Bbbk\text{-}\mathbf{Ab}$$

of a colax functor and a pseudofunctor, we obtain the following:

COROLLARY 8.2.9. *The following is a pseudofunctor:*

$$\begin{aligned}\mathrm{Colax}(I, \mathrm{Mod})\colon \mathrm{Colax}(I, \Bbbk\text{-}\mathbf{Cat}) &\to \mathrm{Colax}(I, \Bbbk\text{-}\mathbf{Ab}).\\ X &\mapsto \mathrm{Mod}\, X.\end{aligned}$$

REMARK 8.2.10.

(1) Let $X\colon I \to \Bbbk\text{-}\mathbf{Cat}$ be a colax functor. Then the colax functor $\mathrm{Mod}\, X$ has the following form: $\mathrm{Mod}\, X\colon I \to \Bbbk\text{-}\mathbf{Ab}$, $(i \xrightarrow{a} j) \mapsto (\mathrm{Mod}\, X(i) \xrightarrow{\text{-}\otimes_{X(i)}\overline{X(a)}} \mathrm{Mod}\, X(j))$, In particular, for $X(i)(\text{-}, x) \in \mathrm{Mod}\, X(i)$ $(i \in I_0, x \in X(i)_0)$, the following holds (cf. (3.1)):

$$((\mathrm{Mod}\, X)(a))(X(i)(\text{-}, x)) = X(i)(\text{-}, x) \otimes_{X(i)} \overline{X(a)} \cong X(j)(\text{-}, X(a)x). \tag{8.1}$$

(2) From the above, the fact that the module category $\mathrm{Mod}\, \mathscr{C}$ of a G-category $\mathscr{C}$ becomes also a G-category is generalized to the setting of the I-colax action. In particular, for a pseudo-G-category $(\mathscr{C}, X)$, this defines a module category $\mathrm{Mod}(\mathscr{C}, X)$ as a pseudo-G-category by setting $\mathrm{Mod}(\mathscr{C}, X) := \mathrm{Mod}\, X$.

(3) Moreover by setting $\mathbf{C}$ in Remark 8.2.3 to be the 2-category $\Bbbk\text{-}\mathbf{Ab}$ of all abelian linear light categories, "Morita equivalence" is defined for $X, X' \in \mathrm{Colax}(I, \Bbbk\text{-}\mathbf{Cat})_0$. Namely, X and X' are said to be *Morita equivalent* if $\mathrm{Mod}\, X$ and $\mathrm{Mod}\, X'$ are equivalent in the 2-category $\mathrm{Colax}(I, \Bbbk\text{-}\mathbf{Ab})$.

(4) By further composing the pseudofunctor that associates the derived category, the "derived category" of each $X \in \mathrm{Colax}(I, \Bbbk\text{-}\mathbf{Cat})_0$ and derived equivalences between them are defined in the same way. (see **[9]** for details).

8.3. Subsequent progress

After this, the pushdown functor between module categories or derived categories is defined and applied as in the case of group actions. For instance, **[9**, Therorem 8.1**]** guarantees that if two colax functors are derived equivalent, then their Grothendieck constructions are also derived equivalent. As a simple application of this theorem to Example 8.1.2, we obtain the following.

COROLLARY 8.3.1. *When $\Bbbk$ is a field, assume that $\Bbbk$-algebras A and A' are derived equivalent. Then the following hold:*

(1) *AQ and $A'Q$ are derived equivalent for all quivers Q.*
(2) *AS and $A'S$ are derived equivalent for all posets S.*
(3) *AG and $A'G$ are derived equivalent for all monoids G.*

Furthermore, the above theorem also allows us to "glue derived equivalences together."

APPENDIX A

Set theory for the foundation of category theory

In this book, we adopt ZFC (Zermelo-Fraenkel set-theory (ZF) with the axiom of choice (C)) as axioms of set theory, and we do not assume the existence of urelements. For any sets x and y, the pair (x, y) is defined as the Kuratowski's pair $(x, y) := \{\{x\}, \{x, y\}\}$, and the set of all maps from x to y is denoted by $\mathrm{Map}(x, y)$ or by y^x. For each set x, the power set (resp. the cardinality) of x is denoted by $\mathscr{P}x$ (resp. $|x|$), and we set $\omega := |\mathbb{N}|$. In the following, the class of all sets (resp. ordinal numbers) is denoted by SET (resp. ORD).

A.1. Universes

Before defining universes, we will introduce tools to control them in terms of ordinal numbers.

DEFINITION A.1.1. A map $\mathsf{V}\colon \mathsf{ORD} \to \mathsf{SET}$ is defined as follows: For each $\alpha \in \mathsf{ORD}$,

$$\mathsf{V}_\alpha := \begin{cases} \emptyset & (\alpha = 0); \\ \mathscr{P}\mathsf{V}_\beta & (\alpha = \beta + 1, \exists \beta \in \mathsf{ORD}); \\ \bigcup_{\beta<\alpha} \mathsf{V}_\beta & (\alpha \text{ is a limit ordinal}). \end{cases}$$

This map is called the *von Neumann hierarchy*.

An element of V_ω is called a *hereditarily finite set*. Since $|\mathsf{V}_0| = 0$, and it holds that $|\mathsf{V}_{n+1}| = 2^{|\mathsf{V}_n|}$ for all $n \in \mathbb{N}$, we have $|A| < \omega$ for all $A \in \mathsf{V}_\omega$ and $|\mathsf{V}_\omega| = \omega$.

The following is well known (see e.g., [**39**, Lemma 7.5.14, 7.5.18] for a proof):

LEMMA A.1.2. *The following hold.*

(1) *The map* V *is monotone, i.e., for any* $\alpha, \beta \in \mathsf{ORD}$, *we have "*$\alpha \leq \beta \Rightarrow \mathsf{V}_\alpha \subseteq \mathsf{V}_\beta$*".*
(2) *It holds that* $\mathsf{SET} = \bigcup_{\alpha \in \mathsf{ORD}} \mathsf{V}_\alpha$.

DEFINITION A.1.3. A *Grothendieck universe* (or a *universe* for short) is a set $\mathfrak{U}$ that satisfies the following conditions:

(1) $x \in \mathfrak{U}, y \in x \Rightarrow y \in \mathfrak{U}$;
(2) $\emptyset \in \mathfrak{U}$;
(3) $x, y \in \mathfrak{U} \Rightarrow \{x, y\} \in \mathfrak{U}$;
(4) $I \in \mathfrak{U}, x_i \in \mathfrak{U}\ (i \in I) \Rightarrow \bigcup_{i \in I} x_i \in \mathfrak{U}$;
(5) $x \in \mathfrak{U} \Rightarrow \mathscr{P}x \in \mathfrak{U}$.

EXAMPLE A.1.4. The set V_ω of all hereditarily finite sets is a universe. However, since each element of V_ω is a finite set, we have $\mathbb{N} \notin \mathsf{V}_\omega$.

REMARK A.1.5. Let $\mathfrak{U}$ be a universe. Then the following are immediate from conditions in Definition A.1.3:

(1) $x \in \mathfrak{U} \Rightarrow x \subseteq \mathfrak{U}$;
(2) $x \subseteq y, y \in \mathfrak{U} \Rightarrow x \in \mathfrak{U}$;
(3) $x, y \in \mathfrak{U} \Rightarrow (x, y) \in \mathfrak{U}$;
(4) $x, y \in \mathfrak{U} \Rightarrow x \cup y, x \times y \in \mathfrak{U}$;
(5) $x, y \in \mathfrak{U} \Rightarrow \mathrm{Map}(x, y) \in \mathfrak{U}$;
(6) $I \in \mathfrak{U}, x_i \in \mathfrak{U}\ (i \in I) \Rightarrow \prod_{i\in I} x_i, \bigsqcup_{i\in I} x_i \in \mathfrak{U}$;
(7) $\emptyset \neq I \in \mathfrak{U}, x_i \in \mathfrak{U}\ (i \in I) \Rightarrow \bigcap_{i\in I} x_i, \in \mathfrak{U}$;
(8) $x \in \mathfrak{U} \Rightarrow x \cup \{x\} \in \mathfrak{U}$;
This implies that $\mathbb{N} \subseteq \mathfrak{U}$. (Note, however, that it is not possible to prove that $\mathbb{N} \in \mathfrak{U}$ by the example above.) Therefore in particular, all finite unions, finite direct products, and finite disjoint unions of elements of $\mathfrak{U}$ belong to $\mathfrak{U}$.
(9) $x \in \mathfrak{U} \Rightarrow \bigcup x \in \mathfrak{U}$.
(10) $x \in \mathfrak{U}, y \subseteq \mathfrak{U}, f : x \to y$ is a surjection $\Rightarrow y \in \mathfrak{U}$.
(11) $\mathfrak{U} \notin \mathfrak{U}$.

PROOF. We omit the proofs since they are all straightforward, but show only (11). Assume contrarily that $\mathfrak{U} \in \mathfrak{U}$ holds. Then by Definition A.1.3 (5), we have $\mathscr{P}\mathfrak{U} \in \mathfrak{U}$. By (1) above, $\mathscr{P}\mathfrak{U} \subseteq \mathfrak{U}$. This shows that $|\mathscr{P}\mathfrak{U}| \leq |\mathfrak{U}|$. However, as long as $\mathfrak{U}$ is a set, Cantor's diagonal argument says that $|\mathfrak{U}| < |\mathscr{P}\mathfrak{U}|$, a contradiction. □

Universes are closely related to the following inaccessible cardinals[1].

DEFINITION A.1.6 (cf. [**44**]). A cardinal α is called an *inaccessible cardinal* if the following two conditions are satisfied:

(1) For any set I and any family $(B_i)_{i\in I}$ of sets, it holds that
$$|I| < \alpha, |B_i| < \alpha\ (i \in I) \Rightarrow \left|\bigcup_{i\in I} B_i\right| < \alpha;$$
(2) For any cardinals β, γ, we have
$$\beta, \gamma < \alpha \Rightarrow \beta^\gamma < \alpha.$$

We set the class of all universes (resp. the class of all inaccessible cardinals) to be Univ (resp. Ina). Then there exists a bijection between them (see [**47**] for a proof) as shown in the following theorem.

THEOREM A.1.7. *The following hold.*

(1) *If α is an inaccessible cardinal, then V_α is a universe.*
(2) *If $\mathfrak{U}$ is a universe, then $|\mathfrak{U}|$ is an inaccessible cardinal.*
(3) *Therefore monotone maps $\mathsf{V}\colon \mathsf{ORD} \to \mathsf{SET}$ and $|\text{-}|\colon \mathsf{SET} \to \mathsf{ORD}$ induce their restrictions*
$$\mathsf{V}\colon \mathsf{Ina} \to \mathsf{Univ}, \quad |\text{-}|\colon \mathsf{Univ} \to \mathsf{Ina},$$
respectively. These are inverse maps to each other. Therefore, Univ *and* Ina *are isomorphic as ordered classes. In particular, since* $\mathsf{Ina} \subseteq \mathsf{ORD}$, *both of them are well-ordered.*

[1] Here we deal with strongly inaccessible cardinals.

REMARK A.1.8. Since $\mathbb{N} \in \mathsf{V}_{\omega+1}$, all universes but V_ω has $\mathbb{N}$ as their element.

In addition to the ZFC, we assume the following axiom.

AXIOM OF UNIVERSE. Every set belongs to a universe.

The axiomatic system that adds this axiom to ZFC is called ZFCU. By Theorem A.1.7, Axiom of universe is equivalent to the following axiom under ZFC:

AXIOM OF INACCESSIBLE CARDINALS. Every cardinal is smaller than an inaccessible cardinal.

Axiom of inaccessible cardinal is known to be independent of the ZFC. Therefore, Axiom of universe is also independent of the ZFC.

DEFINITION A.1.9. A universe having $\mathbb{N}$ as its element is called an *infinite universe*. By the axiom above, an infinite universe exists, and by Remark A.1.8, all universes except for V_ω are infinite universes.

Throughout this book, we fix an infinite universe $\mathfrak{U}$.

REMARK A.1.10. Since $\mathbb{N} \in \mathfrak{U}$, we have $\mathbb{Z}, \mathbb{Q}, \mathbb{R}, \mathbb{C} \in \mathfrak{U}$. Therefore, all structures constructed from these using operations in set theory also belong to $\mathfrak{U}$.

DEFINITION A.1.11. Let A be a set.

(1) A is called a $\mathfrak{U}$-*small* set if $A \in \mathfrak{U}$.
(2) A is called a $\mathfrak{U}$-*class* if $A \subseteq \mathfrak{U}$.
(3) A $\mathfrak{U}$-class that is not a $\mathfrak{U}$-small set is called a *proper* $\mathfrak{U}$-*class*.
(4) A is called an *essentially* $\mathfrak{U}$-*small* set if there exists a $\mathfrak{U}$-small set B such that there exists a bijection $f\colon B \to A$.

In the following, we omit "$\mathfrak{U}$-" if there seems to be no confusion[2].

REMARK A.1.12.

(1) By Remark A.1.5 (1), small sets are classes.
(2) Since $\mathfrak{U} \notin \mathfrak{U}$, $\mathfrak{U}$ is a proper class.
(3) By Remark A.1.5 (10), essentially small classes are small sets.
(4) A class A is a proper class if and only if A is not an element of any classes. Indeed, assume that A is a proper class. If A is an element of some class C, then since $C \subseteq \mathfrak{U}$, we have $A \in \mathfrak{U}$, a contradiction. Since $\mathfrak{U}$ is a class, the converse is obvious. This fact shows, in particular, that *a set is not a class if it has a proper class as its element* .

EXAMPLE A.1.13 (essentially small set that is neither small nor a class). Let C be a proper class (e.g., $C = \mathfrak{U}$). Then the set $\{C\}$ has only one element, and it is essentially small. However, since it has a proper class as its element, it is not a class (therefore not a small set, either) by Remark A.1.12 (4).

Thus we have $\{C\} \in \mathscr{P}^2\mathfrak{U} \setminus \mathscr{P}\mathfrak{U}$, and hence $\mathscr{P}\mathfrak{U} \subsetneq \mathscr{P}^2\mathfrak{U}$. Therefore, for each $k \in \mathbb{N}$, we also have $\mathscr{P}^k\mathfrak{U} \subsetneq \mathscr{P}^{k+1}\mathfrak{U}$.

---

[2]If "$\mathfrak{U}$-" is omitted, then there is a risk of confusion between a class that is not a set and a $\mathfrak{U}$-class. In such a case, we carefully distinguish them.

A.2. Hierarchy of sets

To avoid the set paradox, it is sufficient to set up the three hierarchies defined above:

$$\{x \in \mathsf{SET} \mid x \text{ is a small set}\} \subsetneq \{x \in \mathsf{SET} \mid x \text{ is a class}\} \subsetneq \mathsf{SET}$$

(e.g., cf [**37**][3]). However, this alone does not give us a guarantee that once a construction rises to a non-class set, another larger construction built from it will stay within SET. For instance, even if $\mathscr{C}$ is a small linear category, the object set of the category $\operatorname{Mod}\mathscr{C}$ is a proper class, and hence the object set of the derived category $\mathscr{D}(\operatorname{Mod}\mathscr{C})$ becomes a non-class set. Then the category $\operatorname{Mod}(\mathscr{D}(\operatorname{Mod}\mathscr{C}))$ and/or the category $\mathbf{D}$ whose object "set" is $\{\mathscr{D}(\operatorname{Mod}\mathscr{C}) \mid \mathscr{C} \in \Bbbk\text{-}\mathbf{Cat}_0\}$ may not be constructed within the set. Since this is inconvenient, we will follow Levy [**33**] to hierarchize SET more finely. This is very convenient because it allows us to freely create "large" sets by collecting "many" sets belonging to a certain hierarchy. For example, in the above example, it is shown that $\mathscr{D}(\operatorname{Mod}\mathscr{C})$ is a 2-moderate category, and $\operatorname{Mod}(\mathscr{D}(\operatorname{Mod}\mathscr{C}))$ and $\mathbf{D}$ are 3-moderate category, which fall within the set. In this way, almost all arguments can be carried out within SET.

DEFINITION A.2.1. Let A be a set, and assume that $\mathfrak{U} \subseteq A$. By axiom of universe and Theorem A.1.7, there exists the smallest universe $\mathfrak{U}'$ such that $A \in \mathfrak{U}'$. We define ΨA to be the smallest set $X \in \mathfrak{U}'$ satisfying the condition that

$$X \supseteq A \cup (X \times X) \cup (\bigcup_{I \in A} X^I).$$

Therefore in particular, since $X = \mathfrak{U}$ satisfies this condition for $A = \mathfrak{U}$, we have $\Psi\mathfrak{U} = \mathfrak{U}$.

PROPOSITION A.2.2. *For each set A satisfying $\mathfrak{U} \subseteq A$, there exists ΨA.*

PROOF. For each $X \in \mathfrak{U}'$, we set

$$\tau(X) := A \cup (X \times X) \cup \left(\bigcup_{I \in A} X^I \right).$$

Here since $\mathfrak{U}'$ is a universe, it follows from $A, X \in \mathfrak{U}'$ that $\tau(X) \in \mathfrak{U}'$. Therefore, the above correspondence defines a map $\tau \colon \mathfrak{U}' \to \mathfrak{U}'$, and by this form we see that τ is monotone. By transfinite induction, for each $\alpha \in \mathsf{ORD}$, we define $\tau^\alpha(A) \in \mathfrak{U}'$ as follows:

$$\tau^\alpha(A) := \begin{cases} A & (\alpha = 0); \\ \tau(\tau^\beta(A)) & (\alpha = \beta + 1, \exists \beta \in \mathsf{ORD}); \\ \bigcup_{\beta < \alpha} \tau^\beta(A) & (\alpha \text{ is a limit ordinal}). \end{cases}$$

Again by transfinite induction, it is easy to verify that for any ordinals α, β, we have that $\alpha \leq \beta$ implies $\tau^\alpha(A) \leq \tau^\beta(A)$. By using this, we set

$$\tau^*(A) := \bigcup_{\alpha \in \mathsf{ORD}} \tau^a(A).$$

Here, note that at this stage it is not clear whether $\tau^*(A)$ is a set because ORD is a proper class (not only a proper $\mathfrak{U}$-class but a class that is not a set).

[3]There, a $\mathfrak{U}$-small set (resp. a $\mathfrak{U}$-class, a set) is called a "set" (resp. a "class", a "conglomerate").

Now, since $\tau^?(A)\colon \alpha \mapsto \tau^\alpha(A)$ is a map from a proper class ORD to a set $\mathfrak{U}'$, it is not an injection. Therefore, there exist ordinal numbers $\lambda < \mu$ such that $\tau^\lambda(A) = \tau^\mu(A)$. Here, for each β with $\lambda \le \beta \le \mu$, the inequalities $\tau^\lambda(A) \le \tau^\beta(A) \le \tau^\mu(A) = \tau^\lambda(A)$ show that $\tau^\beta(A) = \tau^\lambda(A)$. In particular, since $\lambda < \lambda + 1 \le \mu$, we have $\tau^\lambda(A) = \tau^{\lambda+1}(A)$. Therefore the set

$$\mathscr{E} := \{\alpha \in \mathsf{ORD} \mid \tau^\alpha(A) = \tau^{\alpha+1}(A)\}$$

is not empty, and hence $\mathscr{E}$ has the smallest ordinal number κ by its well-orderedness. By transfinite induction, it is easy to show that $\tau^\kappa(A) = \tau^\alpha(A)$ for all ordinal number α with $\kappa \le \alpha$. Hence we have

$$\tau^*(A) = \tau^\kappa(A) \in \mathfrak{U}'.$$

Thus we see that $\tau^*(A)$ is a $\mathfrak{U}'$-small set. Since $\kappa \in \mathscr{E}$, we have $\tau^\kappa(A) = \tau^{\kappa+1}(A)$. Thus

$$\tau^\kappa(A) = \tau(\tau^\kappa(A)). \tag{A.1}$$

Therefore the set

$$\mathscr{F} := \{X \in \mathfrak{U}' \mid X \supseteq \tau(X)\} \ (\ni \tau^\kappa(A))$$

is not empty. Hence $\tau^\dagger(A) := \bigcap \mathscr{F}$ $(\subseteq \tau^\kappa(A))$ is defined. It remains to show that $\tau^\dagger(A) \supseteq \tau(\tau^\dagger(A))$ to see that $\tau^\dagger(A)$ satisfies the condition on ΨA. For each $X \in \mathscr{F}$, since $X \supseteq \tau^\dagger(A)$ and τ is monotone, we have $X \supseteq \tau(X) \supseteq \tau(\tau^\dagger(A))$. Therefore $\tau^\dagger(A) = \bigcap \mathscr{F} \supseteq \tau(\tau^\dagger(A))$. Accordingly, $\Psi A = \tau^\dagger(A)$ exists. □

REMARK A.2.3. Similar to the Knaster–Tarski theorem (see e.g., [**28**, Theorem 1.12])[4], in fact it also holds that $\tau^\dagger(A) = \tau^\kappa(A)$ in the proof above. (It is easy to show by transfinite induction on α that $\tau^\dagger(A) \supseteq \tau^\alpha(A)$ for all $\alpha \in \mathsf{ORD}$, which shows the assertion.) Thus ΨA is given by $\Psi A = \tau^\kappa(A)$. Therefore by the equality (A.1), we have

$$\Psi A = \tau(\Psi A) = A \cup (\Psi A \times \Psi A) \cup \left(\bigcup_{I \in A} (\Psi A)^I \right). \tag{A.2}$$

In particular, the set ΨA satisfies the following:

(1) $I \in A \Rightarrow I \in \Psi A$ (in particular, $\mathfrak{U} \subseteq A$ shows that $x \in \mathfrak{U} \Rightarrow x \in \Psi A$);
(2) $x, y \in \Psi A \Rightarrow (x, y) \in \Psi A$;
(3) $I \in A, x_i \in \Psi A$ $(i \in I) \Rightarrow (x_i)_{i \in I} \in \Psi A$.

By the properties of ΨA above, the following is immediate:

PROPOSITION A.2.4. *Let A be a set satisfying $\mathfrak{U} \subseteq A$. Then the following hold.*

(1) *Every subset of A is a subset of ΨA.*
(2) *If $B, C \subseteq \Psi A$, then the following are also subsets of ΨA:*

$$B \sqcup C := \{(0, b) \mid b \in B\} \cup \{(1, c) \mid c \in C\} \text{ and}$$
$$B \times C := \{(b, c) \mid b \in B, c \in C\}.$$

(3) *If $B \subseteq \Psi A, C_b \subseteq \Psi A$ for all $b \in B$, then the following is a subset of ΨA:*

$$\bigsqcup_{b \in B} C_b := \{(b, c) \mid b \in B, c \in C_b\}.$$

---

[4]In our setting, unlike the setting of the Knaster–Tarski theorem, the map τ does not have the form $\mathscr{P}S \to \mathscr{P}S$ $(S \in \mathsf{SET})$, and it is not possible to take $X = S$ as a trivial example of a set X with $\tau(X) \supseteq X$.

(4) *If $I \in A$, $B_i \subseteq \Psi A$ for all $i \in I$, then the following is a subset of ΨA:*

$$\prod_{i \in I} B_i := \{(b_i)_{i \in I} \mid b_i \in B_i \ (\forall i \in I)\}.$$

From here to Section A.5, we let $k \in \mathbb{N}$. Since $\mathfrak{U} \subseteq \mathscr{P}\mathfrak{U}$ and $\mathfrak{U} = \Psi\mathfrak{U}$, we have $\mathfrak{U} \subseteq \mathscr{P}\Psi\mathfrak{U}$. Therefore, since both $\mathscr{P}$ and Ψ are monotone maps from SET to SET, we have the following inclusions:

$$\begin{array}{ccccc} (\mathscr{P}\Psi)^k\mathfrak{U} & \hookrightarrow & (\mathscr{P}\Psi)^{k+1}\mathfrak{U} & & \\ \downarrow & & \downarrow & & \\ \Psi(\mathscr{P}\Psi)^k\mathfrak{U} & \hookrightarrow & \Psi(\mathscr{P}\Psi)^{k+1}\mathfrak{U} & \hookrightarrow & \mathsf{SET} \end{array}$$

Using this, we define the following:

DEFINITION A.2.5.

(1) An element of $(\mathscr{P}\Psi)^k\mathfrak{U}$ is called a *k-class*, and an element of $((\mathscr{P}\Psi)^k\mathfrak{U}) \setminus ((\mathscr{P}\Psi)^{k-1}\mathfrak{U})$ is called a *proper k-class*.
(2) The category of the k-classes (and the maps between them) is denoted by $\mathbf{Class}^k$. Therefore we have $\mathbf{Class}^k_0 = (\mathscr{P}\Psi)^k\mathfrak{U}$.

REMARK A.2.6. The following are immediate from the definition above:

(1) A 0-class is nothing but a small set.
(2) A 1-class is nothing but a class (indeed, $\Psi\mathfrak{U} = \mathfrak{U}$ shows that $(\mathscr{P}\Psi)\mathfrak{U} = \mathscr{P}\mathfrak{U}$).
(3) A k-class is nothing but a subset of $\Psi\mathbf{Class}^{k-1}_0$ for all $k \geq 1$.

By Proposition A.2.4 we have the following, where if $k = 0$, replace "$k - 1$" with 0.

PROPOSITION A.2.7. *The following hold.*

(1) *A subset of a k-class is a k-class* (by Remark A.2.6 (3)).
(2) *A set consisting of k-classes is a $(k+1)$-class.*
(3) *If B, C are k-classes, then both $B \sqcup C$ and $B \times C$ are k-classes.*
(4) *If B is a k-class, and C_b is a k-class for each $b \in B$, then $\bigsqcup_{b \in B} C_b$ is a k-class.*
(5) *If I is a $(k-1)$-class, and B_i is a k-class for each $i \in I$, then $\prod_{i \in I} B_i$ is a k-class.*
(6) *If I is a $(k-1)$-class, and B is a k-class, then* $\mathrm{Map}(I, B)$ *is a k-class.*

PROOF. Every statement is easy to show, and the proof is left to the reader. Note that (6) is obtained by applying (5) to the case that $B_i = B$ $(i \in I)$. □

LEMMA A.2.8. *Let $n \in \mathbb{N}$, $X \in \mathbf{Class}^n_0$, and R be an equivalence relation on X. Then the quotient set X/R satisfies the following:*

$$X/R \in \begin{cases} \mathbf{Class}^0_0 & (n = 0); \\ \mathbf{Class}^{n+1}_0 & (n \geq 1). \end{cases}$$

PROOF. When $n = 0$, since there exists a surjection $\mathfrak{U} \ni X \to X/R \subseteq \mathfrak{U}$, we have $X/R \in \mathfrak{U} = \mathbf{Class}^0_0$.

Next, let $n \geq 1$. For each $x \in X$, denote the equivalence class of x by $[x]$. Then $X/R = \{[x] \mid x \in X\}$. Since $\mathbf{Class}^n_0$ is closed under taking subsets by its construction, $[x] \subseteq X \in \mathbf{Class}_0^n$ implies $[x] \in \mathbf{Class}_0^n$ for all $x \in X$. Therefore, $X/R \subseteq \mathbf{Class}_0^n$. Thus, $X/R \in \mathscr{P}\mathbf{Class}_0^n \subseteq \mathscr{P}\Psi\mathbf{Class}_0^n = \mathbf{Class}_0^{n+1}$. □

A.3. Hierarchy of categories, k-moderate categories

DEFINITION A.3.1. Let $\mathscr{C}$ be a category.

(1) $\mathscr{C}$ is called a *small category* if $\mathscr{C}_0$ and all local morphism sets are small. The 2-category of all small categories is denoted by **Cat**.
(2) $\mathscr{C}$ is called a *light category* if $\mathscr{C}_0$ is a class, and all local morphism sets are small sets. The 2-category of all light categories is denoted by **CAT**.
(3) $\mathscr{C}$ is called a *moderate category* if $\mathscr{C}_0$ and all local morphism sets are classes. The 2-category of all moderate categories is denoted by $\underline{\mathbf{CAT}}$.
(4) More generally, $\mathscr{C}$ is called a *k-moderate category* if $\mathscr{C}_0$ and all local morphism sets are k-classes. The 2-category of all k-moderate categories is denoted by $\underline{\mathbf{CAT}}^k$. $\mathscr{C}$ is called a *properly* k-moderate category if it is k-moderate but not $(k-1)$-moderate.

REMARK A.3.2. A 0-moderate category is nothing but a small category. A 1-moderate category is just a moderate category. A small category is a light category, and a light category is a 1-moderate category.

LEMMA A.3.3. *If $\mathscr{C}$ is a k-moderate category, then $\mathscr{C} \in \Psi\mathbf{Class}_0^k$.*

PROOF. We may set $\mathscr{C} = (\mathscr{C}_0, \mathscr{C}_1, \mathrm{dom}, \mathrm{cod}, \circ, \mathbb{1})$. By assumption, we have $\mathscr{C}_0 \in \mathbf{Class}_0^k$, and since $\mathscr{C}_1 = \bigsqcup_{(x,y)\in\mathscr{C}_0\times\mathscr{C}_0} \mathscr{C}(x,y)$, we have $\mathscr{C}_1 \in \mathbf{Class}_0^k$. Note here that each of the following is a map from a k-class to a k-class: $\mathrm{dom}, \mathrm{cod}\colon \mathscr{C}_1 \to \mathscr{C}_0$, $\circ_{x,y,z}\colon \mathscr{C}(y,z) \times \mathscr{C}(x,y) \to \mathscr{C}(x,z)$ $((x,y,z) \in \mathscr{C}_0 \times \mathscr{C}_0 \times \mathscr{C}_0)$, and $\mathbb{1}\colon \mathscr{C}_0 \to \mathscr{C}_1$. In order to deal with these maps in a unified way, we first show the following.

CLAIM. *If $f\colon A \to B$ is a map, and $A, B \in \mathbf{Class}_0^k$, then $f \in \Psi\mathbf{Class}_0^k$.*

Indeed, at first, when $k = 0$, $A, B \in \mathfrak{U}$ shows that $f \in \mathrm{Map}(A,B) \in \mathfrak{U}$. Therefore, $f \in \mathfrak{U} = \Psi\mathbf{Class}_0^0$, and the claim holds.

Next consider the case where $k \geq 1$. Then $f = (f(x))_{x\in A}$, and $A \in \mathbf{Class}_0^k$. Here since $B \in \mathbf{Class}_0^k = \mathscr{P}\Psi\mathbf{Class}_0^{k-1}$, we have $B \subseteq \Psi\mathbf{Class}_0^{k-1} \subseteq \Psi\mathbf{Class}_0^k$. Therefore, for each $x \in A$, $f(x) \in B$ shows that $f(x) \in \Psi\mathbf{Class}_0^k$. Hence by the property (3) in Remark A.2.3, we have $f = (f(x))_{x\in A} \in \Psi\mathbf{Class}_0^k$. This proves the claim.

By this claim, we have $\mathrm{dom}, \mathrm{cod}, \circ_{x,y,z}, \mathbb{1} \in \Psi\mathbf{Class}_0^k$. Again by the property (3) in Remark A.2.3, we have $\circ = (\circ_{x,y,z})_{(x,y,z)\in\mathscr{C}_0\times\mathscr{C}_0\times\mathscr{C}_0} \in \Psi\mathbf{Class}_0^k$, and also $\mathscr{C}_0, \mathscr{C}_1 \in \mathbf{Class}_0^k \subseteq \Psi\mathbf{Class}_0^k$. Therefore, as $\mathscr{C}$ itself is a finite sequence of these things, we have $\mathscr{C} \in \Psi\mathbf{Class}_0^k$. □

By Proposition A.2.7 we obtain the following.

PROPOSITION A.3.4. *Let $\mathscr{C}$ and $\mathscr{D}$ be categories. Then the functor category* $\mathrm{Fun}(\mathscr{C},\mathscr{D})$ *is*

$$\begin{cases} \textit{small} & (\mathscr{C}, \mathscr{D} \textit{ are small}), \\ \textit{light} & (\mathscr{C} \textit{ is small and } \mathscr{D} \textit{ is light}), \textit{ and} \\ k\textit{-moderate} & (\mathscr{C} \textit{ is } (k-1)\textit{-moderate and } \mathscr{D} \textit{ is } k\textit{-moderate}). \end{cases}$$

PROOF. We only show the last case because the remaining cases are shown similarly. By definition of a functor category, the following holds:

$$\mathrm{Fun}(\mathscr{C},\mathscr{D})_0 \subseteq \bigsqcup_{F_0\in\mathrm{Map}(\mathscr{C}_0,\mathscr{D}_0)} \prod_{(x,y)\in\mathscr{C}_0\times\mathscr{C}_0} \mathrm{Map}(\mathscr{C}(x,y),\mathscr{D}(F_0x,F_0y)).$$

Apply Proposition A.2.7 (3), (6), (5), (6), (4), (1) to this to see that $\mathrm{Fun}(\mathscr{C},\mathscr{D})_0$ is a k-class. For each $E,F\in\mathrm{Fun}(\mathscr{C},\mathscr{D})_0$, the following holds by definition of a functor category:

$$\mathrm{Fun}(\mathscr{C},\mathscr{D})(E,F)\subseteq \prod_{x\in\mathscr{C}_0}\mathscr{D}(Ex,Fx).$$

By applying Proposition A.2.7 (6), (1) to this, we see that $\mathrm{Fun}(\mathscr{C},\mathscr{D})(E,F)$ is a k-class. □

EXAMPLE A.3.5. Let $\mathscr{C}$ be a small $\Bbbk$-category (resp. a light $\Bbbk$-category). Then it is 0-moderate (resp. 1-moderate). Since $\Bbbk$-Mod is a light category, it is also 1-moderate (hence also 2-moderate and 3-moderate). Therefore, by the proposition above, the module category $\mathrm{Mod}\,\mathscr{C} = \mathrm{Fun}(\mathscr{C}^{\mathrm{op}},\Bbbk\text{-}\mathrm{Mod})$ is light (resp. 2-moderate), and the module category $\mathrm{Mod}(\mathrm{Mod}\,\mathscr{C}) = \mathrm{Fun}((\mathrm{Mod}\,\mathscr{C})^{\mathrm{op}},\Bbbk\text{-}\mathrm{Mod})$ is 2-moderate (resp. 3-moderate).

REMARK A.3.6. Let $\mathscr{C}$ be a category with $\mathscr{C}_0$ a proper class. Then $\mathrm{Fun}(\mathscr{C},\mathscr{C})_0$ is a proper 2-class, and $\mathrm{Fun}(\mathscr{C},\mathscr{C})$ has a local morphism set that is a proper 2-class.

Indeed, if $\mathrm{Fun}(\mathscr{C},\mathscr{C})_0$ is a 1-class, then

$$\mathfrak{U}\supseteq \mathrm{Fun}(\mathscr{C},\mathscr{C})_0 \ni \mathbb{1}_{\mathscr{C}} \ni \{\mathbb{1}_{\mathscr{C}_0}\} \ni \mathbb{1}_{\mathscr{C}_0} = \{(x,x)\mid x\in\mathscr{C}_0\} \xrightarrow{\sim} \mathscr{C}_0$$

shows that $\mathscr{C}_0$ is a small set, a contradiction.

It also holds that $\mathrm{Fun}(\mathscr{C},\mathscr{C})(\mathbb{1}_{\mathscr{C}},\mathbb{1}_{\mathscr{C}})$ is not a 1-class. Indeed, otherwise,

$$\mathfrak{U}\supseteq \mathrm{Fun}(\mathscr{C},\mathscr{C})(\mathbb{1}_{\mathscr{C}},\mathbb{1}_{\mathscr{C}}) \ni \mathbb{1}_{\mathbb{1}_{\mathscr{C}}} = \{(x,\mathbb{1}_x)\mid x\in\mathscr{C}_0\} \xrightarrow{\sim} \mathscr{C}_0$$

yields the same contradiction. Therefore, $\mathrm{Fun}(\mathscr{C},\mathscr{C})(\mathbb{1}_{\mathscr{C}},\mathbb{1}_{\mathscr{C}})$ is a proper 2-class.

COROLLARY A.3.7. *The category* $\mathbf{Class}^k$ *is* $(k+1)$*-moderate.*

PROOF. By Proposition A.2.7 (2), $\mathbf{Class}^k_0$ is a $(k+1)$-class. For any $A,B\in\mathbf{Class}^k_0$, we have $(\mathbf{Class}^k)(A,B) = \mathrm{Map}(A,B)$. Here, A is a k-class, and B is a k-class, and hence also a $(k+1)$-class. Therefore by Proposition A.2.7(6), $(\mathbf{Class}^k)(A,B)$ is a $(k+1)$-class. □

A.4. Hierarchy of 2-categories

When $\mathscr{C}$ is a 2-category, if $x,y\in\mathscr{C}_0$ and $f,g\in\mathscr{C}(x,y)_0$, then $\mathscr{C}(x,y)_0$ (resp. $\mathscr{C}(x,y)(f,g)$) is called a *local* 1*-morphism set* (resp. *local* 2*-morphism set*) of $\mathscr{C}$.

DEFINITION A.4.1. Let $\mathscr{C}$ be a 2-category.

(1) $\mathscr{C}$ is said to be *small* if $\mathscr{C}_0$ and all of its local r-morphism sets are small for each $r=1,2$.
(2) $\mathscr{C}$ is said to be *light* if $\mathscr{C}_0$ is a class, and all of its local r-morphism sets are small for each $r=1,2$.
(3) $\mathscr{C}$ is said to be *k-moderate* if $\mathscr{C}_0$ and all of its local r-morphism sets are k-classes for each $r=1,2$, namely if $\mathscr{C}_0$ is a k-class, and categories $\mathscr{C}(x,y)$ are k-moderate for all $x,y\in\mathscr{C}_0$.

PROPOSITION A.4.2. *The following hold.*

(1) *The* 2*-category* **Cat** *is light;*
(2) *The* 2*-category* **CAT** *is* 2*-moderate; and*
(3) *The* 2*-category* $\underline{\mathbf{CAT}}^k$ *is* $(k+1)$*-moderate.*

PROOF. Statement (1) is obvious from definition.

(3) By Lemma A.3.3, $\underline{\mathbf{CAT}}_0^k \subseteq \Psi\mathbf{Class}_0^k$. Thus $\underline{\mathbf{CAT}}_0^k \in \mathscr{P}\Psi\mathbf{Class}_0^k = \mathbf{Class}_0^{k+1}$, and hence $\underline{\mathbf{CAT}}_0^k$ is a $(k+1)$-class. Next for any $\mathscr{C}, \mathscr{D} \in \underline{\mathbf{CAT}}_0^k$, $\underline{\mathbf{CAT}}^k(\mathscr{C}, \mathscr{D}) = \mathrm{Fun}(\mathscr{C}, \mathscr{D})$ is $(k+1)$-moderate by Proposition A.3.4. Therefore, the 2-category $\underline{\mathbf{CAT}}^k$ is $(k+1)$-moderate.

(2) **CAT** is a full 2-subcategory of $\underline{\mathbf{CAT}}^1$. Namely, we have $\mathbf{CAT}_0 \subseteq \underline{\mathbf{CAT}}_0^1$, $\mathbf{CAT}(\mathscr{C}, \mathscr{D}) = \underline{\mathbf{CAT}}^1(\mathscr{C}, \mathscr{D})$ for all $\mathscr{C}, \mathscr{D} \in \mathbf{CAT}_0$, and $\mathbf{CAT}(\mathscr{C}, \mathscr{D})(E, F) = \underline{\mathbf{CAT}}^1(\mathscr{C}, \mathscr{D})(E, F)$ for all $E, F \in \mathbf{CAT}(\mathscr{C}, \mathscr{D})$, where $\underline{\mathbf{CAT}}^1$ is 2-moderate by statement (3). Therefore, **CAT** is also 2-moderate. □

REMARK A.4.3. Let $\mathscr{C}$ be a 1-moderate category with $\mathscr{C}_0$ not small. Then by Remark A.3.6, the local 1-morphism set $\mathrm{Fun}(\mathscr{C}, \mathscr{C})_0$ of $\mathbf{Cat}(\{\mathscr{C}\})$ is not a 1-class, and hence $\mathbf{Cat}(\{\mathscr{C}\})$ cannot be 1-moderate, thus turns out to be a proper 2-moderate 2-category. Therefore even if $\mathscr{U}$ is an essentially small set consisting of light categories, if any of them is not a small category, then $\mathbf{Cat}(\mathscr{U})$ turns out to be a proper 2-moderate 2-category.

PROPOSITION A.4.4. *The following hold.*

(1) *The* 2*-category* $\Bbbk$-**Cat** *is light; and*
(2) *The* 2*-category* $\Bbbk$-**CAT** *is* 2*-moderate.*

PROOF. We show only statement (2) because the similar argument works also for (1). By definition of light $\Bbbk$-categories, we have

$$\begin{aligned}\Bbbk\text{-}\mathbf{Cat}_0 = \{(\mathscr{C}, (+_{x,y}, \cdot_{x,y}))_{(x,y)\in\mathscr{C}_0\times\mathscr{C}_0} \mid \\ &\mathscr{C} \in \mathbf{CAT}_0, (\mathscr{C}(x,y), +_{x,y}, \cdot_{x,y}) \in (\mathrm{Mod}\,\Bbbk)_0 \text{ for all } x, y \in \mathscr{C}_0, \text{ and} \\ &\cdot_{x,y} \colon \mathscr{C}(y,z) \times \mathscr{C}(x,y) \to \mathscr{C}(x,z) \text{ is } \Bbbk\text{-bilinear for all } x, y, z \in \mathscr{C}_0\} \\ \subseteq \bigsqcup_{\mathscr{C}\in\mathbf{CAT}_0} \prod_{x,y\in\mathscr{C}_0} &\mathscr{C}(x,y)^{\mathscr{C}(x,y)\times\mathscr{C}(x,y)} \times \mathscr{C}(x,y)^{\Bbbk\times\mathscr{C}(x,y)}.\end{aligned}$$

Here, since $\mathscr{S}(x,y) := \mathscr{C}(x,y)^{\mathscr{C}(x,y)\times\mathscr{C}(x,y)} \times \mathscr{C}(x,y)^{\Bbbk\times\mathscr{C}(x,y)}$ is small for all $x, y \in \mathscr{C}_0$ and $\mathscr{C}_0$ is a 1-class, $\prod_{x,y\in\mathscr{C}_0} \mathscr{S}(x,y)$ is a 2-class for all $\mathscr{C} \in \mathbf{CAT}_0$. In addition, $\mathbf{CAT}_0$ is a 2-class by Proposition A.4.2. These show that $\bigsqcup_{\mathscr{C}\in\mathbf{CAT}_0} \prod_{x,y\in\mathscr{C}_0} \mathscr{S}(x,y)$ is also a 2-class. Therefore, $\Bbbk$-$\mathbf{CAT}_0$ is a 2-class, as well.

Let $\mathrm{Fgt} \colon \Bbbk\text{-}\mathbf{CAT} \to \mathbf{CAT}$ be the forgetful functor. Then for any $\mathscr{C}, \mathscr{D} \in \Bbbk\text{-}\mathbf{CAT}_0$, we have $(\Bbbk\text{-}\mathbf{CAT})(\mathscr{C}, \mathscr{D}) \subseteq \mathbf{CAT}(\mathrm{Fgt}(\mathscr{C}), \mathrm{Fgt}(\mathscr{D}))$, and the right-hand side is 2-moderate by Proposition A.4.2. Hence $(\Bbbk\text{-}\mathbf{CAT})(\mathscr{C}, \mathscr{D})$ is also 2-moderate. □

A.5. Application 1: The category of colax functors and the derived category

PROPOSITION A.5.1. *If I is a k-moderate category, then* $\mathrm{Colax}(I, \underline{\mathbf{CAT}^k})$ *is a* $(k+1)$*-moderate 2-category. In particular, if I is a small category, then the following hold.*

(1) *Both* $\mathrm{Colax}(I, \Bbbk\text{-}\mathbf{Cat})$ *and* $\mathrm{Colax}(I, \underline{\mathbf{CAT}^0})$ *are 1-moderate 2-categories. Therefore, so is G-*$\mathbf{Cat}$.

(2) *Both* $\mathrm{Colax}(I, \Bbbk\text{-}\mathbf{CAT})$ *and* $\mathrm{Colax}(I, \underline{\mathbf{CAT}^1})$ *are 2-moderate 2-categories.*

PROOF. By definition of colax functors, the following holds:

$$\begin{aligned}
&\mathrm{Colax}(I, \underline{\mathbf{CAT}^k})_0 \\
&\subseteq \bigsqcup_{X \in \mathrm{Fun}(I, \underline{\mathbf{CAT}^k})_0} \left(\prod_{i \in I_0} \mathrm{Fun}(X(i), X(i))(X(\mathbb{1}_i), \mathbb{1}_{X(i)}) \right. \\
&\quad \left. \times \prod_{(b,a) \in I_1 \times_{I_0} I_1} \mathrm{Fun}(X(\mathrm{dom}(a)), X(\mathrm{cod}(b)))(X(ba), X(b) \circ X(a)) \right).
\end{aligned}$$

Here, the category $\mathrm{Fun}(I, \underline{\mathbf{CAT}^k})$ is $(k+1)$-moderate because I is k-moderate and $\underline{\mathbf{CAT}^k}$ is $(k+1)$-moderate. Thus $\mathrm{Fun}(I, \underline{\mathbf{CAT}^k})_0$ is a $(k+1)$-class, and each local 2-morphism set $\mathrm{Fun}(\mathscr{C}, \mathscr{C}')(E, F)$ ($\mathscr{C}, \mathscr{C}' \in \underline{\mathbf{CAT}^k_0}, E, F \in \mathrm{Fun}(\mathscr{C}, \mathscr{C}')_0$) is a $(k+1)$-class as well. Now since both I_0 and $I_1 \times_{I_0} I_1$ are k-classes, the term in parentheses in the right-hand side is a $(k+1)$-class. Therefore, the right-hand side, and hence the left-hand side $\mathrm{Colax}(I, \underline{\mathbf{CAT}^k})_0$ is also a $(k+1)$-class. Next let $X, X' \in \mathrm{Colax}(I, \underline{\mathbf{CAT}^k})_0$. Then by definition of morphisms between colax functors, the following holds:

$$\begin{aligned}
&\mathrm{Colax}(I, \underline{\mathbf{CAT}^k})(X, X') \\
&\subseteq \bigsqcup_{F \in \prod_{i \in I_0} \mathrm{Fun}(X(i), X'(i))_0} \left(\prod_{(i,j) \in I_0^2} \right. \\
&\qquad \left. \bigsqcup_{a \in I(i,j)} \mathrm{Fun}(X(i), X'(j))(X'(a)F(i), F(j)X(a)) \right).
\end{aligned}$$

Here, since I_0 is a k-class and $\mathrm{Fun}(X(i), X'(i))_0$ is a $(k+1)$-class, the set $\prod_{i \in I_0} \mathrm{Fun}(X(i), X'(i))_0$ is a $(k+1)$-class. Similarly, since the term in parenthesis is also a $(k+1)$-class, the right-hand side is a $(k+1)$-class. Therefore, $\mathrm{Colax}(I, \underline{\mathbf{CAT}^k})(X, X')$ is a $(k+1)$-class as well.

Finally, let $(F, \psi), (F', \psi') \in \mathrm{Colax}(I, \underline{\mathbf{CAT}^k})(X, X')_0$. Then by definition of 2-morphisms in $\mathrm{Colax}(I, \underline{\mathbf{CAT}^k})$, the following holds:

$$\begin{aligned}
&\mathrm{Colax}(I, \underline{\mathbf{CAT}^k})(X, X')((F, \psi), (F', \psi')) \\
&\subseteq \prod_{i \in I_0} \underline{\mathbf{CAT}^k}(X(i), X'(i))(F(i), F'(i)).
\end{aligned}$$

Here, since I_0 is a k-class and $\underline{\mathbf{CAT}^k}$ is $(k+1)$-moderate, the right-hand side is a $(k+1)$-class, and $\mathrm{Colax}(I, \underline{\mathbf{CAT}^k})(X, X')((F, \psi), (F', \psi'))$ is a $(k+1)$-class as well. □

Although we do not explain derived categories in the text, we note about the level of moderation of the derived category at the end.

PROPOSITION A.5.2. *If $\mathscr{A}$ is a light abelian category, then its derived category $\mathscr{D}(\mathscr{A})$ is a proper 2-moderate category. Thus, it is beyond the range of the usual categories (i.e., the light categories).*

PROOF. Let $\mathscr{C}(\mathscr{A})$ (resp. $\mathscr{K}(\mathscr{A})$) be the category of chain complexes in (resp. the homotopy category of) $\mathscr{A}$. We first show that $\mathscr{D}(\mathscr{A})_0$ is a proper class. By definition of objects of the derived category, we have

$$\mathscr{D}(\mathscr{A})_0 = \mathscr{C}(\mathscr{A})_0 \subseteq \bigsqcup_{(X^n)_n \in \mathscr{A}_0^{\mathbb{Z}}} \prod_{n\in\mathbb{Z}} \mathscr{A}(X^n, X^{n+1}).$$

Since $\mathscr{A}_0^{\mathbb{Z}}$ is a class, and both $\mathbb{Z}$ and $\mathscr{A}(X^n, X^{n+1})$ are small sets, $\mathscr{D}(\mathscr{A})_0$ is a class. We also have

$$\mathscr{D}(\mathscr{A})_0 \supseteq \{(\cdots \to 0 \to X^0 \to 0 \cdots) \mid X^0 \in \mathscr{A}_0\} \xrightarrow{\text{bijective}} \mathscr{A}_0 \subseteq \mathfrak{U}.$$

Therefore if $\mathfrak{U} \ni \mathscr{D}(\mathscr{A})_0$, then $\mathscr{A}_0 \in \mathfrak{U}$, a contradiction. Hence $\mathscr{D}(\mathscr{A})_0$ has to be a proper class.

We next investigate local morphism sets. Take any $X^{\bullet}, Y^{\bullet} \in \mathscr{D}(\mathscr{A})_0 = \mathscr{K}(\mathscr{A})_0$. Then we first show that $\mathscr{K}(\mathscr{A})(X^{\bullet}, Y^{\bullet})$ is a small set. Since we have $\mathscr{K}(\mathscr{A})(X^{\bullet}, Y^{\bullet}) \in \mathscr{P}(\mathscr{P}(\mathscr{C}(\mathscr{A})(X^{\bullet}, Y^{\bullet})))$ and $\mathscr{C}(\mathscr{A})(X^{\bullet}, Y^{\bullet}) \subseteq \prod_{n\in\mathbb{Z}} \mathscr{A}(X^n, Y^n) \in \mathfrak{U}$, it holds that $\mathscr{K}(\mathscr{A})(X^{\bullet}, Y^{\bullet}) \in \mathfrak{U}$. Thus $\mathscr{K}(\mathscr{A})(X^{\bullet}, Y^{\bullet})$ is a small set. Next we show that $\mathscr{D}(\mathscr{A})(X^{\bullet}, Y^{\bullet})$ is a 2-class. By definition of morphisms of the derived category, the following holds ($[f]$ is the equivalence class of f in $\mathscr{D}(\mathscr{A})(X^{\bullet}, Y^{\bullet})$):

$$\mathscr{D}(\mathscr{A})(X^{\bullet}, Y^{\bullet}) = \{[f] \mid f \in \mathscr{S}(X^{\bullet}, Y^{\bullet})\}, \text{ where}$$
$$\mathscr{S}(X^{\bullet}, Y^{\bullet}) := \bigsqcup_{Z^{\bullet} \in \mathscr{D}(\mathscr{A})_0} \mathscr{K}(\mathscr{A})(Z^{\bullet}, X^{\bullet}) \times \mathscr{K}(\mathscr{A})(Z^{\bullet}, Y^{\bullet}).$$

Here since $\mathscr{D}(\mathscr{A})_0$ is a class, and $\mathscr{K}(\mathscr{A})(Z^{\bullet}, X^{\bullet}) \times \mathscr{K}(\mathscr{A})(Z^{\bullet}, Y^{\bullet})$ is a small set, $\mathscr{S}(X^{\bullet}, Y^{\bullet})$ turns out to be a class. Therefore for each $f \in \mathscr{S}(X^{\bullet}, Y^{\bullet}) \subseteq \mathfrak{U}$, $f \in \mathfrak{U}$ shows that $[f] \in \mathscr{P}\mathfrak{U}$. Hence $\mathscr{D}(\mathscr{A})(X^{\bullet}, Y^{\bullet}) \in \mathscr{P}\mathscr{P}\mathfrak{U} \subset \mathscr{P}\Psi\mathscr{P}\mathfrak{U} = (\mathscr{P}\Psi)^2\mathfrak{U}$. Thus $\mathscr{D}(\mathscr{A})(X^{\bullet}, Y^{\bullet})$ is a 2-class. Finally, take any $X^{\bullet} \in \mathscr{D}(\mathscr{A})$ with $\mathscr{D}(\mathscr{A})(X^{\bullet}, X^{\bullet}) = 0$ (e.g., we can take $X^{\bullet} = (\cdots 0 \to 0 \to 0 \cdots)$ or $X^{\bullet} = (\cdots 0 \to 0 \to \mathscr{A}(\text{-}, x) \xrightarrow{\mathbf{1}} \mathscr{A}(\text{-}, x) \to 0 \to 0 \cdots)$ $(x \in \mathscr{A}_0)$). Then we have

$$\mathscr{D}(\mathscr{A})(X^{\bullet}, X^{\bullet}) = 0 = \{\mathscr{S}(X^{\bullet}, X^{\bullet})\}. \tag{A.3}$$

Here if $\mathscr{S}(X^{\bullet}, X^{\bullet})$ is a small set, then

$$\mathscr{S}(X^{\bullet}, X^{\bullet}) \supseteq \{(Y^{\bullet}, (0,0)) \mid Y^{\bullet} \in \mathscr{D}(\mathscr{A})_0\} \xrightarrow{\text{bijective}} \mathscr{D}(\mathscr{A})_0 \subseteq \mathfrak{U}$$

shows that $\mathscr{D}(\mathscr{A})_0$ is a small set, which contradicts the fact shown above. Therefore, $\mathscr{S}(X^{\bullet}, X^{\bullet})$ is a proper class. Hence by the equality A.3 and Remark A.1.12 (4), $\mathscr{D}(\mathscr{A})(X^{\bullet}, X^{\bullet})$ is not a class. As a consequence, $\mathscr{D}(\mathscr{A})(X^{\bullet}, X^{\bullet})$ is a proper 2-class. Therefore, $\mathscr{D}(\mathscr{A})$ is a 2-moderate category, which is neither light nor 1-moderate. □

A.6. Application 2: Tensor products over a light category

In Chap. 7, we showed the unique existence of tensor product functors over small categories (Theorem 7.1.3), and here we will show that tensor product functors exist uniquely even over light categories.

DEFINITION A.6.1. Let k be a positive integer, $\mathscr{C}, \mathscr{D}$ linear categories, and $F: \mathscr{C} \times \mathscr{D} \to \operatorname{Mod}^k \Bbbk$ a functor of two variables. Then F is called *bilinear* if the map

$$(\mathscr{C} \times \mathscr{D})((x_1, y_1), (x_2, y_2)) = \mathscr{C}(x_1, x_2) \times \mathscr{D}(y_1, y_2) \\ \to \operatorname{Mod}^k \Bbbk(F(x_1, y_1), F(x_2, y_2))$$

induced from F is bilinear for all $(x_1, y_1), (x_2, y_2) \in (\mathscr{C} \times \mathscr{D})_0$. For any bilinear functors $F, F': \mathscr{C} \times \mathscr{D} \to \operatorname{Mod}^k \Bbbk$, a *morphism* from F to F' is defined to be a morphism from F to F' as functors of two variables. (Note that in Chap. 7, the same was defined for $k = 1$.)

We here give a way to compute the colimit $\varinjlim F$ for a functor $F: I \to \operatorname{Mod}^n \mathscr{C}$ from a k-moderate category I with $k \geq 1$ to the *n-class module category* $\operatorname{Mod}^n \mathscr{C} := \operatorname{Fun}(\mathscr{C}^{\mathrm{op}}, \operatorname{Mod}^n \Bbbk)$ of a light linear category $\mathscr{C}$ with $n \in \mathbb{N}$.

PROPOSITION A.6.2. *Let $\mathscr{C}$ be a light category, I a k-moderate category with $k \geq 1$, and $F: I \to \operatorname{Mod}^n \mathscr{C}$ a functor with $n \in \mathbb{N}$. Then there exists the following exact sequence in* $\operatorname{Mod}^m \mathscr{C}$ *for* $m := \max\{n+1, k+2\}$*:*

$$\coprod_{f \in I_1} F(\operatorname{dom}(f)) \xrightarrow{(\sigma_{\operatorname{cod}(f)} \circ F(f) - \sigma_{\operatorname{dom}(f)})_{f \in I_1}} \coprod_{i \in I_0} F(i) \to \varinjlim F \to 0.$$

Therefore in particular, we have $\varinjlim F \in \operatorname{Mod}^m \mathscr{C}$, and a linear functor R: $\operatorname{Mod}^m \mathscr{C} \to \operatorname{MOD} \Bbbk$ *preserves $\varinjlim$ in $(\operatorname{Mod}^n \mathscr{C})^I$ if and only if R is right exact and preserves coproducts with index sets k-classes.*

PROOF. By Proposition A.2.7, it follows from $I_0 \in \mathbf{Class}_0^k$ that

$$\coprod_{i \in I_0} (F(i))(x) \subseteq \prod_{i \in I_0} (F(i))(x) \in \begin{cases} \mathbf{Class}_0^{k+1} & (n \leq k+1) \\ \mathbf{Class}_0^n & (n \geq k+2) \end{cases}$$

for all $x \in \mathscr{C}_0$. Thus we have $\coprod_{i \in I_0} F(i) \in (\operatorname{Mod}^{m-1} \mathscr{C})_0$. Similarly it follows from $I_1 \in \mathbf{Class}_0^1$ that $\coprod_{f \in I_1} F(\operatorname{dom}(f)) \in (\operatorname{Mod}^{m-1} \mathscr{C})_0$. By definition of colimits, it is easy to verify that $\varinjlim F = \operatorname{Coker}(\sigma_{\operatorname{cod}(f)} \circ F(f) - \sigma_{\operatorname{dom}(f)})_{f \in I_1}$, which shows that $\varinjlim F \in \operatorname{Mod}^m \mathscr{C}$ by Lemma A.2.8. Hence the canonical morphism to the cokernel gives us the desired exact sequence in $\operatorname{Mod}^m \mathscr{C}$. □

THEOREM A.6.3. *Let $\mathscr{C}$ be a light category. Then there exists a bilinear functor* $\text{-} \otimes_{\mathscr{C}} ?: \operatorname{Mod} \mathscr{C} \times \mathscr{C}\text{-}\operatorname{Mod} \to \operatorname{Mod}^3 \Bbbk$ *satisfying the following conditions, which is uniquely determined up to isomorphisms and is called the* tensor product functor *over $\mathscr{C}$:*

(1) *For each $N \in \mathscr{C}\text{-}\operatorname{Mod}$,*
   (a) $\text{-} \otimes_{\mathscr{C}} N: \operatorname{Mod} \mathscr{C} \to \operatorname{Mod}^3 \Bbbk$ *is 2-moderate-cocontinuous* (Definition 4.5.5), *which is equivalent to saying that it is right exact, and preserves coproducts with index sets 2-classes by* Proposition A.6.2; *and*

(b) *For each $x \in \mathscr{C}_0$, there exists an isomorphism*

$$s_{x,N} \colon \mathscr{C}(\text{-}, x) \otimes_{\mathscr{C}} N \xrightarrow{\sim} N(x)$$

in $\operatorname{Mod}^3 \Bbbk$ *that is natural in x and in N.*

(2) *For each $M \in \operatorname{Mod} \mathscr{C}$,*

(a) $M \otimes_{\mathscr{C}} \text{-} \colon \mathscr{C}\text{-}\operatorname{Mod} \to \operatorname{Mod}^3 \Bbbk$ *is 2-moderate-cocontinuous; and*

(b) *For each $x \in \mathscr{C}_0$, there exists an isomorphism*

$$t_{x,M} \colon M \otimes_{\mathscr{C}} \mathscr{C}(x, \text{-}) \xrightarrow{\sim} M(x)$$

in $\operatorname{Mod}^3 \Bbbk$ *that is natural in x and in M.*

PROOF. First of all, we explicitly construct the functor

$$\text{-} \otimes_{\mathscr{C}} ? \colon \operatorname{Mod} \mathscr{C} \times \mathscr{C}\text{-}\operatorname{Mod} \to \operatorname{Mod}^3 \Bbbk$$

as follows.

On objects: Let $M \in (\operatorname{Mod} \mathscr{C})_0$ and $N \in (\mathscr{C}\text{-Mod})_0$. Then we set

$$\begin{aligned} M \otimes_{\mathscr{C}} N &:= \left(\coprod_{x \in \mathscr{C}_0} M(x) \otimes_{\Bbbk} N(x) \right) \Big/ I(M, N), \\ I(M, N) &:= \langle mM(f) \otimes n - m \otimes N(f)n \mid x, y \in \mathscr{C}_0, m \in M(y), \\ &\qquad\qquad f \in \mathscr{C}(x, y), n \in N(x) \rangle. \end{aligned} \tag{A.4}$$

For each $x, y \in \mathscr{C}_0, m \in M(x), n \in N(x)$, we set

$$m \otimes_{\mathscr{C}} n := m \otimes n + I(M, N) \in M \otimes_{\mathscr{C}} N$$

Then by definition above, we have

$$m \otimes_{\mathscr{C}} N(f)n = mM(f) \otimes_{\mathscr{C}} n$$

for all $x, y \in \mathscr{C}_0, m \in M(y), f \in \mathscr{C}(x, y), n \in N(x)$. We next verify that $M \otimes_{\mathscr{C}} N \in \operatorname{Mod}^3 \Bbbk$. For each $x \in \mathscr{C}_0$, the set $M(x) \otimes_{\Bbbk} N(x)$ is small, and we have $\coprod_{x \in \mathscr{C}_0} M(x) \otimes_{\Bbbk} N(x) \subseteq \prod_{x \in \mathscr{C}_0} M(x) \otimes_{\Bbbk} N(x) \in \mathbf{Class}_0^2$ by Proposition A.2.7(5). Therefore, $M \otimes_{\mathscr{C}} N \in \mathscr{P}\mathbf{Class}_0^2 \subseteq \mathbf{Class}_0^3$ (see Definition A.2.5). Thus $M \otimes_{\mathscr{C}} N \in \operatorname{Mod}^3 \Bbbk$.

On morphisms: Let $f \colon M \to M'$ be a morphism in $\operatorname{Mod} \mathscr{C}$ and $g \colon N \to N'$ a morphism in $\mathscr{C}$-Mod. Then we define $f \otimes_{\mathscr{C}} g \colon M \otimes_{\mathscr{C}} N \to M' \otimes_{\mathscr{C}} N'$ as follows. It is easy to verify that the map

$$F := \coprod_{x \in \mathscr{C}_0} f_x \otimes_{\Bbbk} g_x \colon \coprod_{x \in \mathscr{C}_0} M(x) \otimes_{\Bbbk} N(x) \to \coprod_{x \in \mathscr{C}_0} M'(x) \otimes_{\Bbbk} N'(x)$$

satisfies $F(I(M, N)) \subseteq I(M', N')$. Therefore we can define $f \otimes_{\mathscr{C}} g$ as the map between $\Bbbk$-modules induced from F:

$$\begin{array}{ccc} \coprod_{x \in \mathscr{C}_0} M(x) \otimes_{\Bbbk} N(x) & \xrightarrow{F} & \coprod_{x \in \mathscr{C}_0} M'(x) \otimes_{\Bbbk} N'(x) \\ {\scriptstyle \text{can.}} \downarrow & & \downarrow {\scriptstyle \text{can.}} \\ M \otimes_{\mathscr{C}} N & \xrightarrow{f \otimes_{\mathscr{C}} g} & M' \otimes_{\mathscr{C}} N' \end{array}$$

It is easy to check that the quiver morphism $\text{-} \otimes_{\mathscr{C}} ? \colon \operatorname{Mod} \mathscr{C} \times \mathscr{C}\text{-}\operatorname{Mod} \to \operatorname{Mod} \Bbbk$ defined above is a bilinear functor.

We now verify that this satisfies the required conditions (1) and (2). Since both are shown similarly, we only show the condition (1). To this end let $N \in (\mathscr{C}\text{-Mod})_0$. We denote by $\mathrm{Mod}^\omega \mathscr{C}$ (resp. $\mathrm{Mod}^\omega \Bbbk$) the full subcategory of $\mathrm{MOD}\,\mathscr{C}$ (resp. $\mathrm{MOD}\,\Bbbk$) with the object set $\bigcup_{n\in\mathbb{N}}(\mathrm{Mod}^n \mathscr{C})_0$ (resp. $\bigcup_{n\in\mathbb{N}}(\mathrm{Mod}^n \Bbbk)_0$). Then first note the following.

CLAIM 1. *The following hold.*

(i) *For each* $n\in\mathbb{N}$ *and each* $M\in(\mathrm{Mod}^n \mathscr{C})_0$*, we have* $M\otimes_{\mathscr{C}} N\in(\mathrm{Mod}^{n+3}\Bbbk)_0$.

(ii) *For each* $n\in\mathbb{N}$ *and each* $V\in(\mathrm{Mod}^n \Bbbk)_0$*, we have* $\mathrm{Hom}_\Bbbk(N,V)\in(\mathrm{Mod}^n \mathscr{C})_0$.

Therefore, it is possible to define the functors $\text{-}\otimes_{\mathscr{C}}N\colon \mathrm{Mod}^\omega \mathscr{C} \to \mathrm{Mod}^\omega \Bbbk$ *and* $\mathrm{Hom}_\Bbbk(N,\text{-})\colon \mathrm{Mod}^\omega \Bbbk \to \mathrm{Mod}^\omega \mathscr{C}$.

Indeed, (i) When $n = 0$, the assertion holds as shown above. Let $n \geq 1$. Then for each $x \in \mathscr{C}_0$, $M(x)\times N(x) \in \mathbf{Class}_0^n$, which shows that $\mathrm{Map}(M(x)\times N(x),\Bbbk) \in \mathbf{Class}_0^{n+1}$. By Lemma A.2.8, as its factor space, we have $M(x)\otimes_\Bbbk N(x) \in \mathbf{Class}_0^{n+2}$. Therefore, $\coprod_{x\in\mathscr{C}_0} M(x)\otimes_\Bbbk N(x) \subseteq \prod_{x\in\mathscr{C}_0} M(x)\otimes_\Bbbk N(x) \in \mathbf{Class}_0^{n+2}$. Again by Lemma A.2.8, as its factor module, we have $M\otimes_{\mathscr{C}}N \in \mathbf{Class}_0^{n+3}$. As a consequence, $M\otimes_{\mathscr{C}} N \in (\mathrm{Mod}^{n+3}\Bbbk)_0$. (ii) It holds that $(\mathrm{Hom}_\Bbbk(N,V))(x) \subseteq \mathrm{Map}(N(x),V) \in \mathbf{Class}_0^n$ for all $x\in\mathscr{C}_0$. Hence $\mathrm{Hom}_\Bbbk(N,V) \in (\mathrm{Mod}^n \mathscr{C})_0$.

CLAIM 2. *There exists the following adjoint:*

$$\mathrm{Mod}^\omega \mathscr{C} \;\underset{\mathrm{Hom}_\Bbbk(N,\text{-})}{\overset{\text{-}\otimes_{\mathscr{C}}N}{\rightleftarrows}}\; \mathrm{Mod}^\omega \Bbbk. \quad (\text{with } \perp)$$

To show this we define a unit $\eta\colon \mathbb{1}_{\mathrm{Mod}^\omega \mathscr{C}} \Rightarrow \mathrm{Hom}_\Bbbk(N,\text{-})\circ(\text{-},\otimes_{\mathscr{C}}N)$ and a counit $\varepsilon\colon (\text{-},\otimes_{\mathscr{C}}N)\circ\mathrm{Hom}_\Bbbk(N,\text{-}) \Rightarrow \mathbb{1}_{\mathrm{Mod}^\omega \Bbbk}$ as follows:

Construction of η: For each $M \in (\mathrm{Mod}^\omega \mathscr{C})_0$ and $x\in\mathscr{C}_0$, we set

$$\eta_{M,x}\colon M(x) \to \mathrm{Hom}_\Bbbk(N(x), M\otimes_{\mathscr{C}} N),\, m \mapsto m\otimes_{\mathscr{C}} \text{-}$$

and set $\eta_M := (\eta_{M,x})_{x\in\mathscr{C}_0}$. Then it is easy to verify that for each morphism $F\colon M\to M'$ in $\mathrm{Mod}^\omega \mathscr{C}$, the following is commutative:

$$\begin{array}{ccc}
M(x) & \xrightarrow{\eta_{M,x}} & \mathrm{Hom}_\Bbbk(N(x), M\otimes_{\mathscr{C}} N) \\
{\scriptstyle F_x}\big\downarrow & & \big\downarrow{\scriptstyle \mathrm{Hom}_\Bbbk(N(x),F\otimes_{\mathscr{C}}N)} \\
M'(x) & \xrightarrow[\eta_{M',x}]{} & \mathrm{Hom}_\Bbbk(N(x), M'\otimes_{\mathscr{C}} N)
\end{array}.$$

Therefore, $\eta := (\eta_M)_{M\in(\mathrm{Mod}^\omega \mathscr{C})_0}$ turns out to be a natural transformation

$$\mathbb{1}_{\mathrm{Mod}^\omega \mathscr{C}} \Rightarrow \mathrm{Hom}_\Bbbk(N,\text{-})\circ(\text{-},\otimes_{\mathscr{C}}N).$$

Construction of ε: For each $V\in(\mathrm{Mod}^\omega \Bbbk)_0$, we define a morphism

$$\varepsilon'_V\colon \coprod_{x\in\mathscr{C}_0} \mathrm{Hom}_\Bbbk(N(x),V)\otimes_\Bbbk N(x) \to V$$

in $\mathrm{Mod}^\omega \Bbbk$ by $(u_x\otimes n_x)_{x\in\mathscr{C}_0} \mapsto \sum_{x\in\mathscr{C}_0} u_x(n_x)$ $(u_x \in \mathrm{Hom}_\Bbbk(N(x),V), n_x\in N(x))$. Then it is easy to check that $\varepsilon'_V(I(\mathrm{Hom}_\Bbbk(N,V),N)) = 0$. Therefore, ε'_V induces a morphism

$$\varepsilon_V\colon \mathrm{Hom}_\Bbbk(N,V)\otimes_{\mathscr{C}} N \to V$$

in $\operatorname{Mod}^\omega \Bbbk$. Then it is immediate that $\varepsilon := (\varepsilon_V)_{V\in(\operatorname{Mod}^\omega \Bbbk)_0}$ turns out to be a natural transformation

$$(\text{-} \otimes_{\mathscr{C}} N) \circ \operatorname{Hom}_k(N, \text{-}) \Rightarrow \mathbb{1}_{\operatorname{Mod}^\omega \Bbbk}.$$

It is also immediate from definition that η, ε satisfy the zigzag equations, i.e., that both of the two diagrams

$$\begin{array}{ccc} \text{-}\otimes_{\mathscr{C}} N & & \\ {\scriptstyle \eta(\text{-})\otimes_{\mathscr{C}} N}\Downarrow & \searrow\!\!\!= & \\ \operatorname{Hom}_\Bbbk(N, \text{-}\otimes_{\mathscr{C}} N)\otimes_{\mathscr{C}} N & \xRightarrow[\varepsilon_{\text{-}\otimes_{\mathscr{C}} N}]{} & \text{-}\otimes_{\mathscr{C}} N, \end{array}$$

$$\begin{array}{ccc} \operatorname{Hom}_\Bbbk(N, \text{-}) & & \\ {\scriptstyle \eta_{\operatorname{Hom}_\Bbbk(N,\text{-})}}\Downarrow & \searrow\!\!\!= & \\ \operatorname{Hom}_\Bbbk(N, \operatorname{Hom}_\Bbbk(N, \text{-})\otimes_{\mathscr{C}} N) & \xRightarrow[\operatorname{Hom}_\Bbbk(N,\varepsilon)]{} & \operatorname{Hom}_\Bbbk(N, \text{-}) \end{array}$$

are commutative. Hence we have an adjoint $\text{-}\otimes_{\mathscr{C}} N \dashv \operatorname{Hom}_\Bbbk(N, \text{-})$. By Theorem 4.4.7, it also holds that there exists an isomorphism

$$\omega_{M,V} \colon \operatorname{Hom}_\Bbbk(M \otimes_{\mathscr{C}} N, V) \xrightarrow{\sim} (\operatorname{Mod}^\omega \mathscr{C})(M, \operatorname{Hom}_\Bbbk(N, V)) \tag{A.5}$$

natural in $M \in (\operatorname{Mod}^\omega \mathscr{C})_0$ and in $V \in (\operatorname{Mod}^\omega \Bbbk)_0$.

Using the claims above we prove statement (1).

(a) By using Proposition 4.5.8 (2), we show that $\text{-}\otimes_{\mathscr{C}} N$ is 2-moderate-cocontinuous. For this let I be a 2-moderate category. We here denote the full subcategory of $(\operatorname{Mod}^\omega \mathscr{C})^I$ whose object set is $\bigcup_{n\in\mathbb{N}} \operatorname{Fun}(I, \operatorname{Mod}^n \mathscr{C})_0$ by $\bigcup_{n\in\mathbb{N}}(\operatorname{Mod}^n \mathscr{C})^I$. Then there exists the following adjoint:

$$\bigcup_{n\in\mathbb{N}}(\operatorname{Mod}^n \mathscr{C})^I \underset{\Delta}{\overset{\varinjlim}{\rightleftarrows}} \operatorname{Mod}^\omega \mathscr{C}, \quad \varinjlim \dashv \Delta.$$

Indeed, since we have $\varinjlim F \in (\operatorname{Mod}^{\max\{n+1,4\}} \mathscr{C})_0$ for all $n \in \mathbb{N}$ and functors F in $(\operatorname{Mod}^n \mathscr{C})^I$ by Proposition A.6.2, it is possible to define the functor

$$\varinjlim \colon \bigcup_{n\in\mathbb{N}}(\operatorname{Mod}^n \mathscr{C})^I \to \operatorname{Mod}^\omega \mathscr{C}.$$

Moreover, since for each $M \in (\operatorname{Mod}^\omega \mathscr{C})_0$, there exists some $n \in \mathbb{N}$ such that $M \in (\operatorname{Mod}^n \mathscr{C})_0$, and then since $\Delta(M) \in ((\operatorname{Mod}^n \mathscr{C})^I)_0$, it is possible to define the functor

$$\Delta \colon \operatorname{Mod}^\omega \mathscr{C} \to \bigcup_{n\in\mathbb{N}}(\operatorname{Mod}^n \mathscr{C})^I.$$

Then it can be shown that $\varinjlim \dashv \Delta$ also in this case, in exactly the same way as in the case where I is a small category.

By the above, we have the following four adjoint pairs:

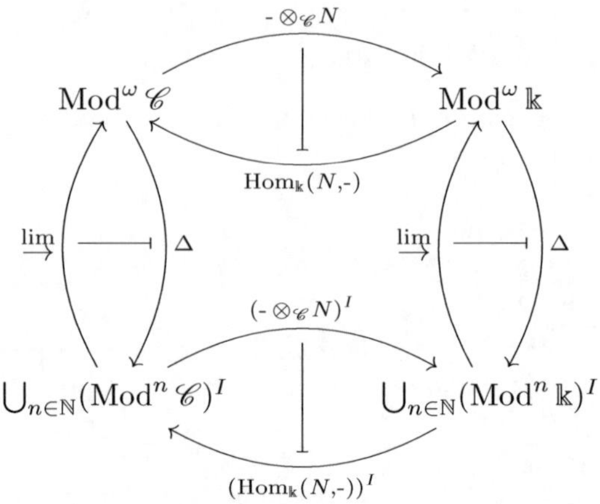

Therefore, it holds that $(\text{-} \otimes_{\mathscr{C}} N) \circ \varinjlim \cong \varinjlim \circ (\text{-} \otimes_{\mathscr{C}} N)^I$ in the diagram above by Proposition 4.4.12 and Lemma 4.5.6. By restricting to $(\operatorname{Mod} \mathscr{C})^I$, this formula holds in the following diagram:

$$\begin{array}{ccc} \operatorname{Mod}^4 \mathscr{C} & \xrightarrow{\text{-} \otimes_{\mathscr{C}} N} & \operatorname{Mod}^7 \Bbbk \\ {\scriptstyle \varinjlim} \uparrow & & \uparrow {\scriptstyle \varinjlim} \\ (\operatorname{Mod} \mathscr{C})^I & \xrightarrow[(\text{-} \otimes_{\mathscr{C}} N)^I]{} & (\operatorname{Mod}^3 \Bbbk)^I. \end{array}$$

Thus $\text{-} \otimes_{\mathscr{C}} N$ is 2-moderate-cocontinuous.

(b) We show that for each $x \in \mathscr{C}_0$, there exists a natural isomorphism $\mathscr{C}(\text{-}, x) \otimes_{\mathscr{C}} N \xrightarrow{\sim} N(x)$ in $\operatorname{Mod}^3 \Bbbk$. By applying the formula (A.5) to $M = \mathscr{C}(\text{-}, x)$, we obtain

$$\begin{aligned} \operatorname{Hom}_{\Bbbk}(\mathscr{C}(\text{-}, x) \otimes_{\mathscr{C}} N, ?) &\xrightarrow{\sim} \operatorname{Mod}^{\omega} \mathscr{C}(\mathscr{C}(\text{-}, x), \operatorname{Hom}_k(N, ?)) \\ &\xrightarrow{\sim} \operatorname{Hom}_{\Bbbk}(N, ?)(x) = \operatorname{Hom}_{\Bbbk}(N(x), ?) \end{aligned}$$

by the Yoneda lemma. Therefore, by Lemma 2.5.12, there exists an isomorphism $\mathscr{C}(\text{-}, x) \otimes_{\mathscr{C}} N \xrightarrow{\sim} N(x)$ in $\operatorname{Mod}^3 \Bbbk$. It is also possible to verify that this is natural in $x \in \mathscr{C}_0$ by the Yoneda lemma. Accordingly, the existence of the tensor product functor $\text{-} \otimes_{\mathscr{C}} ?$ is proved.

Uniqueness: Consider a bilinear functor

$$\text{-} \bar{\otimes}_{\mathscr{C}} ?\colon \operatorname{Mod} \mathscr{C} \times \mathscr{C}\text{-}\operatorname{Mod} \to \operatorname{Mod}^3 \Bbbk$$

having the same property as $\text{-} \otimes_{\mathscr{C}} ?$, and take $N \in (\mathscr{C}\text{-}\operatorname{Mod})_0$. Then by the property (1)(b) of $\text{-} \otimes_{\mathscr{C}} ?$ and of $\text{-} \bar{\otimes}_{\mathscr{C}} ?$, there exists isomorphisms

$$\mathscr{C}(\text{-}, x) \bar{\otimes}_{\mathscr{C}} N \xrightarrow[\sim]{\bar{s}_x} N(x) \xleftarrow[\mathscr{C}]{s_x} (\text{-}, x) \otimes_{\mathscr{C}} N$$

that are natural in $x \in \mathscr{C}_0$.

(i) The case where M is of the form $M = \mathscr{C}(\text{-}, x)$ for some $x \in \mathscr{C}_0$: We set $\phi_{M,N} := s_x^{-1} \circ \bar{s}_x \colon M \bar{\otimes}_{\mathscr{C}} N \xrightarrow{\sim} M \otimes_{\mathscr{C}} N$.

(ii) The case where M is of the form $M = \coprod_{x \in \mathscr{C}_0} \mathscr{C}(\text{-}, x)^{(I_x)}$ for some 2-classes I_x $(x \in \mathscr{C}_0)$: By the property (1)(a) of $\text{-} \otimes_{\mathscr{C}} ?$ and of $\text{-} \bar{\otimes}_{\mathscr{C}} ?$, the vertical canonical morphism are isomorphisms in the diagram

$$\begin{array}{ccccc}
\bigoplus_{x\in\mathscr{C}_0}\mathscr{C}(\text{-},x)^{(I_x)}\bar{\otimes}_{\mathscr{C}}N & \xrightarrow[\sim]{\bigoplus_{x\in\mathscr{C}_0}\bar{s}_x^{(I_x)}} & \bigoplus_{x\in\mathscr{C}_0}N(x)^{(I_x)} & \xleftarrow[\sim]{\bigoplus_{x\in\mathscr{C}_0}s_x^{(I_x)}} & \bigoplus_{x\in\mathscr{C}_0}\mathscr{C}(\text{-},x)^{(I_x)}\otimes_{\mathscr{C}}N \\
\wr\downarrow & & & & \downarrow\wr \\
\left(\bigoplus_{x\in\mathscr{C}_0}\mathscr{C}(\text{-},x)^{(I_x)}\right)\bar{\otimes}_{\mathscr{C}}N & & \overset{\sim}{\dashrightarrow} & & \left(\bigoplus_{x\in\mathscr{C}_0}\mathscr{C}(\text{-},x)^{(I_x)}\right)\otimes_{\mathscr{C}}N
\end{array}.$$

We then define $\phi_{M,N}\colon M\bar{\otimes}_{\mathscr{C}}N\overset{\sim}{\to}M\otimes_{\mathscr{C}}N$ as the unique isomorphism making this diagram commutative. Note that $\phi_{M,N}$ is natural in M for those M having this form. (iii) The general case: First of all we show the following for all $M\in(\operatorname{Mod}\mathscr{C})_0$.

CLAIM 3. *There exists a family $(I_x)_{x\in\mathscr{C}_0}$ of small sets such that we have an epimorphism $\phi_M\colon\coprod_{x\in\mathscr{C}_0}\mathscr{C}(\text{-},x)^{(I_x)}\to M$ in $\operatorname{Mod}^2\mathscr{C}$.*

Indeed, by the Yoneda lemma, for each $x\in\mathscr{C}_0$ there exists an isomorphism

$$\psi_x\colon M(x)\to(\operatorname{Mod}\mathscr{C})(\mathscr{C}(\text{-},x),M)$$

such that $\psi_x(a)(\mathbb{1}_x)=a$ for all $a\in M(x)$. Therefore

$$(\psi_x(a))_{a\in M(x)}\colon\coprod_{a\in M(x)}\mathscr{C}(\text{-},x)\to M$$

becomes an epimorphism $\coprod_{a\in M(x)}\mathscr{C}(x,x)\to M(x)$ in $\operatorname{Mod}\Bbbk$ at each $x\in\mathscr{C}_0$. Thus

$$((\psi_x(a))_{a\in M(x)})_{x\in\mathscr{C}_0}\colon\coprod_{x\in\mathscr{C}_0}\coprod_{a\in M(x)}\mathscr{C}(\text{-},x)\to M$$

is an epimorphism in $\operatorname{Mod}^2\mathscr{C}$ (note that $\coprod_{x\in\mathscr{C}_0}\coprod_{a\in M(x)}\mathscr{C}(y,x)\in\operatorname{Mod}^2\Bbbk$ for all $y\in\mathscr{C}_0$). Hence we can take $I_x:=M(x)$.

CLAIM 4. *There exist a family $(I_x)_{x\in\mathscr{C}_0}$ of small sets and a family $(J_y)_{y\in\mathscr{C}_0}$ of 2-classes such that we have an exact sequence*

$$\coprod_{y\in\mathscr{C}_0}\mathscr{C}(\text{-},y)^{(J_y)}\xrightarrow{F}\coprod_{x\in\mathscr{C}_0}\mathscr{C}(\text{-},x)^{(I_x)}\xrightarrow{E}M\to 0 \tag{A.6}$$

in $\operatorname{Mod}^3\mathscr{C}$.

Indeed, take ϕ_M in Claim 2 as E and set $K:=\operatorname{Ker}\phi_M$. Then noting that $J_y:=K(y)\in\operatorname{Mod}^2\Bbbk$ for all $y\in\mathscr{C}_0$ and $\coprod_{y\in\mathscr{C}_0}\mathscr{C}(z,y)^{(J_y)}\in\operatorname{Mod}^3\Bbbk$ for all $z\in\mathscr{C}_0$, the argument above constructs an epimorphism $\phi_K\colon\coprod_{y\in\mathscr{C}_0}\mathscr{C}(\text{-},y)^{(J_y)}\to K$ in $\operatorname{Mod}^3\mathscr{C}$. Therefore we can take ϕ_K as F.

In the claim above, note that we have isomorphisms

$$P_0:=\coprod_{x\in\mathscr{C}_0}\mathscr{C}(\text{-},x)^{(I_x)}\cong\coprod_{(x,a)\in\bigsqcup_{x\in\mathscr{C}_0}I_x}\mathscr{C}(\text{-},x),$$
$$P_1:=\coprod_{y\in\mathscr{C}_0}\mathscr{C}(\text{-},y)^{(J_y)}\cong\coprod_{(y,b)\in\bigsqcup_{y\in\mathscr{C}_0}J_y}\mathscr{C}(\text{-},y)$$

and that $\bigsqcup_{x\in\mathscr{C}_0}I_x$ is a 1-class and $\bigsqcup_{y\in\mathscr{C}_0}J_y$ is a 2-class. Then by (ii) above, both $\phi_{P_0,N}\colon P_0\bar{\otimes}_{\mathscr{C}}N\to P_0\otimes_{\mathscr{C}}N$ and $\phi_{P_1,N}\colon P_1\bar{\otimes}_{\mathscr{C}}N\to P_1\otimes_{\mathscr{C}}N$ are isomorphisms, and the diagram

$$
\begin{array}{ccccccc}
P_1 \bar{\otimes}_{\mathscr{C}} N & \xrightarrow{F \bar{\otimes}_{\mathscr{C}} N} & P_0 \bar{\otimes}_{\mathscr{C}} N & \xrightarrow{E \bar{\otimes}_{\mathscr{C}} N} & M \bar{\otimes}_{\mathscr{C}} N & \longrightarrow & 0 \\
{\scriptstyle \phi_{P_1,N}} \downarrow \wr & & {\scriptstyle \phi_{P_0,N}} \downarrow \wr & & \downarrow {\scriptstyle \phi_{M,N}} & & \\
P_1 \otimes_{\mathscr{C}} N & \xrightarrow[F \otimes_{\mathscr{C}} N]{} & P_0 \otimes_{\mathscr{C}} N & \xrightarrow[E \otimes_{\mathscr{C}} N]{} & M \otimes_{\mathscr{C}} N & \longrightarrow & 0
\end{array}
\tag{A.7}
$$

is commutative, which has exact rows by the property (1)(a). Therefore, by the universality of cokernels and by the five lemma, we obtain an isomorphism $\phi_{M,N} \colon M \bar{\otimes}_{\mathscr{C}} N \to M \otimes_{\mathscr{C}} N$ that is natural in M. The fact that this is natural in N can be shown in exactly the same way as when $\mathscr{C}$ is a small category (see the last part of the proof of Theorem 7.1.3). □

APPENDIX B

Supplement to the original version

In the original version, due to lack of space some propositions in Sect. 2.3 were turned into exercises or simply cited without proofs. To make up for this, we give answers to those questions and the omitted proofs in this chapter. Namely, in Sect. B.1, we show that the category of *framed* algebras (see Remark B.1.6) is equivalent to a subcategory of the category of finite linear categories, which gives an answer to Exercise 2.3.9. In Sect. B.2, we give answers to Exercises 2.3.15, 2.3.21, 2.3.23, and 2.3.24 with some explanations in order to return exercises to propositions. In Sect. B.3, a proof of the Krull-Schmidt Theorem (Theorem 2.3.16) is given by using a notion of radical maps. In Sect. B.4, we give two proofs of the special form of the Morita Theorem (Theorem 2.3.19). The materials in Sect. B.1–B.4 above cover the omitted parts in Sect. 2.3. Finally, Sect. B.5 is devoted to the definition of left Kan extensions and its relationships with left adjoint functors. This will supplement a background fact of sentences in the beginning of Sect. 7.1 that a left adjoint to the functor $(\text{-}) \circ F \colon \mathscr{A}^{\mathscr{D}} \to \mathscr{A}^{\mathscr{C}}$ between functor categories induced from a functor $F \colon \mathscr{C} \to \mathscr{D}$ and a category $\mathscr{A}$ is consisting of left Kan extensions along F.

B.1. The category of algebras and the category of finite linear categories: an answer to Exercise 2.3.9

In this section, we discuss relationships between the category of algebras and the category of finite linear categories, and give a detailed answer to Exercise 2.3.9. First of all, we add the following remark to the text.

REMARK B.1.1. In what follows, when we consider a complete set E of orthogonal idempotents, we always assume that $E \not\ni 0$.

We recall Proposition 2.3.8 (see Definition 2.3.7 for the definitions of categories $\Bbbk\text{-}\mathbf{Cat}_{\mathrm{f}}$ and $\Bbbk\text{-}\mathbf{Alg}_{\mathrm{coi}}$).

PROPOSITION B.1.2 (Proposition 2.3.8 in the text). *The categories* $\Bbbk\text{-}\mathbf{Alg}_{\mathrm{coi}}$ *and* $\Bbbk\text{-}\mathbf{Cat}_{\mathrm{f}}$ *are equivalent.*

The outline of the proof is given in the text, and Exercise 2.3.9 asks for a detailed proof of it. We give a detailed proof below.

PROOF. We define a functor $\mathrm{Cat} \colon \Bbbk\text{-}\mathbf{Alg}_{\mathrm{coi}} \to \Bbbk\text{-}\mathbf{Cat}_{\mathrm{f}}$ as follows.

On objects: For each $(A, E) \in (\Bbbk\text{-}\mathbf{Alg}_{\mathrm{coi}})_0$, we define $\mathscr{C} = \mathscr{C}_{A,E} = \mathrm{Cat}(A, E)$ by

$$\mathscr{C}_0 := E,\ \mathscr{C}(x, y) := yAx\ (x, y \in \mathscr{C}_0),$$

where the composition is given by the multiplication of A, and $\mathbb{1}_x = x\,(= x1x)$ for all $x \in \mathscr{C}_0$.

On morphisms: For each morphism $f\colon (A,E) \to (A',E')$ in $\Bbbk\text{-}\mathbf{Alg}_{\mathrm{coi}}$, we define a morphism

$$\mathrm{Cat}(f)\colon \mathrm{Cat}(A,E) \to \mathrm{Cat}(A',E')$$

in $\Bbbk\text{-}\mathbf{Cat}_{\mathrm{f}}$, i.e., a functor as follows. On objects, define the map $\mathrm{Cat}(f)_0$ by

$$\begin{array}{ccc} \mathrm{Cat}(A,E) & \xrightarrow{\mathrm{Cat}(f)_0} & \mathrm{Cat}(A',E') \\ \| & \| & \| \\ E & \xrightarrow[f|_E]{} & E'. \end{array}$$

Then we have the following:

$$\text{(B.1)} \qquad f|_E \text{ is injective.}$$

Indeed, take any $x, y \in E$, and assume that $x \neq y$. Then it is enough to show that $f(x) \neq f(y)$. Note that x and y are orthogonal idempotents. Now if $f(x) = f(y)$, then $f(x) = f(x^2) = f(x)f(x) = f(x)f(y) = f(xy) = f(0) = 0$ shows that $E \ni 0$, which contradicts our assumption in Remark B.1.1 above. Therefore, $\mathrm{Cat}(f)_0$ is injective.

Next let $x, y \in E = \mathrm{Cat}(A,E)_0$ and $a \in \mathrm{Cat}(A,E)(x,y) = yAx$. Then we set $\mathrm{Cat}(f)(a) := f(a) \in f(y)A'f(x) = \mathrm{Cat}(A',E')(f(x), f(y))$. Namely, in the category $\mathrm{Cat}(A',E')$ the following equality holds:

$$\begin{array}{ccc} \mathrm{Cat}(f)(x) & \xrightarrow{\mathrm{Cat}(f)(a)} & \mathrm{Cat}(f)(y) \\ \| & \| & \| \\ f(x) & \xrightarrow[f(a)]{} & f(y). \end{array}$$

Here we verify that $\mathrm{Cat}(f)$ is a functor.

Preserving identities: For each $x \in \mathrm{Cat}(A,E)_0$, we have $\mathrm{Cat}(f)(\mathbb{1}_x) = f(x) = \mathbb{1}_{f(x)} = \mathbb{1}_{\mathrm{Cat}(f)(x)}$.

Preserving compositions: If $x \xrightarrow{a} y \xrightarrow{b} z$ is a morphism in $\mathrm{Cat}(A,E)$, i.e., if $x, y, z \in E, a \in yAx, b \in zAy$, then we have $\mathrm{Cat}(f)(b \cdot a) = f(ba) = f(b)f(a) = \mathrm{Cat}(f)(b) \cdot \mathrm{Cat}(f)(a)$. From the above, it is confirmed that $\mathrm{Cat}(f)$ is a morphism in $\Bbbk\text{-}\mathbf{Cat}_{\mathrm{f}}$.

We now verify that $\mathrm{Cat}\colon \Bbbk\text{-}\mathbf{Alg}_{\mathrm{coi}} \to \Bbbk\text{-}\mathbf{Cat}_{\mathrm{f}}$ is a functor.

Preserving identities: It is enough to show that for each $(A,E) \in (\Bbbk\text{-}\mathbf{Alg}_{\mathrm{coi}})_0$, the functor $\mathrm{Cat}(\mathbb{1}_{(A,E)})\colon \mathrm{Cat}(A,E) \to \mathrm{Cat}(A,E)$ is the identity. For each $x \in \mathrm{Cat}(A,E)_0 = E$ we have $\mathrm{Cat}(\mathbb{1}_{(A,E)})(x) = \mathbb{1}_A(x) = x$. In addition, if $x, y \in E, a \in \mathrm{Cat}(A,E)(x,y) = yAx$, then $\mathrm{Cat}(\mathbb{1}_{(A,E)})(a) = \mathbb{1}_A(a) = a$. Therefore, it holds that $\mathrm{Cat}(\mathbb{1}_{(A,E)}) = \mathbb{1}_{\mathrm{Cat}(A,E)}$.

Preserving compositions: Let $(A,E) \xrightarrow{f} (A',E') \xrightarrow{g} (A'',E'')$ be morphisms in $\Bbbk\text{-}\mathbf{Alg}_{\mathrm{coi}}$. We verify that $\mathrm{Cat}(gf) = \mathrm{Cat}(g)\mathrm{Cat}(f)$.

On objects: For each $x \in \mathrm{Cat}(A,E)_0 = E$, we have $\mathrm{Cat}(gf)(x) = (gf)(x) = g(f(x)) = \mathrm{Cat}(g)(\mathrm{Cat}(f)(x))$. Therefore, it holds that $\mathrm{Cat}(gf)_0 = \mathrm{Cat}(g)_0\mathrm{Cat}(f)_0$.

On morphisms: If $x, y \in E, a \in \mathrm{Cat}(A,E)(x,y) = yAx$, then we have $\mathrm{Cat}(gf)(a) = (gf)(a) = g(f(a)) = \mathrm{Cat}(g)(\mathrm{Cat}(f)(a))$. Therefore, $\mathrm{Cat}(gf)_1 = \mathrm{Cat}(g)_1\mathrm{Cat}(f)_1$. By these facts, it holds that $\mathrm{Cat}(gf) = \mathrm{Cat}(g)\mathrm{Cat}(f)$, and it is confirmed that Cat is a functor.

Next, we define a functor $\mathrm{Mat}\colon \Bbbk\text{-}\mathbf{Cat}_{\mathrm{f}} \to \Bbbk\text{-}\mathbf{Alg}_{\mathrm{coi}}$ as follows.

On objects: For each $\mathscr{C} \in \Bbbk\text{-}\mathbf{Cat}_{\mathrm{f}}$, define $(A_{\mathscr{C}}, E_{\mathscr{C}}) = \mathrm{Mat}(\mathscr{C})$ as follows:

$$A_{\mathscr{C}} := \coprod_{x,y\in\mathscr{C}_0} \mathscr{C}(x,y),$$

$$E_{\mathscr{C}} := \{e_{x,x} \mid x \in \mathscr{C}_0\}, \text{where} e_{x,x} := (\mathbb{1}_x \delta_{(j,i),(x,x)})_{j,i\in\mathscr{C}_0} \ (x \in \mathscr{C}_0).$$

In the above, the multiplication is given by the usual matrix multiplication by regarding each element as a matrix $(a_{y,x})_{y,x}$, $a_{y,x} \in \mathscr{C}(x,y)$. Then the identity is given as the matrix $(\mathbb{1}_x\delta_{y,x})_{y,x\in\mathscr{C}_0}$.

On morphisms: For each morphism $F\colon \mathscr{C} \to \mathscr{C}'$ in $\Bbbk\text{-}\mathbf{Cat}_{\mathrm{f}}$, we define a morphism $\mathrm{Mat}(F)\colon \mathrm{Mat}(\mathscr{C}) \to \mathrm{Mat}(\mathscr{C}')$ in $\Bbbk\text{-}\mathbf{Alg}_{\mathrm{coi}}$ as follows. First, we denote the characteristic function of the subset $\mathrm{Im}\, F_0 \subseteq \mathscr{C}'_0$ by χ_F. Namely, $\chi_F\colon \mathscr{C}' \to \{0,1\}$ is a function defined by

$$\chi_F(x) := \begin{cases} 1 & (x \in \mathrm{Im}\, F_0), \\ 0 & (\text{otherwise}). \end{cases}$$

Then for each $(a_{y,x})_{y,x\in\mathscr{C}_0} \in A_{\mathscr{C}}$, we set

$$\mathrm{Mat}(F)((a_{y,x})_{y,x\in\mathscr{C}_0}) := (\chi_F(v)\chi_F(u)F(a_{F^{-1}(v),F^{-1}(u)}))_{v,u\in\mathscr{C}'_0} \in A_{\mathscr{C}'}, \tag{B.2}$$

where we denote the unique element of $F_0^{-1}(w)$ by $F^{-1}(w)$ for all $w \in \mathrm{Im}\, F_0$ by using the fact that $F_0\colon \mathscr{C}_0 \to \mathscr{C}'_0$ is injective. Alternatively, this can be described as follows. Let l, m $(l \le m)$ be the number of objects of $\mathscr{C}, \mathscr{C}'$, respectively. Then we may set

$$\begin{aligned} &\mathscr{C}_0 = \{x_1,\dots,x_l\}, \mathscr{C}'_0 = \{y_1,\dots,y_m\}, \mathrm{Im}\, F_0 = \{y_1,\dots,y_l\}, \\ &\text{where } y_i := F_0(x_i) \ (i = 1,\dots,l). \end{aligned} \tag{B.3}$$

By using this, $\mathrm{Mat}(F)$ can be defined as

$$\mathrm{Mat}(F)((a_{x_j,x_i})_{j,i\in\{1,\dots,l\}}) = \begin{pmatrix} (F(a_{x_j,x_i}))_{j,i} & 0 \\ 0 & 0 \end{pmatrix}. \tag{B.4}$$

Here, we verify that $\mathrm{Mat}(F)$ is a morphism in $\Bbbk\text{-}\mathbf{Alg}_{\mathrm{coi}}$. This can be done by using the form of (B.2), but by using the form of (B.4), this becomes almost obvious as follows: Let $a := (a_{x_j,x_i})_{j,i}, b := (b_{x_j,x_i})_{j,i} \in A_{\mathscr{C}}, t \in \Bbbk$. Then

The linearity of $\mathrm{Mat}(F)$**:**

$$\begin{aligned} \mathrm{Mat}(F)(a+tb) &= \begin{pmatrix} (F(a_{x_j,x_i} + tb_{x_j,x_i}))_{j,i} & 0 \\ 0 & 0 \end{pmatrix} \\ &= \begin{pmatrix} (F(a_{x_j,x_i}) + kF(b_{x_j,x_i}))_{j,i} & 0 \\ 0 & 0 \end{pmatrix} \\ &= \mathrm{Mat}(F)(a) + t\,\mathrm{Mat}(F)(b). \end{aligned}$$

$\mathrm{Mat}(F)$ **preserves multiplication:**

$$\begin{aligned} \mathrm{Mat}(F)(ab) &= \begin{pmatrix} (F(\sum_{k=1}^{l} a_{x_j,x_k}b_{x_k,x_i}))_{j,i} & 0 \\ 0 & 0 \end{pmatrix} \\ &= \begin{pmatrix} ((\sum_{k=1}^{l} F(a_{x_j,x_k})F(b_{x_k,x_i})))_{j,i} & 0 \\ 0 & 0 \end{pmatrix} \\ &= \mathrm{Mat}(F)(a)\,\mathrm{Mat}(F)(b). \end{aligned}$$

Verification of $\mathrm{Mat}(F)(E_{\mathscr{C}}) \subseteq E_{\mathscr{C}'}$**:** For each $i \in \{i, \dots, l\}$, we have

$$\mathrm{Mat}(F)(e_{x_i,x_i}) = e_{y_i,y_i} \in E_{\mathscr{C}'}.$$

By the above, $\mathrm{Mat}(F)$ is a morphism in $\Bbbk\text{-}\mathbf{Alg}_{\mathrm{coi}}$.

We next verify that $\mathrm{Mat}\colon \Bbbk\text{-}\mathbf{Cat}_{\mathrm{f}} \to \Bbbk\text{-}\mathbf{Alg}_{\mathrm{coi}}$ is a functor.

Preserving identities: For each $\mathscr{C} \in (\Bbbk\text{-}\mathbf{Cat}_{\mathrm{f}})_0$, it is obvious from the form of (B.4) that $\mathrm{Mat}(\mathbb{1}_{\mathscr{C}}) = \mathbb{1}_{\mathrm{Mat}(\mathscr{C})}$.

Preserving compositions: Let $\mathscr{C} \xrightarrow{F} \mathscr{C}' \xrightarrow{G} \mathscr{C}''$ be a morphism in $\Bbbk\text{-}\mathbf{Cat}_{\mathrm{f}}$. Set $\mathscr{C}_0, \mathscr{C}'_0, \operatorname{Im} F$ as in (B.3) above, and further set $\mathscr{C}''_0 = \{z_1, \dots, z_n\}$ $(m \le n)$ and $\operatorname{Im} G_0 = \{z_1, \dots, z_m\}, z_i := G(y_i)$ $(i = 1, \dots, m)$. Then it is enough to show the commutativity of the diagram

$$\begin{array}{ccc} \mathrm{Mat}(\mathscr{C}) & \xrightarrow{\mathrm{Mat}(F)} & \mathrm{Mat}(\mathscr{C}') \\ & {\scriptstyle \mathrm{Mat}(GF)} \searrow & \downarrow {\scriptstyle \mathrm{Mat}(G)} \\ & & \mathrm{Mat}(\mathscr{C}''). \end{array}$$

This can be verified by the following calculation: For each $a := (a_{x_j,x_i})_{j,i\in\{1,\dots,l\}} \in A_{\mathscr{C}}$, we have

$$\begin{aligned} \mathrm{Mat}(GF)(a) &= \begin{pmatrix} ((GF)(a_{x_j,x_i}))_{j,i} & 0 & 0 \\ 0 & 0 & 0 \\ 0 & 0 & 0 \end{pmatrix} = \begin{pmatrix} (G(F(a_{x_j,x_i})))_{j,i} & 0 & 0 \\ 0 & 0 & 0 \\ 0 & 0 & 0 \end{pmatrix} \\ &= \mathrm{Mat}(G)(\mathrm{Mat}(F)(a)). \end{aligned}$$

From the above, we see that Mat is a functor.

Finally, we show that Cat and Mat are quasi-inverses of each other.

Definition of a natural isomorphism $\sigma\colon \mathrm{Mat} \circ \mathrm{Cat} \Rightarrow \mathbb{1}_{\Bbbk\text{-}\mathbf{Alg}_{\mathrm{coi}}}$**:** Let $(A, E) \in (\Bbbk\text{-}\mathbf{Alg}_{\mathrm{coi}})_0$, and set $\mathscr{C} := \mathrm{Cat}(A, E)$. Then $\mathrm{Mat}(\mathrm{Cat}(A, E)) = (A_{\mathscr{C}}, E_{\mathscr{C}})$. Here, we define a morphism $\sigma_{(A,E)}\colon (A_{\mathscr{C}}, E_{\mathscr{C}}) \to (A, E)$ as follows. As a map, it is defined as the linear isomorphism $\sigma_{(A,E)}\colon A_{\mathscr{C}} = \coprod_{x,y\in E} yAx \to \bigoplus_{x,y\in E} yAx = A, (a_{y,x})_{y,x\in E} \mapsto \sum_{y,x\in E} a_{y,x}$. Then for each $x \in E$, we have $\sigma_{(A,E)}(e_{x,x}) = \mathbb{1}_x = x \in E$, which shows that $\sigma_{(A,E)}(E_{\mathscr{C}}) \subseteq E$.

σ **preserves multiplication:** Let $a = (a_{y,x})_{y,x\in E}$ and $b = (b_{y,x})_{y,x\in E} \in A_{\mathscr{C}}$. Then we have $\sigma(ab) = \sigma((\sum_{u\in E} a_{y,u} b_{u,x}))_{y,x} = \sum_{y,u,x\in E} a_{y,u} b_{u,x}$, which shows that

$$\begin{aligned} \sigma(a)\sigma(b) &= \left(\sum_{y,v\in E} a_{y,v}\right)\left(\sum_{u,x\in E} b_{u,x}\right) = \sum_{y,v,u,x\in E} a_{y,v} b_{u,x} \\ &\stackrel{*}{=} \sum_{y,v,u,x\in E} \delta_{v,u} a_{y,v} b_{u,x} = \sigma(ab), \end{aligned}$$

where the equality $\stackrel{*}{=}$ follows from the fact that if $v \ne u$, then $vu = 0$, and $a_{y,v} b_{u,x} \in (yAv)(uAx) = 0$. Therefore, $\sigma_{(A,E)}$ is a morphism in $\Bbbk\text{-}\mathbf{Alg}_{\mathrm{coi}}$, and is an isomorphism because it was given as a linear isomorphism.

Naturality of σ**:** Let $f\colon (A, E) \to (A', E')$ be a morphism in $\Bbbk\text{-}\mathbf{Alg}_{\mathrm{coi}}$, and set $\mathscr{C} := \mathrm{Cat}(A, E), \mathscr{C}' := \mathrm{Cat}(A', E'), F := \mathrm{Cat}(f)$. Then it is enough to show that the diagram

$$\begin{array}{ccc} (A_{\mathscr{C}}, E_{\mathscr{C}}) & \xrightarrow{\mathrm{Mat}(F)} & (A_{\mathscr{C}'}, E_{\mathscr{C}'}) \\ {\scriptstyle \sigma_{(A,E)}}\downarrow & & \downarrow{\scriptstyle \sigma_{(A',E')}} \\ (A, E) & \xrightarrow[f]{} & (A', E') \end{array}$$

is commutative. Take $\mathscr{C}_0 = E, \mathscr{C}' = E', \operatorname{Im} F_0 = f(E)$ as in (B.3) above. Then for each $a = (a_{x_j,x_i})_{j,i} \in A_{\mathscr{C}}$, we have

$$\sigma_{(A',E')}(\mathrm{Mat}(F)(a)) = \sigma_{(A',E')}\begin{pmatrix} (F(a_{x_j,x_i}))_{j,i} & 0 \\ 0 & 0 \end{pmatrix} = \sum_{j,i=1}^{l} f(a_{x_j,x_i}) = f(\sigma(a)).$$

Therefore, σ is a natural transformation, and is a natural isomorphism because each entry is an isomorphism. As a consequence, we have $\mathrm{Mat} \circ \mathrm{Cat} \cong \mathbb{1}_{\Bbbk\text{-}\mathbf{Alg}_{\mathrm{coi}}}$.

Definition of a natural isomorphism $\rho\colon \mathbb{1}_{\Bbbk\text{-}\mathbf{Cat}_{\mathrm{f}}} \Rightarrow \mathrm{Cat} \circ \mathrm{Mat}$: Let $\mathscr{C} \in (\Bbbk\text{-}\mathbf{Cat}_{\mathrm{f}})_0$. Then $(A_{\mathscr{C}}, E_{\mathscr{C}}) = \mathrm{Mat}(\mathscr{C})$. Here, we define a morphism $\rho_{\mathscr{C}}\colon \mathscr{C} \to \mathrm{Cat}(A_{\mathscr{C}}, E_{\mathscr{C}})$ in $\Bbbk\text{-}\mathbf{Cat}_{\mathrm{f}}$ as follows.

On objects: For each $x \in \mathscr{C}_0$, $\rho_{\mathscr{C}}(x) := e_{x,x} \in E_{\mathscr{C}} = (\mathrm{Cat}(A_{\mathscr{C}}, E_{\mathscr{C}}))_0$. This map between the objects is obviously bijective, and in particular, is injective.

On morphisms: For each $x, y \in \mathscr{C}_0$, we define a map

$$\rho_{\mathscr{C}}\colon \mathscr{C}(x,y) \to \mathrm{Cat}(A_{\mathscr{C}}, E_{\mathscr{C}})(\rho_{\mathscr{C}}(x), \rho_{\mathscr{C}}(y))$$

as follows. For each $a \in \mathscr{C}(x,y)$, set $\rho_{\mathscr{C}}(a) := (\delta_{(v,u),(y,x)}a)_{v,u\in\mathscr{C}_0} \in e_{y,y}A_{\mathscr{C}}e_{x,x} = \mathrm{Cat}(A_{\mathscr{C}}, E_{\mathscr{C}})(\rho_{\mathscr{C}}(x), \rho_{\mathscr{C}}(y))$. Here, since $e_{y,y}A_{\mathscr{C}}e_{x,x} = e_{y,y}\left(\coprod_{x,y\in\mathscr{C}_0}\mathscr{C}(x,y)\right)e_{x,x}$, we see that this map is bijective. For $\rho_{\mathscr{C}}$ to be a morphism in $\Bbbk\text{-}\mathbf{Cat}_{\mathrm{f}}$ $\rho_{\mathscr{C}}$ must be a functor. When this is the case, $\rho_{\mathscr{C}}$ is also an isomorphism by Exercise 1.4.5.

Preserving identities: For each $x \in \mathscr{C}_0$, the definition of $\rho_{\mathscr{C}}$ shows that $\rho_{\mathscr{C}}(\mathbb{1}_x) = (\delta_{(v,u),(x,x)}\mathbb{1}_x)_{v,u\in\mathscr{C}_0} = e_{x,x}$, which is the identity of $e_{x,x} = \rho_{\mathscr{C}}(x)$ in $\mathrm{Cat}(A_{\mathscr{C}}, E_{\mathscr{C}})$.

Preserving compositions: Let $x \xrightarrow{a} y \xrightarrow{b} z$ be morphisms in $\mathscr{C}$. Then

$$\begin{aligned} \rho_{\mathscr{C}}(b)\rho_{\mathscr{C}}(a) &= (\delta_{(v,u),(z,y)}b)_{v,u\in\mathscr{C}_0}(\delta_{(v,u),(y,x)}a)_{v,u\in\mathscr{C}_0} \\ &= \left(\sum_{w\in\mathscr{C}_0}\delta_{(v,w),(z,y)}b\,\delta_{(w,u),(y,x)}a\right)_{v,u\in\mathscr{C}_0} \end{aligned}$$

$$= (\delta_{(v,u),(z,x)}ba)_{v,u\in\mathscr{C}_0} = \rho_{\mathscr{C}}(ba).$$

Therefore, $\rho_{\mathscr{C}}$ is an isomorphism in $\Bbbk\text{-}\mathbf{Cat}_{\mathrm{f}}$.

Naturally of ρ: Let $F\colon \mathscr{C} \to \mathscr{C}'$ be a morphism in $\Bbbk\text{-}\mathbf{Cat}_{\mathrm{f}}$. Then it is enough to show that the diagram

$$\begin{array}{ccc} \mathscr{C} & \xrightarrow{F} & \mathscr{C}' \\ {\scriptstyle \rho_{\mathscr{C}}}\downarrow & & \downarrow{\scriptstyle \rho'_{\mathscr{C}}} \\ \mathrm{Cat}(\mathrm{Mat}(\mathscr{C})) & \xrightarrow[\mathrm{Cat}(\mathrm{Mat}(F))]{} & \mathrm{Cat}(\mathrm{Mat}(\mathscr{C}')) \end{array}$$

is commutative.

On objects: For each $x \in \mathscr{C}_0$, we have

$$(\mathrm{Cat}(\mathrm{Mat}(F))(\rho_{\mathscr{C}}(x)) = (\mathrm{Mat}(F))(e_{x,x}) = e_{F(x),F(x)} = \rho_{\mathscr{C}'}(F(x)).$$

On morphisms: Let $x, y \in \mathscr{C}_0, a \in \mathscr{C}(x, y)$. Then

$$\begin{aligned}(\mathrm{Cat}(\mathrm{Mat}(F))(\rho_{\mathscr{C}}(a)) &= (\mathrm{Mat}(F))((\delta_{(v,u),(y,x)}a)_{v,u\in\mathscr{C}_0}) \\ &= (\delta_{(q,p),(F(y),F(x))}F(a))_{q,p\in\mathscr{C}'_0} = \rho_{\mathscr{C}'}(F(a)).\end{aligned}$$

Therefore, ρ is a natural isomorphism, and we have $\mathrm{Cat} \circ \mathrm{Mat} \cong \mathbb{1}_{\Bbbk\text{-}\mathbf{Cat}_\mathrm{f}}$. □

DEFINITION B.1.3. We define a subcategory $\Bbbk$-$\mathbf{Cat}'_\mathrm{f}$ of $\Bbbk$-$\mathbf{Cat}_\mathrm{f}$ and a subcategory $\Bbbk$-$\mathbf{Alg}'_\mathrm{coi}$ of $\Bbbk$-$\mathbf{Alg}_\mathrm{coi}$ as follows:

$$\Bbbk\text{-}\mathbf{Cat}'_\mathrm{f} : \begin{cases}(\Bbbk\text{-}\mathbf{Cat}'_\mathrm{f})_0 := (\Bbbk\text{-}\mathbf{Cat}_\mathrm{f})_0, \\ (\Bbbk\text{-}\mathbf{Cat}'_\mathrm{f})_1 := \{F \in (\Bbbk\text{-}\mathbf{Cat}_\mathrm{f})_1 \mid F_0 \text{ is a bijection}\}.\end{cases}$$

$$\Bbbk\text{-}\mathbf{Alg}'_\mathrm{coi} : \begin{cases}(\Bbbk\text{-}\mathbf{Alg}'_\mathrm{coi})_0 := (\Bbbk\text{-}\mathbf{Alg}_\mathrm{coi})_0, \\ (\Bbbk\text{-}\mathbf{Alg}'_\mathrm{coi})_1 := \{f \in (\Bbbk\text{-}\mathbf{Alg}_\mathrm{coi})_1 \mid f \text{ is an algebra homomorphism}\}.\end{cases}$$

LEMMA B.1.4. *For each morphism $f\colon (A, E) \to (A', E')$ in $\Bbbk$-$\mathbf{Alg}_\mathrm{coi}$, the following are equivalent:*

(1) $f \in (\Bbbk\text{-}\mathbf{Alg}'_\mathrm{coi})_1$.
(2) $f(1_A) = 1_{A'}$.
(3) $f(E) = E'$.
(4) $f|_E\colon E \to E'$ *is a bijection.*

PROOF. The equivalence of (1) and (2) is obvious from the definition of algebra isomorphisms.

(2) ⇒ (3). If $f(E) \subsetneq E'$, then $f(1_A) = f(\sum_{x\in E} x) = \sum_{x\in E} f(x) \neq \sum_{y\in E'} y = 1_{A'}$.

(3) ⇒ (4). If (3) holds, then (4) holds because $f|_E$ is injective (B.1).

(4) ⇒ (2). If (4) holds, then we have $f(1_A) = f(\sum_{x\in E} x) = \sum_{x\in E} f(x) = \sum_{y\in E'} y = 1_{A'}$. □

Therefore, Proposition 2.3.8 together with Lemma B.1.4 shows the following:

COROLLARY B.1.5. Cat *and* Mat *restrict to equivalences between* $\Bbbk$-$\mathbf{Cat}'_\mathrm{f}$ *and* $\Bbbk$-$\mathbf{Alg}'_\mathrm{coi}$.

REMARK B.1.6. Let us call a complete set E of orthogonal idempotents in an algebra A a *frame* of A for short, and call a pair (A, E) a *framed algebra*, and also call a linear category with only finitely many objects a *finite linear category*. Then Proposition 2.3.8 says that we can identify a framed algebra with the corresponding finite linear category (namely, there exists a bijection between their isomorphism classes). However, by comparing morphisms, we see that there are much fewer algebra homomorphisms between framed algebras than there are functors between finite linear categories. Namely, they correspond to only those functors that restrict to bijections on the objects. Even if we drop the condition that it send identity to identity, they correspond to only those functors that restrict to injections on the objects. On the other hand, most covering functors appearing in later sections do not restrict to injections on the objects. This shows that it is not possible to give the notion of covering functors by using homomorphisms between algebras. Therefore, to build a covering theory it becomes necessary to consider algebras as linear categories by giving them some frames.

B.2. Answers to some Exercises in Section 2.3

Due to page limits in the original version, we have replaced expositions with exercises so that the reader can at least get a general idea in Section 2.3. In this section, we give answers to those exercises (Exercises 2.3.15, 2.3.21, 2.3.23, and 2.3.24) in order to supplement the explanations given in text.

B.2.a. Answer to Exercise 2.3.15.

EXERCISE 2.3.15. Prove that every finite-dimensional algebra has a complete set of orthogonal primitive idempotents.

An answer. First we show the following:

PROPOSITION B.2.1. *For any finite-dimensional right A-module M ($\neq 0$), the following holds:*

$(P(M))$ *M decomposes into the direct sum of indecomposable modules.*

PROOF. Assume that this statement does not hold. Then there exists some finite-dimensional module M such that $(P(M))$ does not hold. Let M a module with the smallest dimension among them. If M is indecomposable, then $(P(M))$ trivially holds. Therefore M is decomposable, and there exist submodules M_1, M_2 of M such that

$$M = M_1 \oplus M_2, \quad M_1 \neq 0, M_2 \neq 0. \tag{B.5}$$

Here, if both $(P(M_1))$ and $(P(M_2))$ hold, then by substituting those decompositions for M_1 and M_2 in (B.5), respectively, we have $(P(M))$. Therefore, either $(P(M_1))$ or $(P(M_2))$ does not hold. Without los of generality, we may assume that $(P(M_1))$ does not hold. Then by our choice of M, we have $\dim M \leq \dim M_1$. But by (B.5), it holds that $\dim M_1 = \dim M - \dim M_2 < \dim M$, a contradiction. □

By applying Proposition B.2.1 to the right A-module A, we obtain the decomposition

$$A = P_1 \oplus P_2 \oplus \cdots \oplus P_n$$

for some indecomposable submodules $P_1, P_2, \ldots, P_n$. Since $1 \in A$, there exists some $(e_1, e_2, \ldots, e_n) \in P_1 \times P_2 \times \cdots \times P_n$ such that

$$1 = e_1 + e_2 + \cdots + e_n. \tag{B.6}$$

It is enough to show the following.

CLAIM. *Each e_i is a primitive idempotent and any pair of them is orthogonal.*

Let $i \in \{1, 2, \ldots, n\}$. By multiplying e_i from the right to both sides of (B.6), we have

$$e_i = e_1 e_i + e_2 e_i + \cdots + e_n e_i.$$

Since each P_j $(j = 1, 2, \ldots, n)$ is a right module, we have $e_j e_i \in P_j$ $(j = 1, 2, \ldots, n)$. Here since $e_i \in P_i$ and the sum (B.6) is a direct sum, we have

$$e_i = e_i e_i, \quad e_j e_i = 0 \ (j \neq i).$$

Thus each e_i is an idempotent, and e_i and e_j are orthogonal if $i \neq j$. Moreover, by multiplying any $x \in P_i$ from the right to hand sides of (B.6), we have

$$x = e_1 x + e_2 x + \cdots + e_n x.$$

By the same reason, since $e_jx \in P_j$ $(j = 1, 2, \ldots, n)$ and the sum (B.6) is direct, we have

$$x = e_i x.$$

Therefore, $P_i \leq e_iA$. Here, since P_i is a right module, we also have $e_iA \leq P_i$. Hence $P_i = e_iA$. In particular, $P_i \neq 0$ shows that $e_i \neq 0$. Now since P_i is indecomposable, e_i turns out to be primitive by Lemma 2.3.3.

B.2.b. Answer to Exercise 2.3.21.

EXERCISE 2.3.21. Let A be an algebra. Prove that the following are equivalent.

(1) A is local.
(2) For each $a \in A$, either a or $1 - a$ is a unit.

Note here that we implicitly assume that A is not a *zero algebra*. (Also, the base ring $\Bbbk$ does not need to be a field, it is enough to be a commutative ring. For instance, we may take $\Bbbk$ to be the ring $\mathbb{Z}$ of integers, and then A turns out to be just a ring.)

According to Definition 2.3.20, A is called *local* if the sum of a non-unit and a non-unit is again a non-unit, where a unit is, by definition, an element having its inverse. In addition, we call an element *right invertible* (resp. *left invertible*) if it has a right inverse (resp. a left inverse).

An Answer. (1) $\Rightarrow$ (2). Assume that statement (1) holds, and let $a \in A$. If both a and $1-a$ are non-unit, then by applying (1) to $a+(1-a) = 1$, it holds that 1 becomes a non-unit, a contradiction. Therefore statement (2) holds.

(2) $\Rightarrow$ (1). Assume that statement (2) holds, and let $a, b \in A$ be non-units. If $c := a+b$ is a unit, then we obtain $1 = a'+b'$ by multiplying c^{-1} to both sides from the right, where we set $a' := ac^{-1}, b' := bc^{-1}$. By (2), either a' or b' is a unit. We may assume that a' is a unit without loss of generality. Then since $1 = ac^{-1}a'^{-1}$, a is right invertible. By applying the lemma below in our setting that (2) holds (note that the assumption of the lemma is weaker than (2)), we see that a is a unit. □

LEMMA B.2.2. *If for any $a \in A$, either a or $1 - a$ is right invertible, then all right invertible elements of A are units.*

PROOF. Let $a \in A$, and assume that a has a right inverse s, namely that $as = 1$. Then it suffices to show that $sa = 1$. By assumption, either sa or $1 - sa$ has a right inverse.

Case 1. The case where sa has a right inverse $t \in A$. In this case, we have $(*)$ $1 = sat$. By multiplying a from the left, we have $a = asat = at$, which together with$(*)$ shows that $1 = sa$.

Case 2. The case where $1 - sa$ has a right inverse $t \in A$. In this case we have $1 = (1 - sa)t$. By multiplying a from the left, we have $a = a(1 - sa)t = at - asat = at - at = 0$. Then $as = 1$ shows that $0 = 1$, and hence A is a zero ring, a contradiction. □

Actually, this lemma together with its proof shows us the following statement.

PROPOSITION B.2.3. *Let A be an algebra (a ring). Then the following are equivalent.*

(1) *A is local.*
(2) *For any $a \in A$, either a or $1 - a$ is a unit.*

(3) *For any* $a \in A$, *either* a *or* $1 - a$ *is right invertible.*
(4) *For any* $a \in A$, *either* a *or* $1 - a$ *is left invertible.*

Indeed, the equivalence of (1) and (2) is shown in the answer above. It is trivial that (2) $\Rightarrow$ (3). The lemma above shows (3) $\Rightarrow$ (2). The argument in the proof above shows the equivalence of (2) and (4).

The following is immediate from the lemma and the proposition above:

REMARK B.2.4. If an algebra (a ring) A is local, then all right invertible elements and all left invertible elements of A are units.

In addition, we would like to add that the following is also true (by analogy with commutative rings, this is the origin of the name of the *local* ring). In the following, a *proper right (resp. left) ideal* of A is, by definition, a right (resp. left) ideal I with $I \neq A$.

THEOREM B.2.5. *Let A be an algebra (a ring), and set J to be the set of all non-units of A. Then the following are equivalent:*

(8) *A is local.*
(9) *J is an ideal of A.*
(10) *J is the largest proper right ideal of A.*
(11) *J is the largest proper left ideal of A.*
(12) *A has the largest proper right ideal.*
(13) *A has the largest proper left ideal.*

PROOF. (1) $\Rightarrow$ (5). Since J is closed under addition by statement (1), it remains to verify that for any $a \in A$ and $b \in J$, we have $ab, ba \in J$. If $ba \notin J$, then since ba is a unit, there exists some $s \in A$ such that $bas = 1$, which shows that b is right invertible. By Remark B.2.4, b turns out to be a unit, which contradicts $b \in J$. Therefore, we have $ba \in J$. By this argument, we also see that $ab \in J$.

(5) $\Rightarrow$ (6). Assume that statement (5) holds. Then it is trivial that J is a right ideal of A. Let I be any proper right ideal of A. Then it is enough to show that $I \leq J$. Let $x \in I$. If $x \notin J$, then since x is a unit, $xA \ni 1$ shows that $I \geq xA = A$, which contradicts the fact that I is a proper right ideal. Therefore, we must have $x \in J$, and it follows that $I \leq J$.

(6) $\Rightarrow$ (8). This is trivial.

(8) $\Rightarrow$ (1). Assume that A has the largest proper right ideal I. Then it is enough to show statement (3). Let $a \in A$. If neither a nor $1 - a$ is right invertible, then $aA \neq A$ and $(1 - a)A \neq A$, i.e., these are proper right ideals of A, and hence we have $aA, (1 - a)A \leq I$. Then $1 = a + (1 - a) \in aA + (1 - a)A \leq I$ shows that $I = A$, which contradicts the fact that I is a proper right ideal.

Similarly the implications (5) $\Rightarrow$ (7) $\Rightarrow$ (9) $\Rightarrow$ (1) can be proved. $\square$

B.2.c. Answer to Exercise 2.3.23.

EXERCISE 2.3.23. Let A be an algebra, and $0 \neq M \in (\operatorname{Mod} A)_0$.

(1) Assume that $M_1, M_2 \leq M, M = M_1 \oplus M_2$. Then for each $i = 1, 2$ show that $e_i \colon M \xrightarrow{\pi_i} M_i \hookrightarrow M$ is an idempotent in $\operatorname{End}_A(M)$ and that $\mathbb{1}_M = e_1 + e_2$, where π_i is the i-th projection $m_1 + m_2 \mapsto m_i$ $(m_1 \in M_1, m_2 \in M_2)$.
(2) Let $e \in \operatorname{End}_A(M)$ be an idempotent. Then show that $M = e(M) \oplus (\mathbb{1}_M - e)(M)$ as right A-modules.

(3) Prove that the following are equivalent:
 (a) M is indecomposable.
 (b) The set of idempotents in $\operatorname{End}_A(M)$ is equal to $\{0, \mathbb{1}_M\}$.

The statements above also hold for any ring A in general.

An answer.

(1) For each $m = m_1 + m_2 \in M$ with $m_1 \in M_1, m_2 \in M_2$, we have $e_1(e_1(m)) = e_1(m_1) = e_1(m_1 + 0) = m_1 = e_1(m)$. Hence $e_1^2 = e_1$, and e_1 is an idempotent in $\operatorname{End}_A(M)$. The same proof works for e_2.

Since $m = m_1 + m_2 = e_1(m) + e_2(m) = (e_1 + e_2)(m)$, we have $\mathbb{1}_M = e_1 + e_2$.

(2) Let $e \in \operatorname{End}_A(M)$ be an idempotent: $e^2 = e$. First we show that $M = e(M) + (\mathbb{1}_M - e)(M)$. The inclusion $(\supseteq)$ is obvious because both $e(M)$ and $(\mathbb{1}_M - e)(M)$ are submodules of M. Let $m \in M$. Then $m = e(m) + (m - e(m)) \in e(M) + (\mathbb{1}_M - e)(M)$, which shows the converse inclusion.

Next we show that $e(M) \cap (\mathbb{1}_M - e)(M) = 0$. Let x be on the left-hand side. Then $x \in e(M)$ and $x \in (\mathbb{1}_M - e)(M)$ shows that $x = e(m)$ and $x = m' - e(m')$ for some $m \in M$ and $m' \in M$. Since $e^2 = e$, the former shows $e(x) = e^2(m) = e(m) = x$, and the latter shows $e(x) = e(m') - e^2(m') = e(m') - e(m') = 0$. Hence $x = 0$.

(3) (a) $\Rightarrow$ (b). Assume (a) and let $e \in \operatorname{End}_A(M)$ be an idempotent. By (2), we have $M = e(M) \oplus (\mathbb{1}_M - e)(M)$. Since M is indecomposable, either $e(M) = 0$ or $(\mathbb{1}_M - e)(M) = 0$. Therefore, $e = 0$ or $\mathbb{1}_M - e = 0$, and hence $e \in \{0, \mathbb{1}_M\}$.

(b) $\Rightarrow$ (a). Assume (b) and suppose that $M = M_1 \oplus M_2$ for some $M_1, M_2 \le M$. Define e_1, e_2 as in (1) above. Then by (1), both of them are idempotents in $\operatorname{End}_A(M)$, and $\mathbb{1}_M = e_1 + e_2$. By (2), we have either $e_1 = 0$ or $e_1 = \mathbb{1}_M$. The former shows that $M_1 = 0$, and the latter shows that $e_2 = 0$ and $M_2 = 0$. □

To be more precise, the following proposition holds.

PROPOSITION B.2.6. *Let A be an algebra, and $0 \ne M \in (\operatorname{Mod} A)_0$, and set*

$$\mathscr{M} := \{(M_1, M_2) \mid M = M_1 \oplus M_2;\ M_1, M_2 \le M\},$$
$$\mathscr{E}' := \{(e_1, e_2) \mid \mathbb{1}_M = e_1 + e_2;\ e_1, e_2 \text{ are idempotents in } \operatorname{End}_A(M)\}, \text{ and}$$
$$\mathscr{E} := \{e \in \operatorname{End}_A(M) \mid e^2 = e\}.$$

Then there exist the following bijections between them:

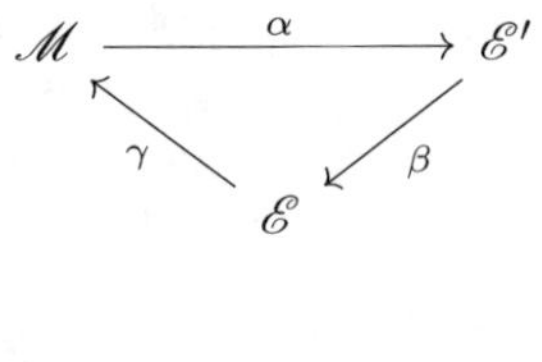

,

where α, β, γ are defined by

- $\alpha\colon \mathscr{M} \to \mathscr{E}'$, $(M_1, M_2) \mapsto (e_1, e_2)$ *(as defined in* (1) *above),*
- $\beta\colon \mathscr{E}' \to \mathscr{E}$, $(e_1, e_2) \mapsto e_1$, *and*
- $\gamma\colon \mathscr{E} \to \mathscr{M}$, $e \mapsto (\operatorname{Im} e, \operatorname{Ker} e)$,

and their inverses are defined by

- $\alpha^{-1} = \gamma \circ \beta\colon (e_1, e_2) \mapsto (\operatorname{Im} e_1, \operatorname{Im} e_2)$,
- $\beta^{-1} = \alpha \circ \gamma\colon e \mapsto (e, \mathbb{1}_M - e)$, *and*
- $\gamma^{-1} = \beta \circ \alpha\colon (M_1, M_2) \mapsto e_1$.

REMARK B.2.7. If $M = M_1 + M_2$, then it also holds that $M = M_2 + M_1$. Therefore, $(M_1, M_2) \in \mathscr{M}$ implies $(M_2, M_1) \in \mathscr{M}$. Similarly, $(e_1, e_2) \in \mathscr{E}$ implies $(e_2, e_1) \in \mathscr{E}$.

The proof is direct, and is left to the reader as an exercise. We now remark that statement (3) follows from the proposition above. First note the following two characterizations of (a) and (b):

i) (a) $\iff \mathscr{M} = \{(M, 0), (0, M)\}$.
ii) (b) $\iff \mathscr{E} = \{\mathbb{1}_M, 0\}$.
Then the bijection γ in Proposition B.2.6 shows that
iii) $\mathscr{M} = \{(M, 0), (0, M)\} \iff \mathscr{E} = \{\mathbb{1}_M, 0\}$.

Hence i), ii), and iii) shows the equivalence (a) $\iff$ (b).

REMARK B.2.8 (Relationship with Definition 2.5.17(2) (splitting idempotents)). Recall that an idempotent $e\colon x \to x$ in a category $\mathscr{C}$ is said to *split* if we have $e = \sigma\pi$ for some pair $y \xrightarrow{\sigma} x \xrightarrow{\pi} y$ of morphisms in $\mathscr{C}$ with $y \in \mathscr{C}_0$ satisfying $\pi\sigma = \mathbb{1}_y$. By the general theory (Proposition 2.5.21), we already know that all idempotents in Mod A split. This is proved alternatively by the bijective correspondences above as follows: Let $0 \neq M \in (\operatorname{Mod} A)_0$, and e be an idempotent in $\operatorname{End}_A(M)$. Then since $e = \beta\alpha\gamma(e)$, we have $M = \operatorname{Im} e \oplus \operatorname{Ker} e$. Set π_1 as the first projection with respect to this decomposition, and $\sigma_1\colon \operatorname{Im} e \hookrightarrow M$ as the inclusion. Then $e = \sigma_1\pi_1$, and also it is trivial that $\pi_1\sigma_1 = \mathbb{1}_{\operatorname{Im} e}$. Thus e splits.

We now deal with the above discussion in terms of category theory. Since Mod A is idempotent complete for any algebra A by Proposition 2.5.21 as mentioned above, we consider an idempotent complete linear category $\mathscr{D}$ in general.

PROPOSITION B.2.9. *Let $\mathscr{D}$ be an idempotent complete linear category, and $M \in \mathscr{D}_0$. We set*

$$\mathscr{S} := \{(s, r) \mid L \xrightarrow{s} M \xrightarrow{r} L, rs = \mathbb{1}_L, \exists L \in \mathscr{D}_0\},$$
$$\mathscr{E} := \{e \in \mathscr{D}(M, M) \mid e^2 = e\}$$

and define an equivalence relation $\sim$ on $\mathscr{S}$ as follows: For any $(s, r), (s', r') \in \mathscr{S}$, we write $(s, r) \sim (s', r')$ if and only if there exists an isomorphism $u\colon L \to L'$ such that the following diagram commutes:

$$\begin{array}{ccccc} L & \xrightarrow{s} & M & \xrightarrow{r} & L \\ {\scriptstyle u}\downarrow & & \| & & \downarrow{\scriptstyle u} \\ L' & \xrightarrow[s']{} & M & \xrightarrow[r']{} & L' \end{array}.$$

For each $(s, r) \in \mathscr{S}$, we denote by $[s, r]$ the equivalence class of (s, r). Then we have a bijection

$$\varepsilon\colon \mathscr{S}/{\sim} \to \mathscr{E}, \ [s, r] \mapsto sr.$$

PROOF. It is easy to check that the relation $\sim$ on $\mathscr{S}$ is an equivalence relation. The fact that $sr \in \mathscr{E}$ for all $(s, r) \in \mathscr{S}$ follows from the equalities $srsr = s\mathbb{1}_L r = sr$. Moreover, if we have $(s, r) \sim (s', r')$ by an isomorphism u that makes the diagram above commutative, then since $s' = su^{-1}$ and $r' = ur$, it holds that $s'r' = su^{-1}ur = sr$. This shows that the map ε is well-defined. Since $\mathscr{D}$ is idempotent complete, ε is surjective.

It remains to show that ε is injective. Let $(s_1, r_1), (s_1', r_1') \in \mathscr{S}$, and assume that $s_1r_1 = s_1'r_1'(=: e)$. Then it is enough to show that $(s_1, r_1) \sim (s_1', r_1')$. Now by definition of $\mathscr{S}$, there exist some $L_1, L_1' \in \mathscr{D}_0$ and some morphisms $L_1 \xrightarrow{s_1} M \xrightarrow{r} L_1$, $L_1' \xrightarrow{s_1'} M \xrightarrow{r_1'} L_1'$ such that $r_1s_1 = \mathbb{1}_{L_1}$, $r_1's_1' = \mathbb{1}_{L_1'}$. Since $\mathbb{1}_M - e$ is also an idempotent, it splits, and hence there exist some morphisms $L_2 \xrightarrow{s_2} M \xrightarrow{r_2} L_2$ such that $\mathbb{1}_M - e = s_2r_2$, $r_2s_2 = \mathbb{1}_{L_2}$. By Lemma 2.5.19, $(L_i \xrightarrow{s_i} M \xrightarrow{r_i} L_i)_{i=1}^2$ is a direct sum system.

We here show that s_1 is a kernel morphism of $\mathbb{1}_M - e$. First we have $(\mathbb{1}_M - e)s_1 = s_1 - s_1r_1s_1 = s_1 - s_1 = 0$. Next assume that a morphism $f: X \to M$ satisfies $(\mathbb{1}_M - e)f = 0$. Then since $s_2r_2f = 0$ and $\mathbb{1}_M = s_1r_1 + s_2r_2$, we have $f = s_1r_1f + s_2r_2f = s_1(r_1f)$. Thus f factors through s_1. To show the uniqueness of r_1f, suppose that $g: X \to L_1$ satisfies $f = s_1g$. Then by composing with r_1 from the left, we have $r_1f = r_1s_1g = g$, which shows the uniqueness of r_1f. As a consequence, s_1 is a kernel morphism of $\mathbb{1}_M - e$.

Set $L_2' := L_2, s_2' := s_2, r_2' := r_2$. Then $(L_i' \xrightarrow{s_i'} M \xrightarrow{r_i'} L_i')_{i=1}^2$ is also a direct sum system. Therefore, the same argument shows that s_1' is also a kernel morphism of $\mathbb{1}_M - e$. Hence by the universality of the kernel, there exists an isomorphism $u: L_1 \to L_1'$ such that $s_1 = s_1'u$. Then since $s_1r_1 = s_1'r_1'$, we have $s_1'ur_1 = s_1'r_1'$, which shows that $ur_1 = r_1'$ by composing with r_1' from the left. Thus we have $(s_1, r_1) \sim (s_1', r_1')$. □

It is also possible to give an answer for Exercise 2.3.23 (3) by using the proposition above.

LEMMA B.2.10. *Let $\mathscr{D}$ be an idempotent complete linear category, and $0 \neq M \in \mathscr{D}_0$. Then the following are equivalent:*

(1) *M is indecomposable.*
(2) *For each $(s, r) \in \mathscr{S}$, $(s, r) = (0, 0)$ or s, r are inverses of each other.*
(3) *$\mathscr{E} = \{0, \mathbb{1}_M\}$.*

PROOF. (1) ⇒ (2). Assume statement (1), and let $(s_1, r_1) \in \mathscr{S}$. Set $M_1 := \mathrm{dom}(s_1) = \mathrm{cod}(r_1)$. Then by Lemma 2.5.19, there exists a direct sum system $(M_i \xrightarrow{s_i} M \xrightarrow{r_i} M_i)_{i=1}^2$. In particular, we have $M \cong M_1 \amalg M_2$. Since M is indecomposable, we have $M_1 = 0$ or $M_2 = 0$. In the former case, $(s_1, r_1) = (0, 0)$. In the latter case, since $s_2r_2 = 0$, the equality $\mathbb{1}_M = s_1r_1 + s_2r_2$ shows that $\mathbb{1}_M = s_1r_1$. This together with $\mathbb{1}_M = r_1s_1$ shows that s, r are inverses of each other.

(2) ⇒ (3). By using the bijection in Proposition B.2.9, statement (2) implies that $\mathscr{E} = \mathrm{Im}\,\varepsilon = \{0, \mathbb{1}_M\}$.

(3) ⇒ (1). Assume statement (3), and suppose that $M \cong M_1 \amalg M_2$. By Proposition 1.5.14, there exists a direct sum system $(M_i \xrightarrow{s_i} M \xrightarrow{r_i} M_i)_{i=1}^2$. Since $s_1r_1 \in \mathscr{E}$, statement (3) shows that $s_1r_1 = 0$ or $s_1r_1 = \mathbb{1}_M$. In the former case, by composing with r_1 from the left, we have $r_1 = 0$, and hence $\mathbb{1}_{M_1} = r_1s_1 = 0$, which shows that $M_1 = 0$. In the latter case, since $\mathbb{1}_M = s_1r_1 + s_2r_2$, we have $s_2r_2 = 0$. Then the same argument shows that $M_2 = 0$. Thus M is indecomposable. □

B.2.d. Answer to Exercise 2.3.24.

EXERCISE 2.3.24. Prove the following statements for an algebra A and a finite-dimensional right A-module M:

(1) If $f \in \operatorname{End}_A(M)$, then there exists some $n \geq 1$ such that $M = \operatorname{Ker} f^n \oplus \operatorname{Im} f^n$.
(2) If M is indecomposable, then every $f \in \operatorname{End}_A(M)$ is either an isomorphism or nilpotent.
(3) If M is indecomposable, then $\operatorname{End}_A(M)$ is a local algebra.

An Answer.

(1) First of all, note that we have $\operatorname{Im} f^i = f^i(M)$ and $\operatorname{Ker} f^i = f^{-i}(0)$ for all integers $i \geq 1$. Since $M \geq f(M)$, it holds that $f^i(M) \geq f^{i+1}(M)$ for all integers $i \geq 1$. Namely, we have a decreasing sequence

$$M \geq f(M) \geq f^2(M) \geq \cdots . \tag{B.7}$$

Since M is finite-dimensional, there exists some integer $m \geq 1$ such that

$$f^m(M) = f^{m+1}(M) = f^{m+2}(M) = \cdots .$$

In particular, we have $f^m(M) = f^{2m}(M)$.

On the other hand, since $0 \leq f^{-1}(0)$, for all integers $i \geq 1$, we have $f^{-i}(0) \leq f^{-(i+1)}(0)$. Thus there exists an increasing sequence

$$0 \leq f^{-1}(0) \leq f^{-2}(0) \leq \cdots . \tag{B.8}$$

Again since M is finite-dimensional, there exists some integer $n \geq 1$ such that

$$f^{-n}(0) = f^{-(n+1)}(0) = f^{-(n+2)}(0) = \cdots .$$

In particular, we have $f^{-n}(0) = f^{-2n}(0)$. Reset n as $\max\{n, m\}$. Then we have both $f^n(M) = f^{2n}(M)$ and $f^{-n}(0) = f^{-2n}(0)$.

We show that $M = \operatorname{Ker} f^n \oplus \operatorname{Im} f^n$ for this n. First we check that $M = \operatorname{Ker} f^n + \operatorname{Im} f^n$. Since both terms in the right-hand side are submodules of M, it is obvious that (LHS) $\geq$ (RHS). To show the inverse inclusion, let $x \in M$. Then since $f^n(x) \in f^n(M) = f^{2n}(M)$, there exists some $y \in M$ such that $f^n(x) = f^{2n}(y) = f^n(f^n(y))$, which shows that $f^n(x - f^n(y)) = 0$. Hence $x - f^n(y) \in \operatorname{Ker} f^n$, and we have $x \in \operatorname{Ker} f^n + f^n(y) \leq \operatorname{Ker} f^n + \operatorname{Im} f^n$. This shows that (LHS) $\leq$ (RHS). Finally, we show that this sum is a direct sum, i.e., that $\operatorname{Ker} f^n \cap \operatorname{Im} f^n = 0$. For each element x of the left-hand side, $x \in \operatorname{Im} f^n$ shows that there exists some $y \in M$ such that $x = f^n(y)$. Since $x \in \operatorname{Ker} f^n$, we have $f^n(x) = 0$. Hence $f^{2n}(y) = 0$. But then we have $y \in \operatorname{Ker} f^{2n} = \operatorname{Ker} f^n$, which shows that $0 = f^n(y) = x$, as desired. Therefore, we have $M = \operatorname{Ker} f^n \oplus \operatorname{Im} f^n$.

(2) Assume that M is indecomposable. Then for the $n \geq 1$ above, we have $M = \operatorname{Ker} f^n \oplus \operatorname{Im} f^n$, which shows that either $\operatorname{Ker} f^n = 0$ or $\operatorname{Im} f^n = 0$. In the former case, the increasing sequence (B.8) shows that $\operatorname{Ker} f = 0$, and that f is a monomorphism. Then the equality $\dim \operatorname{Ker} f + \dim \operatorname{Im} f = \dim M$ shows that $\dim \operatorname{Im} f = \dim M$, and that $\operatorname{Im} f = M$. Thus f is also an epimorphism, and hence an isomorphism. In the latter case, it holds that $f^n = 0$, and that f is nilpotent.

(3) Note in general that for each $f \in \operatorname{End}_A(M)$, f is a unit in $\operatorname{End}_A(M)$ if and only if f is an isomorphism. Now assume that M is indecomposable, and let $f, g \in \operatorname{End}_A(M)$ be non-units. Then it is enough to show that $f + g$ is also a non-unit. Alternately, assume that $f + g$ is a unit, and set $h := (f + g)^{-1}$. Then

we have $fh + gh = \mathbb{1}_M$. If gh is a unit, then $g = (gh)(f+g)$ becomes a unit as a product of units, a contradiction. Thus gh is not a unit, and hence nilpotent by statement (2). Therefore, $(gh)^n = 0$ for some $n \geq 1$. This yields the equalities

$$(\mathbb{1}_M - gh)(1 + gh + (gh)^2 + \cdots + (gh)^{n-1}) = \mathbb{1}_M, \text{ and}$$
$$(1 + gh + (gh)^2 + \cdots + (gh)^{n-1})(\mathbb{1}_M - gh) = \mathbb{1}_M,$$

which shows that $fh = \mathbb{1}_M - gh$ becomes a unit of $\operatorname{End}_A(M)$. Thus $f = (fh)(f+g)$ becomes also a unit as a product of units, a contradiction. Therefore, $f+g$ must be a non-unit. □

REMARK B.2.11. Let A be an algebra, and M a finite-dimensional right A-module (not necessarily indecomposable). Then the argument used in statement (2) above shows that the following are equivalent:

(a) f is a monomorphism.
(b) f is an epimorphism.
(c) f is an isomorphism.

Indeed, the finite-dimensionality of M shows that $\dim M = \dim \operatorname{Im} f + \dim \operatorname{Ker} f$, from which we see that $\operatorname{Ker} f = 0 \iff M = \operatorname{Im} f$. Thus f is a monomorphism if and only if f is an epimorphism, which immediately shows the assertion above.

This gives an alternative proof of the fact that g is a unit if gh is a unit in statement (3) above. Indeed, if gh is a unit, then it is an epimorphism, which shows that g is an epimorphism, hence an isomorphism, thus a unit.

REMARK B.2.12. In statement (3), if $\operatorname{End}_A(M)$ were commutative, then the binomial theorem would show that the sum of nilpotent elements is again nilpotent, and the longer proof above would become superfluous. But $\operatorname{End}_A(M)$ is not necessarily commutative in general; it is not possible to use this argument.

B.3. Proof of the Krull–Schmidt Theorem

Throughout this section A is an algebra, and all A-modules are assumed to be finite-dimensional right A-modules. Here we give a proof of the Krull–Schmidt Theorem using a notion of radical maps in $\operatorname{mod} A$.

DEFINITION B.3.1. Let X, Y be A-modules, and $f \in \operatorname{Hom}_A(X, Y)$. Then f is called a *radical map* if hfg is not an isomorphism for all $g \in \operatorname{Hom}_A(X', X)$, $h \in \operatorname{Hom}_A(Y, Y')$ and all indecomposable A-modules X', Y'. The set of all radical maps in $\operatorname{Hom}_A(X, Y)$ is denoted by $\operatorname{rad}_A(X, Y)$.

LEMMA B.3.2. rad_A *has the following properties.*

(1) rad_A *is an ideal of the category* $\operatorname{mod} A$, *i.e., for any* X, X_1, Y, $Y_1 \in (\operatorname{mod} A)_0$, *the following hold:*
   (a) $\operatorname{rad}_A(X, Y)$ *is closed under addition; and*
   (b) $\operatorname{Hom}_A(Y_1, Y) \circ \operatorname{rad}_A(X_1, Y_1) \circ \operatorname{Hom}_A(X, X_1) \subseteq \operatorname{rad}_A(X, Y)$.
   In particular, $\operatorname{rad}_A(X, X)$ *is an ideal of the algebra* $\operatorname{End}_A(X)$ *and we can consider the factor algebra* $E(X) := \operatorname{End}_A(X)/\operatorname{rad}_A(X, X)$. *Moreover,* $\operatorname{rad}_A(X, Y)$ *is a right* $\operatorname{End}_A(X)$*-submodule of* $\operatorname{Hom}_A(X, Y)$, *and* $\operatorname{Hom}_A(X, Y)/\operatorname{rad}_A(X, Y)$ *becomes a right* $E(X)$*-module.*
(2) $\operatorname{rad}_A(X, Y \oplus Y') \cong \operatorname{rad}_A(X, Y) \oplus \operatorname{rad}_A(X, Y')$ *as right* $\operatorname{End}_A(X)$*-modules for all* $X, Y, Y' \in (\operatorname{mod} A)_0$.

(3) *Let X, Y be indecomposable A-modules. Then*

$$\operatorname{rad}_A(X,Y) = \{f \in \operatorname{Hom}_A(X,Y) \mid f \text{ is not an isomorphism}\}.$$

In particular, $E(X)$ is a skew-field, and $\operatorname{Hom}_A(X,Y) = \operatorname{rad}_A(X,Y)$ if $X \not\cong Y$. Hence we have an isomorphism of right $E(X)$-modules:

$$\text{(B.9)} \qquad \operatorname{Hom}_A(X,Y)/\operatorname{rad}_A(X,Y) \cong \begin{cases} E(X) & (\text{if } X \cong Y), \\ 0 & (\text{if } X \not\cong Y). \end{cases}$$

PROOF. (1) Let X', Y' be indecomposable A-modules, and let $g \in \operatorname{Hom}_A(X', X)$ and $h \in \operatorname{Hom}_A(Y, Y')$.

(a) Let $f, f' \in \operatorname{Hom}_A(X,Y)$. Assume that $\alpha := h(f+f')g \in \operatorname{Hom}_A(X', Y')$ is an isomorphism, and set $h' := \alpha^{-1}h \in \operatorname{Hom}_A(Y, X')$. Then $\mathbb{1}_{X'} = h'(f+f')g = h'fg + h'f'g$. Here, since $f, f' \in \operatorname{rad}_A(X,Y)$, neither $h'fg$ nor $h'f'g$ is an isomorphism in the local algebra $\operatorname{End}_A(X')$ (see Exercise 2.3.24(3)), which shows that $\mathbb{1}_{X'}$ is not an isomorphism, a contradiction. Hence $h(f+f')g$ is not an isomorphism. Thus $f + f' \in \operatorname{rad}_A(X,Y)$.

(b) Let $b \in \operatorname{Hom}_A(Y_1, Y), f \in \operatorname{rad}_A(X_1, Y_1), a \in \operatorname{Hom}_A(X, X_1)$. Then $h(bfa)g = (hb)f(ag)$ is not an isomorphism by definition of radical maps, which shows that $bfa \in \operatorname{rad}_A(X,Y)$.

(2) By (1) above, for each homomorphism $f\colon Y \to Y'$ in $\operatorname{mod} A$, we can define an $\operatorname{End}_A(X)$-homomorphism $\operatorname{rad}_A(X,f)\colon \operatorname{rad}_A(X,Y) \to \operatorname{rad}_A(X,Y')$ as the restriction of $\operatorname{Hom}_A(X,f)\colon \operatorname{Hom}_A(X,Y) \to \operatorname{Hom}_A(X,Y')$. This defines a linear functor $\operatorname{rad}_A(X,\text{-})\colon \operatorname{mod} A \to \operatorname{mod} \operatorname{End}_A(X)$. Hence it preserves direct sums.

(3) Let X, Y be indecomposable A-modules.

($\subseteq$). Let $f \in \operatorname{rad}_A(X,Y)$. Then $f = \mathbb{1}_Y f \mathbb{1}_X$ is not an isomorphism by definition of $\operatorname{rad}_A(X,Y)$.

($\supseteq$). Assume that $f \in \operatorname{Hom}_A(X,Y)$ is not an isomorphism, and let X', Y', g, h be as in (1) above. If hfg is an isomorphism, then g becomes a section and h becomes a retraction. But since X, X', Y, Y' are indecomposable, both g and h turn out to be isomorphisms, which shows that f is also an isomorphism, a contradiction. Therefore, hfg is not an isomorphism, and hence $f \in \operatorname{rad}_A(X,Y)$. □

THEOREM B.3.3 (The Krull–Schmidt Theorem). *Let M be a finite-dimensional A-module. Then*

(1) *M decomposes into the direct sum of indecomposable modules, i.e., there exists indecomposable modules $M_1, \dots, M_m$ such that $M = \bigoplus_{i=1}^m M_i$.*
(2) *The above decomposition is unique in the sense that if $M = \bigoplus_{j=1}^n N_j$ is another such decomposition, then $m = n$ and there exists some permutation σ of $\{1, \dots, n\}$ such that $N_i \cong M_{\sigma(i)}$ for all $i = 1, \dots, n$.*

PROOF.

(1) was already proved in Proposition B.2.1.

(2) Choose a complete set $\{X_1, \dots, X_n\}$ of representatives of isoclasses of $\{M_1, \dots, M_m\}$, and set $d_i := |\{j \in \{1, \dots, m\} \mid M_j \cong X_i\}|$ for all $i = 1, \dots, n$. Then we have $M \cong \bigoplus_{i=1}^n X_i^{(d_i)}$, and $X_i \not\cong X_j$ if $i \neq j$. Let $i \in \{1, \dots, n\}$. It is enough to show that the number d_i is uniquely determined by M. By Lemma B.3.2

we have the following isomorphisms of right $E(X_i)$-modules:

$$\begin{aligned}\operatorname{Hom}_A(X_i, M)/\operatorname{rad}_A(X_i, M) &\cong \operatorname{Hom}_A(X_i, \bigoplus_{j=1}^{n} X_j^{(d_j)})/\operatorname{rad}_A(X_i, \bigoplus_{j=1}^{n} X_j^{(d_j)}) \\ &\cong \bigoplus_{j=1}^{n}(\operatorname{Hom}_A(X_i, X_j)^{(d_j)} \Big/ \bigoplus_{j=1}^{n} \operatorname{rad}_A(X_i, X_j)^{(d_j)} \\ &\cong \bigoplus_{j=1}^{n}(\operatorname{Hom}_A(X_i, X_j)/\operatorname{rad}_A(X_i, X_j))^{(d_j)} \\ &\underset{(B.9)}{\cong} (\operatorname{Hom}_A(X_i, X_i)/\operatorname{rad}_A(X_i, X_i))^{(d_i)} \\ &= E(X_i)^{(d_i)}.\end{aligned}$$

Hence $\dim_{E(X_i)} \operatorname{Hom}_A(X_i, M)/\operatorname{rad}_A(X_i, M) = d_i$.

□

REMARK B.3.4. In the above, $\operatorname{Hom}(X, \text{-})/\operatorname{rad}_A(X, \text{-}) = (\operatorname{mod} A/\operatorname{rad}_A)(X, \text{-})$ is a linear functor $\operatorname{mod} A \to \operatorname{mod} E(X)$. The proof above tells us that if X is indecomposable, then the function $\dim_{E(X)}(\operatorname{mod} A/\operatorname{rad}_A)(X, \text{-})\colon (\operatorname{mod} A)_0 \to \mathbb{N}$ gives each module M the multiplicity of X in the indecomposable decomposition of M.

B.4. Proofs of Morita's Theorem

We here give proofs of Theorem 2.3.19 in two ways. The first one is elementary; it only requires knowledge in Chapter 2 and a fundamental fact on tensor products over an algebra A that for each right A-module M, the map $M \otimes_A A \to M, m\otimes a \mapsto ma$ is an isomorphism with the inverse $M \to M \otimes_A A, m \mapsto m \otimes 1$. If the reader is not yet familiar with tensor products over an algebra, then we recommend reading Sect. 7.1 – 7.3 regarding all algebras A as the category $\operatorname{Cat}(A, \{1\})$. The second proof explains a relationship between the first and the Hom-tensor adjoint. After that we extend this theorem to (one half of) the full version of the Morita theorem by generalizing the second proof. For a full explanation on the Morita theorem we refer the reader to further reading such as [**1**, §22].

THEOREM 2.3.19 (Morita). Let A be a finite-dimensional algebra and e a basic idempotent of A. Then the functor

$$\operatorname{Hom}_A(eA, \text{-})\colon \operatorname{Mod} A \to \operatorname{Mod} eAe$$

is en equivalence.

FIRST PROOF. We identify the functor $\operatorname{Hom}_A(eA, \text{-})$ with $(\text{-}) \cdot e$ by the usual way. Let $M, M' \in (\operatorname{Mod} A)_0$. Then note that

$$\phi := (\text{-}) \cdot e\colon \operatorname{Hom}_A(M, M') \to \operatorname{Hom}_{eAe}(Me, M'e)$$

is defined by $\phi(f) := f|_{Me}$ for all $f \in \operatorname{Hom}_A(M, M')$. First, $(\text{-})\cdot e$ is dense because for each $N \in (\operatorname{Mod} eAe)_0$, we have $N \otimes_{eAe} eA \in (\operatorname{Mod} A)_0$, and $(N \otimes_{eAe} eA)e = N \otimes_{eAe} (eAe) \cong N$. It remains to show that ϕ is bijective. For this we define its inverse $\psi\colon \operatorname{Hom}_{eAe}(Me, M'e) \to \operatorname{Hom}_A(M, M')$. Let $E = \{e_{11}, \dots, e_{1a_1}, \dots, e_{n1}, \dots, e_{na_n}\}$ be a complete set of orthogonal primitive idempotents in A, where as before $e_{ij} \cong e_{pq}$ in $\operatorname{Cat}(A, E)$ if and only if $i = p$. Then we may take $e := e_1 + \cdots + e_n$, where

$e_i := e_{i1}$ for all $i = 1, \dots, n$. For any $i = 1, \dots, n$ and $j = 1, \dots, a_i$, since $e_{ij} \cong e_i$ in $\operatorname{Cat}(A, E)$, there exist $s_{ij} \in e_{ij}Ae_i$ and $r_{ij} \in e_iAe_{ij}$ such that $s_{ij}r_{ij} = e_{ij}$ and $r_{ij}s_{ij} = e_i$. We then define ψ by

$$[\psi(g)](m) := \sum_{i,j} g(me_{ij}s_{ij})r_{ij} \quad (g \in \operatorname{Hom}_{eAe}(Me, M'e), m \in M).$$

We now verify that $\psi = \phi^{-1}$. For any $g \in \operatorname{Hom}_{eAe}(Me, M'e)$ and $m \in M$, since $e \cdot e_{ij} = \delta_{j,1}e_i$, we have

$$\begin{aligned}[(\phi \circ \psi)(g)](me) = [\psi(g)](me) &= \sum_{i,j} g(mee_{ij}s_{ij})r_{ij} = \sum_i g(me_is_{i1})r_{i1} \\ &= \sum_i g(me_is_{i1}r_{i1}) = g(me).\end{aligned}$$

Therefore, $\phi \circ \psi = \mathbb{1}_{\operatorname{Hom}_{eAe}(Me,M'e)}$. For any $f \in \operatorname{Hom}_A(M, M')$ and $m \in M$, by setting $g := \phi(f)$, we have

$$\begin{aligned}[(\psi \circ \phi)(f)](m) &= \sum_{i,j} g(me_{ij}s_{ij})r_{ij} = \sum_{i,j} f(me_{ij}s_{ij})r_{ij} = \sum_{i,j} f(me_{ij}s_{ij}r_{ij}) \\ &= \sum_{i,j} f(me_{ij}e_{ij}) = f(m).\end{aligned}$$

Therefore, $\phi \circ \psi = \mathbb{1}_{\operatorname{Hom}_A(M,M')}$. $\square$

We now give the second proof, which explains a relationship between the proof above and the Hom-tensor adjoint given by the eAe-A-bimodule eA.

SECOND PROOF. Regard $A = \operatorname{Cat}(A, \{1\})$ and $eAe = \operatorname{Cat}(eAe, \{e\})$, and consider the eAe-A-bimodule eA. Then by Proposition 7.3.2, we have the following adjoint:

$$\operatorname{Mod} eAe \underset{R:=\operatorname{Hom}_A(eA,\text{-})}{\overset{L:=\text{-}\otimes_{eAe}eA}{\rightleftarrows}} \operatorname{Mod} A \quad (L \dashv R)$$

with the unit $\eta\colon \mathbb{1}_{\operatorname{Mod} eAe} \Rightarrow R \circ L$ and the counit $\varepsilon\colon L \circ R \Rightarrow \mathbb{1}_{\operatorname{Mod} A}$ defined in its proof. It is enough to show that they are natural isomorphisms. Recall that η is defined as $\eta_N\colon N \to \operatorname{Hom}_A(eA, N \otimes_{eAe} eA), x \mapsto x \otimes (\text{-}) = (a \mapsto x \otimes a)$ for all $N \in (\operatorname{Mod} eAe)_0$. Then the diagram

$$\begin{array}{ccccc} N & \xrightarrow{\eta_N} & \operatorname{Hom}_A(eA, N \otimes_{eAe} eA) & x & \longmapsto & x \otimes (\text{-}) \\ \cong\downarrow & & \downarrow\cong & \downarrow & & \downarrow \\ N \otimes_{eAe} eAe & \underset{\cong}{\longleftarrow} & (N \otimes_{eAe} eA)e & x \otimes e = x \otimes (ee) & \longleftarrow & (x \otimes e)e = x \otimes e \end{array}$$

is commutative on the left as is confirmed on the right, where the maps without names are natural isomorphisms. Hence η is a natural isomorphism. Next, recall that ε is defined by $\varepsilon_M\colon \operatorname{Hom}_A(eA, M) \otimes_{eAe} eA \to M$, $f \otimes ea \mapsto f(ea)$ for all $M \in (\operatorname{Mod} A)_0$. Let E be a complete set of orthogonal primitive idempotents in A and take $s_{ij} \in e_{ij}Ae_i$ as in the first proof so that the right multiplication of them give isomorphisms $Ae_{ij} \xrightarrow{\sim} Ae_i$, which altogether induce an isomorphism $A = \bigoplus_{i,j} Ae_{ij} \xrightarrow{\sim} \bigoplus_{i=1}^n Ae_i^{(a_i)}$ as left A-modules. By applying $eA \otimes_A$ -

and $M \otimes_A$ -, this yields isomorphisms $eA = \bigoplus_{i,j} eAe_{ij} \xrightarrow{\sim} \bigoplus_{i=1}^n eAe_i^{(a_i)}$ and $\bigoplus_{i,j} Me_{ij} \xrightarrow{\sim} \bigoplus_{i=1}^n Me_i^{(a_i)}$. Then we have the diagram

$$\begin{array}{ccc} \mathrm{Hom}_A(eA, M) \otimes_{eAe} eA & \xrightarrow{\varepsilon_M} & M \\ \cong\downarrow & & \downarrow\cong \\ Me \otimes_{eAe} \left(\bigoplus_{i=1}^n eAe_i^{(a_i)}\right) & & \bigoplus_{i,j} Me_{ij} \\ \cong\downarrow & & \downarrow\cong \\ \bigoplus_{i=1}^n (Me \otimes_{eAe} (eAe)e_i)^{(a_i)} & \xrightarrow{\cong} & \bigoplus_{i=1}^n Mee_i^{(a_i)} \end{array},$$

where the maps without names are given by natural isomorphisms and isomorphisms induced from s_{ij}. Its commutativity is checked in the following diagram:

$$\begin{array}{ccc} f \otimes ea & \longmapsto & f(ea) = \sum_{i,j} f(e)ae_{ij} \\ \downarrow & & \downarrow \\ f(e) \otimes (eas_{i1}e_i, \dots, eas_{ia_i}e_i)_{i=1}^n & & (f(e)eae_{i1}, \dots, f(e)eae_{ia_i})_{i=1}^n \\ \downarrow & & \downarrow \\ (f(e) \otimes eas_{1i}ee_i, \dots, f(e) \otimes eas_{ia_i}ee_i)_{i=1}^n & \longmapsto & (f(e)eas_{i1}ee_i, \dots, f(e)eas_{ia_i}ee_i)_{i=1}^n \end{array}.$$

Hence ε is a natural isomorphism. □

REMARK B.4.1. The last commutativity check corresponds to the verification of $\psi = \phi^{-1}$ in the first proof. This can be avoided as follows. By Corollary 4.4.9 (1), the fact that η is a natural isomorphism shows that L is fully faithful. The last diagram without commutativity shows that L is dense. Therefore, L is an equivalence, say with a quasi-inverse L'. Then since $L \dashv R$ and $L \dashv L'$, we have $R \cong L'$ by Proposition 4.4.10 (1), and hence $R = \mathrm{Hom}_A(eA, \text{-})$ is an equivalence.

Finally, we explain how the proof above can be generalized to show the following theorem, where the base ring $\Bbbk$ can be any commutative ring.

THEOREM B.4.2 (Morita). *Let A be an algebra (not necessarily finite-dimensional) and P a finitely generated projective right A-module. Assume that P is a* generator *in* $\mathrm{Mod}\,A$, *i.e., that there exists an epimorphism $P^{(m)} \to A$ in* $\mathrm{Mod}\,A$ *for some $m \in \mathbb{N}$. Set $B := \mathrm{End}_A(P)$, and regard P as a B-A-bimodule. Then the functor*

$$\mathrm{Hom}_A(P, \text{-})\colon \mathrm{Mod}\,A \to \mathrm{Mod}\,B$$

is an equivalence.

Note that for $P = eA$ in Theorem 2.3.19, we may have $B = eAe$, and eA satisfies the conditions on P above. Thus the theorem above generalizes Theorem 2.3.19. Before giving the proof, we prepare some statements.

LEMMA B.4.3. *Let $\mathscr{C}$ and $\mathscr{D}$ be linear categories, E and F linear functors $\mathscr{C} \to \mathscr{D}$, and $\alpha\colon E \Rightarrow F$ a natural transformation. For $x_1, x_2 \in \mathscr{C}_0$, assume that we have a direct sum x of x_1 and x_2 in $\mathscr{C}$. Then α_x is an isomorphism if and only if so are both α_{x_1} and α_{x_2}.*

PROOF. Let $(x_i \xrightarrow{\sigma_i} x \xrightarrow{\pi_i} x_i)_{i=1}^2$ be a direct sum system. Then we have the diagram

$$\text{(B.10)} \qquad \begin{array}{ccccc} E(x_1) & \underset{E(\pi_1)}{\overset{E(\sigma_1)}{\rightleftarrows}} & E(x) & \underset{E(\sigma_2)}{\overset{E(\pi_2)}{\rightleftarrows}} & E(x_2) \\ \downarrow{\scriptstyle \alpha_{x_1}} & & \downarrow{\scriptstyle \alpha_x} & & \downarrow{\scriptstyle \alpha_{x_2}} \\ F(x_1) & \underset{F(\pi_1)}{\overset{F(\sigma_1)}{\rightleftarrows}} & F(x) & \underset{F(\sigma_2)}{\overset{F(\pi_2)}{\rightleftarrows}} & F(x_2) \end{array},$$

in which the right arrows and down arrows (resp. left arrows and down arrows) form a commutative diagram because α is a natural transformation.

$(\Rightarrow)$. Assume that α_x is an isomorphism. Set

$$\alpha_i := E(\pi_i) \circ \alpha_x^{-1} \circ F(\sigma_i) \quad (i = 1, 2).$$

Then by the commutativity above, we have $\alpha_i \circ \alpha_{x_i} = \mathbb{1}_{E(x_i)}$ and $\alpha_{x_i} \circ \alpha_i = \mathbb{1}_{F(x_i)}$ for all $i = 1, 2$, and hence both α_{x_1} and α_{x_2} are isomorphisms.

$(\Leftarrow)$. Assume that both α_{x_1} and α_{x_2} are isomorphisms. Set

$$\alpha' := E(\sigma_1) \circ \alpha_{x_1}^{-1} \circ F(\pi_1) + E(\sigma_2) \circ \alpha_{x_2}^{-1} \circ F(\pi_2).$$

Then by the commutativity above, we see that α' is the inverse of α_x, and hence α_x is an isomorphism. □

REMARK B.4.4. Let A be an algebra, and consider the case that $\mathscr{C} = \operatorname{Mod} A$ in Lemma B.4.3. Let $X \in (\operatorname{Mod} A)_0$. Then we denote by $\operatorname{add} X$ the full subcategory of $\operatorname{Mod} A$ with the object set

$$\{M \in (\operatorname{Mod} A)_0 \mid M \text{ is a direct summand of } X^{(n)} \text{ for some } n \in \mathbb{N}\}.$$

Note that by this lemma, if α_X is an isomorphism, then so is α_Y for all $Y \in (\operatorname{add} X)_0$. In particular, if α_A is an isomorphism, then so is α_P for all finitely generated projective A-modules because $\operatorname{add} A$ coincides with the category of all finitely generated projective A-modules.

PROPOSITION B.4.5. *Let A and B be algebras, and U a B-A-bimodule. For each $P \in (\operatorname{Mod} A)_0$ and $X \in (\operatorname{Mod} B)_0$, define*

$$\alpha_{P,X} \colon X \otimes_B \operatorname{Hom}_A(P, U) \to \operatorname{Hom}_A(P, X \otimes_B U)$$

by $[\alpha_{P,X}(x \otimes f)](p) := x \otimes f(p)$ for all $x \in X, f \in \operatorname{Hom}_A(P,U), p \in P$. Then the family $\alpha := (\alpha_{P,X})_{(P,X)\in(\operatorname{Mod} A)_0\times(\operatorname{Mod} B)_0}$ is a natural transformation in P and X. If P is a finitely generated projective module, then $\alpha_{P,X}$ is an isomorphism for all $X \in (\operatorname{Mod} B)_0$.

PROOF. It is easy to check that α is a natural transformation, and the verification is left to the reader. For $P = A$ we have the following commutative diagram:

$$\begin{array}{ccc} X \otimes_B \operatorname{Hom}_A(A, U) & \xrightarrow{\alpha_{A,X}} & \operatorname{Hom}_A(A, X \otimes_U) \\ & {\scriptstyle X\otimes_A \phi_U}\searrow\cong \qquad \cong\swarrow{\scriptstyle \phi_{X\otimes_B U}} & \\ & X \otimes_B U & \end{array},$$

where $\phi_Y \colon \operatorname{Hom}_A(A, Y) \to Y,\ f \mapsto f(1)$ is the isomorphism in Proposition 2.3.4 for all $Y \in (\operatorname{Mod} A)$. Therefore, $\alpha_{A,X}$ is an isomorphism, and hence so is $\alpha_{P,X}$ for all finitely generated projective A-modules P by Remark B.4.4. □

PROPOSITION B.4.6. *Let A be an algebra, and P a right A-module, and set $B := \operatorname{End}_A(P)$. For each $X \in (\operatorname{Mod} A)_0$, define*

$$\beta_X : \operatorname{Hom}_A(P, X) \otimes_B P \to X$$

by $\beta_X(f \otimes p) := f(p)$ for all $f \in \operatorname{Hom}_A(P, X)$, $p \in P$. Then the family $\beta := (\beta_X)_{X \in (\operatorname{Mod} A)_0} : \operatorname{Hom}_A(P, \text{-}) \otimes_B P \Rightarrow \mathbb{1}_{\operatorname{Mod} A}$ is a natural transformation. If $X \in (\operatorname{add} P)_0$, then β_X is an isomorphism.

PROOF. Again the verification of the naturality of β is left to the reader. For $X = P$, the map β_P coincides with the usual isomorphism $B \otimes_B P \xrightarrow{\sim} P$. Hence the assertion follows by Remark B.4.4. □

We are now in a position to give a proof of the theorem.

PROOF OF THEOREM B.4.2. We show that the functor $\text{-} \otimes_B P : \operatorname{Mod} B \to \operatorname{Mod} A$ is a quasi-inverse of $\operatorname{Hom}_A(P, \text{-})$. Let $N \in (\operatorname{Mod} B)_0$. Then

$$N \cong N \otimes_B \operatorname{Hom}_A(P, P) \overset{\text{B.4.5}}{\cong} \operatorname{Hom}_A(P, N \otimes_B P) \text{ natural in } N.$$

Let $M \in (\operatorname{Mod} A)_0$. Then since there exists an epimorphism $P^{(m)} \to A$ in $\operatorname{Mod} A$ for some $m \in \mathbb{N}$, the projectivity of A shows that $A \in \operatorname{add} P$. Hence

$$\begin{aligned}\operatorname{Hom}_A(P, M) \otimes_B P \cong \operatorname{Hom}_A(P, M \otimes_A A) \otimes_B P &\overset{\text{B.4.5}}{\cong} M \otimes_A \operatorname{Hom}_A(P, A) \otimes_B P \\ &\overset{\text{B.4.6}}{\cong} M \otimes_A A \cong M \text{ natural in } M.\end{aligned}$$

□

REMARK B.4.7. As is easily seen, the isomorphisms in the proof above coincide with the components η_N and ε_M of the unit and the counit of the adjoint given by the bimodule P, respectively.

B.5. Left Kan extensions and tensor products

In Chapter 7, we introduced tensor products over a small category, where we used terminology of left Kan extensions without any explanations. In this section, we give a definition of left Kan extensions and explain relationships with left adjoint functors. This will make clear the fact that a left Kan extension is a special type of tensor product in the linear category case (see the beginning of Sect. 7.1). Here also, we use string diagrams to aid understanding. For categories $\mathscr{A}$ and $\mathscr{C}$, we denote by $\mathscr{A}^{\mathscr{C}}$ the functor category $\operatorname{Fun}(\mathscr{C}, \mathscr{A})$.

DEFINITION B.5.1. Let $F : \mathscr{C} \to \mathscr{D}$ be a functor between categories, $\mathscr{A}$ a category, and $M \in (\mathscr{A}^{\mathscr{C}})_0$. Then a *left Kan extension* of M along F is a pair $(\hat{M}, \theta)$ of a functor $\hat{M} \in (\mathscr{A}^{\mathscr{D}})_0$ and a natural transformation $\theta : M \Rightarrow \hat{M} \circ F$

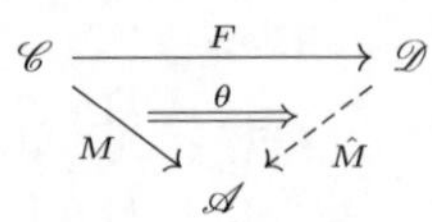

with the following universal property: For any natural transformation $\sigma : M \Rightarrow S \circ F$, there exists a unique natural transformation $\tau : \hat{M} \Rightarrow S$ such that $\sigma = (\tau \circ F) \bullet \theta$.

The last equation is illustrated by the following string diagram:

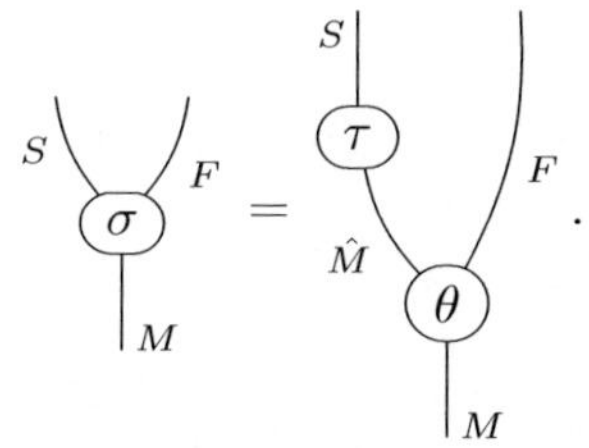

In the sequel, we fix the setting in the definition above, and set $F^{\bullet}$ as the functor $(\text{-}) \circ F \colon \mathscr{A}^{\mathscr{D}} \to \mathscr{A}^{\mathscr{C}}$.

PROPOSITION B.5.2. *The following hold:*

(1) *Assume that there exists a left adjoint L to $F^{\bullet}$ with a unit $\eta \colon \mathbb{1}_{\mathscr{A}^{\mathscr{C}}} \Rightarrow F^{\bullet} \circ L$. Then the pair $(L(M), \eta_M)$ turns out to be a left Kan extension of M along F.*

(2) *Conversely, assume that for each $M \in \mathscr{A}^{\mathscr{C}}$, there exists a left Kan extension $(\hat{M}, \theta_M)$ of M along F. Then the set $\{(\hat{M}, \theta_M) \mid M \in \mathscr{A}^{\mathscr{C}}\}$ is extended to a left adjoint to $F^{\bullet}$ in the sense that there exists a left adjoint L to $F^{\bullet}$ with a unit η such that $(\hat{M}, \theta_M) = (L(M), \eta_M)$ for all $M \in \mathscr{A}^{\mathscr{C}}$.*

PROOF. (1) First, for each $M \in \mathscr{A}^{\mathscr{C}}$, $\eta_M \colon M \to F^{\bullet}(L(M)) = L(M) \circ F$ is a morphism in the category $\mathscr{A}^{\mathscr{C}}$, thus a natural transformation. Next, let $S \in \mathscr{A}^{\mathscr{D}}$ and $\sigma \colon M \Rightarrow S \circ F$ be a natural transformation. Then since $S \circ F = F^{\bullet}(S)$, the morphism $\sigma \colon M \Rightarrow F^{\bullet}(S)$ in $\mathscr{A}^{\mathscr{C}}$ uniquely corresponds to some $\tau \colon L(M) \Rightarrow S$ in $\mathscr{A}^{\mathscr{D}}$ such that $\sigma = F^{\bullet}(\tau) \bullet \eta_M = (\tau \circ F) \bullet \eta_M$ by the adjoint. This means that the pair $(L(M), \eta_M)$ is a left Kan extension of M along F.

(2) We first define a functor $L \colon \mathscr{A}^{\mathscr{C}} \to \mathscr{A}^{\mathscr{D}}$ as follows. For each $M \in (\mathscr{A}^{\mathscr{C}})_0$, set $L(M) := \hat{M}$. For each morphism $f \colon M \Rightarrow N$ in $\mathscr{A}^{\mathscr{C}}$, there exists a unique natural transformation $\tau \colon L(M) \Rightarrow L(N)$ such that $\theta_N \bullet f = (\tau \circ F) \bullet \theta_M$ by the universal property of θ_M. Then we set $L(f) := \tau$, and the following equality holds:

(B.11)

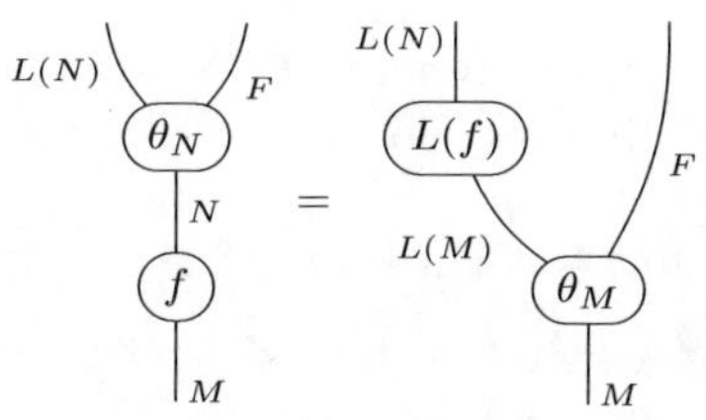

It is easy to see that L is a functor by using the universality of θ_M for all $M \in \mathscr{A}^{\mathscr{C}}$.

Next, we define a natural isomorphism $\omega \colon \mathscr{A}^{\mathscr{D}}(L(\text{-}), ?) \to \mathscr{A}^{\mathscr{C}}(\text{-}, F^{\bullet}(?))$ as follows. For any $M \in (\mathscr{A}^{\mathscr{C}})_0$ and $N \in (\mathscr{A}^{\mathscr{D}})_0$, define

$$\omega_{M,N} \colon \mathscr{A}^{\mathscr{D}}(L(M), N) \to \mathscr{A}^{\mathscr{C}}(M, F^{\bullet}(N))$$

by $\alpha \mapsto (\alpha \circ F) \bullet \theta_M$ for all $\alpha \in \mathscr{A}^{\mathscr{D}}(L(M), N)$, and set $\omega := (\omega_{M,N})_{(M,N) \in (\mathscr{A}^{\mathscr{C}} \times \mathscr{A}^{\mathscr{D}})_0}$. Then the universality of θ_M shows the bijectivity of $\omega_{M,N}$. To show that $\omega_{M,N}$ is natural in M and in N, let $f \colon M' \Rightarrow M$ be in $\mathscr{A}^{\mathscr{C}}$ and $g \colon N \Rightarrow N'$ in $\mathscr{A}^{\mathscr{D}}$. Then it is enough to show that the following are commutative:

$$\begin{array}{ccc} \mathscr{A}^{\mathscr{D}}(L(M),N) & \xrightarrow{\omega_{M,N}} & \mathscr{A}^{\mathscr{C}}(M,f^{\boldsymbol{\cdot}}(N)) \\ {\scriptstyle \mathscr{A}^{\mathscr{D}}(L(f),N)}\downarrow & & \downarrow{\scriptstyle \mathscr{A}^{\mathscr{C}}(f,F^{\boldsymbol{\cdot}}(N))} \\ \mathscr{A}^{\mathscr{D}}(L(M'),N) & \xrightarrow[\omega_{M',N}]{} & \mathscr{A}^{\mathscr{C}}(M',f^{\boldsymbol{\cdot}}(N)) \end{array} \quad \text{and} \quad \begin{array}{ccc} \mathscr{A}^{\mathscr{D}}(L(M),N) & \xrightarrow{\omega_{M,N}} & \mathscr{A}^{\mathscr{C}}(M,f^{\boldsymbol{\cdot}}(N)) \\ {\scriptstyle \mathscr{A}^{\mathscr{D}}(L(M),g)}\downarrow & & \downarrow{\scriptstyle \mathscr{A}^{\mathscr{C}}(M,F^{\boldsymbol{\cdot}}(g))} \\ \mathscr{A}^{\mathscr{D}}(L(M),N') & \xrightarrow[\omega_{M,N'}]{} & \mathscr{A}^{\mathscr{C}}(M,f^{\boldsymbol{\cdot}}(N')) \end{array}.$$

Let $\alpha \in \mathscr{A}^{\mathscr{D}}(L(M),N)$. Then the former is commutative if $(\alpha \circ F) \bullet \theta_M \bullet f = ((\alpha \bullet L(f)) \circ F) \bullet \theta_{M'}$. By (B.11) this is verified as follows:

$(LHS) =$ 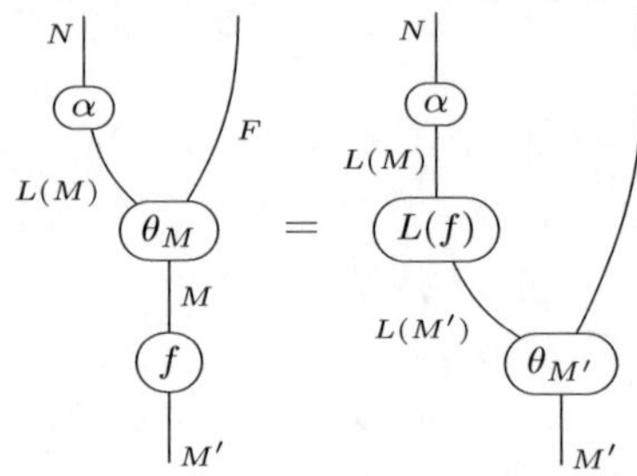 $= (RHS)$.

The latter is commutative if $F^{\boldsymbol{\cdot}}(g) \bullet (\alpha \circ f) \bullet \theta_M = ((g \bullet \alpha) \circ F) \bullet \theta_M$, which is verified by $(LHS) = (g \circ F) \bullet (\alpha \circ f) \bullet \theta_M = (RHS)$.

By Theorem 4.4.7, the natural isomorphism ω defines a unit η by $\eta_M := \omega_{M,L(M)}(\mathbb{1}_{L(M)}) = (\mathbb{1}_{L(M)} \circ F) \bullet \theta_M = \theta_M$ for all $M \in \mathscr{A}^{\mathscr{C}}$. Hence we have $(\hat{M}, \theta_M) = (L(M), \eta_M)$ for all $M \in \mathscr{A}^{\mathscr{C}}$. □

EXERCISE B.5.3. In the above, after the definition of L, give an alternative proof (instead of defining ω as above) by defining a unit η (as $(\theta_M)_{M \in \mathscr{A}^{\mathscr{C}}}$) and a counit ε suitably, and by checking that they are natural transformations and satisfy the zigzag equations.

By applying Proposition B.5.2 (1) to Chapter 7, we obtain the following:

COROLLARY B.5.4. *Let $F\colon \mathscr{C} \to \mathscr{D}$ be a linear functor between small linear categories $\mathscr{C}$ and $\mathscr{D}$. Consider the $\mathscr{C}$-$\mathscr{D}$-bimodule ${}_F\mathscr{D} := \mathscr{D}(\text{-}, F(?))$. Then the pair $(M \otimes_{\mathscr{C}} {}_F\mathscr{D}, \eta_M)$ is a left Kan extension of M along F, where $\eta_M\colon M \to (M \otimes_{\mathscr{C}} {}_F\mathscr{D}) \circ F$ is defined by $\eta_{M,x}\colon M_x \to M \otimes_{\mathscr{C}} {}_F\mathscr{D}_{F(x)}$, $m \mapsto m \otimes \mathbb{1}_{F(x)}$ $(m \in M_x, x \in \mathscr{C}_0)$.*

PROOF. By Lemma 7.3.4, the functor $F^{\boldsymbol{\cdot}} \cong \operatorname{Mod}\mathscr{D}({}_F\mathscr{D}, ?)$ has a left adjoint $\text{-} \otimes_{\mathscr{C}} {}_F\mathscr{D}$ with a unit η, and the assertion follows. Since the isomorphism above is given by the Yoneda lemma, the unit defined in the proof of Proposition 7.3.2 gives us the precise form of η as follows: For each $M \in (\operatorname{Mod}\mathscr{C})_0$ and $x \in \mathscr{C}_0$,

$$\begin{aligned} \eta_{M,x}\colon M_x &\to (\operatorname{Mod}\mathscr{D})({}_x({}_F\mathscr{D}), M \otimes_{\mathscr{C}} {}_F\mathscr{D}) \to M \otimes_{\mathscr{C}} {}_F\mathscr{D}_{F(x)} \\ m &\mapsto \qquad\qquad m \otimes \text{-} \qquad\qquad \mapsto m \otimes \mathbb{1}_{F(x)}. \end{aligned}$$

□

Bibliography

[1] Frank W. Anderson and Kent R. Fuller, *Rings and categories of modules*, 2nd ed., Graduate Texts in Mathematics, vol. 13, Springer-Verlag, New York, 1992, DOI 10.1007/978-1-4612-4418-9. MR1245487

[2] Hideto Asashiba, *A covering technique for derived equivalence*, J. Algebra **191** (1997), no. 1, 382–415, DOI 10.1006/jabr.1997.6906. MR1444505

[3] Hideto Asashiba, *The derived equivalence classification of representation-finite selfinjective algebras*, J. Algebra **214** (1999), no. 1, 182–221, DOI 10.1006/jabr.1998.7706. MR1684880

[4] Hideto Asashiba, *Derived and stable equivalence classification of twisted multifold extensions of piecewise hereditary algebras of tree type*, J. Algebra **249** (2002), no. 2, 345–376, DOI 10.1006/jabr.2001.9063. MR1901163

[5] Hideto Asashiba, *A generalization of Gabriel's Galois covering functors and derived equivalences*, J. Algebra **334** (2011), 109–149, DOI 10.1016/j.jalgebra.2011.03.002. MR2787656

[6] Hideto Asashiba, *Representations of quivers*, Lecture Notes at Shizuoka University, 2014, `https://wwp.shizuoka.ac.jp/asashiba/hideto-asashibas-website/quiver3-1/`.

[7] Hideto Asashiba, *A generalization of Gabriel's Galois covering functors II: 2-categorical Cohen-Montgomery duality*, Appl. Categ. Structures **25** (2017), no. 2, 155–186, DOI 10.1007/s10485-015-9416-9. MR3638358

[8] Hideto Asashiba, *Derived equivalences of actions of a category*, Appl. Categ. Structures **21** (2013), no. 6, 811–836, DOI 10.1007/s10485-012-9284-5. MR3127372

[9] Hideto Asashiba, *Gluing derived equivalences together*, Adv. Math. **235** (2013), 134–160, DOI 10.1016/j.aim.2012.10.021. MR3010054

[10] Hideto Asashiba and Mayumi Kimura, *Presentations of Grothendieck constructions*, Comm. Algebra **41** (2013), no. 11, 4009–4024, DOI 10.1080/00927872.2012.699572. MR3169503

[11] Hideto Asashiba, *Smash products of group weighted bound quivers and Brauer graphs*, Comm. Algebra **47** (2019), no. 2, 585–610, DOI 10.1080/00927872.2018.1487562. MR3935367

[12] Hideto Asashiba, *Categories and representation theory (with a focus on 2-categoriacal covering theory)*, SGC library series **155**, Saiensu-sha, 2019, in Japanese.

[13] Hideto Asashiba, *Cohen-Montgomery duality for pseudo-actions of a group*, Bull. Iranian Math. Soc. **47** (2021), no. 3, 767–842, DOI 10.1007/s41980-020-00413-6. MR4249177

[14] Ibrahim Assem, Daniel Simson, and Andrzej Skowroński, *Elements of the representation theory of associative algebras. Vol. 1*, London Mathematical Society Student Texts, vol. 65, Cambridge University Press, Cambridge, 2006. Techniques of representation theory, DOI 10.1017/CBO9780511614309. MR2197389

[15] Aslak Bakke Buan, Robert Marsh, Markus Reineke, Idun Reiten, and Gordana Todorov, *Tilting theory and cluster combinatorics*, Adv. Math. **204** (2006), no. 2, 572–618, DOI 10.1016/j.aim.2005.06.003. MR2249625

[16] K. Bongartz and P. Gabriel, *Covering spaces in representation-theory*, Invent. Math. **65** (1981/82), no. 3, 331–378, DOI 10.1007/BF01396624. MR643558

[17] Francis Borceux, *Handbook of categorical algebra. 3*, Encyclopedia of Mathematics and its Applications, vol. 52, Cambridge University Press, Cambridge, 1994. Categories of sheaves. MR1315049

[18] Jianmin Chen, Xiao-Wu Chen, and Shiquan Ruan, *The dual actions, equivariant autoequivalences and stable tilting objects* (English, with English and French summaries), Ann. Inst. Fourier (Grenoble) **70** (2020), no. 6, 2677–2736. MR4245628

[19] Claude Cibils and Eduardo N. Marcos, *Skew category, Galois covering and smash product of a k-category*, Proc. Amer. Math. Soc. **134** (2006), no. 1, 39–50, DOI 10.1090/S0002-9939-05-07955-4. MR2170541

[20] M. Cohen and S. Montgomery, *Group-graded rings, smash products, and group actions*, Trans. Amer. Math. Soc. **282** (1984), no. 1, 237–258, DOI 10.2307/1999586. MR728711

[21] P. Deligne, *Action du groupe des tresses sur une catégorie* (French, with English summary), Invent. Math. **128** (1997), no. 1, 159–175, DOI 10.1007/s002220050138. MR1437497

[22] Vladimir Drinfeld, Shlomo Gelaki, Dmitri Nikshych, and Victor Ostrik, *On braided fusion categories. I*, Selecta Math. (N.S.) **16** (2010), no. 1, 1–119, DOI 10.1007/s00029-010-0017-z. MR2609644

[23] P. Gabriel, *The universal cover of a representation-finite algebra*, Representations of algebras (Puebla, 1980), Lecture Notes in Math., vol. 903, Springer, Berlin-New York, 1981, pp. 68–105. MR654725

[24] P. Gabriel and A. V. Roĭter, *Representations of finite-dimensional algebras*, Algebra, VIII, Encyclopaedia Math. Sci., vol. 73, Springer, Berlin, 1992, pp. 1–177. With a chapter by B. Keller. MR1239447

[25] R. Gordon, A. J. Power, and Ross Street, *Coherence for tricategories*, Mem. Amer. Math. Soc. **117** (1995), no. 558, vi+81, DOI 10.1090/memo/0558. MR1261589

[26] Edward L. Green, *Graphs with relations, coverings and group-graded algebras*, Trans. Amer. Math. Soc. **279** (1983), no. 1, 297–310, DOI 10.2307/1999386. MR704617

[27] Alexander Grothendieck, *Revêtements étales et groupe fondamental. Fasc. I: Exposés 1 à 5*, Institut des Hautes Études Scientifiques, Paris, 1963. Troisième édition, corrigée; Séminaire de Géométrie Algébrique, 1960/61. MR0217087

[28] David Harel, Dexter Kozen, and Jerzy Tiuryn, *Dynamic logic*, Foundations of Computing Series, MIT Press, Cambridge, MA, 2000. MR1791342

[29] John M. Howie, *Fundamentals of semigroup theory*, London Mathematical Society Monographs. New Series, vol. 12, The Clarendon Press, Oxford University Press, New York, 1995. Oxford Science Publications. MR1455373

[30] David Hughes and Josef Waschbüsch, *Trivial extensions of tilted algebras*, Proc. London Math. Soc. (3) **46** (1983), no. 2, 347–364, DOI 10.1112/plms/s3-46.2.347. MR693045

[31] G. M. Kelly and Ross Street, *Review of the elements of 2-categories*, Category Seminar (Proc. Sem., Sydney, 1972/1973), Springer, Berlin, 1974, pp. 75–103. Lecture Notes in Math., Vol. 420. MR0357542

[32] Tom Leinster, *Basic category theory*, Cambridge Studies in Advanced Mathematics, vol. 143, Cambridge University Press, Cambridge, 2014, DOI 10.1017/CBO9781107360068. MR3307165

[33] P. B. Levy, *Formulating categorical concepts using classes*, `arXiv:1801.08528`.

[34] Saunders MacLane, *Categories for the working mathematician*, Graduate Texts in Mathematics, Vol. 5, Springer-Verlag, New York-Berlin, 1971. MR0354798

[35] D. Marsden, *Category theory using string diagrams*, `arXiv:1401.7220v2` [math.CT].

[36] R. Martínez-Villa and J. A. de la Peña, *The universal cover of a quiver with relations*, J. Pure Appl. Algebra **30** (1983), no. 3, 277–292, DOI 10.1016/0022-4049(83)90062-2. MR724038

[37] D. Murfet, *Foundations for category theory*, Available at `therisingsea.org`, (2006).

[38] H. Nakaoka, *Techniques for category theory* (Homological algebra in abelian and triangulated categories), Nihon Hyoronsha, (2015).

[39] Alexander von Eye and Eun-Young Mun, *Log-linear modeling*, John Wiley & Sons, Inc., Hoboken, NJ, 2013. Concepts, interpretation, and application. MR3235585

[40] Idun Reiten and Christine Riedtmann, *Skew group algebras in the representation theory of Artin algebras*, J. Algebra **92** (1985), no. 1, 224–282, DOI 10.1016/0021-8693(85)90156-5. MR772481

[41] C. Riedtmann, *Algebren, Darstellungsköcher, Überlagerungen und zurück* (German), Comment. Math. Helv. **55** (1980), no. 2, 199–224, DOI 10.1007/BF02566682. MR576602

[42] Christine Riedtmann, *Representation-finite self-injective algebras of class A_n*, Representation theory, II (Proc. Second Internat. Conf., Carleton Univ., Ottawa, Ont., 1979), Lecture Notes in Math., vol. 832, Springer, Berlin, 1980, pp. 449–520. MR607169

[43] Christine Riedtmann, *Representation-finite self-injective algebras of class D_n*, Compositio Math. **49** (1983), no. 2, 231–282. MR704393

[44] A. Tarski, *Über unerreichbare Kardinalzahlen*, Fundamenta Mathematicae, **30** (1938), 68–89.

[45] TheCasters: *String diagrams, 1 – 4*, YouTube videos, `https://www.youtube.com/watch?v=USYRDDZ9yEc`.

[46] Josef Waschbüsch, *Universal coverings of self-injective algebras*, Representations of algebras (Puebla, 1980), Lecture Notes in Math., vol. 903, Springer, Berlin-New York, 1981, pp. 331–349. MR654721

[47] N. H. Williams, *On Grothendieck universes*, Compositio Math. **21** (1969), 1–3. MR244035

Index

Selected Published Titles in This Series

271 **Hideto Asashiba,** Categories and Representation Theory, 2022
270 **Leslie Hogben, Jephian C.-H. Lin, and Bryan L. Shader,** Inverse Problems and Zero Forcing for Graphs, 2022
267 **Dragana S. Cvetković Ilić,** Completion Problems on Operator Matrices, 2022
266 **Kate Juschenko,** Amenability of Discrete Groups by Examples, 2022
265 **Nabil H. Mustafa,** Sampling in Combinatorial and Geometric Set Systems, 2022
264 **J. Scott Carter and Seiichi Kamada,** Diagrammatic Algebra, 2021
263 **Vugar E. Ismailov,** Ridge Functions and Applications in Neural Networks, 2021
262 **Ragnar-Olaf Buchweitz,** Maximal Cohen–Macaulay Modules and Tate Cohomology, 2021
261 **Shiri Artstein-Avidan, Apostolos Giannopoulos, and Vitali D. Milman,** Asymptotic Geometric Analysis, Part II, 2021
260 **Lindsay N. Childs, Cornelius Greither, Kevin P. Keating, Alan Koch, Timothy Kohl, Paul J. Truman, and Robert G. Underwood,** Hopf Algebras and Galois Module Theory, 2021
259 **William Heinzer, Christel Rotthaus, and Sylvia Wiegand,** Integral Domains Inside Noetherian Power Series Rings, 2021
258 **Pramod N. Achar,** Perverse Sheaves and Applications to Representation Theory, 2021
257 **Juha Kinnunen, Juha Lehrbäck, and Antti Vähäkangas,** Maximal Function Methods for Sobolev Spaces, 2021
256 **Michio Jimbo, Tetsuji Miwa, and Fedor Smirnov,** Local Operators in Integrable Models I, 2021
255 **Alexandre Boritchev and Sergei Kuksin,** One-Dimensional Turbulence and the Stochastic Burgers Equation, 2021
254 **Karim Belabas and Henri Cohen,** Numerical Algorithms for Number Theory, 2021
253 **Robert R. Bruner and John Rognes,** The Adams Spectral Sequence for Topological Modular Forms, 2021
252 **Julie Déserti,** The Cremona Group and Its Subgroups, 2021
251 **David Hoff,** Linear and Quasilinear Parabolic Systems, 2020
250 **Bachir Bekka and Pierre de la Harpe,** Unitary Representations of Groups, Duals, and Characters, 2020
249 **Nikolai M. Adrianov, Fedor Pakovich, and Alexander K. Zvonkin,** Davenport–Zannier Polynomials and Dessins d'Enfants, 2020
248 **Paul B. Larson and Jindrich Zapletal,** Geometric Set Theory, 2020
247 **István Heckenberger and Hans-Jürgen Schneider,** Hopf Algebras and Root Systems, 2020
246 **Matheus C. Bortolan, Alexandre N. Carvalho, and José A. Langa,** Attractors Under Autonomous and Non-autonomous Perturbations, 2020
245 **Aiping Wang and Anton Zettl,** Ordinary Differential Operators, 2019
244 **Nabile Boussaïd and Andrew Comech,** Nonlinear Dirac Equation, 2019
243 **José M. Isidro,** Jordan Triple Systems in Complex and Functional Analysis, 2019
242 **Bhargav Bhatt, Ana Caraiani, Kiran S. Kedlaya, Peter Scholze, and Jared Weinstein,** Perfectoid Spaces, 2019
241 **Dana P. Williams,** A Tool Kit for Groupoid C^*-Algebras, 2019
240 **Antonio Fernández López,** Jordan Structures in Lie Algebras, 2019
239 **Nicola Arcozzi, Richard Rochberg, Eric T. Sawyer, and Brett D. Wick,** The Dirichlet Space and Related Function Spaces, 2019

For a complete list of titles in this series, visit the AMS Bookstore at **www.ams.org/bookstore/survseries/**.